# Soil Nitrogen Uses and Environmental Impacts

# Advances in Soil Science

Series Editors: Rattan Lal and B. A. Stewart

## Published Titles

**Methods for Assessment of Soil Degradation**
Rattan Lal, Winfried E. H. Blum, C. Valentin, Bobby A. Stewart

**Soil Processes and the Carbon Cycle**
R. Lal, J. M. Kimble, R. F. Follett, and B. A. Stewart

**Global Climate Change and Cold Regions Ecosystems**
R. Lal, J. M. Kimble, and B. A. Stewart

**Assessment Methods for Soil Carbon**
R. Lal, J. M. Kimble, R. F. Follett, and B. A. Stewart

**Soil Erosion and Carbon Dynamics**
E.J. Roose, R. Lal, C. Feller, B. Barthès, and B. A. Stewart

**Soil Quality and Biofuel Production**
R. Lal and B. A. Stewart

**Food Security and Soil Quality**
R. Lal and B. A. Stewart

**World Soil Resources and Food Security**
R. Lal and B. A. Stewart

**Soil Water and Agronomic Productivity**
R. Lal and B. A. Stewart

**Principles of Sustainable Soil Management in Agroecosystems**
R. Lal and B. A. Stewart

**Soil Management of Smallholder Agriculture**
R. Lal and B. A. Stewart

**Soil-Specific Farming: Precision Agriculture**
R. Lal and B. A. Stewart

**Soil Phosphorus**
R. Lal and B. A. Stewart

**Urban Soils**
R. Lal and B. A. Stewart

For more information about this series, please visit:
https://www.crcpress.com/Advances-in-Soil-Science/book-series/CRCADVSOILSCI

# Soil Nitrogen Uses and Environmental Impacts

Edited by
## Rattan Lal
## B.A. Stewart

CRC Press
Taylor & Francis Group
Boca Raton London New York

CRC Press is an imprint of the
Taylor & Francis Group, an **informa** business

Explanation for equation on book cover: GY is grain yield, T is transpiration, ET is evapotranspiration, HI is the harvest index, and TR is the transpiration ratio. In other words, GY is proportional to the evapotranspiration and it represents the total water use by the crops. The grain yield, and thus the food security, depends on the amount of water available for plants to grow.

CRC Press
Taylor & Francis Group
6000 Broken Sound Parkway NW, Suite 300
Boca Raton, FL 33487-2742

First issued in paperback 2021

© 2018 by Taylor & Francis Group, LLC
CRC Press is an imprint of Taylor & Francis Group, an Informa business

No claim to original U.S. Government works

ISBN 13: 978-1-03-209565-3 (pbk)
ISBN 13: 978-1-138-62636-2 (hbk)

**Library of Congress Cataloging-in-Publication Data**

Names: Lal, R., editor. | Stewart, B. A. (Bobby Alton), 1932- editor.
Title: Soil nitrogen uses and environmental impacts / editors: Rattan Lal and B. A. Stewart.
Other titles: Advances in soil science (Boca Raton, Fla.)
Description: Boca Raton, FL : CRC Press, Taylor & Francis Group, 2018. |
Series: Advances in soil science | Includes bibliographical references and index.
Identifiers: LCCN 2018001400 | ISBN 9781138626362 (hardback : alk. paper)
Subjects: LCSH: Nitrogen in agriculture. | Soils--Nitrogen content.
Classification: LCC S587.5.N5 S65 2018 | DDC 631.8/4--dc23
LC record available at https://lccn.loc.gov/2018001400

**Visit the Taylor & Francis Web site at**
**http://www.taylorandfrancis.com**

**and the CRC Press Web site at**
**http://www.crcpress.com**

# Contents

# Preface

Being an important component of protein and other organic molecules, nitrogen (N) is an essential element for all organisms. Following oxygen, carbon, and hydrogen, N is the most common element in plant and animal tissues. In natural ecosystems, N is in short supply and is an important control of numerous ecosystem functions. While Earth's atmosphere contains predominantly N (78%), it is nonreactive and unavailable to plants. The nonreactive N in the atmosphere is converted into the reactive form available to plants by biological N fixation (BNF) and lightning. The amount of N fixed globally is estimated to be 5–10 Tg/yr by lightning and 90–140 Tg/yr by BNF. The inert N is fixed as nitric oxide (NO) by lightning and ultraviolet rays, and as ammonia ($NH_3$), nitrites ($NO_2$), and nitrates ($NO_3$) through BNF by soil microorganisms. Two kinds of soil microorganisms involved in BNF are: (1) nonsymbiotic or free-living including Cyanobacteria and Azotobacter, and (2) symbiotic (mutualistic) bacteria including Rhizobium (associated with leguminous plants) and Azospirillum species (associated with cereal grasses). The industrial process of N fixation by the Haber–Bosch process, into reactive N (e.g., nitrates, ammonia, or nitrites), was discovered in World War I in 1913 ($N_2 + 4H_2 \rightarrow 2NH_3$, $\Delta H = -92.22$ KJ/mol).

The world demand for N fertilizer (Tg, million ton) was 108, 110, 112, 113, 115, and 116 in 2011, 2012, 2013, 2014, 2015, and 2016, respectively. In comparison, the global demand for $P_2O_5$ (Tg) was 41, 42, 43, 43.5, 44, and 45 and that of $K_2O$ (Tg) was 28, 29, 29.5, 31, 32, and 33 for 2011, 2012, 2013, 2014, 2015, and 2016, respectively. Thus, the increase in global demand over the 6-year period (Tg/yr) is 1.3 for N, 0.7 for $P_2O_5$, and 0.8 for $K_2O$ (FAO 2016). Therefore, industrial production of N fertilizer has strongly impacted the global N cycle. Similar to water ($H_2O$) and carbon (C), N compounds also cycle through soil, water, plants, and air. The global N cycle is a continuous sequence of events by which atmospheric N and nitrous compounds in the soil are converted, as by nitrification and N fixation, into substances that can be utilized by green plants, and substances returning to the air and soil because of decay of plants and denitrification. The natural N cycle has been drastically disturbed by the industrial process of fertilizer production with dire consequences to climate change through emission of nitrous oxide ($N_2O$) with global warming potential 310 times that of $CO_2$, and transport of $NO_3^-$ into surface runoff and groundwater causing eutrophication (e.g., algal bloom) and severe adverse impacts on human and animal health. There has been a strong increase in the use of synthetic N fertilizers, especially since 1990, and fertilizer use is projected to increase because of the increasing demand for food production to feed the projected world population of 9.7 billion by 2050. Therefore, this volume is specifically devoted to availability, production, and recycling of N (in soils, air, water, and plants), with impact on climate change and water quality, and management of N in agroecosystems in the context of maximizing the use efficiency and minimizing the risks of leakage of reactive N ( $NO_3^-$, $N_2O$) into the environment.

The 15-chapter book describes the global N cycle, the need and environmental consequences of its use as an essential fertilizer for food production, and the socioeconomic and policy implications of its use and misuse. The challenge is to reconcile the need for food production with that of reducing the environmental footprint of using the reactive N. The book also provides the scientific basis of N management and describes recent advances in slow release formulations to reduce the environmental impact.

The editors thank all the authors for their outstanding contributions, and for sharing their knowledge and experiences. Despite the busy schedules and numerous commitments, preparation of the manuscripts in a timely manner by all authors is greatly appreciated. The editors also thank the editorial staff of Taylor & Francis for their help and support in publishing this volume. The office staff of the Carbon Management and Sequestration Center provided support in the flow of manuscripts between authors and editors and made valuable contributions, and their help and support is greatly appreciated. In this context, special thanks are due to Ms. Laura Conover who formatted the text and prepared the final submission. It is a challenging task to thank by listing names of all those who contributed in one way or another to bringing this volume to fruition. Thus, it is important to build on the outstanding contributions of numerous soil scientists, agricultural engineers, and technologists whose research is cited throughout the book.

# Editors

**Rattan Lal, Ph.D.,** is a Distinguished University Professor of Soil Science and Director of the Carbon Management and Sequestration Center, The Ohio State University, and an Adjunct Professor at the University of Iceland. His current research focus is on climate-resilient agriculture, soil carbon sequestration, sustainable intensification, enhancing use efficiency of agroecosystems, and sustainable management of soil resources of the tropics. He received honorary degrees of Doctor of Science from Punjab Agricultural University (2001), the Norwegian University of Life Sciences, Aas (2005), Alecu Russo Balti State University, Moldova (2010), Technical University of Dresden, Germany (2015), and University of Lleida, Spain (2017). He was president of the World Association of the Soil and Water Conservation (1987–1990), the International Soil Tillage Research Organization (1988–1991), the Soil Science Society of America (2005–2007), and is President Elect of International Union of Soil Science. He was a member of the Federal Advisory Committee on U.S. National Assessment of Climate Change-NCADAC (2010–2013), member of the SERDP Scientific Advisory Board of the US-DOE (2011–), Senior Science Advisor to the Global Soil Forum of Institute for Advanced Sustainability Studies, Potsdam, Germany (2010–), member of the Advisory Board of Joint Program Initiative of Agriculture, Food Security and Climate Change (FACCE-JPI) of the European Union (2013–), and Chair of the Advisory Board of the Institute for Integrated Management of Material Fluxes and Resources of the United Nations University (UNU-FLORES), Dresden, Germany (2014–2017). Prof. Lal was a lead author of IPCC (1998–2000). He has mentored 106 graduate students and 54 postdoctoral researchers, and hosted 156 visiting scholars. He has authored/co-authored 818 refereed journal articles, has written 19 and edited/co-edited 65 books. For 3 years (2014, 2015, 2016), Reuter Thomson listed him among the world's most influential scientific minds and having citations of publications among the top 1% of scientists in agricultural sciences.

**B.A. Stewart, Ph.D.,** is Director of the Dryland Agriculture Institute and a distinguished professor of Agriculture at West Texas A&M University, Canyon, Texas. He is a former director of the USDA Conservation and Production Laboratory at Bushland, Texas; past president of the Soil Science Society of America; and member of the 1990–1993 Committee on Long-Range Soil and Water Policy, National Research Council, National Academy of Sciences. He is a fellow of the Soil Science Society of America, American Society of Agronomy, Soil and Water Conservation Society, a recipient of the USDA Superior Service Award, a recipient of the Hugh Hammond Bennett Award of the Soil and Water Conservation Society, and an honorary member of the International Union of Soil Sciences in 2008. In 2009, Dr. Stewart was inducted into the USDA Agriculture Research Service Science Hall of Fame. Dr. Stewart is very supportive of education and research on dryland agriculture. The B.A. and Jane Ann Stewart Dryland Agriculture Scholarship Fund was established at West Texas A&M University in 1994 to provide scholarships for undergraduate and graduate students with a demonstrated interest in dryland agriculture.

# Contributors

**Benjamin Babst**
Arkansas Forest Resource Center
University of Arkansas
Monticello, Arkansas

**S. Babu**
ICAR Research Complex for NEH Region
Umiam, Meghalaya, India

**Marianne Bechmann**
Norwegian Institute of Bioeconomy Research
Ås, Norway

**Barbara J. Cade-Menun**
Agriculture and Agri-Food Canada
Saskatchewan, Canada

**Mark S. Coyne**
Department of Plant and Soil Sciences
University of Kentucky
Lexington, Kentucky

**Anup Das**
ICAR Research Complex for NEH Region
Umiam, Meghalaya, India

**Otto C. Doering III**
Department of Agricultural Economics
Purdue University
West Lafayette, Indiana

**Richard A. Ferrieri**
Missouri Research Reactor Center
University of Missouri
Columbia, Missouri

**Benjamin M. Gramig**
Department of Agricultural and Consumer
  Economics
University of Illinois
Urbana, Illinois

**Eliot Herman**
University of Arizona
Tucson, Arizona

**Peter R. Hobbs**
College of Agriculture and Life Sciences
Cornell University
Ithaca, New York

**Dawoon Jeong**
Department of Agricultural Economics
Purdue University
West Lafayette, Indiana

**Rattan Lal**
Carbon Management and Sequestration Center
The Ohio State University
Columbus, Ohio

**Jayanta Layek**
ICAR Research Complex for NEH Region
Umiam, Meghalaya, India

**Yang Liu**
Liaoning Academy of Agricultural Sciences
Shenyang, China

**Jiafa Luo**
AgResearch
Hamilton, New Zealand

**Shou-Tian Ma**
College of Agronomy and Biotechnology
China Agricultural University
Beijing, China

**Brian G. McConkey**
Agriculture and Agri-Food Canada
Saskatchewan, Canada

**R.S. Meena**
Benaras Hindu University
Varanasi, India

**Minh-Long Nguyen**
Group One Consultancy Ltd
Auckland, New Zealand

**Anne Falk Øgaard**
Norwegian Institute of Bioeconomy Research
Ås, Norway

**Rajendra Prasad**
Indian Agricultural Research Institute
New Delhi, India

**Chao Pu**
College of Agronomy and Biotechnology
China Agricultural University
Beijing, China

**Bert F. Quin**
Quin Environmentals (NZ) Limited
Auckland, New Zealand

**Carl J. Rosen**
Department of Soil, Water, and Climate
University of Minnesota–Twin Cities
St. Paul, Minnesota

**Amit Roy**
International Fertilizer Development Center
Muscle Shoals, Alabama

**Matthew D. Ruark**
Department of Soil Science
University of Wisconsin
Madison, Wisconsin

**D. Sarkar**
Bidhan Chandra Krishi Viswavidyalaya
Mohanpur, Nadia, India

**Kimberley D. Schneider**
Agriculture and Agri-Food Canada
Ontario, Canada

**Michael J. Schueller**
Missouri Research Reactor Center
University of Missouri
Columbia, Missouri

**Rogério P. Soratto**
College of Agricultural Sciences
São Paulo State University
São Paulo, Brazil

**B.A. Stewart**
Amarillo, Texas

**Arumugam Thiagarajan**
Agriculture and Agri-Food Canada
Saskatchewan, Canada

**Henry F. Wilson**
Agriculture and Agri-Food Canada
Manitoba, Canada

**Jian-Fu Xue**
College of Agriculture
Shanxi Agricultural University
Taigu, Shanxi, China

**Gulab Singh Yadav**
ICAR Research Complex for NEH Region
    Tripura Centre
Lembucherra, Agartala, India

**Hai-Lin Zhang**
College of Agronomy and Biotechnology
China Agricultural University
Beijing, China

**Xin Zhao**
College of Agronomy and Biotechnology
China Agricultural University
Beijing, China

# 1 Benefits and Unintended Consequences of Synthetic Nitrogen Fertilizers

*B.A. Stewart*

## CONTENTS

## 1.1 INTRODUCTION

Nitrogen (N) is the most important essential element for crop production because it is required in large amounts and is nearly always the first nutrient that becomes limiting after an ecosystem is converted to cropland. However, since the development of synthetic N that requires only energy to combine hydrogen from the energy with nitrogen gas ($N_2$) from the air to produce ammonia ($NH_3$), N fertilizers have become readily available. Economics have constrained the use of N fertilizers in some cases, particularly in developing countries, but they have greatly increased crop yields and especially so in favorable rainfall areas and on irrigated lands. As will be discussed in more detail later, it has been estimated that synthetic N is responsible for feeding more than 50% of the world's population of 7.4 billion level of today. Without the use of synthetic N, the world population could not have increased to more than about one-half of the 7.4 billion level of today (Erisman et al. 2008; Smil 1999, 2001).

This chapter will look at some of the history of N fertilizer and its importance. Although it is a partial review, the author will also inject some personal viewpoints. I started my career after receiving a B.S. degree as a soil scientist jointly employed by USDA and Oklahoma Experiment Station. The project was to conduct wheat fertilization studies in western Oklahoma and the first publication summarized the first synthetic N studies conducted on farmer fields (Eck and Stewart 1954). We also conducted the first study in Oklahoma that directly applied anhydrous ammonia into the soil (Eck et al. 1957). After completing an M.S. degree while working at Stillwater, I transferred to the newly established USDA Agricultural Research Service Nitrogen Laboratory in Ft. Collins, CO where I also pursued a Ph.D. in soil science. It was during that period that I first became acquainted with possible unintended consequences of using synthetic N. Barry Commoner (1971) raised concerns that fertilizer N applied to soil in the U.S. Corn Belt was causing high levels of nitrate in river waters that subsequently led to methoglobinemia (blue-baby syndrome) in infants. Commoner concluded that the country's N cycle was seriously out of balance, and this led to controversy between agriculturalists and environmentalists. About the same time, large concentrated animal feeding operations were becoming common raising more questions about N, particularly nitrates, polluting surface and groundwater. The USDA Agricultural Research Service Nitrogen Laboratory was one of the early investigators to address this issue (Stewart et al. 1967). The U.S. Environmental Protection Agency (EPA) was established in 1970 and this increased the attention of the effect of N fertilizers and other chemicals on the environment. I chaired a five-scientist committee to author a two-volume manual *Control of Water Pollution from Cropland* jointly published by the USDA Agricultural Research Service and the Office of Research and Development of the Environmental Protection Agency (Stewart et al. 1975). Later, as a member of the Committee on Long-Range Soil and Water Conservation organized by the National Research Council, I was one of the authors of the book *Soil and Water Quality: An Agenda for Agriculture* (National Research Council 1993). From 1997 to 2003, I was a member of the Farmlands Work Group of The John Heinz III Center for Science, Economics and the Environment that had the goal of developing indicators to measure the state of the nation's ecosystems. In addition to the farmlands group, they had working groups for coasts and oceans, forests, fresh water, grasslands and shrublands, and urban and suburban areas. A comprehensive (Heinz Center 2002) report was published listing indicators that could be measured over time to monitor changes in the state of various ecosystems.

The activities above were listed to show that while I am an agriculturalist that believes strongly the world's food and fiber needs cannot be met without the use of synthetic N, I am also passionate about protecting the environment. As a young scientist in the 1950s studying the benefits of N fertilizer on wheat (*Triticum aestivum*) production, there was no thought given to what happened to the N that was not used by the plant. The recommended rate of N fertilizer was mostly determined by economics and adding more than an adequate amount was often considered good insurance in case favorable weather conditions resulted in a bumper crop. As environmental issues began to surface, there developed serious conflicts between agriculturalists and environmentalists. Today, while almost everyone recognizes that the use of synthetic N has resulted in serious unintended consequences, there is little agreement on developing meaningful policies. Developing policies is particularly difficult because it is an international problem because use of synthetic N increases greenhouse gases (GHGs) as well as affects water quality. Furthermore, phosphorus (P) is in many cases causing more environmental problems than N and the two are closely linked. For approximately every 5 units of N removed from cropland with cereal grains and many other crops, there is about 1 unit of P removed. While N is nearly always the first limiting essential nutrient for crop production, P is nearly always the second limiting nutrient. Thus, the use of N fertilizer is closely followed simultaneously by the use of P. As N use increases, P use also increases. Stewart et al. (2016) estimated that 1 Mg P was added for every 5.8 Mg N worldwide in 2010. Phosphorus is readily adsorbed by soil particles and causes serious eutrophication problems in surface waters. Also, unlike $N_2$ that occupies 78% of the atmosphere that is an endless supply for producing synthetic N, P fertilizer is derived from phosphate rock located in mines that is a finite resource. About 90% of the world's known reserves are controlled by five countries: Morocco, Jordan, South Africa, the United States, and China (Amundson et al. 2015).

The objective of this chapter is (1) to take a brief look at the history of the development of synthetic N; (2) evaluate the impact its use has had on food production and how dependent worldwide food security is on its use; (3) assess the seriousness of the environmental problems that have resulted from unintended consequences of its use; and (4) discuss strategies and policies that can alleviate some of the problems. Having spent a long career intimately involved with both sides of this issue, I think I understand many of the challenges ahead. Because it seems clear that we cannot live without using synthetic N, we have no choice but to develop strategies and practices that minimize damage to the environment.

## 1.2　BRIEF HISTORY OF DEVELOPMENT OF SYNTHETIC N

Nitrogen has long been known as the essential plant nutrient that is the first to limit production. Justus von Liebig (1803–1873) established the "Law of the Minimum" that states that plant growth is limited by the element, or the compound, that is present in the soil in the least adequate amount. Water is often the most limiting factor but with continued cropping, soil fertility declines and an essential plant nutrient becomes deficient and limits growth. In most cases, N is the first nutrient that becomes deficient. Then, according to the Law of the Minimum, the addition of N will restore production until another essential nutrient becomes deficient. Phosphorus is commonly the second nutrient to limit growth. Smil (2001) reviewed some of the early experiments and wrote an excellent summary of the findings. Smil (2001) stated that Jean-Baptiste Boussingault, in 1838, was the first researcher to demonstrate that legumes could restore N to soil. However, he also reported that the Chinese used leguminous green manures in the fifth century B.C., and the practice was almost certainly even more ancient.

By the early 1800s, however, it was becoming increasingly clear that there was a great need for N fertilizers. Europeans began importing guano (solidified bird excrement) and sodium nitrate from South America (Smil 2001) but it became apparent that supplies of these N sources would be insufficient.

Coal generally contains between 1 and 1.6% N derived from the decomposition of proteins that were present in the biomass that was eventually transformed by pressure and heat to produce the solid fuel (Smil 2001). During combustion, most of the N is oxidized to nitrate and lost to the atmosphere. However, when burned in the absence of air to produce coke, part of the N is released as ammonia and can be converted to ammonium sulfate and recovered for use as a fertilizer.

Despite these advances, there was great concern among many scientists that there was a great need to find a reliable, long-term source of fixed nitrogen. William Crookes (1832–1919) addressed the British Association in 1898 on the wheat problem. Although it caused a furor among economists and politicians and was criticized by agricultural experts, it became one of the best-known speeches at the turn of the century (Smil 2001). Crookes concluded that if the low wheat yield continued, the rising demand would result in a global wheat deficiency as early as 1930. He stated, "the fixation of nitrogen is vital to the progress of civilized humanity."

The first attempt at synthesizing $NH_3$ from $N_2$ and $H_2$ using high temperature dates to 1788, and more than a dozen chemists published results of similar experiments during the nineteenth century (Smil 2001). It was not until 1909, however, that Fritz Haber, a chemist with BASF, perfected an apparatus to clearly demonstrate the potential of high-pressure synthesis with gas recirculation. Smil (2001) reported that Haber's third key patent for ammonia synthesis—No. 238450 filed on September 14, 1909, and issued on September 28,1911, generally known as the high-pressure patent—stressed the catalysis under pressure and high temperature. Haber received the Nobel Prize for Chemistry in 1919 for this achievement that is considered one of the most important inventions of all times.

Carl Bosch, an engineer with BASF, was charged with converting Haber's bench-top apparatus into a large-scale commercial reality and he did so at an impressive pace. On September 9, 1913, a plant was finished that was synthesizing more than 10 t (metric ton = 1000 kg = Mg) of $NH_3$ a day (Smil 2001). By 1914, the plant was synthesizing 20 Mg a day with plans to expand to 40 Mg a day. However, World War I broke out and instead of continuing as a fertilizer plant, the $NH_3$ was converted to compounds used to make munitions. Carl Bosch was also awarded with a Nobel Prize in Chemistry in 1931 that he shared with Friedrich Bergius in recognition of their contributions to the invention and development of chemical high-pressure methods.

The process is today known as the Haber–Bosch process and is by far the primary source of N fertilizers worldwide. However, even though a large-scale plant was developed in 1914, it was almost two decades later before synthesized ammonia became the dominant means of providing fixed N for agriculture, nor did the commercial availability of fertilizers developed by the process lead to a rapid increase in average application rates of N fertilizer (Smil 2001). The first ammonia plant in the United States was established in Syracuse, NY in 1928 (Roberts and Dibb 2009). World War II occurred from 1939 to 1945 and it was during this period that there was a rapid increase in the amount of N fixed by the Haber–Bosch process but most of it was used to produce munitions. At the beginning of World War II, the U.S. government built 10 new ammonia plants to produce munitions (Genzel and Reinhardt 2017). All the plants were in the interior parts of the United States and mostly along natural gas pipelines so that there was an energy source for producing the hydrogen to combine with atmospheric N. By the end of World War II, plants in the United States were producing 730,000 Mg of ammonia and had a capacity of producing 1.6 million Mg or Tg (Genzel and Reinhardt 2017). Therefore, when the war ended, there was a huge excess of ammonia being produced that was no longer needed for making bombs and this was the beginning of the widespread use of nitrogen fertilizer. The estimated amounts of N fixed by the Haber–Bosch process from 1920 to 2014 are shown in Figure 1.1. Clearly, production began to increase following the end of World War II, but it took a few years for the transition from munitions being the dominant uses to agriculture use. With the ready availability of N fertilizer at relatively low cost, more and more producers began using for the first time and found marked increases in yields. The average yields of corn (*Zea mays*) and wheat for U.S. farmers from 1866 to 2016 increased only marginally until after the end of World War II and then yields increased dramatically (Figure 1.2). The increased yields were the result of several factors, but N fertilizer was perhaps the most important single factor. Removal of N as a limiting factor also allowed for other factors to become more important. Improved cultivars and cultural practices along with better harvesting machines all contributed, and yields are continuing to increase although the rate of increase is slowing.

Similar increases occurred in Europe following World War II and the rest of the world followed rather quickly. Bumb and Baanante (1996) reported that the worldwide use of fertilizer in 1959/1960 was 27.4 Tg of which 24.7 Tg were used in developed countries. Nitrogen usage was

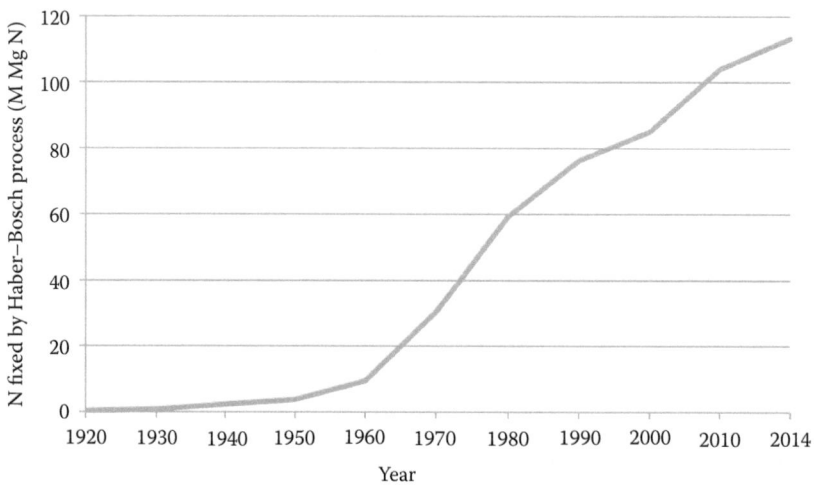

**FIGURE 1.1**  Nitrogen fixed by Haber–Bosch process for N fertilizer (million Mg N year$^{-1}$) from 1920 to 2014. (Data from Smil, V. 1999. Detonator of the population explosion. *Nature* 400:415; and FAOSTAT. 2017. Food and Agriculture Organization Corporate Statistical Database Statistics Division. Food and Agriculture Organization of the United Nations. Rome. Available at http://faostat3.fao.org/home/E.)

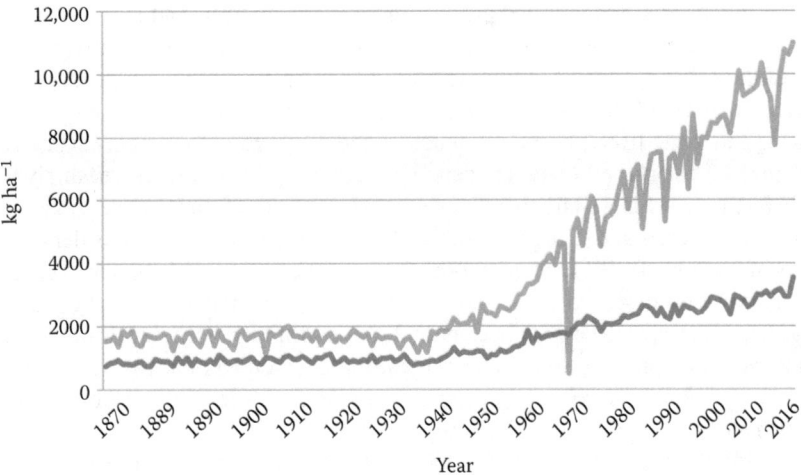

**FIGURE 1.2** U.S. corn (top line) and wheat (bottom line) yields (kg ha$^{-1}$) from 1866 to 2016. (Data from USDA. 2017. ERS Feed Outlook. Economic Research Service, USDA. Washington, DC. Available at: https://www.ers.usda.gov/publications/pub-details/?pubid=82089.)

9.5 Tg. By 1989/1990, the usage had increased to 143.6 Tg and 81.3 Tg was used in developed countries and total N usage was 79.2 Tg. They forecasted by 2020 that N usage would be 115.3 Tg and total fertilizer would be 208 Tg with 121.6 Tg used in developing countries. The forecasted amounts for N were fairly close to the N amounts fixed by the Haber–Bosch process shown in Figure 1.1. It is evident that N fertilizer has been a key factor, and probably the most important factor, for increasing food production, particularly cereal grains that provide much of the world's food. Although the world has over 50,000 edible plants, 60% of the world's energy intake is provided by rice (*Oryza sativa*), maize, and wheat (FAO 2017). Figure 1.3 shows increases in world

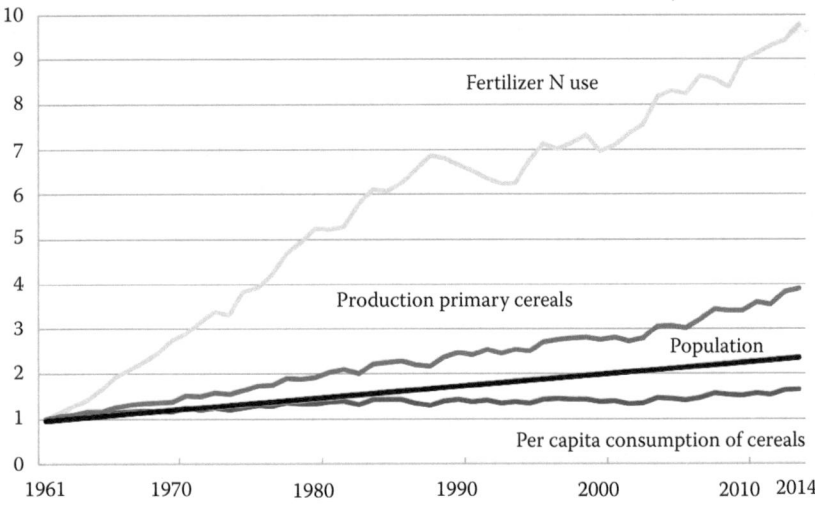

**FIGURE 1.3** Relative increases in world use of N fertilizer, production of primary cereals (wheat, maize, rice), world population, and kg capita$^{-1}$ yr$^{-1}$ consumption of primary cereals (1961 values were 11.6 Mt N; 643 Mt wheat, maize, and rice; 3.08 billion population; and 209 kg capita$^{-1}$ yr$^{-1}$ primary cereals). (Data from FAOSTAT. 2017. Food and Agriculture Organization Corporate Statistical Database Statistics Division. Food and Agriculture Organization of the United Nations. Rome. Available at http://faostat3.fao.org/home/E.)

population; N fertilizer use; combined production of maize, wheat, and rice; and kg per capita of maize, wheat, and rice between 1961 and 2014. While the world population increased 2.3 times, production of the primary cereals increased almost four times, and the kg per capita of the cereal grains increased from 209 to about 355. However, worldwide N fertilizer usage increased almost 10-fold during the same time period and much of the N usage is for cereals. Therefore, the data shown in Figure 1.3 clearly indicates that cereal production is becoming increasingly dependent on additional N fertilizer. This will likely increase even more in the future because maize is becoming the dominant cereal crop and it requires more N because the yields are considerably higher than those for wheat and rice. In 1961, wheat production was the greatest followed by rice and then corn although they were all essentially the same (Figure 1.4). From 1961 until about 2000, all three increased at similar rates. However, since about 2000, maize has increased dramatically compared to wheat and rice. The primary reason for this dramatic change is increasing prosperity in many of the developing countries resulting in people changing their diets to include more animal-based protein. Worldwide, about 35% of maize grain produced in 2007 was fed to animals and 17% was used to make ethanol and other fuels. However, 48% of the world's grain supply was consumed directly by humans and accounts for 48% of their calories so essentially all of the rice and most of the wheat are consumed directly by people (Worldwatch Institute 2016). The main driver today, however, is the growing demand for animal-based protein that requires more maize mostly used for animal feed. As recently as 2000, the amounts of maize, wheat, and rice were essentially equal, but in 2014, maize production accounted for 41% of the total and rice and wheat remained almost equal at slightly less than 40% each (Figure 1.4). This trend is almost certainly to increase more in the future and therefore, with population continuing to increase at the same time as the per capita consumption of grain is increasing, the demand for grain will increase significantly requiring ever increasing amounts of N fertilizer.

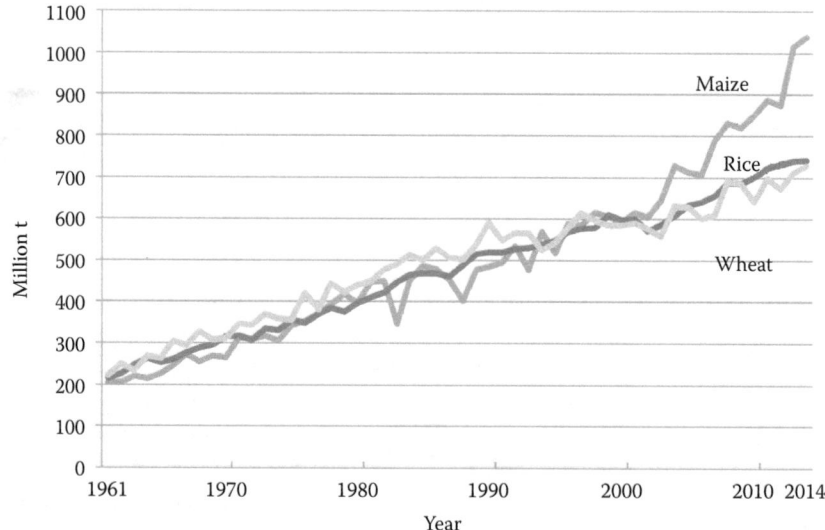

**FIGURE 1.4** Worldwide production (million tonnes) of maize, rice, and wheat from 1961 to 2014. (Data from FAOSTAT. 2017. Food and Agriculture Organization Corporate Statistical Database Statistics Division. Food and Agriculture Organization of the United Nations. Rome. Available at http://faostat3.fao.org/home/E.)

## 1.3 DEPENDENCY OF WORLDWIDE FOOD SECURITY ON SYNTHETIC NITROGEN

Although no one can say for sure how dependent the world population is on synthetic N fixed by the Haber–Bosch process, there is overwhelming evidence as presented above in Section 1.1 that the world population could not have reached 7.4 billion in 2016 and certainly could not increase to the expected 9.7 billion by 2050 without N fertilizer. Most agricultural scientists agree that we cannot survive without it. Smil (2001) wrote a book that carefully detailed the history of N and crop production with particular emphasis on the importance of the Haber–Bosch process. He made perhaps the most compelling analysis ever regarding how dependent world food production is on synthetic N. He made his analysis during the late 1990s when the world population was 5.745 billion. An excerpt from his book is as follows:

> The range of our dependence on the Haber–Bosch synthesis of ammonia thus is as follows: for about 40% of the humanity it now provides the very means of survival; only half as many people as are alive today could be supplied by prefertilizer agriculture with very basic, overwhelmingly vegetarian, diets; and prefertilizer farming could provide today's average diets to only about 40% of the existing population.
>
> The only way this global dependence on nitrogen fertilizer could be lowered would be to adopt an unprecedented degree of sharing and restraint. Because the world's mean daily protein supply of nearly 75 g/capita is well above the needed medium, equitable distribution of available food among the planet's 6 billion people content to subsist on frugal, but adequate, diets would provide enough protein even if the global food harvests were to be some 10% lower than they are today.
>
> In such a world the populations of affluent countries would have to reduce their meat consumption as hundreds of millions of people would have to revert to simpler diets containing more cereals and legumes. Needless to say, the chances of this dietary transformation, running directly against the long-term trend of global nutritional transitions, are extremely slim. But even in such an altruistic and frugal world ammonia synthesis would still have to supply at least 1/3 of all nitrogen assimilated by the global food harvest!

Since the late 1990s when Smil concluded that approximately 40% of the world's 5.7 billion people survived only because of synthetic N, there have been 1.7 billion people added to the planet and projections are that there will be another 2.3 billion added by 2050 (UN 2017). Furthermore, nearly all the increase will be in developing countries. The total amount of cropland in the world is about 1400 million hectare (M ha) and this has increased only about 10% since 1961 and there is limited opportunity for increase. In 2014, 568 M ha of cropland was used for maize, wheat, and rice (FAOSTAT 2017). As the demand for grain increases, most of the increased production will have to come from increased yields rather than from increased area and this can only be done by further increasing the use of synthetic N. Not only will the increased number of people require more grain, the amount per capita will likely increase at a faster rate than population growth rate because of increasing prosperity leading to changing diets to include more animal-based protein. The per capita consumption of meat and milk and dairy products is increasing dramatically in the developing countries where the greatest increase in the number of people is occurring (Table 1.1). In 2017, approximately 1 billion of the 7.4 billion people on the Earth live in the more developed countries. In 2050, it is estimated there will be about 9.5 billion people and approximately 8.5 billion will live in the less developed countries (Haub 2012). Many of these less developed countries are rapidly developing and as incomes rise, diets change quickly to include more eggs, milk, meat, and other animal-based proteins that require more grain, particularly maize, for animal feed.

Based on present data and forecasts, there appears no way that enough food and fiber for the present and future world populations can be produced without using huge amounts of N fertilizer. Without it, a large portion of the population can likely not survive. Thus, the only valid question is: How can we use it in ways that eliminate or minimize damage to the environment?

**TABLE 1.1**

**Consumption of Meat and Milk and Dairy Food Products from 1964 to 1999 and Projected Consumption in 2015 and 2030**

| Years | 1964/1966 | 1974/1976 | 1984/1986 | 1994/1996 | 1997/1999 | 2015 | 2030 |
|---|---|---|---|---|---|---|---|
| **Milk and Dairy Food Products: kg/capita of Whole Milk Equivalent** | | | | | | | |
| World | 74 | 75 | 78 | 77 | 78 | 83 | 90 |
| Developing countries | 28 | 30 | 37 | 42 | 45 | 55 | 66 |
| Industrial countries | 186 | 191 | 212 | 212 | 212 | 217 | 221 |
| **Food Consumption of Meat: kg/capita of Carcass Weight Equivalent** | | | | | | | |
| World | 24.2 | 27.4 | 30.7 | 34.8 | 38.4 | 41.3 | 45.3 |
| Developing countries | 10.2 | 11.4 | 15.5 | 22.7 | 25.5 | 31.6 | 36.7 |
| Industrial countries | 61.5 | 73.5 | 80.7 | 86.2 | 88.2 | 96.7 | 100.1 |

*Source:* Data from Bruinsma, J. (Ed.) 2003. *World Agriculture: Towards 2015/2030: An FAO Perspective.* Earthscan Publications Ltd., London. Available at http://www.fao.org/3/a-y4252e.pdf.

## 1.4  UNINTENDED CONSEQUENCES OF N FERTILIZER USAGE

### 1.4.1  Earliest Concerns

Although the Haber–Bosch process that made N fertilizer readily available and generally inexpensive relative to its benefit is considered one of the most important inventions of all time, its use has resulted in serious environmental problems. As discussed above, early day scientists fully recognized that N was seriously limiting crop production, particularly wheat that was the primary cereal grain for human consumption. Its widespread use, particularly in the United States and Europe following World War II, resulted in sudden and significant increases in yields (Figure 1.2) and there was little or no concern about possible negative effects. As a young scientist in the mid-1950s conducting experiments on farmer fields where no fertilizer had ever been applied, I marveled at the smiles on farmer faces as they reaped wheat harvests greater than ever. They often wanted to add more and more N and we included treatments with as much as 270 kg N ha$^{-1}$ even though grain yields were usually below 2000 kg ha$^{-1}$ (Eck and Stewart 1954). The only constraint considered was when the cost of the fertilizer was considered to be greater than the expected benefit. It was not until the 1960s that concerns became common and Barry Commoner of St. Louis University was perhaps the most vocal critic that later authored a widely read book *The Closing Circle* (Commoner 1971). Commoner stated that the widespread use of N fertilizer in the U.S. Corn Belt was resulting in high concentrations of nitrate in water supplies that caused methemoglobinemia (blue-baby syndrome) in infants. The last thing the agricultural community wanted to hear was that the use of N fertilizer was causing environmental problems, and this led to widespread conflicts between agriculturalists and environmentalists that lasted for several years.

In 1976, following the publication of a manual *Control of Water Pollution from Cropland* published jointly by the USDA and EPA (Stewart et al. 1975), I spoke at the Annual Meeting of the Fertilizer Institute. It was a memorable experience because even though I tried to emphasize the agricultural viewpoint as much as possible, very little I said was considered favorably. The good news today is that the agricultural community and environmental community are unified and fully recognize that there are serious environmental problems associated with the use of N fertilizers. The International Plant Nutrition Institute (IPNI) whose members include most of the world's fertilizer companies is actively involved in addressing fertilizer associated environmental issues.

Needless to say, the problem is complex and difficult with no easy solutions. Even more so, the problem almost defies a solution. Refer again to Liebig's "Law of the Minimum" that states yield is proportional to the amount of the most limiting nutrient, whichever nutrient it may be. Clearly, N was the most limiting factor for crop production on most cropland fields. When N fertilizer became readily available, N was no longer the limiting factor and in most cases, water, or the lack of water, became the limiting factor. The amount of water available for crop production is highly variable making it difficult to determine how much N fertilizer should be applied. This is not only true for areas that depend entirely on precipitation, but for irrigated areas as well because most irrigated areas receive some precipitation. Worldwide, 35% of the water used by irrigated crops comes from growing season precipitation (Molden 2007). Therefore, water is often limited even in irrigated fields and temperatures are often higher in years when precipitation is low exacerbating the limited water. Since farmers do not want N to be the limiting factor, and the amount of water available cannot be accurately predicted, excess N fertilizer is often applied.

An even greater factor that leads to excess N in the environment is the response curve between added N fertilizer and crop yield. Figure 1.5 illustrates a well-established response between crop yield and added rates of N fertilizer. Although the y-axis shows sufficiency index, it can be thought of as yield because the maximum yield occurs when the sufficiency index reaches 1. In the example shown in Figure 1.5, the crop yield was about 70% of the maximum when no fertilizer N was added. When 50 kg ha$^{-1}$ was added, the yield was increased to about 85% of the maximum. The addition of another 50 kg ha$^{-1}$ increased the yield to 95%, and another 50 kg ha$^{-1}$ increased the yield to approximately 99% of maximum, but it took an additional 20 kg ha$^{-1}$ to reach the maximum yield. This clearly shows that added N is used less efficiently with every additional unit of N applied. Unless the cost of N becomes a major constraint, farmers tend to add enough N to prevent it from being the limiting factor. To make matters even worse, as improved seed and other management practices lead to higher yields, then the amount of N required to reach a sufficiency index of 1 becomes higher, so the amount of N not fully utilized by the crop becomes greater and greater. The excess N is then available for leaching, lost from the field with runoff, or lost to the atmosphere as ammonia or nitrous oxide which is a major greenhouse gas.

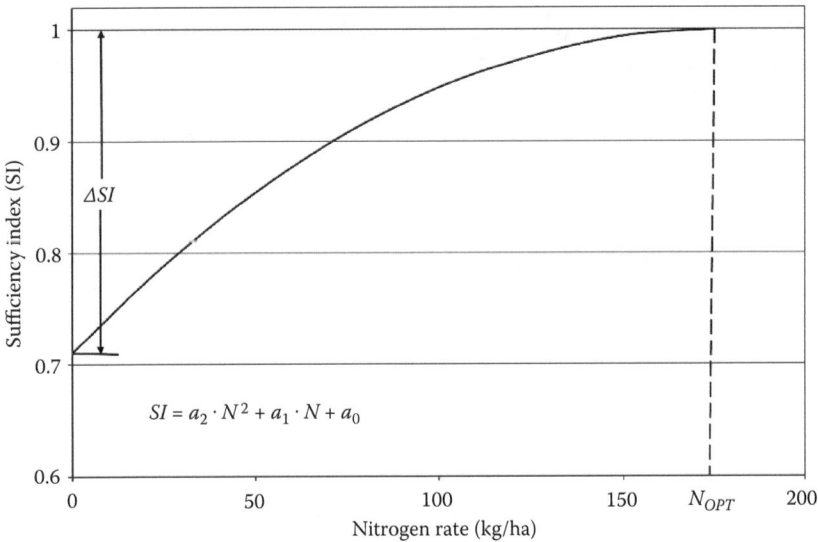

**FIGURE 1.5**   Generalized response curve between nitrogen sufficiency index and rate of nitrogen fertilizer. (From Holland, K.H. and J.S. Schepers. 2010. Derivation of a variable rate nitrogen application model for in-season fertilization of corn. *Agronomy J.* 102:1415–1424. doi:10.2134/agronj2010.0015.)

### 1.4.2 Phosphorus Effect

The close linkage between the use of N fertilizer and P fertilizer has also had a significant negative effect on the environment. The first essential nutrient that becomes limiting following the conversion of grassland or forest land to cropland is in most cases N. However, when N inputs are added as a single input, it is generally not many years before P becomes a limiting factor. This is particularly true for grain crops because cereal grains contain about 1 unit of P for every 5 units of N. Prior to widespread use of synthetic N, decomposition of soil organic matter (SOM), recycling of crop residues, green manure crops, and animal manures supplied most of the N and P used by the crops. The N to P ratios in these sources are in the realm of 4 to 5 units of N for each unit of P. Himes (1997) reported that SOM is about 58% C, and the C/N ratio is about 12 to 1, and the C/P ratio is about 50 to 1. Therefore, the N to C ratio in SOM is about 4 to 1. The uptake of N and P in most crops is also in this range so once large amounts of synthetic N were added, it was necessary to begin adding chemical P fertilizers. The primary source was rock phosphate treated with acid to increase the availability of the P for use by crops. The data in Table 1.2 show a close relationship between the additions of N and P fertilizers. It appears that the ratios of N to P added with fertilizers tend to widen somewhat with time. This is likely because adding P fertilizer usually results in build-up of plant available P over time and the amount can be decreased. However, the total amounts of N and P added from 1961 to 2014 were very close to the 1 unit of P for every 5 units of N.

Phosphorus is also a major environmental concern and since it is so closely linked to N, increased N use has had a double effect on degrading the environment. In some ways, P can be considered more damaging to the environment than N. Phosphorus is attached to soil particles and when soil particles are eroded by wind or water, the attached P go with the particles. Also, as the amount of plant available P in the soil is increased, runoff water contains more dissolved P, and this can cause eutrophication of surface waters. Phosphorus is usually more limiting than N for eutrophication of surface waters, but if runoff water contains increased concentrations of both N and P, the rate of eutrophication can be severe.

Perhaps the greatest concern about the effects of N and P on the environment is simply the enormous amounts that have been added over time and that the amounts added will continue to increase. Between 1961 and 2014, there were approximately 470 Tg of N and 96 Tg of P added to U.S. cropland (FAOSTAT 2017). These amounts were added to the environment from outside sources in that the N came from the atmosphere and the P from mines. To gain some perspective as to the amount of these added nutrients, the United States has roughly 160 M ha of cropland. If we assume that the 0–15 cm of topsoil of all the cropland averages 2% SOM, the total amount of N and P in the topsoil

---

**TABLE 1.2**

**Use of Synthetic N and P Fertilizers in the United States and Worldwide from 1961 to 2014**

| Years | 1961 | 1980 | 2000 | 2014 | 1961 to 2014 |
|---|---|---|---|---|---|
| United States | | million tonnes N or P | | | |
|    N fertilizer use | 3.1 | 10.8 | 10.5 | 9.1 | 469.5 |
|    P fertilizer use | 1.1 | 2.2 | 1.8 | 1.9 | 95.8 |
| Worldwide | | | | | |
|    N fertilizer use | 11.6 | 60.8 | 80.8 | 113.3 | 3543.7 |
|    P fertilizer use | 4.8 | 13.6 | 14.3 | 20.5 | 731.3 |

*Source:* Data obtained from FAOSTAT. 2017. Food and Agriculture Organization Corporate Statistical Database Statistics Division. Food and Agriculture Organization of the United Nations. Rome. Available at http://faostat3.fao.org /home/E.

would be 2166 kg N and 542 kg P based on the values of Himes (1997) presented above. Between 1961 and 2014, an average of 2938 kg of N and 600 kg of P have been added to each hectare of U.S. cropland which is more than the total amount of organic N and organic P in the 15-cm layer of topsoil. Furthermore, the SOM content of much of the cropland is lower now than when fertilizer additions began. There are some positive signs, particularly in the United States, of better managing fertilizer usage. Data in Table 1.2 show that the amounts of both N and P were less in 2014 than in 1980 even though yield of corn, the largest user of fertilizer, has increased dramatically during the same period. Reduced tillage and no-tillage systems have also become more widely used that have increased the SOM content by sequestering more carbon (Islam and Reeder, 2014).

While the amounts of fertilizer in the United States peaked during the 1970s, worldwide use is continuing to increase significantly. In 1961, the United States used 26% of the N and 23% of the P (Table 1.2). In contrast, the United States used only 8% of the N and 9% of the P in 2014. There are approximately 1400 M ha of cropland worldwide and about 160 M ha in the United States (FAOSTAT 2017). Somewhat surprising, the average amount of N fertilizer applied to total cropland in the world was 80 kg ha$^{-1}$ compared to 57 kg ha$^{-1}$ for U.S. cropland even though there are vast areas, particularly in Africa, that receive little or no fertilizer (FAOSTAT 2017).

### 1.4.3 TRANSLOCATION OF NUTRIENTS

Although there have been many practices developed that increase the efficient use of fertilizers, there have been other practices that increase N and P concerns for the environment. Perhaps of greatest concern is the translocation of nutrients contained in crops produced in one area but moved to another area for consumption. Corn for grain is the most widely produced crop in the United States both in terms of area and production. While it is grown in many states, most of it is grown in the Corn Belt located in the north-central part of the United States. The soils in the Corn Belt are highly fertile and the climate is favorable. Therefore, with added fertilizers, improved hybrids, and good cropping practices, the yields have increased almost every year since the 1950s. This has been achieved with no significant increase in the rates of N since the mid-1970s (Figure 1.6). Much of the corn, however, is fed to animals. In 2016, almost 40% of the U.S. corn crop was fed to animals

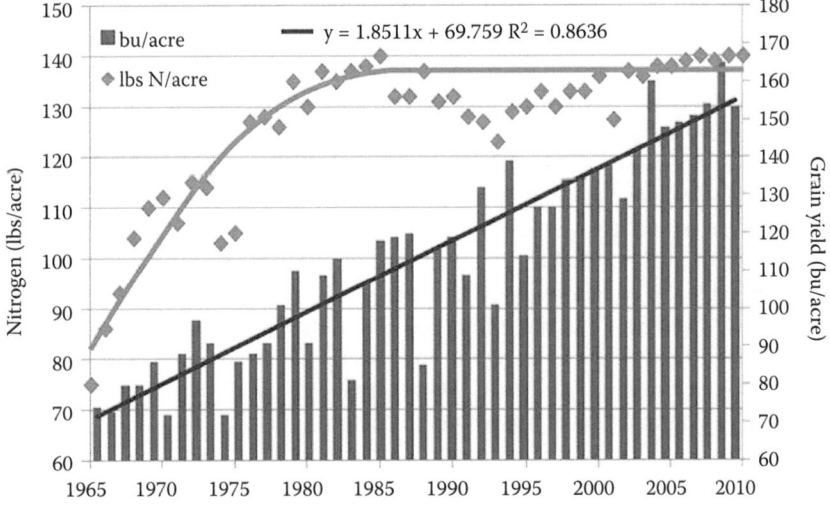

**FIGURE 1.6** Historical grain yields and nitrogen application rates on corn acres in the United States from 1966 to 2014. (From DeBruin, J. 2013. Nitrogen uptake in corn. Crop Insights, DuPont Pioneer, Johnston, IA. Available at: https://www.pioneer.com/home/site/us/agronomy/library/n-uptake-corn/.).

(USDA 2017). While this has been true for many years, the thing that has changed dramatically in the past few decades is where the animals are being fed the corn. Historically, the animals were fed on or very near the land where the corn was grown, and the manure produced by the animals was spread on the land to recycle the nutrients. Much of the N and P consumed by growing animals is excreted. Lory et al. (2006) stated that growing cattle excrete 50–70% of N and 65–75% of the P consumed, and van Heugten and van Kempen (2017) reported that finishing swine animals excreted 50–70% of N and 55–80% of P consumed. In the past few decades, large concentrated animal feeding operations (CAFOs) have become common and large percentages of dairy cows and beef, swine, and poultry animals fed for meat production are being fed large amounts of corn in areas often far removed from where the corn is grown. In simple terms, historically the animals were taken to the corn while in many cases now the feed is being taken to the animal. A single CAFO can have from 50 to 100,000 beef animals, millions of chickens, and as many as 10,000 or more dairy cows. Corn is moved by truck or sometimes in unit trains that have 110 cars filled with corn grain. The CAFOs are often located in warmer and sometimes drier areas where the animals will perform more efficiently, but in many cases, there is limited cropland. Thus, there are huge amounts of manure produced and even when there is enough area to spread the manure, it is often spread at higher rates than needed so the potential for environmental damage is increased. Then, the farmers in the Corn Belt add more chemical fertilizer that contains N fixed from the atmosphere and P taken from mines, so large amounts of N and P are introduced into the environment every successive year.

The translocation of nutrients is not simply a U.S. issue, but a worldwide issue, as large facilities for feeding animals are rapidly spreading to many countries as demand for meat, dairy products, and eggs continues to rapidly increase. Many countries are importing large amounts of maize and soybeans (*Glycine max*) for animal feed. Even European countries that are highly productive agricultural countries import large amounts. Therefore, Europe uses large amounts of N fertilizer and in addition imports large amounts of N and P with the maize and soybeans. Excess N is a major concern in Europe and led to an extensive European Nitrogen Assessment (Sutton et al. 2011). More than 100 scientists selected by the European Science Foundation contributed to the book that assessed the effect of N fertilizer on greenhouse balance, soil quality, water quality, and air quality was having on ecosystems and biodiversity. A cost-benefit economic analysis estimated that N losses in Europe resulted in environmental costs of U.S. \$75 to \$350 billion $yr^{-1}$. The highest societal costs were on the environment and human health.

The Netherlands serves as an example of adverse effects of excess N and some of the common factors that lead to excess N. Again, translocation of N from other areas is a major factor. Although the Netherlands is a small country, it is second only to the United States in the world for the export of agricultural products (Logan 2016). The Netherlands produces large amounts of meat, dairy products, and eggs but roughly 70% is exported. Maize and soybeans are among the main feedstuffs and they are largely imported. From 2014 to 2016, 5 Tg of maize were imported compared to only 200,000 Mg produced (Logan 2016). Also, most of the soybeans is imported. In 2009, the Netherlands was the second largest importer of soybeans in the world (Dutch Soy Coalition 2009). As discussed above, from 50 to 70% of N in feed consumed by animals is excreted so there are large amounts of N added to the Netherlands environment each year. Schwägerl (2016) stated that intensive animal farming, fertilizer use, and industrial N emissions have resulted in 200 kg $ha^{-1}$ $yr^{-1}$ of excess N. Atmospheric deposition of ammonia and nitrates has also greatly increased in areas where large amounts of N are present (Smil 2001). He reported that over 80 kg N $ha^{-1}$ $yr^{-1}$ was occurring in the Netherlands toward the end of the twentieth century.

These examples show that even if crops are produced with management practices that utilize N fertilizer efficiently for growing crops that are used for animal feed, the movement of the crops to feed animals in other areas often results in environmental degradation of the areas where the animals are located.

## 1.4.4 WATER QUALITY CONSEQUENCES

The record is clear and largely undisputed that the greatly increased use of N fertilizer produced by the Haber–Bosch process has resulted in degradation of water quality. This is not only true for the United States, but also worldwide. Countless numbers of research papers and reviews have been written about the negative effects of excess N and P on water quality at local sites, watersheds, countries, and the Earth. Among them are "Human Alteration of the Global Nitrogen Cycle: Causes and Consequences" (Vitousek 1997); "Excess Nitrogen in the Environment" (UNEP 2014); "Massive Nitrogen Pollution Accompanies China's Growth" (Narayanan 2013); and "Reactive Nitrogen: Too Much of a Good Thing?" (Galloway et al. 2002).

Johan Rockström, Executive Director of Stockholm Resilience Centre in Sweden, and colleagues developed planetary boundaries for guiding human development on a changing planet (Steffen et al. 2015). The idea of planetary boundaries was conceived in 2007 based on assuming the Holocene Epoch, the current period of geologic time that began about 11,000 years ago at the end of the last Ice Age, has been a time of relative climate stability (Ruddiman 2003). This allowed people to grow crops, domesticate animals, and launch industrial and communications revolutions. But now, Rockström and his colleagues believe that climatic stability is under threat. Of the nine boundaries they thought should not be crossed, four had already been crossed by 2015. These were climate change where 350 ppm $CO_2$ should not exceed 350 but passed 400 in 2016; biodiversity should be maintained about 90% and has decreased to about 80%; addition of N and P to crops and ecosystems should not exceed 62 Tg and 11 Tg, respectively, but are about 150 Tg and 22 Tg; and original forests should be maintained above 75% and they are down to 62%. The other boundaries pertain to emission of aerosols, stratospheric ozone depletion, ocean acidification, freshwater use, and dumping of organic pollutants (Steffen et al. 2015). Rockström and colleagues are concerned about each boundary but also about how one boundary possibly interacts with other boundaries that could possibly lead to tipping points that could have serious effects on changing the climate.

Excess N use can degrade the environment in several ways. The atmosphere can be affected by ammonia volatilization and denitrification that emits nitrous oxide ($N_2O$), a GHG. The most serious concerns, however, have been focused on water quality issues. Excess P is also primarily a water quality issue. This makes excess nutrients more of a local issue than a global issue like increasing $CO_2$. An agricultural practice that contributes $CO_2$ to the atmosphere affects the entire world but has no immediate impact locally. In contrast, excess nutrients can affect only a local site, a small watershed, a river basin, a river, or some part of the ocean. Therefore, the manner in how the public reacts to environmental concerns differs greatly.

As already stated, the agricultural community, as well as the environmental community, agrees that the widespread use of chemical fertilizers is affecting water quality. And while fertilizer use efficiency has increased because of improved technologies, the problems in some cases become greater simply because of the cumulative amounts of N and P being added to the environment. The agricultural community is almost universally opposed to regulations of any kind, and with water quality problems being somewhat localized, there have been few attempts to regulate the application of N and P although many guidelines have been established. The likely possibility of regulations will increase if and when localized water quality problems are perceived by the local residents as a serious health hazard. In the United States, two recent incidents are of interest.

### 1.4.4.1 Toledo, Ohio

On August 3, 2014, Toledo, the second largest city in Ohio, issued a "Do Not Drink" advisory for 400,000 residents served by the city water system. Chemical tests confirmed the presence of unsafe levels of the algal toxin Microcystin in the plant's treated water (Spear 2014). Harmful algal blooms in Lake Erie, one of the five Great Lakes of the United States, had become so highly concentrated that the toxin in the water entering the city treatment facility reached unsafe levels. The filtering system did not remove the toxin and boiling the water did not affect the toxin, so the residents were without drinking water and also advised not to bathe in the water. The cause of the algal bloom was largely considered excess P from fertilized cropland in the watershed of the lake. Even if N is limited, P can cause algal blooms because there are algae that can fix N from the atmosphere. However, if there are both N and P present in the runoff water, the algae blooms can occur much faster and more concentrated. This again shows the close linkage between N and P fertilizers and increased use of N fertilizer is almost always accompanied by increased use of P fertilizer. To date, this incident has not resulted in widespread agreement on how to address the problem, but this is an example of the kind of localized incidents that could cause citizens to demand to action, particularly if the problem occurs multiple times.

### 1.4.4.2   Des Moines, Iowa

The city of Des Moines, Iowa gets water from the Raccoon River for treatment and use by the nearly 500,000 residents. The U.S. Safe Water Drinking Standards requires that the nitrates cannot be greater than 45 ppm (10 ppm $NO_3$-N). In recent years, the concentration in water pumped from the rivers frequently exceeds this amount making it necessary for the treatment plant to use expensive methods to remove some of the nitrates. The cause of the nitrates is considered extensive use of N fertilizer and the high concentration of swine production facilities in Calhoun, Buena Vista, and Sac counties. In 2015, the Des Moines Water Works filed one of the first lawsuits in the United States for pollution from agricultural watersheds. Although the suit was scheduled to go to trial in 2016, it was postponed, and on January 27, 2017, the Iowa Supreme Court ruled against the Des Moines Water Works lawsuit (Maricle 2017). Again, this illustrates the difficulty, but most likely not the end, of regulating the use of chemical fertilizers to control or prevent degradation of water quality.

### 1.4.4.3   Phosphorus Fertilizer Bans

Even though the increased use of P fertilizer is directly linked to the increased use of N fertilizer, in many ways P is considered a more immediate concern than N. Based on a library review conducted for a legislative review, there are at least 11 states that have bans on P fertilizer use or sales (Miller 2012). Miller stated, "in general, these states prohibit P fertilizer application unless it is for (1) curing a lack of necessary P, (2) establishing new turf, or (3) repairing turf. Many states exempt agricultural lands and production, commercial or sod farms, gardening, or golf courses from the ban. And many prohibit applying fertilizer (not only P fertilizer) on impervious, frozen, or saturated surfaces, or within a certain distance of a water body." Therefore, even though these states have bans, P fertilizer use by crop producers, which are by far the primary users, is little affected.

## 1.5   WHY DOES THE NITROGEN PROBLEM DEFY A SOLUTION?

As stated in the Introduction, I applied synthetic N to cropland for the first time in 1953 as a graduate student. For every succeeding year, I have interacted with N as a scientist, a farmer, a water skier dodging algae blooms, a member of environmental task forces to address N problems, and an organizer of conferences and workshops to improve the efficient use of N. The widespread use of synthetic N began around 1950 with little thought of unintended consequences of its use. However, for more than 50 years, it has been well documented that excess N is associated with water quality, GHGs, and perceived as a health hazard. Countless experiments and computer simulation models have addressed the problem, and while there have been many outstanding accomplishments, the problems associated with the use of synthetic N are considered by many to become more serious with each succeeding year.

The author's view of some of the reasons that the N problem defies a solution is stated in this section. However, the generalized response curve between nitrogen sufficiency index and rate of N fertilizer by Holland and Schepers (2010) shown in Figure 1.1 is used as a resource. A statement often heard at conferences is "If producers would cut the rate of N applied by 10%, the pollution caused by excess N could be reduced by 50%." While I might quibble a little over the numbers, the statement is basically valid. This is in effect saying the farmer can largely solve the problem in a simple manner. The reason the farmer does not do this is that the amount of N needed by the crop is not known because the farmer does not know how much water will be available for the crop. The amount of N does not determine the yield; the amount of water transpired by the crop is the primary factor that determines the yield. The yield of a grain crop can be expressed by the following equation (Stewart and Peterson 2014).

$$GY = ET \times T/ET \times 1/TR \times HI \qquad (1.1)$$

where GY is kg ha$^{-1}$ of dry grain yield; ET is kg ha$^{-1}$ of evapotranspiration (water use by evaporation from soil surface and transpiration by the crop between seeding and harvest); T/ET is the fraction of evapotranspiration transpired by the crop; TR is the transpiration ratio (number of kilograms of water transpired to produce 1 kg of aboveground biomass); and HI is the harvest index (kg dry grain/kg aboveground dry biomass). This equation applies to all situations where grain crops are produced. A similar equation is presented by Fischer et al. (2014).

Nitrogen is not a factor in Equation 1.1, yet it is clear that N affects yield but only as it affects one of the other factors. The factors included in the equation have specific numbers for every situation. Phosphorus, diseases, insects, and countless other factors also affect grain yield but again only as they affect one of the factors included in the equation. For example, N affects plant vigor which can influence rooting depth to extract more water. It can affect how quick a plant canopy develops to shade the ground which will reduce evaporation from the soil surface and increase the T/ET factor. It can increase grain filling that will increase the HI. In the final analysis, however, the equation clearly shows that it is water that determines yield, and the amount of water available for a crop is beyond the control of the farmer, even if the crop is irrigated. This is because it is only the amount of water transpired by the growing crop that determines the amount of biomass produced by photosynthesis, and this is affected not only by the amount of water available but on other climatic factors such as temperature, radiation, humidity, and wind. Most irrigated crops also get a significant amount of their water from growing season rainfall and the amount of rainfall for a given year also affects humidity and other climatic factors. Even as complexing as not knowing how much water will be available to the crop is not knowing the amount of N supplied to the growing crop from mineralization of soil organic matter and in some cases from deposition with rainfall. Therefore, farmers never know with certainty how much N fertilizer will be required.

Even when farmers make good estimates of how much N the soil will supply, they still face perhaps the hardest choice of all and that is to decide how much fertilizer N to add based on the response curve shown in Figure 1.1. The curve clearly shows that the amount of yield increase becomes less with each succeeding unit of N fertilizer added until the maximum yield is obtained. This is the basis for the statement that a 10% reduction will reduce pollution by 50%. The farmer does not know what the maximum yield is going to be, so the first decision is to estimate the yield. This is difficult at best. The 1961 to 2014 maize grain yields for the United States and the world are shown in Figure 1.7. The yield from one year to the next can be quite large and in most cases the main reason is differences in weather. Also, with improved technologies such as improved cultivars and management practices that increase yields, the variation between years becomes greater. Because N is usually the first limiting factor other than water, most farmers want to make sure they have enough N available to fully utilize the water. If we assume a farmer reduces the estimated fertilizer requirement by 10%, and the yield is reduced by 5%, the profit of the farmer may be reduced by 20 or 30% or even more. This is because even though the reduced yield is rather small, it may have been a large part of the profit because the only added cost was the added N. All other input costs remain the same as well as harvest costs and fixed costs, so the farmer is very reluctant to allow N to limit production even for a "normal" year and most farmers always want to be able to capitalize on that good year that occurs every so often. Therefore, it is not realistic to expect a

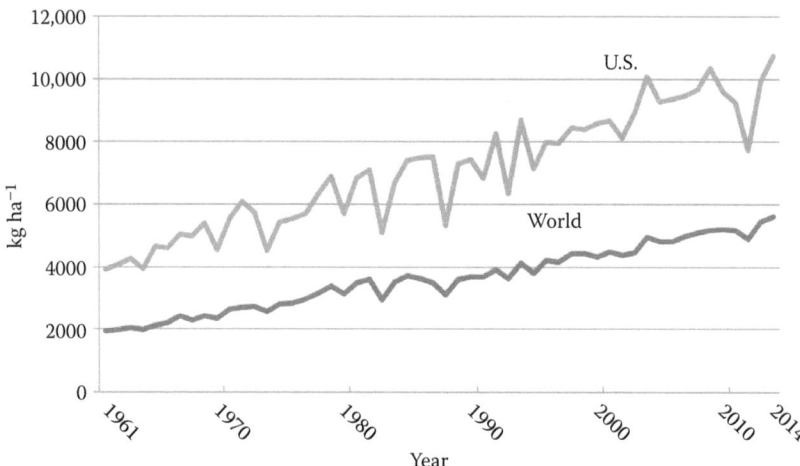

**FIGURE 1.7**    Yield of maize grain (kg ha⁻¹) for United States and the world from 1961 to 2014.

farmer to voluntarily cut back on an input that is so important to potential profit. Granted, some farmers set yield goals that are not realistic, particularly when the price of N fertilizer is relatively inexpensive. A yield goal equal to the average yield for the last 5 years is considered by some as realistic, but some farmers set a goal higher than they have ever achieved in the past.

There are many strategies that have been developed that have greatly increased the use efficiency of N fertilizer. The International Plant Nutrition Institute has promoted the 4-Rs program for several years that focuses on increasing fertilizer use efficiency. The 4-Rs are to add the right product at the right place at the right amount at the right time. Therefore, there have been major advances, but the right amount is still an enormous challenge as discussed above.

The increasing practice of taking feed to the animals rather than taking animals to the feed as was historically the practice is another concern that has no easy answer. Furthermore, it is likely to become even more serious in the future as the demand for animal protein increases worldwide. The economy of scale makes concentrated animal feeding operations more efficient, which leads to less expensive products that consumers demand. This is not likely to change unless at some point in the future a system is developed for factoring in environmental costs. The primary reason that these concentrated feeding operations are making excess N problems worse is that there is less efficient recycling of the manure back to crop production. Thus, more N must be added each year as fertilizer and this is the most overwhelming of all the causes of excess N. In 2014, there was about 115 M t of N and 20 M t of P worldwide with chemical fertilizer. Assuming an average N content of the grain as 1.7% and an average P content of 0.33, the N and P removed with maize, wheat, and rice grains in 2014 would be approximately 43 Tg of N and 8 Tg of P (calculated from data presented in Figure 1.3), which is less than 40% of that added. More than 50% of the world's fertilizer is added to wheat, rice, and maize (FAO, 2000). Thus, substantially more N and P are being added each year to these cereal crops as fertilizer, and much of the N and P in the grain come from natural sources. The process is repeated every succeeding year and the added N is coming from the air and the P is coming from mines so enormous amounts of these elements are added to the environment every year. Perhaps slowly, but surely, some of these excess nutrients have negative effects on the environment. The obvious solution is to put more emphasis on recycling, but the present trends of production make this difficult.

The record is clear that synthetic N fertilizer has been the single most important factor that has allowed food production, particularly rice, wheat, and maize that account for more than half the world's calories, to increase fourfold in 50 years. Without N fertilizer, the world population likely could not have increased from 3 to 7 billion during the same period. More importantly, with the world population expected to reach 9.7 billion by 2050 and an increasing number of them demanding more animal-based protein, the continued use of huge amounts of fixed N and mined P will be required that will make the problem of excess nutrients in the environment even more challenging. Therefore, the problem will continue to defy a solution even though essentially all agricultural, environmental, scientific, and public organizations agree a solution is needed. The difficulty is that there is not one solution because the problems are somewhat localized both regarding cause and effect. Also, while most people agree on the problem, there is not agreement on how serious the problems are. In many cases, the perception of a problem may be greater than the problem.

While the scientific community needs to continue developing more efficient methods for applying and utilizing fertilizer, science cannot develop a solution. Likely, solutions will come only after people in localized areas become concerned enough to demand change. They must also be willing to accept the consequences of what these changes may bring. Success in localized areas may eventually spread to other areas, but the solution for each area will be different.

## 1.6 CONCLUSIONS

The Haber–Bosch process that synthesizes $NH_3$ from $N_2$ and $H_2$ using high temperature is without question one of the most significant achievements ever. This resulted in readily available fertilizer at relatively affordable cost that allowed rapid and significant increases in food production. Between 1961 and 2014, the world population increased from 3 billion to almost 7.4 billion people. However, largely because of synthetic N fertilizer, the production of rice, wheat, and maize increase almost fourfold and these three cereals account for more than half of the world's food calories. It is unlikely that the population could have increased to this level without N fertilizer. This accomplishment did not come without damage to the environment. Some of the added N leached through the soil into groundwater resulting in health concerns, particularly for infants. Some also reached lakes and rivers causing algal growth, and some became denitrified emitting nitrous oxides into the atmosphere that added to greenhouse gases. Thus, while the early discovered benefits of N fertilizer were celebrated, the unintended consequences are now considered serious problems that seem to defy solutions. Nitrogen was the most limiting element for crop production but when N fertilizer removed this constraint, phosphorus became limiting in most soils leading to a demand for P fertilizer. The demand was met by obtaining rock phosphate from mines and treating it with acid to make the P available to plants. Between 1961 and 2014, there were 470 Tg of N and 96 Tg of P used as fertilizer. In many cases, excess P is considered more damaging to the environment than N. Both N and P occur naturally but synthetic N comes from the atmosphere and fertilizer N comes from mines, so all the N and P from these sources are new inputs to the environment. These huge amounts of N and P are added each succeeding year, so the historical N and P cycles are being seriously affected. In addition, more than half of fertilizer is added to rice, wheat, and maize. With the increasing demand for animal-based protein, maize has become the most highly produced cereal because of its high use for animal feed. In the United States, about 40% of the maize produced is fed directly to animals and another 40% is used to produce ethanol for fuel but about 1/3 of that ends up as dried distillers grain that is used for animal feed. Because of efficiencies gained from feeding animals in large concentrated feeding operations, increasing amounts of maize are being transported from areas where it is grown to where the animals are concentrated. In many cases, these are far removed and often located in areas where there is little or no cropland. From 50 to 70% of the N and P consumed by the animals is excreted but not recycled efficiently if at all. This means that even if the N and P fertilizer used to produce the maize was used efficiently, the result is that much of the N and P ultimately often lead to degrading the environment. This not only happens in localized areas, but even in countries such as the Netherlands and Denmark where huge amounts of grain are imported to feed animals and then the products exported leaving the manure behind.

Even though there have been outstanding achievements in improving the use efficiency of fertilizer, the problem of excess nutrients in the environment is becoming more serious. The problem will not be solved by science. Although bad in some ways, it is good in others that the while the problems occur worldwide, they tend to be localized. Thus, solutions will likely occur only when public citizens of a localized area become so concerned that they demand action and are willing to accept the consequences that result. If and when such actions occur that are successful, they could serve as catalysts for other areas to follow that would slowly, but surely, lessen the negative impacts of N and P fertilizers on the environment. In reality, there is no other choice because the world population has reached the level that it does not appear possible to produce enough food without synthetic N.

## REFERENCES

Amundson, R., A.S. Berhe, J.W. Hopmans, C. Olson, A.E. Sztein, and D.L. Sparks. 2015. Soil and human security in the 21st century. *Science* 348 (6235). doi:10.1126/science.1261071.

Bruinsma, J. (Ed.). 2003. *World Agriculture: Towards 2015/2030: An FAO Perspective*. Earthscan Publications Ltd., London. Available at http://www.fao.org/3/a-y4252e.pdf.

Bumb, B.L. and C.A. Baanante. 1996. World trends in fertilizer use and projections to 2020. International Food Policy Research Institute, 2020 Brief 38. Available at http://ageconsearch.umn.edu/bitstream/16353/1/br38.pdf.

Commoner, B. 1971. *The Closing Circle*. Knopf, New York, NY.

DeBruin, J. 2013. Nitrogen uptake in corn. *Crop Insights*, DuPont Pioneer, Johnston, IA. Available at: https://www.pioneer.com/home/site/us/agronomy/library/n-uptake-corn/.

Dutch Soy Coalition. 2009. Strategies for reducing the negative impacts of soy production: Replacing soy in animal feed. Available at: http://141.105.120.208/dsc/wp-content/uploads/2014/04/Factsheet-1-Strategies-for-reducing-the-negative-effects-of-soy-production-Responsible-soy-production.pdf.

Eck, H.V. and B.A. Stewart. 1954. Wheat fertilization studies in western Oklahoma. Bulletin B-432. Oklahoma Agricultural Experiment Station and USDA Agricultural Research Service. Stillwater, OK.

Eck, H.V., B.A. Stewart, and G.O. Boatwright. 1957. Anhydrous ammonia and ammonium nitrate fertilizer for wheat. Bulletin B-493. Oklahoma Agricultural Experiment Station, Stillwater, OK.

Erisman, J.A., M.A. Sutton, J. Galloway, Z. Klimont, and W. Winiwarter. 2008. How a century of ammonia synthesis changed the world. *Nature Geoscience* 1:363–369.

FAO. 2000. Fertilizer requirements in 2015 and 2030. Food and Agriculture Organization of the United Nations. Available at: ftp://ftp.fao.org/agl/agll/docs/barfinal.pdf.

FAO. 2017. Dimensions of need—An atlas of food and agriculture. Food and Agriculture Organization of the United Nations, Rome. Available at: http://www.fao.org/docrep/u8480e/u8480e07.htm.

FAOSTAT. 2017. Food and Agriculture Organization Corporate Statistical Database Statistics Division. Food and Agriculture Organization of the United Nations. Rome. Available at http://faostat3.fao.org/home/E.

Fischer, R.A., D. Byerlee, and G.O. Edmeades. 2014. Crop yields and global security: Will yield increase continue to feed the world? *ACIAR Monograph* No. 158. Australian Centre for International Agricultural Research, Canberra.

Galloway, J.N., E.B. Cowling, S.P. Seiltzinger, and R.H. Socolow. 2002. Reactive nitrogen: Too much of a good thing? *Ambio* 31:60–63. Available at: https://www.jstor.org/stable/4315216?seq=1#page_scan_tab_contents.

Genzel, B. and C. Reinhardt. 2017. Postwar fertilizer explodes. Wessels Living History Farm, York, NE. Available at http://www.livinghistoryfarm.org/farmingthe40s/crops_04.html.

Haub, C. 2012. Fact Sheet: World Population Trends 2012. Population Reference Bureau, Washington, DC. Available at: http://www.prb.org/publications/Datasheets/2012/world-population-data-sheet/fact-sheet-world-population.aspx.

Heinz Center. 2002. The State of the Nation's Ecosystems: Measuring the Lands, Waters, and Living Resources of the United States. The Heinz Center for Science, Economics and the Environment. Washington, DC.

Himes, F.L. 1997. Nitrogen, sulfur, and phosphorus and the sequestering of carbon. In: R. Lal, J.M. Kimble, R.F. Follett, and B.A. Stewart (eds.), *Soil Processes and the Carbon Cycle. Advances in Soil Science*, CRC Press LLC, Boca Raton, FL, pp. 315–319.

Holland, K.H. and J.S. Schepers. 2010. Derivation of a variable rate nitrogen application model for in-season fertilization of corn. *Agronomy J.* 102:1415–1424. doi:10.2134/agronj2010.0015.

Islam, R. and R. Reeder. 2014. No-till and conservation agriculture in the United States: an example from the David Brandt farm, Carroll, Ohio. *International Soil and Water Conservation Research* 2:97–107.

Logan, C. 2016. Focus on The Netherlands. The Grain and Grain Processing Information Site. Available at: http://www.world-grain.com/Departments/Country-Focus/Country-Focus-Home/Focus-on-the-Netherlands.aspx.

Lory, J.A., K.C. Olson, and C. Zumbrunnen. 2006. Calculating fertilizer value of supplemental feed for cattle on pasture. MU Guide G 2083. MU Extension, University of Missouri, Columbia. Available at: http://extension.missouri.edu/explorerpdf/agguides/ansci/g2083.pdf.

Maricle, K. 2017. Supreme Court rules against Des Moines Water Works over nitrates. Available at: http://whotv.com/2017/01/27/supreme-court-rules-against-des-moines-water-works-lawsuit-over-nitrates/.

Miller, K.L. 2012. State laws banning phosphorus fertilizer use. OLR Research Report 2012-R-0076. Available at: https://www.cga.ct.gov/2012/rpt/2012-R-0076.htm.

Molden, D. 2007. Water for Food, Water for Life. International Water Management Institute and Earthscan, London. Available at http://www.iwmi.cgiar.org/assessment/files_new/synthesis/Summary_SynthesisBook.pdf.

Narayanan, N. 2013. Massive nitrogen pollution accompanies China's growth. *Scientific American*. Available at: https://www.scientificamerican.com/article/massive-nitrogen-pollution-accompanies-chinas-growth/.

National Research Council. 1993. *Soil and Water Quality: An Agenda for Agriculture*. National Academy Press, Washington, DC.

Roberts, T.L. and D.W. Dibb. 2009. Fertilizer use in North America: Types and amounts. In: Rattan, L. (ed.), *Agricultural Sciences*, Volume II. Eolss Publishers Co. Ltd., Oxford, United Kingdom, pp. 176–189.

Ruddiman, W.F. 2003. The anthropogenic greenhouse ERA began thousands of years ago. *Climate Change* 61:261–293.

Schwägerl, C. 2016. With too much of a good thing, Europe tackles excess nitrogen. *Yale Environment* 360. Available at: http://e360.yale.edu/features/with_too_much_of_a_good_thing_europe_tackles_excess_nitrogen.

Smil, V. 1999. Detonator of the population explosion. *Nature* 400:415.

Smil, V. 2001. *Enriching the Earth: Fritz Haber, Carl Bosch, and the Transformation of World Food Production*. MIT Press, Cambridge, MA.

Spear, S. 2014. Toxic algae bloom leaves 500,000 without drinking water in Ohio. *EcoWatch*. Available at: http://www.ecowatch.com/toxic-algae-bloom-leaves-500-000-without-drinking-water-in-ohio-1881940537.html.

Steffen, W., K. Richardson, J. Rockström, S.E. Cornell, I. Fetzer, E.M. Bennett, R. Biggs et al. 2015. Planetary boundaries: Guiding human development on a changing planet. *Science* 347. 1259855 DOI: 10.1126/science.1259855.

Stewart, B.A. and G.A. Peterson. 2014. Managing green water in dryland agriculture. *Agronomy Journal* 106:1–10. doi:10.2134/agronj14.0038.

Stewart, B.A., F.G. Viets, Jr., G.L. Hutchinson, and W.D. Kemper. 1967. *Environmental Science and Technology* 1:736–739.

Stewart, B.A., D.A. Woolhiser, W.H. Wischmeier, J.H. Caro, and M.H. Frere. 1975. Control of water pollution from cropland: Volume I: A manual for guideline development. Report No. ARS H-5-1. USDA Agricultural Research Service and Office of Research and Development, Environmental Protection Agency, Washington, DC.

Stewart, B.A., P. Pokhrel, and M. Bhandari. 2017. Effects of phosphorus fertilizer on U.S. agriculture and the environment. In: R. Lal and B.A. Stewart (Eds.), *Soil Phosphorus. Advances in Soil Science*. CRC Press, Boca Raton, FL, pp. 23–42.

Sutton, M.A., C.M. Howard, J.W. Erisman, G. Billen, A. Bleeker, P. Grennfelt, H. van Grinsven, and B. Grizzetti (Eds.). 2011. *The European Nitrogen Assessment: Sources, Effects and Policy Perspectives*. Cambridge University Press, New York. Available at: http://www.nine-esf.org/node/360/ENA-Book.html.

UNEP. 2014. *Excess Nitrogen in the Environment*. 2014 Year Book, United Nations Environmental Programme, Nairobi, Kenya. Available at: http://hqweb.unep.org/yearbook/2014/PDF/chapt1.pdf.

UN. 2017. World population projected to reach 9.7 billion by 2000. United Nations Department of Economics and Social Affairs. Available at: http://www.un.org/en/development/desa/news/population/2015-report.html.

USDA. 2017. ERS Feed Outlook. Economic Research Service, USDA. Washington, DC. Available at: https://www.ers.usda.gov/publications/pub-details/?pubid=82089.

van Heugten, E. and T. van Kempen. 2017. Understanding and applying nutrition concepts to reduce nutrient excretions in swine. North Carolina Cooperative Extension Service, Raleigh. Available at: https://projects.ncsu.edu/project/swine_extension/nutrition/environ/concepts.pdf.

Vitousek, P.M., J. Aber, R.W. Howarth, G.E. Linens, P.A. Matson, D.W. Schindler, W.H. Schlesinger, and G.D. Tilman. 1997. Human alteration of the global nitrogen cycle: Causes and consequences. *Issues in Ecology* 1, Spring 1997. Available at: http://www.esa.org/esa/documents/2013/03/issues-in-ecology-issue-1.pdf.

Worldwatch Institute. 2016. Grain harvest sets record, but supplies still tight. Available at: http://www.world-watch.org/node/5359.

# 2 Nitrogen Fixation by Pulse Crops and the Use of Nitrogen Isotopic Techniques to Measure the Fixation Capacity

*Minh-Long Nguyen*

## CONTENTS

## 2.1 INTRODUCTION

Pulse crops, also referred to as grain legumes, are annual crops that are harvested solely for dry grain, thus excluding crops harvested green for food (green peas, green beans, etc.) which are classified as vegetable crops. Pulses include field peas (*Pisum sativum* L.), dried or split peas (*Pisum sativum* L.), pigeon peas (*Cajanus cajan* L.), chickpeas (*Cicer arietinum* L.), lentils (*Lens culinaris*), beans (*Phaseolus vulgaris* L.), including the common bean, kidney beans, haricot beans, pinto beans, snap beans and navy beans, faba beans (*Vicia faba* L.), mung beans (*Vigna radiata* L.), bambara beans (*Vigna subterranea* (L.) Verdc.), and lupins (e.g., *Lupinus albus* L., *Lupinus mutabilis*, *L. angustifolius*, *L. luteus*) (FAO 2016a, b; http://iyp2016.org/).

The total annual world production of pulses (2011–2013) is around 72.3 million tons (Mt) and global pulse demand has continued to increase over the past 30 years from 42 Mt in 1980–1981, to 66 Mt in 2011. Over the period between 1981 and 2013, the area under pulse production increased from 63 million hectare (Mha) to 80 Mha. The main producing countries include India, Myanmar, China, Canada, Australia, the United States, and African countries including Ethiopia, the United Republic of Tanzania, Nigeria, and Niger (FAO 2014; Anglade et al. 2015; Joshi and Rao 2016; OECD-FAO 2016).

Pulses are known to have a low carbohydrate, but high protein and fiber content, thus providing a healthy, protein-rich diet with a low glycemic index which can address the hidden hunger affecting more than 2 billion people caused by deficiencies in protein and micronutrients, compared to the 795 million affected by hunger (low calorie intake) (FAO 2016a). These protein-rich crops have many other benefits compared to continuous cereal-based systems, particularly when they are included in an arable cropping rotation (Lemke et al. 2007; Gan et al. 2009, 2011a–c, 2015; Kopke and Nemecek 2010; Chianu et al. 2011; FAO 2016b). These include: (1) an improvement in soil fertility, especially in the availability of soil nitrogen (N) via biological nitrogen fixation (BNF), (2) an enhancement of soil organic carbon (SOC) pool because of the input of a high quality (low C:N ratio) biomass of pulse residues, and (3) an increase in soil aggregation and structure because of improvement in SOC and polysaccharides. The increase in SOC can increase nutrient and water holding capacity, reduce soil degradation, and minimize nutrient leaching losses. Additionally, pulses require less water for growth and production than many high protein agricultural products such as chicken, pork, and beef (Hoekstra and Chapagain 2008; Schlink et al. 2010; FAO 2016b). For example, the water footprints required to produce a kilogram (kg) of beef, pork, chicken, and soybeans are 43, 18, 11, and 5 times higher than that for pulses. Thus, pulses are ideal for dry climatic conditions and low water input agriculture (Marcellos et al. 1998; Herridge et al. 1998, 2008; Angadi et al. 2008; Cutforth et al. 2009).

Pulses, similar to other leguminous plants such as clover and alfalfa, have properties that can use atmospheric nitrogen ($N_2$) through a process called BNF. It is a symbiotic relationship between pulses/ leguminous crops and specific soil bacteria (*Rhizobia*). Soon after the crop germinates, *Rhizobia* enter the root hairs of pulses/leguminous crops and induce nodule formation (Figure 2.1). The crops absorb atmospheric $N_2$ and provide energy, nutrients, and water to the *Rhizobia* in the nodules (Hardarson and Atkins 2003; Fustec et al. 2010). The soil *Rhizobia* in turn converts atmospheric $N_2$ into ammonium $\left( NH_4^+ \right)$, which is available for use by plants for growth. BNF can supply 50–80% of plant N uptake in a pulse crop, and also contributes some residual N to the following cereal crop which is grown in pulse- arable rotation (Figure 2.2). It is estimated that some 190 Mha of grain legumes globally contribute around 5–7 Mt of N to soils (i.e., $5–7 \times 10^9$ kg or 5–7 teragram (Tg)) (FAO 2016a).

The amount of $N_2$ fixed by pulses varies, depending on the type of pulse crop. The faba bean is known to be a stronger N fixer than chickpeas, lentils, and common beans (102–123 kg N ha$^{-1}$ yr$^{-1}$ vs. 50–86 kg N ha$^{-1}$ yr$^{-1}$, respectively; Walley et al. 2007; Herridge et al. 2008; Gan et al. 2010). Besides crop types, the presence of the appropriate *Rhizobium* strain and optimum environmental conditions such as adequate soil moisture and adequate plant-available soil phosphorus (P) and sulfur (S) for pulse growth (Singh and Sekhon 2007), can also influence the amount of $N_2$ fixed

**FIGURE 2.1    (See color insert.)** Nodules on faba bean. (With permission from Dr. M. Unkovich; School of Agriculture, Food and Wine, The University of Adelaide, Australia; http://www.soilquality.org.au/factsheets /legumes-and-nitrogen-fixation-south-australia.)

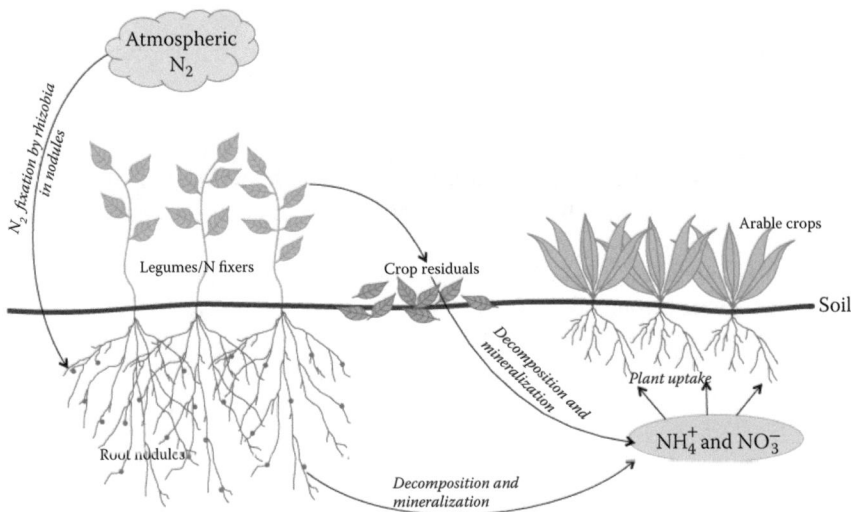

**FIGURE 2.2  (See color insert.)** Inputs of biological nitrogen fixation from grain (pulses) and forage legumes to arable crops via decomposition and subsequent mineralization of dead nodules and crop residues.

by pulses (Herridge 2009; Gan et al. 2011a–c, Seymour et al. 2015). The estimated amounts of $N_2$ fixed by chickpeas is lower under drought than under adequate moisture conditions (19–24 and 60–80 kg N ha$^{-1}$, respectively; Carranca et al. 1999; Unkovich and Pate 2000). Only a small amount of N is required as a starter for pulses where soil available N (soil ammonium and predominantly soil nitrate) is extremely low (Peoples et al. 2009a). Urea application that provides 10 or more kg N ha$^{-1}$ can suppress nodulation and hence reduce BNF. Optimum soil nitrate $\left( NO_3^- \right)$ levels for BNF are reported to be below 50–56 kg N ha$^{-1}$ in the top 100 cm soil depth (Voisin et al. 2002; Herridge et al. 2015). Virtually no BNF is fixed when soil $NO_3^-$ in the topsoil (100 cm depth) is at 200–350 kg N ha$^{-1}$ (Doughton et al. 1993; Schwenke et al. 1998; Seymour et al. 2015). Intensive soil cultivation can also lower BNF, since it can increase soil organic N (SON) mineralization to $NO_3^-$ which can reduce BNF (https://grdc.com.au/research-and-development /grdc-update-papers/2009/09/nitrogen-fixation-benefits-of-pulse-crops).

The average amount of $N_2$ fixed in an above-ground biomass by pulse crops (shoots including grains) is 100–200 kg N ha$^{-1}$ yr$^{-1}$ (Evans et al. 2001; Crews and Peoples 2004, 2005; Peoples et al. 2001; Herridge 2009; Anglade et al. 2015; Seymour et al. 2015). The amount of $N_2$ fixed (kg N ha$^{-1}$ yr$^{-1}$) by a pulse is directly proportional to its growth, which is estimated to be 20–25 kg N fixed per Mg (1000 kg = metric tonne = Mg) of above-ground dry matter produced (Peoples et al. 2001; Unkovich et al. 2010). It is the product of the amount of above-ground biomass production (kg dry matter ha$^{-1}$ yr$^{-1}$) by pulse crops and the proportion of N derived from atmosphere (%Ndfa). Anglade et al. (2015) proposed that the amount of $N_2$ fixed is strongly related to N yield (N in grain and in shoot DM).

The percentage of N derived from the atmosphere (%Ndfa) is crucial for determining the amount of $N_2$ fixed by pulse crops and hence how much of the remaining N is coming from soil or fertilizer sources to achieve optimum yields. Pulse crops can derive up to 50–80% (mean of 75%) of their N requirements from the atmosphere through BNF with a range from as low as 0%, to as high as 80%, partly depending on the effectiveness of the symbiotic relationships and environmental conditions as stated above (Peoples et al. 2001; Anglade et al. 2015). Published data suggests that %Ndfa is higher in faba beans (71–85%) than in chickpeas, field peas, mung beans, and lentils (38–60%) (Walley et al. 2007; Herridge et al. 2008; Anglade et al. 2015).

The N-fixing ability of pulses can also reduce the reliance of the following cereal crop on N fertilizer application because of the residual soil N contribution from the pulse crops to soils following grain harvest. This residual soil N is contributed not only from pulse crop residue remaining after grain harvest but also from below-ground ground contributions from pulse crops. Field studies suggest that N associated with the nodules and roots of pulse crops may represent 30–60% of the total amount of N taken up by pulse crops (e.g., Rochester et al. 1998; Khan et al. 2002, 2003; McNeill and Fillery 2008; Peoples et al. 2001, 2015). Thus, both above-ground and below-ground pulse crop residues can contain significant amounts of fixed N even after a large amount of N is removed in grain at harvest. However, only 2–40% of N from pulse residues is made available to succeeding crops the following year through decomposition and mineralization processes (Stevenson and van Kessel 1996; Lupwayi and Kennedy 2007), depending on factors such as the C:N ratios of residues, soil microbial biomass activity, soil moisture, and soil temperature (Ross et al. 2008; Arcand et al. 2014). Ross et al. (2008) estimated that 44% of the N from faba beans was retained as soil N, 40% was taken up by crops, and 16% was lost to the environment. Walley et al. (2007) studies in the Northern Great Plains in Canada have reported that the extent of N contribution from pulses to a subsequent crop is difficult to predict, but pulse crops with relatively high levels of $N_2$ fixation (60–80%), such as faba beans and field peas, are more likely to contribute positively to the overall N economy of a cropping system than pulses with lower levels of $N_2$ fixation (38–53%) such as desi and kabuli chickpeas and common beans, particularly when a cropping system is evaluated over a long term.

Studies conducted in Australia (Marcellos et al. 1993; Dalal et al. 1998; Weston et al. 2002; Herridge 2009) have also reported the rotational benefits of pulses. Cereals grown after pulses (e.g., wheat following chickpeas) commonly yield 0.5–1.5 t/ha more than cereals grown after cereals (e.g., wheat-wheat rotation). This increase was attributed to a higher level of soil nitrate (equivalent to 35–40 kg nitrate-N ha$^{-1}$ in the top 120 cm soil profile) after pulse cropping than in a continuous wheat crop rotation.

The extent of the residual benefit of pulses on soil $NO_3^-$ and subsequent crops may last for one or more growing seasons, depending on many factors such as the decomposition rate of pulse residues, the mineralization of soil organic N, and N losses via leaching, denitrification, and volatilization. Marcellos et al. (1993) reported an average yield benefit of 46% for wheat (3.2 t ha$^{-1}$) after chickpea crops versus 2.2 t ha$^{-1}$ for wheat after wheat crops for the residual first wheat crop in northern New South Wales, Australia. However, there were no beneficial effects from chickpeas on soil $NO_3^-$ and yields in a second residual wheat crop for five of the studied six sites. In contrast, Ross et al. (2015) in their studies in Alberta, Canada, showed that pulse crops can improve the yield and quality (protein level) of a second subsequent crop. In year 3, increases in barley grain yields averaged 11% (0.33 Mg ha$^{-1}$) and increases in seed N yields averaged 11% (7.2 kg N ha$^{-1}$) after a 1-year (YR1) faba bean crop, when compared with YR1 canola or barley without added N. Gan et al. (2015) in their 3-year crop growing studies (repeated for five cycles from 2005 to 2011) in Canada also reported a long residual effect of pulses on cereal crops, depending on the seasonal conditions under which the crops were grown. When averaged over the five cycles, the available soil N in the top 120 cm soil profile when durum wheat was planted was 76.3 kg ha$^{-1}$ in the fields following pulses, which was 57.5% greater than in the fields following cereal crops. In a 3-year cropping sequence, the pulse system increased total grain production by 35.5%, improved protein yield by 50.9%, and enhanced fertilizer-N use efficiency by 33.0% over the summer-fallow system.

By reducing the requirement for N fertilizers, pulse crops can reduce $N_2O$ emissions and the energy use associated with the production and use of N fertilizers, thus contributing to the reduction of agricultural impacts on climate change (Anglade et al. 2015; FAO 2016b). The effectiveness of fertility maintenance depends upon the balance between (1) N inputs from above-ground and below-ground pulse crops and from N fertilizer application to arable crops and (2) N outputs through harvested products (grains), straws out of farm gates, the return of crop residues and N losses via drainage beyond the plant rooting zone, denitrification, and volatilization. To construct an N balance, one of the important aspects is to quantify the BNF of different grain legumes under different rotations and environmental conditions. The N rhizodeposition from the pulse rooting zone, which is mainly the senescence and decay of pulse roots and nodules, and the exudation of N compounds by living roots is also an important N pool for subsequent cereal crops (Carlsson and Huss-Danell 2003; Crews and Peoples 2004, 2005; Walley et al. 2007; McNeill and Fillery 2008). In addition, the interaction between N rhizodeposition and soil organic matter (OM) and soil micro-organisms may enhance the turnover of plant residues and the plant-availability of soil nutrients for subsequent crops (Fustec et al. 2010). Herridge et al. (2008) in their comprehensive review of BNF have reported that the amount of pulse's below-ground N (BGN) accounted for 22–68% of the total plant N for pulses such as faba beans, chickpeas, mung beans, lupins, peas, and pigeon peas. To account for BGN when calculating $N_2$ fixation, Herridge et al. (2008) used a multiplication factor of 2.0 for chickpeas (assuming 50% of plant N is below-ground) and 1.4 for the remainder of the pulses (assuming 30% BGN). A subsequent review by Anglade et al. (2015) separated the amount of pulse's below-ground N (BGN), into two components: (1) below-ground plant-derived N (BGP-N, i.e., the amount of total plant N that is below-ground) and (2) N depositions (Ndfr) via exudates and decaying root cells and hyphate. The proportions of total plant N as BGP-N and Ndfr for a pulse crop ranged from 11 to 56% (mean of 26% with a standard deviation [SD] of 10%) and 1.4 to 51% (mean of 20% with SD of 13%), respectively. With these data, Lassaletta et al. (2014) and Anglade et al. (2015) have recommended that a multiplication factor of 1.3 is used to account for the input of BGN from a pulse crop. The total $N_2$ fixed by pulses is a product of this factor with the above-ground N (estimated from grain yield and harvest index which is the ratio of harvested grain to the total above-ground biomass production) through a linear regression model (Anglade et al. 2015). Herridge et al. (2008) in their global review for (pulse and forage legumes) have also reported a similar proportion of the total plant N as BGN with a multiplication factor of 1.4.

The purpose of this chapter is to highlight the role of nitrogen stable ($^{15}$N) isotopic techniques to quantify BNF by pulse crops. The review will also briefly mention the use of non-isotopic conventional techniques to assess BNF.

## 2.2 QUANTIFICATION OF FIELD BIOLOGICAL NITROGEN FIXATION

### 2.2.1 METHODS FOR QUANTIFYING BIOLOGICAL NITROGEN FIXATION

Biological nitrogen fixation of pulse crops can be quantified by methods such as acetylene reduction assay (ARA), nodule observations, xylem sap analysis (ureide production), total N difference (TND), and $^{15}$N stable isotopic methods (natural abundance and isotope dilution methods) (Peoples et al. 1989, 2002; Peoples and Herridge 1990; Hardarson and Atkins 2003; Anglade et al. 2015). This review focuses on the use of $^{15}$N isotopic methods to quantify BNF and hence the principles and procedures for other methods are only briefly mentioned here. Further details can be viewed in published research from many authors (e.g., Unkovich and Pate 2000; Herridge et al. 2008; Peoples et al. 2002, 2009b).

#### 2.2.1.1 Acetylene Reduction Assay (ARA)

This method measures the rate of conversion of added acetylene ($C_2H_4$) to ethylene ($C_2H_4$) by the $N_2$-fixing enzyme (nitrogenase), which exists in pulse crops. This enzyme is not only involved in BNF but also in the reduction of added $C_2H_4$ to $C_2H_4$. The ethylene ($C_2H_4$) produced can be analyzed using gas chromatography and converted into $N_2$ fixation by multiplying with a conversion factor of three ($C_2H_2$:$N_2$ ratio of 3:1), which represents the theoretical stoichiometric relationship between $C_2H_2$ reduction and $N_2$ fixation (Hardy et al. 1973; Roskowski 1981). However, there is evidence that experimental ratios may vary from the theoretical 3:1 ratio (Peoples and Herridge 1990). In addition, this ARA method can only provide a spot BNF measurement, instead of an integrated measurement over a growing season of a pulse crop.

### 2.2.1.2 Nodule Observation

The indices of nodulation, number of nodules, fresh and dry weight of nodules, and visual appraisal of the leghemoglobin content (degree of red pigmentation) in nodules may provide a rough guide to the level of $N_2$ fixation within a single cultivar (Hardarson and Atkins 2003).

### 2.2.1.3 Xylem Sap Analysis

The xylem sap technique can be used to measure BNF for plant species that produce significant quantities of ureide in plant sap as a product of $N_2$ fixation. The proportion of N as ureide in the xylem sap in these plants is directly proportional to the amount of $N_2$ fixation (Herridge et al. 1996). Although this technique is simple, relatively inexpensive, and provides an instantaneous assay of $N_2$ fixation, both the composition of the sap and the rate of BNF must be calibrated against an independent method of measurement (Herridge et al. 1996).

### 2.2.1.4 Nitrogen Balance or the Difference Method

The total N-balance method is based on the principal that the soil-plant system will accumulate N over time if there is an input of $N_2$ fixation. However, changes in soil and plant N may be influenced not only by $N_2$ fixation but also by other external N inputs such as rainfall and dry deposition and N losses from the plant-soil system through ammonia volatilization, denitrification, leaching, and so on (Peoples and Herridge 1990; Giller and Merckx 2003). A simple variation of N balance for quantifying $N_2$ fixation is the difference in total N uptake (kg N ha$^{-1}$) between $N_2$-fixing plants and neighboring non-$N_2$-fixing (i.e., reference) plants (Unkovich et al. 2008; Chemining'wa et al. 2013) as shown below:

$$\% \text{ Ndfa} = \left[(\text{Total N uptake in } N_2\text{-fixing plants} - \text{Total N in reference plants})/ \right.$$
$$\left. \text{Total N uptake in } N_2\text{-fixing plants}\right] \times 100$$

where total N uptake (kg N ha$^{-1}$) is the product of dry matter (DM) yield and N content (%) in plants.

The main assumption of this total N difference (TND) method is that the $N_2$-fixing plants assimilate the same amount of soil mineral N as the neighboring non-$N_2$-fixing plants. However, this method can be influenced by soil N status and the differences between $N_2$-fixing and non-$N_2$-fixing plants in root morphology and rooting depth to exploit soil N (Herridge et al. 1995; Chalk 1998).

### 2.2.1.5 Nitrogen Stable (Nitrogen-15) Isotopic Techniques

Both $^{15}N$ isotopic dilution and $^{15}N$ natural abundance are widely used to estimate BNF (Unkovich et al. 2008). The principle of these techniques is that $N_2$-fixing plants such as pulses have different amounts of $^{15}N$ in their tissues than non-fixing plants. Based on this difference in the $^{15}N$ signatures, it is possible to estimate BNF based on $^{15}N$ analyses from the $N_2$-fixing plant and a non-$N_2$-fixing (reference) plant. These isotopic techniques will be reviewed in the next section (Section 2.2.2).

### 2.2.2 BASIC PRINCIPLES OF NITROGEN STABLE ISOTOPES

The nitrogen (N) atom has two stable isotopes $^{14}N$ and $^{15}N$ $\left(^{14}_{7}N \text{ and } ^{15}_{7}N\right)$, which have the same number of protons (Z = 7) but different atomic weights (A of 14 and 15, respectively) and a different number of neutrons (N, defined as the difference between A and Z, of 7 and 8, respectively). These isotopes are stable, since they do not emit any radiation. The heavy $^{15}N$ (atomic weight of 15) accounts for only 0.3663% of naturally occurring N in the atmosphere, while the light $^{14}N$ (atomic weight of 14) makes up the vast majority (99.63%) of naturally occurring N. $N_2$-fixing plants, which are able to fix atmospheric $N_2$, contain lower atom % $^{15}N$ values (and hence have lower $^{15}N/^{14}N$ ratios) when compared with those of non-N-fixing plants, which derive all of their N from soil (Herridge et al. 2008). This is because of N transformation processes which tend to result in soil and biological material (e.g., plant tissues) having higher Atom% $^{15}N$ and hence higher $^{15}N/^{14}N$ ratio than the atmospheric N (Unkovich et al. 2008). The change in $^{15}N$ concentrations in soil or plant samples with respect to atmospheric $^{15}N$ is expressed as $\delta^{15}N$ in parts per thousand (‰) or per mil.

Isotopic data obtained from isotopically enriched studies are usually reported as atom $^{15}N$ percent (Atom% $^{15}N$) or atom percent excess (Atom% $^{15}N$ excess). In contrast, isotopic data obtained from natural abundance studies using high precision mass spectrometers which can discriminate the differences in the natural abundances of $^{15}N$ between $N_2$-fixing and non-fixing (reference) plants, are expressed as the ratio (R) of the $^{15}N$ to $^{14}N$ in the sample (R Sample) compared to the same ratio ($^{15}N{:}^{14}N$) in an international standard (R Standard), using the "delta" ($\delta$) notation ($\delta^{15}N$).

### 2.2.2.1  Notation of $^{15}N$ Isotopic Data

The $^{15}N$ abundance in samples is determined by an isotope ratio mass spectrometer (IRMS) coupled on-line to an elemental analyzer (EA) or by spectroscopic techniques such as wavelength-scanned cavity ring-down spectroscopy (WS-CRDS) combined with a combustion module (Nguyen et al. 2011). Data obtained is expressed as Atom% $^{15}N$ or $\delta^{15}N$ notion. Before describing the $^{15}N$ isotopic dilution and $^{15}N$ natural abundance methods, the isotopic notations are discussed.

#### 2.2.2.1.1  Atom% $^{15}N$

Atom percent $^{15}N$ (Atom% $^{15}N$) is the absolute number of atoms of $^{15}N$ in 100 atoms of total N. Thus, Atom% $^{15}N$ in a sample is notated as follows:

$$\text{Atom\% }^{15}N \text{ sample} = \left[ ^{15}N/(^{14}N + ^{15}N) \right] \times 100.$$

Atmospheric N is an internationally accepted reference standard and the Atom% $^{15}N$ of a reference standard (subsequently referred to as standard) is known to be 0.3663 (Nguyen et al. 2011). This means the Atom% $^{14}N$ in an atmospheric standard is 99.6337% (the difference between 100 and 0.3663) and the isotopic ratio (Atom%$^{15}N$/Atom% $^{14}N$) in a standard is 0.0036765 (i.e., 0.3663/99.6337).

#### 2.2.2.1.2  Atom% $^{15}N$ excess

The Atom% $^{15}N$ atom excess in a sample is calculated as the difference between the Atom% $^{15}N$ in a sample and the background natural abundance (i.e., Atom% $^{15}N$ of a reference standard of 0.3663).

#### 2.2.2.1.3  Delta ($\delta$) $^{15}N$

The change in $^{15}N$ concentrations in samples (e.g., soils or plants) with respect to atmospheric $^{15}N$ (atmospheric N as a standard) is expressed in delta notation $\delta^{15}N$, which is the deviation of $^{15}N$ abundance of the sample in relation to the standard as $\delta^{15}N$ per mil (‰) (i.e., molecules per thousand). Thus, $\delta$ $^{15}N$ in sample, the isotopic ratio of $^{15}N$ to $^{14}N$ in the sample (R sample) compared to the $^{15}N/^{14}N$ isotopic ratio in a standard (R standard), is calculated as follows:

$$\delta \ ^{15}N \text{ sample (‰)} = \left[ (R \text{ sample} - R \text{ standard/R standard}) \right]$$
$$\times 1000 = \left[ (R \text{ sample/R standard}) - 1 \right] \times 1000$$

It can be rewritten as:

$$\delta\ ^{15}N\ sample\ (\permil) = \left[(^{15}N/^{14}N\ sample/^{15}N/^{14}N\ standard) - 1\right] \times 1000$$

$$\delta\ ^{15}N\ sample\ (\permil) = \left[(Atom\%\ ^{15}N\ in\ sample - Atom\%\ ^{15}N\ in\ standard)\right.$$
$$\left./Atom\%\ ^{15}N\ in\ standard\right] \times 1000$$

where R is the $^{15}N/^{14}N$ ratio of the sample or standard. R standard (atmospheric N) is 0.003676 (i.e., 0.3663/99.6337). Samples with ratios of $^{15}N/^{14}N$ greater than 0.003676 have positive $\delta$ values and those with ratios of $^{15}N/^{14}N$ lower than 0.003676 have negative $\delta$ values. A sample with a $\delta\ ^{15}N$ value of +5‰ means that a sample has a $^{15}N/^{14}N$ ratio greater than the standard ratio (R standard) by 5 parts-per-thousand or 0.5% more $^{15}N$ in the sample relative to the standard.

### 2.2.2.2   Conversion of Atom% $^{15}$N to Delta ($\delta$) $^{15}$N and Vice Versa

*2.2.2.2.1   From $\delta^{15}N$ (‰) to Atom% $^{15}N$*

Atom% $^{15}N$ in samples can be obtained from $\delta^{15}N$ (‰) using the following equation:

$$Atom\%\ ^{15}N\ in\ a\ sample = 100/\left[(1/A) + 1\right]$$

where $A = [(\delta^{15}N\ sample/1000) + 1] \times R$ standard and R standard = Atom% $^{15}N$ in a standard/ Atom% $^{14}N$ in a standard, or R standard = 0.3663/(100 – 0.3663) = 0.0036765.

For example, the Atom% $^{15}N$ abundance of a sample with $\delta^{15}N$ of –30‰ and +30‰ is 0.35535 and 0.377251, respectively. They can be calculated as follows:

a. Atom% $^{15}N$ in a sample with $\delta^{15}N$ of –30‰

$$A = \left[(-30/1000) + 1\right] \times 0.0036765 = [-0.03 + 1] \times 0.0036765 = 0.003566205.$$

Thus, atom% $^{15}N$ in the sample = 100/[(1/0.003566205) +1] = 100/281.4101 = 0.35535.
b. Atom% $^{15}N$ in a sample with $\delta^{15}N$ of +30‰

$$A = \left[(30/1000) + 1\right] \times 0.0036765 = [0.03 + 1] \times 0.0036765 = 0.003786795.$$

Thus, atom% $^{15}N$ in the sample = 100/(1/0.003786795) + 1= 100/265.075572985 = 0.377251.

*2.2.2.2.2   From Atom% $^{15}N$ to $\delta^{15}N$ (‰)*

Delta ($\delta$) $^{15}N$ (‰) can be obtained from Atom% $^{15}N$ using the following equation:

$$\delta\ ^{15}N\ (\permil)\ in\ a\ sample = \left[(Atom\%\ ^{15}N\ in\ a\ sample/A) - 1\right] \times 1000$$

where A = R standard × (100 – Atom% $^{15}N$ in a sample) and R standard = Atom% $^{15}N$/Atom% $^{14}N$ = 0.3663/(100 – 0.3663) = 0.003676467.

For example, $\delta^{15}N$ (‰) for a sample with 0.371 Atom% $^{15}N$ is 12.878‰. This is calculated as follows:

$$\delta^{15}N \ (‰) \text{ in a sample} = \left[(0.371/A)-1\right] \times 1000$$

where A = 0.003676467 × (100 – 0.371) = 0.003676467 × 99.629 = 0.36628273.
Thus, $\delta^{15}N$ (‰) in the sample = [(0.371/0.36628273) – 1] × 1000 = 12.878‰.

### 2.2.3 THE $^{15}N$ DILUTION METHOD

This method involves growing both $N_2$-fixing and non-fixing plants in soil that has been enriched with the same amount of $^{15}N$ labeled fertilizer, which has a higher Atom% $^{15}N$ than that in the atmospheric N standard of 0.3663 Atom% (Hardarson and Danso 1990; IAEA, 1998, 2001, 2008; Unkovich et al. 2008). The $^{15}N$ enrichment in $N_2$-fixing plants is diluted by atmospheric $N_2$ fixation, while the $^{15}N$ enrichment of the non-fixing plants (often termed as control or reference plants) is assumed to be equal to the $^{15}N$ enrichment of the N derived from the soil in the $N_2$ fixed plants (Unkovich et al. 2008). In this respect, the $^{15}N$ enrichment of the soil N needs to be relatively constant over time, or soil N uptake by the $N_2$-fixing and non-fixing plants needs to be the same. However, the mineralization of unlabeled N can contribute to a decline in $^{15}N$ enrichment of plant available soil N with time (Viera-Vargas et al. 1995). In addition, soil mineral N uptake patterns vary among plant species, making it difficult to choose an ideal reference plant (Unkovich and Pate 2000; Peoples and Herridge 1990).

The degree of soil $^{15}N$ enrichment in a field can vary considerably and is influenced by a number of physical and biochemical factors (Hauggaard-Nielsen et al. 2010). The isotopic dilution (ID) method may still be favorable for soils where the $\delta^{15}N$ of plant-available soil N is less than 2‰ or where high-precision mass-spectrometers are not available. The inclusion of more than one reference plant has been suggested to increase the accuracy of the method (Viera-Vargas et al. 1995; Boddey et al. 2000; Cadisch et al. 2000; Unkovich et al. 2008).

The proportion of N derived from a $^{15}N$-labeled fertilizer is:

$$\% \ Ndff = (\text{Atom\% } ^{15}N \text{ excess in plant sample} / \text{Atom\% } ^{15}N \text{ excess of fertilizer}) \times 100.$$

The proportion of N fixed from the atmosphere (%Ndfa) can be calculated from the following equation:

$$Ndfa\% = 100 \times \left[ 1 - (Ndff\% \text{ of the } N_2\text{-fixing plants/} \right.$$
$$\left. Ndff\% \text{ of the non-fixing plants)} \right]$$

$$Ndfa\% = 100 \times \left[ 1 - (\text{Atom\% } ^{15}N \text{ excess in } N_2\text{-fixing plants/} \right.$$
$$\left. \text{Atom\% } ^{15}N \text{ excess in non-fixing plants)} \right]$$

where Atom% $^{15}N$ excess = Atom% $^{15}N$ – 0.3663 (i.e., $^{15}N$ abundance of atmospheric $N_2$).

The above equation is derived from the following relationships in $N_2$-fixing (F) and non-fixing (NF) plants:

a. Non-fixing (NF) plants:
   The proportion of N derived from a $^{15}N$-labeled fertilizer in NF plants ($\%Ndff_{NF}$):

$$\%Ndff_{NF} = (\text{Atom\% } ^{15}N \text{ excess in NF plant samples}$$
$$/\text{Atom\% } ^{15}N \text{ excess in fertilizer}) \times 100.$$

Thus, the proportion of plant N derived from soil in NF plants ($\%Ndfs_{NF}$):

$$\%Ndfs_{NF} = 100 - \%Ndff_{NF}$$

b. $N_2$-fixing (F) plants:
   The proportion of N derived from a $^{15}N$-labeled fertilizer in F plants ($\%Ndff_F$):

$$\%Ndff_F = (\text{Atom\% } ^{15}N \text{ excess in F plant samples}/$$
$$\text{Atom\% } ^{15}N \text{ excess in fertilizer}) \times 100. \tag{2.1}$$

Thus, the proportion of plant N derived from atmosphere ($\%Ndfa$) in F plants:

$$\%Ndfa = 100 - (\%Ndff_F + \%Ndfs_F)$$

where $\%Ndfs_F$ is the proportion of plant N derived from soils in F plants.
The relationship between F and NF plants can be expressed as below:

$$(\%Ndff_F/\%Ndfs_F) = (\%Ndff_{NF}/\%Ndfs_{NF}).$$

Thus, the proportion of N derived from soils in F plants is

$$\%Ndfs_F = \left[(\%Ndff_F/\%Ndff_{NF}) \times \%Ndfs_{NF} = \left[(\text{Atom\% } ^{15}N \text{ excess in F plant samples/}\right.\right.$$
$$\text{Atom\% } ^{15}N \text{ excess of fertilizer}) / (\text{Atom\% } ^{15}N \text{ excess in NF plant samples/}$$
$$\left.\text{Atom\% } ^{15}N \text{ excess of fertilizer}\right] \times \%Ndfs_{NF}.$$

The above equation can be rearranged as follows:

$$\%Ndfs_F = (\text{Atom\% } ^{15}N \text{ excess in F plant samples}$$
$$/\text{Atom\% } ^{15}N \text{ excess in NF plants}) \times \%Ndfs_{NF}.$$

$$\%Ndfs_F = \left[(\text{Atom\% } ^{15}N \text{ excess in F plant samples/Atom\% } ^{15}N \text{ excess in NF plants}\right]$$
$$\times \left[100 - (\text{Atom\% } ^{15}N \text{ excess in NF plant samples}\right. \tag{2.2}$$
$$\left./\text{Atom\% } ^{15}N \text{ excess in fertilizer}) \times 100\right].$$

The proportion of %Ndfa (%Ndfa = 100 – (%Ndff$_F$ + %Ndfs$_F$)) can be rewritten by replacing the terms %Ndff$_F$ and %Ndfs$_F$ with Atom% $^{15}$N excess in F and NF plants as shown in Equations 2.1 and 2.2 as follows:

$$\%\text{Ndfa} = 100 - 100 \times (\text{Atom\% }^{15}\text{N excess in plant samples in F plants/}$$
$$\text{Atom\% }^{15}\text{N excess in plant samples in NF plants}).$$

$$\%\text{Ndfa} = 100\left[1 - (\text{Atom\% }^{15}\text{N excess in plant samples in F plants/}\right.$$
$$\left.\text{Atom\% }^{15}\text{N excess in plant samples in NF plants})\right].$$

The amount of N$_2$ fixed/fixation is calculated as follows:

$$\text{N fixed} = (\%\text{Ndfa} \times \text{Total N in fixing crop})/100$$

where the amount of N$_2$ fixed and the total N in a fixing crop are expressed as kg N ha$^{-1}$ and the total N yield in a N$_2$-fixing crop (kg N ha$^{-1}$) = Dry matter yield (kg ha$^{-1}$) × N% content in dry matter.

## 2.2.4 $^{15}$N NATURAL ABUNDANCE (NA) METHOD

This method relies on naturally occurring differences in $^{15}$N abundance between plant-available soil N and that of atmospheric N$_2$ (Danso et al. 1993; Unkovich et al. 2008). Because of isotopic discrimination effects during soil N transformations (e.g., mineralization, nitrification, denitrification, and ammonia volatilization), soils tend to have higher $^{15}$N natural abundance than the atmosphere. In N$_2$-fixing plant tissues, N becomes diluted by the lower natural $^{15}$N abundance of fixed atmospheric N$_2$. The amount of N$_2$ fixation is calculated by the differences in $^{15}$N abundance between the N$_2$-fixing plants obtaining N from both atmospheric N$_2$ through BNF and soil N and non-N$_2$-fixing reference plants obtaining N only from the soil, after accounting for isotopic fractionation (the B value) between $^{14}$N and $^{15}$N during the process of N$_2$ fixation in the above-ground shoot of the N$_2$-fixing plants grown under N$_2$-free conditions.

The first assumption of the NA method is that the N$_2$-fixing plants and reference plants are accessing the same soil N pool. This assumption requires that the two plants are grown near one another and also that the two plants have similar rooting morphology and similar planting and harvesting times. A second assumption is that there is no discrimination between $^{14}$N and $^{15}$N in the plants' uptake and metabolism of N (Unkovich et al. 2008). Including more than one reference plant, such as in-field weeds, has been suggested to increase the accuracy of the NA method (Boddey et al. 2000; Unkovich et al. 2008). Cadisch et al. (2000) have shown that there was a good agreement in the amount of N$_2$ fixed by groundnuts (*Arachis hypogaea*) over an average of 2-year study period (21–24 kg N ha$^{-1}$) between the natural $^{15}$N abundance method and $^{15}$N dilution method, where an appropriate reference plant (e.g., non-nodulating groundnuts and/or maize) was available. However, the $^{15}$N dilution method was much more sensitive to a difference in planting time between the reference and fixing plants, compared to the natural abundance $^{15}$N method. The authors concluded that under the relatively high plant available $^{15}$N conditions in this soil ($\delta$ $^{15}$N ‰ of 6–10‰), the $^{15}$N natural abundance method is a viable alternative to measure N$_2$ fixation of N$_2$-fixing plants under field conditions.

The proportion of N in the fixing plant derived from atmospheric N (%Ndfa) (Shearer and Kohl 1986; Cadisch et al. 2000; Unkovich et al. 2008) is calculated as follows:

$$\%\text{Ndfa} = \left[ (\delta\,^{15}\text{N in reference plants} - \delta\,^{15}\text{N in N}_2\text{-fixing plants})/ \right.$$
$$\left. (\delta\,^{15}\text{N in reference plants} - \text{B}) \right] \times 100$$

where $\delta\,^{15}\text{N}$ (‰) = [($^{15}\text{N}/^{14}\text{N}$ sample/$^{15}\text{N}/^{14}\text{N}$ standard) $-1$] $\times$ 1000 and B is the $\delta\,^{15}\text{N}$‰ value of $N_2$-fixing plants when grown in the absence of soil mineral N (sand-culture greenhouse experiments) with $N_2$ as the sole source of N. The published B values reported for a range of grain legumes by Unkovich et al. (2008) are approximately $-0.50$ to $-1.75$‰.

Pauferro et al. (2010) have proposed that the $^{15}\text{N}$ natural abundance technique is one of those most easily applied "on farm" to evaluate the contribution of BNF from $N_2$-fixing plants. Instead of using the traditional technique to determine the B value from $N_2$-fixing plants grown in a glasshouse with an N-free soil medium and controlled irrigation with sterile water, Pauferro et al. (2010) have proposed an alternative approach to estimate the B value. In this approach, the proportion of N derived from BNF (%Ndfa) was estimated in separate pots, to which a small quantity of enriched $^{15}\text{N}$ ammonium sulfate was added. The %Ndfa was then used with the $^{15}\text{N}$ natural abundance data of the $N_2$-fixing legume and reference plants, to determine the B value as follows:

$$\text{B value (‰)} = \left[ (\delta\,^{15}\text{N in reference plants} \times \%\text{Ndfa}) - 100\,(\delta\,^{15}\text{N in reference plants} \right.$$
$$\left. -\delta\,^{15}\text{N in N}_2\text{-fixing plants}) \right]/\%\text{Ndfa}.$$

where $\delta\,^{15}\text{N}$ is the variation (‰) of $^{15}\text{N}$ natural abundance of reference plants and of $N_2$-fixing plants and % Ndfa is calculated using the formula shown in the isotopic dilution (ID) technique.

Total above-ground N fixed (kg N ha$^{-1}$) was calculated by multiplying biomass N yield and %Ndfa as follows:

$$\text{N fixed} = (\%\text{Ndfa} \times \text{N yield})/100.$$

where N yield is = DM yield (kg DM ha$^{-1}$) $\times$ N content (%) in DM.

## 2.3   SUMMARY AND CONCLUSIONS

Quantification of BNF is important to evaluate the sustainability of soil fertility in grain legume-based farming systems (Herridge et al. 2008; Anglade et al. 2015). This depends upon the balance of N-fixing legumes and N depleting non-legumes in the rotation. Many authors (e.g., Hardarson and Danso 1993; Hardarson and Atkins 2003; Herridge et al. 2008) have provided a comprehensive review of different BNF assessment techniques. Although the ureide method can be readily applied in the field (Unkovich et al. 2008), the snapshot information obtained using this colorimetric technique needs to be calibrated against quantitative data obtained from techniques such as $^{15}\text{N}$ dilution (Herridge and Peoples 1990). In addition, ureide measurements need to be repeated over a growing season. Use of the $^{15}\text{N}$ dilution technique is limited because of the cost associated with $^{15}\text{N}$-enriched fertilizers. In addition, the decline in $^{15}\text{N}$ enrichment of plant-available soil N with time and the non-uniform $^{15}\text{N}$ distribution with soil depth can lead to substantial errors if the temporal and spatial N uptake of the non-fixing reference plants differs from that of the $N_2$-fixing plants (Cadisch et al. 2000; Unkovich et al. 2008).

The cost associated with the use of [15]N-enriched fertilizers is avoided when using the [15]N natural abundance method, provided that the soil and plant samples obtained can be analyzed in laboratories with an automated continuous flow stable isotope ratio mass spectrometer which allows for a large number of analyses per day at a high level of precision ($\delta$ [15]N precision of 0.2‰) and a reasonable cost (Cadisch et al. 2000; Unkovich et al. 2008). The natural abundance technique has been successfully employed in agricultural ecosystems to quantify BNF, since many different non $N_2$-fixing reference plants were found to have relatively minor variations in their [15]N natural abundance (Unkovich et al. 1994; Boddey et al. 2000). Pauferro et al. (2010) have shown that the [15]N natural abundance technique is one of those most easily applied "on farm" to evaluate the contribution of biological $N_2$ fixation (BNF) from legume crops, provided that an accurate estimate of the B value (the [15]N abundance of the N accumulated in the plants derived from solely from BNF) is conducted. Unkovich (2013) has identified the need to reliably measure [15]N isotope discrimination and proposes approaches to improve the quantification of BNF using the [15]N natural abundance method.

Anglade et al. (2015), in their review, have proposed the use of a linear regression model that is based on the observed linear relationship between the reviewed %Ndfa values obtained from a range of BNF assessment techniques including the [15]N isotopic dilution and natural abundance techniques and shoot DM yields produced from 33 organic farms in France to estimate BNF for grain (and also forage) legumes in northern Europe. This equation considers harvested grain dry matter yield (kg ha$^{-1}$ yr$^{-1}$), grain N uptake (kg N ha$^{-1}$ yr$^{-1}$), the N harvest index, which is defined as the ratio between harvested material (grain N) to the total above-ground production (total shoot N), and the BGN multiplicative factor of 1.3. The total BNF in lentil, field pea, and faba bean shoots for 33 organic farms in France was estimated to be 40, 85, and 127 kg N ha$^{-1}$ yr$^{-1}$, respectively. Considering the below-ground N (BGN) contributions to total N fixation for pulse crop legumes, the total BNF in shoots and roots of lentils, field peas, and faba beans was 52, 111, and 165 kg N ha$^{-1}$ yr$^{-1}$, respectively. Unkovich et al. (2000) have also explored the complimentary use of a linear equation model which considers the published %Nfda and harvested N yield when estimating BNF. However, they have also cautioned the need for careful consideration of the influence of factors such as the availability of plant available soil N and DM yield on BNF under local growing conditions.

Although the BNF values estimated by both [15]N isotopic dilution and the [15]N natural abundance method may be influenced by factors such as isotopic discrimination, the size of the plant-available soil N pool and the difference in rooting depths between $N_2$-fixing and reference plants (Unkovich et al. 1994; Walley et al. 2001; Unkovich 2013), these methods can provide valuable information on the extent of BNF by pulse crops and its impacts on soil N sustainability in an arable cropping rotation. For example, Knight (2012) has successfully used the [15]N isotopic dilution technique to elucidate the impacts of BNF by field peas under different crop rotations in Canada, while Pauferro et al. (2010) were able to quantify the extent of N fixation in soybeans in both the South and Cerrado regions of Brazil, using the [15]N isotopic dilution technique. Several other publications (e.g., IAEA 1998, 2001, 2008; Abi-Ghanem et al. 2012) have also highlighted the usefulness of the [15]N isotopic dilution technique to assist in varietal breeding and the strain selection of grain legumes for enhanced agricultural N fixation.

Increasing emphasis on the use of pulse crops to provide food protein, enhance the adaptation of agricultural systems to drought and reduce the reliance of fertilizer N for cereal crops in pulse-based cropping rotations highlights the need to quantify BNF by pulse crops using [15]N isotopic techniques. These techniques are extremely useful in areas where previous data on BNF are not available but where pulse crops are gaining in popularity and credibility as suitable alternative growing crops which are capable of responding to changes in growing conditions resulting from climate change.

## ACKNOWLEDGMENTS

The assistance of Ms. E. Thom and Dr. J. Luo in drawing Figure 2.2 is acknowledged.

## REFERENCES

Abi-Ghanem, R., Carpenter-Boggs, L., Smith, J. L., and Vandemark, G. J. 2012. Nitrogen fixation by U.S. and Middle Eastern chickpeas with commercial and wild Middle Eastern inocula. *International Scholarly Research Network (ISRN) Soil Science* doi:10.5402/2012/981842.

Angadi, S. V., McConkey, B. G., Cutforth, H. W., Miller, P. R., Ulrich, D., Selles, F., Volkmar, K. M., Entz, M. H., and Brandt, S. A. 2008. Adaptation of alternative pulse and oilseed crops to the semiarid Canadian Prairie: Seed yield and water use efficiency. *Canadian Journal of Plant Science* 88: 425–438.

Anglade, J., Billen, G., and Garnier J. 2015. Relationships for estimating N₂ fixation in legumes: Incidence for N balance of legume-based cropping systems in Europe. *Ecosphere* 6 (3), Article 37: 1–24.

Arcand, M. M., Knight, J. D., and Farrell, R. E. 2014. Differentiating between the supply of N to wheat from above and belowground residues of preceding crops of pea and canola. *Biology and Fertility of Soils* 50: 563–570.

Boddey, R. M., Peoples, M. B., Palmer, B., and Dart, P. J. 2000. Use of the ¹⁵N natural abundance technique to quantify biological nitrogen fixation by woody perennials. *Nutrient Cycling in Agroecosystems* 57: 235–270.

Cadisch, G., Hairiah, K., and Giller, K. E. 2000. Applicability of the natural ¹⁵N abundance technique to measure N₂ fixation in *Arachis hypogaea* grown on an Ultisol. *Netherlands Journal of Agricultural Science* 48: 31–45.

Carlsson, G., and Huss-Danell, K. 2003. Nitrogen fixation in perennial forage legumes in the field. *Plant and Soil* 253: 353–372.

Carranca, C., De Varennes, A., and Rolston, D. 1999. Biological nitrogen fixation by fababean, pea and chickpea, under field conditions, estimated by the ¹⁵N isotope dilution technique. *European Journal of Agronomy* 10: 49–56.

Chalk, P. M. 1998. Dynamics of biologically fixed N in legume-cereal rotation: A review. *Australian Journal of Agricultural Research* 49: 303–316.

Chemining'wa, G. N, Mwangi, P. W., Mburu, M. W. K., and Mureithi, J. G. 2013. Nitrogen fixation potential and residual effects of selected grain legumes in a Kenyan soil. *International Journal of Agronomy and Agricultural Research* 3 (2): 14–20.

Chianu, J. N., Nkonya, E. M., Mairura, F. S., Chianu, J., and Akinnifesi, F. K. 2011. Biological nitrogen fixation and socioeconomic factors for legume production in sub-Saharan Africa: A review. *Agronomy for Sustainable Development* 31: 139–154.

Crews, T. E., and Peoples, M. B. 2004. Legume versus fertilizer sources of nitrogen: Ecological tradeoffs and human needs. *Agriculture, Ecosystems and Environment* 102: 279–297.

Crews, T. E., and Peoples, M. B. 2005. Can the synchrony of nitrogen supply and crop demand be improved in legume and fertilizer-based agroecosystems? A review. *Nutrient Cycling in Agroecosystems* 72: 101–120.

Cutforth, H. W., Angadi, S. V., McConkey, B. G., Entz, M. H., Ulrich, D., Volkmar, K. M., Miller, P. R., and Brandt, S. A. 2009. Comparing plant water relations for wheat with alternative pulse and oilseed crops grown in the semiarid Canadian prairie. *Canadian Journal of Plant Science* 89: 826–835.

Dalal, R. C., Strong, W. M., Doughton, J. A., Weston, E. J., Cooper, J. E., Wildermuth, G. B., Lehane, K. J., King, A. J., and Holmes, C. J. 1998. Sustaining productivity of a Vertisol at Warra, Queensland, with fertilisers, no-tillage or legumes 5. Wheat yields, nitrogen benefits and water-use efficiency of chickpea-wheat rotation. *Australian Journal of Experimental Agriculture* 38: 489–501.

Danso, S. K. A., Hardarson, G., and Zapata, F. 1993. Misconceptions and practical problems in the use of ¹⁵N soil enrichment techniques for estimating N₂ fixation. *Plant and Soil* 152: 25–52.

Doughton, J. A., Vallis, I., and Saffina, P. G. 1993. Nitrogen fixation in chickpea. 1. Influence of prior cropping or fallow, nitrogen fertilizer and tillage. *Australian Journal of Agricultural Research* 44: 1403–1413.

Evans, J., McNeill, A. M., Unkovich, M. J., Fettell, N. A., and Heenan, D. P. 2001. Net nitrogen balances for cool-season grain legume crops and contributions to wheat nitrogen uptake: A review. *Australian Journal of Experimental Agriculture* 41: 347–359.

FAO (Food and Agricultural Organization of the United Nations). 2014. Food Outlook: Biannual report on global food markets. pp. 63–67, http://www.fao.org/3/a-i3751e.pdf.

FAO. 2016a. International Year of Pulses (IYP)-Pulses and climate change. #IYP2016 © FAO 2016 I5384E /1/02.16, fao.org/pulses-2016.

FAO. 2016b. International Year of Pulses (IYP)-Nutritional benefits of pulses. #IYP2016 © FAO 2016 I5384E /1/02.16, fao.org/pulses-2016.

FAO. 2016c. Pulses Contribute to Food Security. #IYP2016 © FAO 2016 I5387E/1/02.16, fao.org/pulses-2016.

Fustec, J., Lesuffleur, F., Mahieu, S., and Cliquet, J-B. 2010. Nitrogen rhizodeposition of legumes. A review. *Agronomy for Sustainable Development* 30: 57–66.

Gan, Y., Campbell, C. A., Janzen, H. H., Lemke, R. L., Basnyat, P., and McDonald, C. L. 2010. Nitrogen accumulation in plant tissues and roots and N mineralization under oilseeds, pulses, and spring wheat. *Plant and Soil* 332: 451–461.

Gan, Y., Hamel, C., O'Donovan, J. T., Cutforth, H., Zentner, R. P., Campbell, C. A., Niu, Y., and Poppy, L. 2015. Diversifying crop rotations with pulses enhances system productivity. *Scientific Reports* 5, Article number 14625, 14 pp.

Gan, Y., Liang, C., Hamel, C., Cutforth, H., and Wang, X. Y. 2011a. Strategies for reducing the carbon footprint of field crops for semiarid areas. A review. *Agronomy for Sustainable Development* 31: 643–656.

Gan, Y., Liang, C., Liu, L. P., Wang, X. Y., and McDonald, C. L. 2011c. C:N ratios and carbon distribution profile across rooting zones in oilseed and pulse crops. *Crop and Pasture Science* 62: 496–503.

Gan, Y., Liang, C., Wang, X. Y., and McConkey, B. 2011b. Lowering carbon footprint of durum wheat by diversifying cropping systems. *Field Crops Research* 122: 199–206.

Gan, Y.T., Campbell, C. A., Liu, L., Basnyat, P., and McDonald, C. L. 2009. Water use and distribution profile under pulse and oilseed crops in semiarid northern high latitude areas. *Agricultural Water Management* 96: 337–348.

Giller, K. E., and Merckx, R. 2003. Exploring the boundaries of $N_2$-fixation in cereals and grasses: A hypothetical and experimental framework. *Symbiosis* 35: 3–17.

Hardarson, G., and Atkins, C. 2003. Optimising biological $N_2$ fixation by legumes in farming systems. *Plant and Soil* 252: 41–54.

Hardarson, G., and Danso, S. K. A. 1990. Use of $^{15}$N methodology to assess biological nitrogen fixation. In: G. Hardarson, editor, Use of nuclear techniques in studies of soil-plant relationships; training course series no.2. International Atomic Energy Agency (IAEA). Vienna, Austria. pp. 129–159.

Hardarson, G., and Danso, S. K. A. 1993. Methods for measuring biological nitrogen fixation in grain legumes. *Plant and Soil* 152: 19–23.

Hardy, R. W. F., Bums, R. C., and Holstein, R. D. 1973. Applications of the acetylene-ethylene assay for measurement of nitrogen fixation. *Soil Biology and Biochemistry* 5: 47–81.

Hauggaard-Nielsen, H., Ambus, P., and Jensen, E. S. 2003. The comparison of nitrogen use and leaching in sole cropped versus intercropped pea and barley. *Nutrient Cycling in Agroecosystems* 65: 289–300.

Herridge D. F., Marcellos, H., Felton, W. L., Turner, G. L., and Peoples, M. B. 1998 Chickpea in wheat-based cropping systems of northern New South Wales III. Prediction of $N_2$ fixation and N balance using soil nitrate at sowing and chickpea yield. *Australian Journal of Agricultural Research* 49: 409–418.

Herridge D. F., Peoples, M. B., and Boddey, R. M. 2008. Global inputs of biological nitrogen fixation in agricultural systems. *Plant Soil* 311: 1–18.

Herridge, D., Manning, B., and McCaffery, D. 2015 *Nitrogen Benefits of Chickpea and Faba Bean.* July 2015 Primefact 1163, Australian Government—Grains Research and Development, NSW Department of Primary Industries.

Herridge, D. F. 2009. *Nitrogen Fixation Benefits of Pulse Crops.* Australian Government – Grains Research and Development, NSW Department of Primary Industries (https://grdc.com.au/Research-and-Development /GRDC-Update-Papers/2009/09/Nitrogen-Fixation-Benefits-of-Pulse-Crops).

Herridge, D. F., and Peoples, M. B. 1990. Ureide assay for measuring nitrogen-fixation by nodulated soybean calibrated by N-15 methods. *Plant Physiology* 93: 495–503.

Herridge, D. F., Marcellos, H., Felton, W. I., Turner, G. I., and Peoples, M. B. 1995. Chickpea increases soil-N fertility in cereal systems through nitrate sparing and $N_2$ fixation. *Soil Biology and Biochemistry* 27: 545–551.

Herridge, D. F., Palmer, B., Nurhayati, D. P., and Peoples, M. B. 1996. Evaluation of the xylem ureide method for measuring N2 fixation in six tree legume species. *Soil Biology and Biochemistry* 28: 281–289.

Hoekstra, A. Y., and Chapagain, A. 2008. *Globalization of Water: Sharing the Planet's Freshwater Resources.* Wiley-Blackwell.

IAEA (International Atomic Energy Agency). 1998. *Improving Yield and Nitrogen Fixation of Grain Legumes in the Tropics and Sub-tropics of Asia,* IAEA TECDOC 1027, IAEA, Vienna.

IAEA. 2001. *Use of Isotope and Radiation Methods in Soil and Water Management and Crop Nutrition.* IAEA Training Course Series 14, IAEA, Vienna.

IAEA. 2008. *Guidelines on Nitrogen Management in Agricultural Systems.* IAEA Training Course Series 29, IAEA, Vienna.

Joshi, P. K., and Rao, P. P. 2016. *Global and Regional Pulse Economies: Current Trends and Outlook.* International Food Policy Research Institute (IFPRI) Discussion Paper Series No. 01544. Washington, DC.

Khan, D. F., Peoples, M. B., Chalk, P. M., and Herridge, D. F. 2002. Quantifying below-ground nitrogen of legumes. 2. A comparison of $^{15}N$ and non-isotopic methods. *Plant and Soil* 239: 277–289.

Khan, D. F., Peoples, M. B., Schwenke, G. D., Felton, W. L., Chen, D., and Herridge, D. F. 2003. Effects of below-ground nitrogen on N balances of field-grown faba bean, chickpea and barley. *Australian Journal of Agricultural Research* 54: 333–340.

Knight, J. D. 2012. Frequency of field pea in rotations impacts biological nitrogen fixation. *Canadian Journal of Plant Science* 92: 1005–1011.

Kopke, U., and Nemecek, T. 2010. Review: Ecological services of faba bean. *Field Crops Research* 115: 217–233.

Lassaletta, L., Billen, G., Grizzetti, B., Anglade, J., and Garnier, J. 2014. *Supplementary Material 1 Methods.* Supplementary information to the paper (Suppl. 1): 50-year trends in nitrogen use efficiency of world cropping systems: the relationship between yield and nitrogen input to cropland. *Environmental Research Letters* 9 (10): 1–12.

Lemke, R. L., Zhong, Z., Campbell, C. A., and Zentner, R. 2007. Can pulse crops play a role in mitigating greenhouse gases from North American agriculture? *Agronomy Journal* 99: 1719–1725.

Lupwayi, N. Z., and Kennedy, A. C. 2007. Grain legumes in northern Great Plains: Impacts on selected biological soil processes. *Agronomy Journal* 99: 1700–1709.

Marcellos, H., Felton, W. L., and Herridge, D. F. 1993. Crop productivity in a chickpea-wheat rotation. *Proc. 7th Australian Agronomy Conference, Australian Society of Agronomy*, pp. 276–278.

Marcellos, H., Felton, W. L., and Herridge, D. F. 1998. Chickpea in wheat-based cropping systems of northern New South Wales. I. $N_2$ fixation and influence on soil nitrate and water. *Australian Journal of Agricultural Research* 49: 391–400.

McNeill, A. M., and Fillery, I. R. P. 2008. Field measurement of lupin belowground nitrogen accumulation and recovery in the subsequent cereal-soil system in a semi-arid Mediterranean-type climate. *Plant and Soil* 302: 297–316.

Nguyen, L., Zapata, F., Lal, R., and Dercon, G. 2011. Role of nuclear and isotopic techniques in sustainable land management. In Lal, R., and Stewart, B. A. (Eds.), *World Soil Resources and Food Security, Advances in Soil Science*, pp. 345–418. CRC Press, Boca Raton, FL.

OECD-FAO (The Organization for Economic Cooperation and Development—Food and Agricultural Organization of the United Nations). 2016. *Pulses*, pp. 81–82. In OECD-FAO Agricultural Outlook 2016-2025: Special focus: Sub-Saharan Africa.

Pauferro, N., Guimarães, A. P., Jantalia, C. P., Urquiaga, S., Alves, B. J. R., and Boddey, R. M. 2010. $^{15}N$ natural abundance of biologically fixed $N_2$ in soybean is controlled more by the Bradyrhizobium strain than by the variety of the host plant. *Soil Biology and Biochemistry* 42: 1694–1700.

Peoples, M. B., Bowman, A. M., Gault, R. R., Herridge, D. F., McCallum, M. H., McCormick, K. M., Norton, R. M., Rochester, I. J., Scammell, G. J., and Schwenke, G. D. 2001. Factors regulating the contributions of fixed nitrogen by pasture and crop legumes to different farming systems of eastern Australia. *Plant and Soil* 228: 29–41.

Peoples, M. B., and Herridge, D. F. 1990. Nitrogen fixation by legumes in tropical and subtropical agriculture. *Advances in Agronomy* 44: 155–223.

Peoples, M. B., Boddey, R. M., and Herridge, D. F. 2002. Quantification of nitrogen fixation. In Leigh, G. J. (Ed.), *Nitrogen Fixation at the Millennium*. Elsevier, Amsterdam, pp. 357–389.

Peoples, M. B., Brockwell, J., Herridge, D. F., Rochester, I. J., Aves, B. J. R., Urquiaga, S., Boddey, R. et al. 2009a. The contributions of nitrogen-fixing crop legumes to the productivity of agricultural systems. *Symbiosis* 48: 1–17.

Peoples, M. B., Faizah, A. W., Rerkasem, B., and Herridge, D. F., Eds. 1989. Methods for evaluating nitrogen fixation by nodulated legumes in the field. Canberra, *Australian Centre for International Agricultural Research (ACIAR) Monograph No. 11*, vii + 76pp.

Peoples, M. B., Unkovich, M. J., and Herridge, D. F. 2009b. Measuring symbiotic nitrogen fixation by legumes. Nitrogen fixation in crop production. *Agronomy Monograph* 52: 125–170.

Peoples, M., Swan, T., Goward, L., Hunt, J., Li, G., Harris, R., Ferrier De-Anne et al. 2015. *Legume Effects on Soil N Dynamics—Comparisons of Crop Response to Legume and Fertiliser N*. Australian Government Grains Research and Development Corporation (GRDC) Newsletter. (https://grdc.com.au/Research -and-Development/GRDC-Update-Papers/2015/02/Legume-effects-on-soil-N-dynamics--comparisons -of-crop-response-to-legume-and-fertiliser-N).

Rochester, I. J., Peoples, M. B., Constable, G. A., and Gault, R. R. 1998. Faba beans and other legumes add nitrogen to irrigated cotton cropping systems. *Australian Journal of Experimental Agriculture* 38: 253–260.

Roskowski, P. 1981. Comparative $C_2H_2$ reduction and $^{15}N_2$ fixation in deciduous woody litter. *Soil Biology and Biochemistry* 13: 83–85.

Ross, S. M., King, J. R., Williams, C. M., Strydhorst, S. M., Olson M. A., Hoy, C. F., and Lopetinsky, K. J. 2015. The effects of three pulse crops on a second subsequent crop. *Canadian Journal of Plant Science* 95: 779–786.

Ross, S. M., Izaurralde, R. C., Janzen, H. H., Robertson, J. A., and McGill, W. B. 2008. The nitrogen balance of three long-term agroecosystems on a boreal soil in western Canada. *Agriculture, Ecosystems and Environment* 127: 241–250.

Schlink, A. C., Nguyen, M. L., and Viljoen, G. J. 2010. Water requirements for livestock production: A global perspective. *Plurithematic Issue of the Scientific and Technical Review (Revue Scientifique et Technique-International Office of Epizootics)* 29: 603–619.

Schwenke, G. D., Peoples, M. B., Turner, G. L., and Herridge, D. F. 1998. Does nitrogen fixation of commercial, dryland chickpea and faba bean crops in north-west New South Wales maintain or enhance soil nitrogen? *Australian Journal of Experimental Agriculture* 38: 61–70.

Seymour, N., McKenzie, K., and Krosch, S. 2015. *Fixing More Nitrogen in Pulse Crops*. Grains Research Development Corporation (GRDC), Australian Government (https://grdc.com.au/Research-and -Development/GRDC-Update-Papers/2015/07/Fixing-more-nitrogen-in-pulse-crops).

Shearer, G., and Kohl, D. H. 1986. $N_2$ fixation in field settings: Estimations based on natural $^{15}N$ abundance. *Australian Journal of Plant Physiology* 13: 699–756.

Singh, G., and Sekhon, H. S. 2007. Effect of phosphorus and sulphur application on yield and yield attributing characters in pigeon pea—Wheat cropping sequence. *Journal of Food Legumes* 20: 212–213.

Stevenson, F.C., and C. Van Kessel. 1996. The nitrogen and non-nitrogen rotational benefits of pea to succeeding crops. *Canadian Journal of Plant Science* 76: 735–745.

Unkovich M. J., and Pate, J. S. 2000. An appraisal of recent field measurements of symbiotic $N_2$ fixation by annual legumes. *Field Crops Research* 65: 211–228.

Unkovich M. J., Baldock, J., and Peoples, M. B. 2010. Prospects and problems of simple linear models for estimating symbiotic $N_2$ fixation by crop and pasture legumes. *Plant and Soil* 329: 75–89.

Unkovich, M. J. 2013. Isotope discrimination provides new insight into biological nitrogen fixation. *New Phytologist* 198: 643–646.

Unkovich, M. J., Herridge, D. F., Peoples, M. B., Cadisch, G., Boddey, R. M., Giller, K. E., Alves B. J. R., and Chalk, P. M. 2008. Measuring plant-associated nitrogen fixation in agricultural systems. *Australian Centre for International Agricultural Research (ACIAR) Monograph* No. 136. 258 pp., Canberra, SA, Australia.

Unkovich, M. J., Pate, J. S., Sanford, P., and Armstrong, E. L. 1994. Potential precision of the delta-N-15 natural-abundance method in-field estimates of nitrogen-fixation by crop and pasture legumes in south-west Australia. *Australian Journal of Agricultural Research* 45: 119–132.

Viera-Vargas, M. S., Oliveira, O. C., Souto, C. M., Cadisch, G., Urquiaga, S., and Boddey, R. M. 1995. Use of different $^{15}N$ labelling techniques to quantify the contribution of biological $N_2$ fixation to legumes. *Soil Biology and Biochemistry* 9: 1185–1192.

Voisin, A., Salon, C., Munier-Jolain, N. G., and Ney, B. 2002. Quantitative effects of soil nitrate, growth potential and phenology on symbiotic nitrogen fixation of pea (*Pisum sativum L.*). *Plant and Soil* 243: 31–42.

Walley, F. L., Clayton, G. W., Miller, P. R., Carr, P. M., and Lafond, G. P. 2007. Nitrogen economy of pulse crop production in the Northern Great Plains. *Agronomy Journal* 99: 1710–1718.

Walley, F. L., Fu, G. M., van Groenigen, J. W., and van Kessel, C. 2001. Short-range spatial variability of nitrogen fixation by field-grown chickpea. *Soil Science Society of America Journal* 65: 1717–1722.

Weston, E. J., Dalal, R. C., Strong, W. M., Lehane, K. J., Cooper, J. E., King, A. J., and Holmes, C. J. 2002. Sustaining productivity of a Vertisol at Warra, Queensland, with fertilisers, no-tillage or legumes. 6. Production and nitrogen benefits from annual medic in rotation with wheat. *Australian Journal of Experimental Agriculture* 42: 961–969.

# 3 The Role of Nitrogen Stable Isotopes to Investigate Soil Nitrogen Transformations and Cycling in Agricultural Systems

*Minh-Long Nguyen, Jiafa Luo, and Bert F. Quin*

## CONTENTS

## 3.1  INTRODUCTION

The current world population of 7.3 billion is expected to reach 8.5 billion by 2030 and 9.7 billion in 2050, according to a report by the United Nations Department of Economic and Social Affairs (UNDESA 2015). Most of this increase in population will occur in developing countries, where the majority depend upon agriculture for their livelihoods (Lal 2000; Smil 2011). Such increase will place great pressure on global food security and hence agricultural production. Nitrogen (N) is one of the major nutrients for arable and pastoral production in many parts of the world since N is an essential component of all proteins and of deoxyribonucleic acid (DNA) in plants and animals (Crews and Peoples 2004; Parfitt et al. 2008; Peoples et al. 2009; Smil 2011).

Nitrogen demand is met by the inputs from inorganic (chemical/synthetic N fertilizers) sources, biological N fixation (BNF) from leguminous crop in arable cropping rotation systems or in mixed legume-grass pastures, animal and green manure, residual soil N from previous fertilizer N applications and atmospheric N deposition. Smil (2011) highlighted the importance of N fertilizers in worldwide food production. He estimated that inorganic N fertilizers are responsible for around 50% of the world's food production and the remaining 50% comes mostly from BNF, organic recycling (manures and crop residues), and atmospheric deposition. Smil (2011) suggested that by 2025 more than half of the world's food production will still depend on inorganic N fertilizers, and this share will keep rising for at least several more decades. The Food and Agriculture Organisation (FAO) of the United Nations has documented the substantial increase in the world N fertilizer demand from 85.6 teragrams (Tg) per year in 2000 to 108.2 Tg in 2011 and then to 109.9 Tg in 2012 with an expected increase to 118.2 Tg in 2019 to meet the world's food demand (FAO 2012, 2016; http://www.fao.org/3/a-i5627e.pdf). However, the benefits of inorganic N fertilizer inputs to enhance food production come at environmental and social costs on water quality, human health, biodiversity, and climate change (e.g., Sutton et al. 2011; Velthof et al. 2011; Fowler et al. 2013; Gu et al. 2013; Jones et al. 2014; Liu et al. 2016). These costs are brought about by (1) an elevated nitrate $\left( NO_3^- \right)$ concentration in waters through leaching losses, runoff, and soil erosion, (2) the volatilization of ammonia ($NH_3$) and its consequent impact on ecosystems via atmospheric deposition, and (3) the emissions of nitrous oxide ($N_2O$), one of the most important greenhouse gases (GHGs) that can have potential impact on climate change and variability (e.g., Sutton et al. 2011; Thomson et al. 2012; Fowler et al. 2013; Gu et al. 2013; Jones et al. 2014; Liu et al. 2016). The efficiency of N fertilizer application in agroecosystems is often between 35 and 50% of the N applied being taken up by plants for their growth (Smil 1999, 2011; Crews and Peoples 2004; Canfield et al. 2010) although some studies have shown that N recoveries can be as low as 20–30% (Fan et al. 2007) or as high as 65% (Schepers and Mosier 1991). These losses are dependent on soil types, agronomic practices, and environmental conditions such as annual rainfall and time of N fertilizer applications (Smil 2011). On a global basis, Smil (2011) has estimated that the overall N efficiency of the global food system is no more than 15%.

To minimize N impacts on environment and enhance N fertilizer use efficiency, it is necessary to have detailed knowledge of the processes involved in N cycling and losses in agricultural systems. Information obtained would help to tighten the N cycle with a reduction in N losses through leaching, volatilization, and denitrification (Galloway et al. 2008; Liu et al. 2013; Suddick et al. 2013; Cui et al. 2014; Stevens et al. 2015).

This chapter will provide an overview of current knowledge on N cycle, sources of N inputs, processes that influence soil N pools, soil N dynamics, fertilizer N use efficiency (referred to as FNUE), the uptake of N by plants and soil microorganisms, and N losses from the N cycle. The use of $^{15}N$ stable isotopic and $^{15}N$ natural abundance techniques to provide a better understanding of the N transformations in soil-plant systems will then be reviewed since information obtained from these isotopic techniques can help to assess the relative importance of management factors that affect FNUE, N uptake from crop residues, and N losses to the environment. This information will be valuable in formulating measures to mitigate N losses, minimize synthetic N inputs, and enhance FNUE, particularly in areas where changes in land uses and climate may impact on various N transformation processes such as ammonium $(NH_4^+)$ oxidation and immobilization (Barnard et al. 2006; Shibata et al. 2015; Liu et al. 2016).

## 3.2  AN OVERVIEW OF SOIL NITROGEN POOLS, PROCESSES, AND TRANSFORMATIONS

### 3.2.1  NITROGEN IN NATURAL SOIL-PLANT ECOSYSTEMS

Nitrogen is unique among the key building blocks for both soil microorganisms and plants, in that by far the biggest reservoir of N that is accessible to some organisms is in the atmosphere, of which the very stable gas dinitrogen ($N_2$) makes up 78%, or $3.7 \times 10^9$ Tg (Sorai et al. 2007). Estimates of total soil N range from $0.2–1.4 \times 10^5$ Tg (Batjes 1996; Lerman et al. 2004), less than 0.01% of that present in the atmosphere.

In the early stages of evolution of life, some microorganisms developed the ability to "fix" $N_2$ from the atmosphere into the $NH_4^+$ form. This requires considerable energy, in the form of 16 molecules of the nucleotide adenosine triphosphate (ATP) per molecule of $N_2$ (Ward 2012). The energy for the formation of ATP comes either from free-living $N_2$-fixing microorganisms directly, or more efficiently from plants, known as legumes, which host the N-fixing bacteria (Zehr et al. 2003). The sole enzyme known to be responsible for N fixation is nitrogenase, a complex enzyme comprised of two separate groups of proteins with separate functions, and whose genealogy indicates evolution from a common ancestor (Raymond et al. 2004).

The N fixed by N-fixers is assimilated into amino acids, protein, and enzymes (Ward 2012). This becomes available for the growth of non-fixing microorganisms and microfauna in the soil, and to plants, through death and decomposition (mineralization), and through excretion of amino acids and other products, particularly in the rhizosphere of plants (Compant et al. 2010). Many soil microorganisms have also developed the ability to pass directly into plant roots by various methods; some then colonize specific plant organs. In doing so, many confer health advantages to the plant (Compant et al. 2010).

A group of microorganisms called nitrifiers oxidize the $NH_4^+$-N to nitrite; another group rapidly oxidize the nitrite to $NO_3^-$, the most easily utilizable form of N for many plants. For many years Nitrosomonas and Nitrobacter were regarded as the most important $NH_3$ and nitrite oxidizing bacteria, respectively; however, recent genomic research has demonstrated that many different microorganisms carry the genomic coding for nitrification and other N processes, often in the same microorganism (Nelson et al. 2016). As well as being utilized by most plants, nitrate is utilized by a group of soil microorganisms, the denitrifiers. Most denitrifiers that exist in natural ecosystems reduce $NH_4^+$ to $N_2O$; others reduce this to $N_2$ gas, which escapes to the atmosphere, thus completing the N cycle (Ward 2012). The total amount of N incorporated into soil microbes is estimated to represent about half of the total terrestrial biotic N (including plants and animals), of which estimates range from $0.5–3 \times 10^4$ Tg (Lerman et al. 2004; Sorai et al. 2007).

Non-microbial soil N can be divided into various categories. These were traditionally classified according to solubility in various extractant components: (1) a small inorganic pool (typically 2–200 kg N/ha) consisting mainly of $NH_4^+$ adsorbed onto clay particles and nitrate dissolved in the soil water, (2) a larger labile organic pool made up largely of decomposing plant material or detritus (above and below the soil surface), microbial litter, animal excreta and remains, and various organic molecules derived from all these, making up 30–80% of soil N, and (3) another large pool consisting of more stable organic N, still commonly referred to as humus or humic acid. Humus has for many decades been widely considered to be the end result of all the microbial decomposition processes, producing unique high molecular weight compounds that are very resistant to further change (Whitehead and Tinsley 1963; Brady and Well 1999; Berg and McClaugherty 2007). This view has been convincingly challenged in recent years; Lehmann and Kleber (2015) argued that soil organic

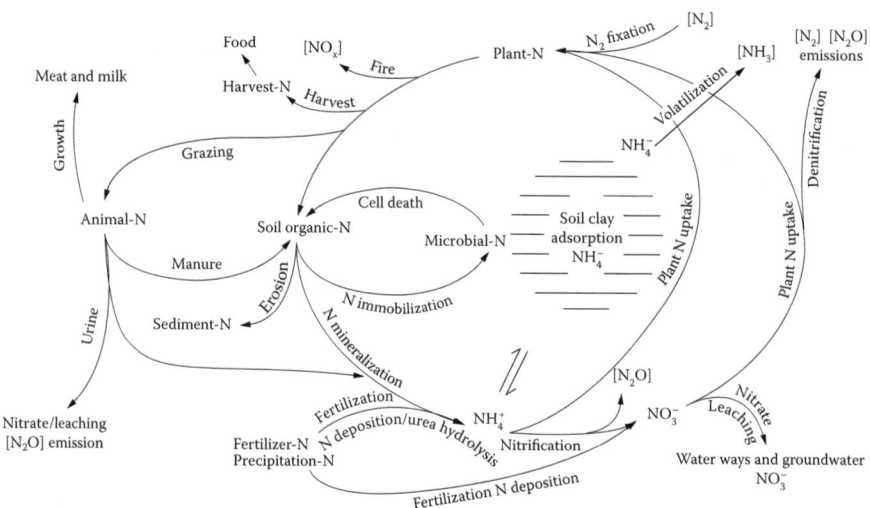

**FIGURE 3.1**   The N cycle from a soil perspective.

matter (SOM) is instead a continuum of progressively decomposing organic compounds, with their size being dependent on the source and stage of decomposition. Other authors have previously spoken of the benefits that certain high molecular weight molecules bring to soils, rather than referring to humus formation (Oades 1984).

Regardless of their origin and ultimate fate, the larger molecules, and whether they are classified as a stable, resistant group, or alternatively as what we suggest be called "saslom" (soil aggregate-stabilizing large organic molecules), they play a very important role in all soils. By assisting aggregation of soil particles, they improve soil structure, rainfall infiltration ability, and water-holding capacity. They also increase the soil cation exchange capacity (CEC), thereby improving the retention of not just the major cations (positively charged ions) but many trace elements as well (Oades 1984; Kong et al. 2007). Larger soil organisms, from the very small nematodes through to worms and beetles, also pay a vital role in maintaining soil health (van Groenigen et al. 2015).

Natural soil-plant ecosystems evolve to be very efficient recyclers of nutrients; N is no different in this regard (Fowler et al. 2013). Losses of $NO_3^-$ and $NH_4^+$ from natural ecosystems are negligible, as are emissions of the GHG-$N_2O$. Even in natural grasslands grazed by what commonly used to be huge herds of migrating ruminant animals, N losses are low. The average protein content of the native grasses dominant in different seasons of the year and migratory areas are just adequate for the species that evolved on them (Beeson 1946), meaning there is little excess N to be excreted as urea and ureic acid in the urine, and therefore very low losses of N as $NH_3$ volatilization, $N_2O$ emission, and $NO_3^-$ leaching.

Figure 3.1 illustrates the processes driven by soil microorganisms, within the greater terrestrial N cycle (elaborated from Robertson and Groffman 2015). Note that the lower loop has come about largely because of anthropogenic activities, particularly in terms of industrial N fixation.

### 3.2.2   Effects of Agriculture on Soil N Cycling

The development of increasingly sophisticated, sensitive, and precise chemical and biochemical analytical technologies and measurement protocols in the last 20 years, and the growth of interdisciplinary research, have led to much greater knowledge of the processes and transformations taking place within and between groups of microorganisms (Butterbach-Bahl et al. 2013). This perspective has also greatly reduced earlier soil-science focus on chemically or physically defined soil N "pools." Increasing effort is being put into studying the effects of these microbial processes and transformations on soil nutrient cycling efficiency by soils, and losses from these processes to the environment, particularly in respect of water quality and global warming. It is becoming increasingly important to quantify these effects under different environmental conditions with the use of the $^{15}N$ isotope (Barraclough 1995; Robinson 2001; Peoples et al. 2015).

Increasing $NO_3^-$ leaching and GHG emissions and their effects on the environment are essentially a result of the rapid adoption of the single most important industrial invention in agricultural production ever; the Haber–Bosch industrial N fixation process of 1919. In this process, atmospheric N is forced to react with hydrocarbons from natural gas in a catalyzed reaction under pressure to form ammonia, which is then reacted with carbon dioxide ($CO_2$) to form fertilizer urea and/or with other chemicals to form ammonia-based N fertilizers (Heffer and Prud'homme 2016).

The effects of this invention have been truly immense, both good and bad. From the commencement of industrial-scale manufacturing in the 1940s, improvements in manufacturing efficiency and the discovery of major natural gas deposits around the world have seen massive increases in urea production worldwide, reaching over 100 Tg annually in 2010 (Fowler et al. 2013). Urea is easily the cheapest form of fertilizer N to produce, convenient to handle, store, distribute, and spread. Straight urea makes up 57% of the world's fertilizer N; urea combined with other ammonia-based fertilizers represents almost 90% of total fertilizer N production (Heffer and Prud'homme 2016).

The increased use of fertilizer N has facilitated a massive increase in agricultural production, both directly through increased plant growth and reduced dependency on legumes to maintain soil N supply, and indirectly through opening the opportunity for the breeding of higher growth-potential plant varieties that require frequent inputs of fertilizer N. This has allowed agriculture to feed a rapidly increasing human population, but it has brought about some increasingly concerning side-effects; in particular, the soil-plant system in agriculture has become much more "leaky" with respect to N (Figure 3.1). Despite increased rates of soil microbial N processes and recycling, increased losses of $NH_3$ and $N_2O$ to the atmosphere and $NO_3^+$ leaching are occurring globally (Galloway et al. 2004). Reductions in soil biota diversity and in total soil N levels are increasingly being reported (Dise et al. 2011). It is important to understand (1) how the various climatic and management factors affect soil microbial processes and emissions, (2) the lifespan of the various emissions (Fowler et al. 2013), and (3) their implications for future agricultural production, soil health, environmental quality, global warming, and changes in climate.

### 3.2.3  MICROBIAL N PROCESSES IN AGRICULTURAL SOILS THAT CAN RESULT IN N LOSSES TO THE ENVIRONMENT

The great majority of soil N processes and transformations are carried out by soil microorganisms, particularly bacteria and to a lesser extent by archaea and fungi. The hydrolysis of urea to $NH_4^+$ by the ubiquitous *urease* enzyme occurs both within bacteria (and in another related form in fungi and plants), and via extracellular bacterial urease in the soil (Saggar et al. 2009). Other soil N processes are believed to occur only within microorganisms (Peoples et al. 2015).

Table 3.1 lists the major soil abiotic and microbial N processes that can directly or indirectly lead to N losses; the microorganisms, enzymes, and chemical reactions involved; and the soil conditions that favor these losses. The information given has been collated from many of the cited references, but particularly from Peoples et al. 2004; Saggar et al. 2009; Ward 2012; Robertson and Groffman 2015.

Without natural or anthropogenic interference, soil microorganisms and fauna will evolve ecosystems that promote efficient recycling within the soil-plant system and minimize the losses by the pathways illustrated in Figure 3.1 and listed in Table 3.1. Whenever these systems are disturbed, the system can be overloaded with N, either temporarily or long-term, which can greatly increase the usually small amounts lost through one or more of the processes described. The increasing $NO_3^-$ leaching and $NH_3$ and $N_2O$ emissions taking place throughout most of the world's agriculture are essentially a result of soil-plant systems being overloaded with more fertilizer mineral N than crop uptake can keep up with, and that microorganisms can efficiently recycle within the soil root zone.

## TABLE 3.1

### Important Soil Microorganism and Autonomous Processes that Can Lead Directly to Nitrate Leaching or Emissions of Ammonia, Nitrous Oxide, or Dinitrogen Gases

| Description of Process | Organisms Involved | Enzymes | Chemical Reactions | N Forms Lost | Soil Conditions that Increase Losses |
|---|---|---|---|---|---|
| *Urea(fertilizer/urine) hydrolysis* | Abiotic and biotic | Urease | $(NH_2)_2CO + 2H_2O \rightarrow (NH_4)_2CO_3 \rightarrow$ $NH_4^+ + NH_3 + CO_2 + OH^-$ | $NH_3$ | Higher pH(6); low CEC ($<10^*$); warm; wind; drying all push $NH_4^+/NH_3$ equilibrium to right |
| *$N_2$ fixation* | Rhizobia (plant-hosted) and free-living bacteria | Nitrogenase I/II | $N_2 + 5H_2O \rightarrow 2\ NH_4^+ + 2OH^- + 1.5O_2$ (Provides N for reactions below) | None directly | All fixed N is usually used for growth of plants and micro-organisms |
| *Mineralization/ ammonification* | | | | | |
| - degradation of detritus protein | Many AOB, AOA fungi | Proteinases | org-N $+ H_2O \rightarrow NH_3 +$ org-OH (+ energy) | $NH_3$ | $NH_3$ losses mainly via mineralization of manure on soil surface; pH>6; CEC<$10^*$ |
| - deamination of protein | Many AOB, AOA | Deaminases | $R\text{-}NH_2 + H_2O \rightarrow NH_3 + R\text{-}OH$ (+energy) | $NH_3$ | |
| *Nitrification* | Autotrophic AOB | $NH_3$ monooxygenase | (i) $2\ NH_4^+ + 3O_2 \rightarrow 2NO_2^- + 2H_2O + 4H^+$ (ii) $2\ NO_3^- + O_2 \rightarrow 2\ NO_3^-$ | $NO_3^-$ | - excess rainfall/irrigation; rapid drainage |
| *Anammox* | AOB (planctomycetes) | Known, notnamed | $NH_4^+ + NO_2^- \rightarrow N_2 + 2H_2O$ | $N_2$ | - occurs only under anaerobic conditions |
| *Denitrification* | | | | | |
| - partial denitrification | FABs and fungi | $NO_3^-/NO_2^-/NO$ red | $2\ NO_3^- - 2\ NO_2^- \rightarrow 2NO - N_2O$ | $N_2O$ | -anaerobic; inside soil aggregates; high $NO_3^-$ |
| - full denitrification | FAB | $N_2O$ reductase | $N_2O \rightarrow N_2$ | $N_2$ | - as above, plus very low $O_2$ |
| - co-denitrification | Fungi? | Not known | $R\text{-}NH_2 + NO \rightarrow N_2/N_2O$ | $N_2, N_2O$ | - high mineral N and pH; high org C-N turnover |

*Source:* Ward, B. 2012. The global nitrogen cycle. In *Fundamentals of geobiology*. A. H. Knoll, D. E. Canfield, and K. O. Konhauser, Eds. Blackwell Publishing Ltd.; Cameron, K. C., Di, H. J., and J. L. Moir. 2013. Nitrogen losses from the soil/plant system: a review. *Annals of Applied Biology* 162: 145–173; Robertson, G. P., and P. M. Groffman. 2015. Nitrogen transformations. In *Soil microbiology, ecology and biochemistry*, 4th ed., pp. 421–446. Burlington, MA: Academic Press; and van Groenigen, J. W., Huygens, D., Boeckx, P., Kuyper, Th. W., Lubbers, M. T., and Groffman, P. M. 2015. The soil N cycle: new insights and key challenges. *Soil* 1: 235–256.

*Note:* AOB: ammonia-oxidizing bacteria; AOA: ammonia-oxidizing archaea; FAB: facultative anaerobic bacteria; Anammox; anaerobic ammonium oxidation using $NO_2^-$ as the oxidant; R: protein.
* CEC units in meq/100 g soil.

### 3.2.4   N in Arable Systems

The conversion of natural ecosystems to arable farming involves soil cultivation and creates considerable risks such as soil erosion and a decline in properties such as soil structure, infiltration rates, biological diversity, and water-holding capacity (Low 1972). Cultivation increases the mineralization of soil organic N (SON) to plant-available soil inorganic N forms ($NH_4^+$ and $NO_3^-$), giving good crop yields initially (Doran 1980), but the SON content and crop yields can quickly decline without N inputs from leguminous crops or application of organic manures. Careful selection of crops, crop rotation, and management of crop residues plays a critical role in maintaining SON content (Wienhold and Tanaka 2001). Zero or minimal tillage and the elimination of residue burning have increasingly being recognized as practical and cost-effective measures to maintain stable soil N pools (Busari et al. 2015). In addition, the return of crop residues can help to reduce water losses from the soil by evaporation, moderate soil temperature, promote soil microbial activity, and enhance nitrogen mineralization (Hobbs et al. 2008).

While the world's need for food continues to increase, the availability of suitable new land for conversion to cropping is declining; humanity's ability to continue to increase food production at the required rate faces serious challenges. Inadequate rainfall and/or declining availability and quality of water for irrigation, changing temperature and rainfall patterns, deteriorating soil structure, nutrient runoff and leaching, and increasing particulate and GHG pollution of the atmosphere all present challenges to sustainable agriculture.

Increasing demand for food also places increasing demands on soils already under cropping. New varieties offer increased potential yield but place more demands on N supply, so that farmers are progressively incentivized—willingly or not—into the use of increasing amounts of fertilizer N and less reliance on leguminous crop rotations. Many soils under cropping are simply not getting the management, study, and inputs they need to be truly sustainable; SON levels are believed to be slowly declining in many situations under intensive arable cropping (Lal 2004). In the Mississippi Basin, although decades of soil loss through erosion by over-cropping are slowly being reversed, and SON levels are increasing again due to fertilizer N inputs and improved crop husbandry, huge losses in inorganic and organic N are still being lost in rivers to the U.S. Gulf, causing enormous damage to aquatic life (Mitsch et al. 2001; Van Meter et al.2016).

Soil biological biodiversity has also been shown to be gradually declining under many cropping situations, particularly monocultures (Ge et al. 2008). Little is known about the ability of soil organisms to cope with what are likely to be increasingly frequent and severe episodes of flooding, water shortages, and temperature extremes.

### 3.2.5   N in High-Density Confinement or Housed Meat and Dairy Systems

The increasing affluence of at least part of the population in many emerging economies has greatly increased the demand for meat and dairy products. Demand was traditionally met by the farming of both native and intensive, fertilized pastures (with small animal houses in inclement winter climates). However, increasing demand has led to the increasing construction of what can be best described as meat and dairy production "factories," where millions of chickens, hundreds of thousands of pigs, or tens of thousands of beef cattle or milking cows are confined in high-density outside pens (as in beef feedlots) or housed essentially full-time (as in chicken, pork, and dairy production).

The surrounding land controlled by the operators of these "factories" is almost invariably incapable of growing more than a small fraction of the feed consumed by the animals; the great majority is brought in from considerable distance, in many cases from other regions and countries. While this has reduced feed costs considerably due to low land and labor costs, it has come at the cost of increasingly serious threats to the environment, particularly in the area surrounding the animal production unit, and especially affecting $NH_3$ losses to the atmosphere. Ammonia is lost from excreta directly; from manure during storage; and during the application of manure to land. Losses as $NH_3$ range from 25–50% of the N in feed depending on the type of operation and its design (Hristov et al. 2011).

The chemical processes involved in the formation of lethal particulate air pollutants from the reaction of $NH_3$ emissions from high density animal confinement or housing, with industrial acidic air pollutants such as $NO_3^-$ and other N species, have been elucidated in recent years. While technologies such as "acid waterfalls" and application of alum, lignite coal, or urease inhibitors to the surfaces of pens can reduce these losses by up to 70% (Shi et al. 2001; Chen et al. 2015), they come with cost and practical limitations, and of course produce their own by-products. The 2% N content maximum in ammonia-saturated lignite (compared to 46% N in urea) places significant transport, storage, and spreading barriers to its proposed use as an N fertilizer.

Furthermore, while the animal feed may be cost-effectively brought in from great distances, transportation costs of low nutrient content manure and legislative restriction reasons often mean that the entire manure and effluent output is applied to the relatively small areas of surrounding land either owned by the enterprise or by local farmers. The result is often excessive soil nutrient levels, deteriorating soil structure and microbial health, increasing $NO_3^-$ leaching, $NH_3$ volatilization, and $N_2O$ emissions. The effects of excessive application of chicken and beef manures onto cropping land in Chesapeake Bay in Maryland is a well-documented example (Boesch et al. 2001).

For these and consumer-preference reasons, there is increasing interest in more traditional animal farming, particularly on grazed pastures for dairy cows and beef cattle. This is discussed in the next section.

### 3.2.6 N in Grazed Pastures

The cycling of N in developed or "intensively grazed" pastures differs greatly from N cycling in natural pastures grazed by migrating or semi-contained herds of native grazing animals. These herds follow the seasonal growth of different species that on average contain just adequate protein for survival. On the other hand, developed pastures in subtropical and particularly temperate climates, contain much higher protein levels than do native pastures, generally considerably more than dietary requirements (Misselbrook et al. 2013). Most of the excess N is excreted in the urine, mainly in the form of urea. Early quantitative or semi-quantitative detailed N cycles for grazed cattle and sheep (Ball et al. 1979; Quin 1982) emphasized the extent of N losses from urine patches as $NO_3^-$ leaching, $NH_3$ volatilization, and denitrification. The potential of $^{15}N$ isotope research to develop a deeper understanding of N cycling processes was quickly realized (Goh 1982).

As a generalization, N losses from developed pastures (in produce and to the environment) were matched by BNF by the legume component of pastures (Ball 1979). Perennial ryegrass/clover mixes have been the most widely used forages for intensive grazing in temperate climates. However, pastoral farmers have come under increasing pressure, both from the need to increase production in the face of increasing competition and from consumer pressure to reduce prices, to use more fertilizer N. Originally used by farmers only as a seasonal strategy to meet any unexpected feed deficits (Field and Ball 1978; Eckard and Franks 1998), fertilizer N has increasingly come to be used as part of standard practice on intensively grazed farms throughout the world, typically being applied 5–8 times per year at rates of 20–60 kg N/ha, or 150–400 kg N/ha annually (Reid 1993; Clark and Harris 1996). Increases in total farm production by 30% or more can be achieved compared to the nil-N grass/clover system. On most farms that have adopted this practice, the legume component of the sward has declined from a typical 20–40% to less than 15% of the sward, due to the increased shading by ryegrass. Very good pasture management is required to maintain clover in fertilizer-N fed pastures (Chapman et al. 2014).

Under intensive grazing, the soil-plant system has become progressively more "leaky." Environmentally adverse losses occur as $NO_3^-$ leaching, $NH_3$ volatilization, and $N_2O$ emissions (Saggar et al. 2009). Nitrate leaching increases with fertilizer N use (Ledgard et al. 1999; Jarvis

2000; Cameron et al. 2013). The use of urease and nitrification inhibitors can reduce losses (Zaman et al. 2008). The majority (>70%) of all N losses from grazed pastures typically come from urine patches, where the N application rate ranges from 300–1200 kg N/ha, much more than can be recovered by the pasture or efficiently recycled by soil microorganisms (Saggar et al. 2009, Cameron et al. 2013; Selbie et al. 2015). Excessive treading by grazing animals can also reduce infiltration rates and soil aeration, resulting in lower pasture growth and high denitrification losses.

### 3.2.7 Maintenance of Optimum Soil Inorganic N Levels

The optimum size of the soil inorganic pool (comprised principally of $NO_3^-$-N and $NH_4^+$-N) can be described as one that permits close to the highest possible crop yield or pasture growth being attained, while restricting losses of all undesirable forms of N from the system ( $NO_3^-$ e leaching, $NH_3$ volatilization, and $N_2O$ emissions) to levels that are assessed as being environmentally sustainable.

Meeting this objective requires a combination of agronomy, fertilizer, crop husbandry, soil fertility, soil biology, soil management, and environmental knowledge, and the adoption of (1) the most efficient water use systems, (2) the most efficient fertilizer application strategies (frequency and rates of the various nutrients and trace elements), and (3) development and use of the most efficient forms of each nutrient, especially N and phosphorus (P). Several of these factors are often absent. Lack of technology, resources, technical knowledge, and capital often mean compromises are made which minimize input costs at the expense of yields and/or of adverse effects on the environment.

The concentration of $NO_3^-$ in the soil fluctuates considerably both seasonally and short term due to many reasons; the most important are soil conditions, especially moisture, oxygen supply and temperature, inputs in fertilizer and organic N inputs such as manure and plant residues, and plant uptake. Ammonium tends to be less variable because (1) it is bound to negatively charged clay particles and organic matter, and (2) it is rapidly reutilized for growth by microorganisms and some plants, as $NH_4^+$ requires the least energy of any N species for conversion to amino acids and thence proteins and enzymes (Ward 2012). Micro-scale changes in moisture, $NH_4^+$-N, $NO_3^-$-N, soil pH, and other factors can result in large variability in the concentrations and activity of microorganisms, especially the nitrifiers and denitrifiers, over very small distances (Parkin 1987).

A problem unique to grazed pastures is the deposition of very high rates of N (300–1200 kg N/ha) in urine patches, generally far beyond the ability of the pasture to recover, and/or soil microorganisms to recycle internally. Urine patches have consequently been found to be the largest sources of both total $NO_3^-$ leaching and $N_2O$ emissions from intensively grazed dairy pastures (Saggar et al. 2009). Recent $^{15}N$ research indicates that the great majority of gaseous N losses from urine patches may be in the form of $N_2$, by the process of co-denitrification (Selbie et al. 2015). However, while this is an important knowledge which explains the "missing N" found in several previous N balance studies on pastures, it does not change the fact that significant losses of $N_2O$ are occurring. In addition, the $N_2$ lost from the farm represents a substantial economic loss. Solutions to this problem include (1) the breeding of grasses that maintain high levels of growth and energy content, but contain lower protein levels, more in line with animal dietary requirements (Gardiner et al. 2016) and (2) the detection of fresh urine patches on a daily basis (while the N is still largely present as urea) and their treatment with appropriate N inhibitors and growth stimulants (Saggar et al. 2009; Bates et al. 2015). Option (1) can result in less excess N to be voided in the urine; option (2) can increase pasture N recovery and growth, and reduce fertilizer N inputs.

### 3.2.8 Maintaining Soil Organic N Levels

Increasing the size of SOM levels in agricultural soils has many benefits, including increased water-holding capacity and crop growth, increased resistance to erosion, and increased N use efficiency, with their side-effects such as reduced fertilizer N requirements and $NO_3^-$ leaching, and increased carbon sequestration. However, most studies have focused on soil C pools rather than on N content.

Partly as a result, there are considerable differences in conclusions in the literature regarding the changes in SON with time under different farming and climatic systems.

Long-term studies at Rothamsted and elsewhere under cropping (Christensen et al. 2008; Ludwig et al. 2010) have demonstrated that initially low organic N contents in soils (in either previously over-cropped or in undeveloped land) can, over periods of 10–30 years, be increased and then sustained by the use of crop-legume rotations, retention of crop and root residues, and appropriate fertilizer and/or manure applications. This can also be accomplished in grazed pastures, by ensuring that the soil fertility status, particularly available P, is non-limiting for the optimum (40–50%) clover component of the sward (Harris et al. 1997; Chapman et al. 2014). Long-term SON levels and C/N ratios (0–30 cm depth) in both cropping and intensive pasture systems are typically 0.15–1.0% N and 9–15, respectively (Batjes 1996). However, SON levels can decline rapidly under cropping if inputs through BNF, fertilizer N, crop residues, and organic manures are insufficient, or if erosion or prolonged water saturation allow excessive denitrification. Even where previously declining SON and soil organic carbon (SOC) contents in over-cropped soils have started increasing again through improved husbandry and nutrient inputs, soil N pools (organic and inorganic) can still be very leaky, as found in the Mississippi Delta, where large river-born N losses to the U.S. Gulf continue (Howarth 2008; Van Meter et al. 2016). The increasing adoption of zero-tillage cropping, improved crop rotations, greater water-use efficiency, and optimum fertilizer N strategies help offset these losses.

Likewise, the increasingly common practice of converting forest soils directly to high growth potential, fertilizer N-fed crops and forages, without a phase of leguminous crop or legume-based pasture, can limit the rate of accumulation of SON (B.F. Quin, pers. comm.). Possible factors include (1) over-stimulation of mineralization with liming and fertilizer inputs, (2) insufficient microbial populations to recycle excess fertilizer N, and (3) the uneven distribution of excreta from grazing animals, and particularly urine.

### 3.2.9  Helping Soil Microorganisms to Maintain Tight N Cycles

Much progress has been made in the last 15 years in the use of genotyping to determine which microorganisms carry the genetic codes to carry out the various N cycling processes (Nelson et al. 2016). For example, large numbers of strains were found to carry the genes for fixation of $N_2$ gas, including 402 strains of bacteria, and smaller numbers of archaea and fungi (Nelson et al. 2016). Far less is known about how many of these are active N fixers in any given habitat, and what factors trigger the activation of their encoded N-fixing capability. However, their existence suggests that many soil microorganisms have evolved through climates and environments in which N fixing and cycling rates were much higher than exist today.

It has been demonstrated that more fertile soils in given types of ecosystems support much higher numbers of bacteria (but not archaea) involved in N processes than less fertile soils (Di et al. 2009). In wider terms, the ratios of the different microorganisms present in a particular ecosystem stay surprisingly constant (Nelson et al. 2016). It seems logical to assume that the microbial populations present in different soils are likely to rearrange their activity levels over time to reduce N losses from soils, provided system managements remain consistent.

The increasing references being made in the literature to "N saturation" of soils, especially in a forestry ecosystem as a consequence of increased atmospheric N deposition (Gundersen and Bashkin 1994; Niu et al. 2016) illustrate how quickly anthropogenic effects can overload the soil biological system present. Biodiversity is also threatened by N deposition (Dise et al. 2011). There is a need for multidisciplinary research involving soil chemists, biochemists, microbiologists, fertilizer chemists, N-15 specialists, geneticists, and ecosystem and N-process modelers, to greatly enhance our understanding of how we can increase the N utilization efficiency (FNUE) of crops from the currently estimated 25–40% to 80% in ways that are truly environmentally and economically sustainable (Barraclough et al. 2010; Hawkesford 2014). A range of approaches has been proposed to tighten the N cycle (Crews and Peoples 2004; Canfield et al. 2010; Smil 2011; Jones et al.

2014), including: (1) minimizing, wherever possible, human-made soil conditions and inputs that overload both the ability of crops to recover fertilizer N efficiently and the ability of soil microorganisms to process and internally recycle excess inorganic soil N, (2) finding the most cost- and biologically efficient means of matching fertilizer N rate of supply to the N uptake needs of individual crops on individual soils, (3) optimizing crop rotations, and (4) breeding or developing genetically engineered varieties for improved N use efficiency.

### 3.2.10 Conclusions

While the world's need for food continues to increase, the availability of suitable new land for conversion to cropping is declining; our ability to continue to increase food production at the required rate faces serious challenges. Inadequate rainfall and/or declining availability and quality of water for irrigation, changing temperature and rainfall patterns, deteriorating soil structure, nutrient runoff and leaching, and increasing particulate and GHG pollution of the atmosphere all present challenges to sustainable agriculture. It is therefore important to understand N cycling in agricultural systems and to identify factors that may influence N transformation and losses.

The N cycle, particularly agricultural N cycle, is influenced by inputs, outputs, and the interactions between different processes such as plant uptake, nitrification, mineralization, immobilization, volatilization, and denitrification. Inputs such as wet and dry deposition and management practices that influence fertilizer application, the return of excreta (urine and dung) from grazing animals, the land application of agricultural wastewater, and the extent of BNF by legumes can affect the extent of N losses from the cycle. Although the use of legumes in agriculture may reduce reliance on inorganic N fertilizers through the BNF from legume, the transfer of N from legumes to grass in mixed legume-grass pasture or to subsequent crops in rotation depends on many factors that influence mineralization-immobilization and losses. Understanding the N cycle and its driving factors is important for evaluating the N dynamics and their effects on agricultural ecosystems and for determining adaptable measures to tighten the N cycle, enhance FNUE, and minimize N losses to the environment.

[15]N stable isotopic techniques, which will be reviewed in the next section (Section 3.3), may be used in conjunction with the conventional non-isotopic techniques to provide an understanding of the interactions between different N transformation processes under different management and environmental conditions and to investigate management practices that can be put in place to enhance FNUE and minimize the potential N losses to the environment.

## 3.3  BASIC PRINCIPLES OF NITROGEN STABLE ISOTOPIC TECHNIQUES

The nitrogen (N) atom has two stable isotopes, [14]N and [15]N. These isotopes are stable, since they do not undergo any radiation. The heavy [15]N (atomic weight of 15) and the light [14]N (atomic weight of 14) account for 0.3663% and 99.6337% of the atmospheric N (in the form of dinitrogen $N_2$), respectively (Mariotti 1983, 1984; IAEA 2001, 2008). Although atmospheric N is abundant in the diatomic N2 form (99%), it cannot be used directly by plants, except for the leguminous crops that have developed a symbiotic relationship with $N_2$-fixing bacteria (Galloway et al. 2003; IFIA 2007).

There are two commonly used stable N isotope methods: the [15]N dilution method and the [15]N-natural abundance method to investigate FNUE, N transformations or sources-sinks in N cycle (Bedard-Haughn et al. 2003; IAEA 2008). The use of these techniques in quantifying BNF has been reviewed in Chapter 2 and hence they are not discussed here. Only brief information on the notation units of [15]N stable isotopes, which has been stated in Chapter 2, is given below before the review of the [15]N isotopic techniques in assessing N transformations, determining FNUE, and identifying N sources-sinks is presented.

With the [15]N dilution method, also termed [15]N-tracer method, [15]N-labeled inorganic N (i.e., N fertilizers such as urea) or organic N (i.e., manure or crop residue) is added to soils to investigate the flow of $^{15}$N through the various soil and plant N pools. The $^{15}$N enrichment of a given N pool is dependent on (1) the amount of $^{15}$N applied, (2) the sizes and the turnover rates of the individual N pools, and (3) the interactions between these pools. The change in $^{15}$N enrichment in different soil N pools over time can be used to calculate the turnover rate of N in the various soil N pools.

In contrast to the $^{15}$N dilution method, the natural abundance method does not involve the use of $^{15}$N-enriched materials. Instead, it is based on the differences in natural $^{15}$N abundance or the differences in the $^{15}$N:$^{14}$N ratios between N sources and sinks to follow the flow of N through the soil-plant systems (Bedard-Haughn et al. 2003; Murphy et al. 2003).

For $^{15}$N dilution studies, $^{15}$N concentration in samples (e.g., soil or plant samples) are expressed as atom percent $^{15}$N (atom% $^{15}$N) or atom% $^{15}$N excess. Atom% $^{15}$N is the absolute number of atoms of $^{15}$N in 100 atoms of total N. Thus atom% $^{15}$N in a sample is notated as follows:

$$\text{Atom\% } {}^{15}\text{N sample} = {}^{15}\text{N/(Total N)} \times 100 = \left[ {}^{15}\text{N/({}^{15}\text{N} + {}^{15}\text{N})} \right] \times 100 \tag{3.1}$$

Atom% $^{15}$N excess is defined as atom% $^{15}$N in sample after correcting for $^{15}$N background abundance. Thus, atom% $^{15}$N excess is calculated as the difference between atom% $^{15}$N in a sample and the atom% $^{15}$N in a globally accepted standard (atmospheric N$_2$), which is 0.3663. Thus, it can be expressed as follows:

$$\text{Atom\% } {}^{15}\text{N excess} = \text{Atom\% } {}^{15}\text{N sample} - 0.3663 \tag{3.2}$$

For natural abundance studies, where no $^{15}$N-labeled materials are added to the N cycle, $^{15}$N concentrations in studied samples are expressed in terms of $^{15}$N:$^{14}$N ratios (expressed as $^{15}$N/$^{14}$N). The changes in these ratios (R) in studied samples, relative to that of the atmospheric N$_2$ (subsequently referred to as a standard) are expressed as delta ($\delta$) values, which are given in parts per thousand or per mil (‰). The $^{15}$N/$^{14}$N ratios in the standard is constant at 0.003676 (i.e., 0.3663/99.6337) and hence $\delta$ $^{15}$N in the standard is 0‰.

The $\delta$ value in a sample is calculated from the following equation (Equation 3.3) taking into account the isotope ratios (R) for the sample and for the standard:

$$\delta\,{}^{15}\text{N in a sample (‰)} = \left[ (\text{R sample/R standard}) - 1 \right] \times 1000 \tag{3.3}$$

Equation 3.3 can be expressed as:

$$\delta\,{}^{15}\text{N in a sample (‰)} = \left\{ \left[ ({}^{15}\text{N/}^{14}\text{N sample})/({}^{15}\text{N/}^{14}\text{N standard}) \right] - 1 \right\} \times 1000 \tag{3.4}$$

Equation 3.4 can be rewritten as:

$$\delta\,{}^{15}\text{N sample (‰)} = \left[ (\text{Atom\% } {}^{15}\text{N in sample} - \text{Atom\% } {}^{15}\text{N in standard}) \right.$$
$$\left. /\text{Atom\% } {}^{15}\text{N in standard} \right] \times 1000 \tag{3.5}$$

Samples with ratios of $^{15}$N/$^{14}$N greater than that in the atmosphere (i.e., R standard = 0.003676) have positive $\delta$ $^{15}$N values and those with ratios of $^{15}$N/$^{14}$N lower than that in the atmosphere have negative $\delta$ $^{15}$N values. A sample with a $\delta$ $^{15}$N value of +5‰ means that a sample has a $^{15}$N/$^{14}$N ratio greater than the standard ratio (R standard) by 5 parts-per-thousand or 0.5% more $^{15}$N in the sample relative to the standard.

## 3.3.1  $^{15}$N DILUTION METHOD

The $^{15}$N dilution method, also known as the $^{15}$N-enriched method, is valuable for the determination of: (1) the rate of N cycling/transformations in the various N pools, (2) the extent of N losses from the applied N sources and hence the N use efficiency (NUE) of these sources, and (3) the fate of applied N in studied ecosystems under different management practices [e.g., conservation tillage (Castellanos et al. 1998), irrigation (Delgado et al. 1996), and the use of slow-release fertilizers and nitrification inhibitors (Hauck et al. 1994; Delgado and Mosier 1996; Shoji et al. 2001)]. The extent of N enrichment (% atomic excess) introduced to applied materials (e.g., fertilizers, manure, or crop residues) can range from 3 to 98%, depending on environmental conditions in which studies are conducted and the aims of these studies to determine NUE or to act as a tracer (IAEA 2001, 2008). The applied $^{15}$N can be traced in each or all of the components of the soil–plant system, such as plant N uptake, soil mineralization-immobilization, leaching losses, $NH_3$ volatilization, denitrification, and codenitrification (IAEA 2001, 2008; Bedard-Haughn et al. 2013).

The extent of $^{15}$N enrichment or depletion in various N pools (e.g., soil $NH_4^+$ or $NO_3^-$ pool) after the addition of $^{15}$N-labeled materials allows the determination of the rate of N cycling in the various N pools and the extent of N losses from the studied ecosystems (Hauck and Bremner, 1976; IAEA 2001, 2008). For example, the addition of $^{15}$N-labeled ammonium sulfate ($^{15}NH_4)_2SO_4$ to label soil $NH_4^+$ pool can be used to determine a gross N mineralization rate from SON to $NH_4^+$ using the $^{15}NH_4^+$ pool dilution technique (Kirkham and Bartholomew 1954; Davidson et al. 1991; Hart et al. 1994). The main principle of this technique is based on a change in the size of soil $NH_4^+$ pool and a dilution of $^{15}NH_4^+$ abundance due to mineralized $^{14}$N after the addition of $(^{15}NH_4)_2SO_4$. It is assumed that the soil $NH_4^+$ pool is uniformly labeled and that any losses from soil $NH_4^+$ pool have the same $^{15}$N abundance (i.e., the same $^{15}N/(^{14}N + ^{15}N)$ ratio) as the whole pool. Thus, processes removing $NH_4^+$ from the soil $NH_4^+$ pool will remove $^{15}$N and $^{14}$N in proportion to their presence in the pool and they will not alter the $^{15}$N abundance of the pool. The unlabeled ($^{14}NH_4^+$) released from SON dilutes the $^{15}$N enrichment of the soil $NH_4^+$ pool, whereas consumptive processes are assumed to remove $^{15}$N and $^{14}$N at equal rates from the soil $NH_4^+$ pool (Murphy et al. 2003). Any consumption of $NH_4^+$ would only change the size of the $NH_4^+$ pool, not the $^{15}$N abundance, so gross N mineralization (GNM) rate (mg N/kg soil. day) over an experimental period is calculated as follows (Murphy et al. 2003; IAEA 2008; Bedard-Haughn et al. 2013):

$$GNM = \left\{[NH_4^+]t_o - [NH_4^+]t_1\right\}/(t_1 - t_o) \times$$
$$\log\left\{(APEt_o/APEt_1)\right\}/\log\left\{[NH_4^+]t_o/[NH_4^+]t_1\right\} \tag{3.6}$$

where, $t_o$ and $t_1$ are times (days) after the addition of $(^{15}NH_4)_2SO_4)$, and APE stands for atom% $^{15}$N excess in the soil $NH_4^+$ at $t_o$ and $t_1$ ($APEt_o$ and $APEt_1$, respectively) and soil $NH_4^+$ concentrations at $t_o$ and $t_1$ are $\left[NH_4^+\right]t_o$ and $\left[NH_4^+\right]t_1$, respectively (Griffin 2007; IAEA 2008).

Similarly, gross N nitrification (GNN; mg N/kg soil. day) is calculated by measuring the size and $^{15}$N abundance of the $NO_3^-$ pool at two times ($t_o$ and $t_1$) after the addition of $K^{15}NO_3$ to label soil $NO_3^-$ pool. The GNM equation stated above is used to calculate GNN by substituting the $NH_4^+$ values in the GNM equation with the $NO_3^-$ values (i.e., with atom% $^{15}$N excess in the soil $NO_3^-$ pool and with soil $NO_3^-$ concentrations $\left[NO_3^-\right]$ at $t_o$ and $t_1$) (Barraclough 1991; Hart et al. 1994; Murphy et al. 2003; Bedard-Haughn et al. 2013).

There are many assumptions regarding the use of isotopic dilution technique such as: (1) no isotopic discrimination between $^{15}$N and $^{14}$N during soil N transformation processes, (2) no remineralization of labeled mineral N that has been immobilized over a study period; (3) the added $^{15}$N is uniformly mixed within a soil and distributed homogenously in soil N pool, and (4) an equilibrium between added $^{15}$N and indigenous $^{14}$N pools (Kirkham and Bartholomew 1954; Hart et al. 1994).

Various researchers (e.g., Follett 2001; Murphy et al. 2003; Luxhøi et al. 2004) have reported errors associated with gross N mineralization estimates even over a short-term (1–2 days) experimental period, resulting from non-homogenous labeling of $^{15}N$ in soils using a single needle injection technique. Bedard-Haughn et al. (2013) have reported some negative values of gross N mineralization and nitrification rates within the 24-h incubation period in their study. They have attributed these to a violation of one or more assumptions of the isotope dilution method such as the violation of the immobilization and remineralization assumption, where the added $^{15}N$ is remineralized within the incubation period. Another possible reason is a dissimilatory nitrate reduction to ammonium (DNRA) and the effect of nonuniformity of $^{15}N$ addition.

The extent of $^{15}N$ recovery in aboveground and belowground plant biomass pools, as well as in soil $NH_4^+$, soil $NO_3^-$, and soil microbial biomass N pool, was also calculated using the isotopic dilution principle (Bedard-Haughn et al. 2003; IAEA 2008). The proportion of N in a particular pool derived from $^{15}N$ labeled material (e.g., fertilizers), expressed as %Ndff is calculated as follows:

$$\%Ndff = (APE\ in\ a\ sample/APE\ in\ added\ ^{15}N\ labeled\ material) \times 100 \tag{3.7}$$

where APE is atom% $^{15}N$ excess in a particular pool (e.g., plant) and in the labeled material. Thus, the amount of N in a particular pool (e.g., plant N uptake) derived from $^{15}N$ labeled material (e.g., fertilizers) is

$$Ndff\ (kg\ n/ha) = \%Ndff \times Total\ plant\ N\ uptake \tag{3.8}$$

where Ndff and %Ndff are the amount (kg N/ha) and the proportion (%) of N in plants (kg derived from a $^{15}N$-labeled fertilizer. Total plant N uptake (kg N/ha) is the product between a harvested dry matter (DM), expressed as kg DM/ha and an N content in DM.

Thus, N use efficiency from labeled fertilizer can be calculated as follows:

$$NUE = Ndff/Amount\ of\ N\ in\ applied\ ^{15}N\text{-labeled fertilizer} \tag{3.9}$$

Although the $^{15}N$ dilution method is valuable in determining the fate of N fertilizers and other N sources such as manure and crop residues, there are some limitations associated with the cost of using $^{15}N$ enriched materials. The cost increases with an increase in $^{15}N$ enrichment (IAEA 2008) and the use of a substantial amount of $^{15}N$ labeled materials at a landscape scale (Bedard-Haughn et al. 2003).

### 3.3.2 $^{15}N$ NATURAL ABUNDANCE APPROACH

The $^{15}N$ natural abundance (NA) method relies on natural differences in $^{15}N$ abundance (i.e., $\delta^{15}N$ values) between different N pools in soil-plant systems to provide insight into the main N transformation processes in the N cycle of agricultural farms (Robinson 2001; Bedard-Haughn et al. 2003). Many factors including isotopic fractionation associated with soil N transformation processes can influence the variability of $\delta^{15}N$ values, making it difficult to interpret the $\delta^{15}N$ data obtained (Robinson 2001; Bedard-Haughn et al. 2003). They have recommended that factors affecting $\delta^{15}N$ variability should be taken into account when considering the use of $\delta^{15}N$ values of soils and plants as an integrated indicator of N cycle processes.

Isotopic fractionation is the separation of $^{15}N$ and $^{14}N$ isotopes during the biological-chemical-physical processes because of the difference in their atomic weights (atomic weights for $^{14}N$ and $^{15}N$ of 14 and 15, respectively; Högberg 1997; Robinson 2001; Bedard-Haughn et al. 2003; Evans

2007). Light isotope ($^{14}$N) reacts faster than heavy isotope ($^{15}$N) during chemical-physical processes (e.g., $NH_4^+$ exchange and diffusion; Högberg 1997) and there is also a discrimination against $^{15}$N during micro-biological processes such as denitrification, nitrification, and volatilization (Högberg 1997; Robinson 2001; Bedard-Haughn et al. 2003). Because of this isotopic fraction, there is a depletion of $^{15}$N in products and an enrichment of $^{15}$N in a residual substrate. The increase in $\delta\ ^{15}$N in a substrate does not occur if a transformation from a substrate to a product is complete and all N atoms in a substrate are converted to products. In this case, the $\delta\ ^{15}$N value in a product will be the same as that in a substrate (Robinson 2001; Bedard-Haughn et al. 2003; Sebilo et al. 2003, 2006; Lehmann et al. 2004). For example, if nitrification converts all $NH_4^+$ into $NO_3^-$, as normally is the case under aerobic conditions, there would be no isotopic fractionation at the end of the nitrification process (Robinson 2001; Bedard-Haughn et al. 2003; Sebilo et al. 2003, 2006). Both $N_2O$ and $N_2$ are simultaneously emitted from $NO_3^-$ rich sediments following microbially mediated denitrification. Denitrifiers preferentially reduce $N_2O$ molecules that contain the lighter $^{14}$N stable isotope, which leads to increasing $^{15}$N enrichment of the remaining $N_2O$ (Ostrom et al. 2007; Vieten et al. 2007). Theoretically, the larger the reduction of $N_2O$ by microorganisms, the higher the $\delta\ ^{15}$N-$N_2O$ signature of $N_2O$ emitted into the atmosphere.

The isotopic fractionation factor (Högberg 1997; Robinson 2001; Bedard-Haughn et al. 2003; Menyailo and Hungate 2006), expressed as $\alpha$, is the ratio of rate constants ($^{14}$k and $^{15}$k) for molecules containing $^{14}$N and $^{15}$N isotopes as shown below:

$$\alpha = {}^{14}k/{}^{15}k \tag{3.10}$$

In an open system with an unlimited supply of a substrate, the isotopic fractionation factor, expressed as $\beta$, can be calculated as:

$$\beta = \left[ ({}^{15}N/{}^{14}N)\ \text{subtrate}/({}^{15}N/{}^{14}N)\ \text{product} \right] \tag{3.11}$$

This isotopic discrimination can lead a $^{15}$N enrichment in a substrate as follows:

$$\varepsilon = (\alpha - 1) \times 1000 \tag{3.12}$$

where $\varepsilon$ is defined as a $^{15}$N enrichment factor.

A range of isotopic fractionation values ($\alpha$, $\beta$, and $\varepsilon$) associated with various N processes has been provided by Robinson (2001) and Bedard-Haughn et al. (2003). They are summarized in Table 3.2. As shown from Table 3.2, the $\beta$ value for denitrification is 1.0185. This indicates that the products from denitrification of $NO_3^-$ ($N_2O$ or $N_2$) become depleted in $^{15}$N by 18.5‰ while the remaining (unreacted) $NO_3^-$ is enriched with $^{15}$N. Sebilo et al. (2003) has reported a similar enrichment factor ($\varepsilon$) of 18.5‰ for denitrification process under conditions where $NO_3^-$ diffusion to a reaction site is not limiting. For volatilization, the $\beta$ value of 1.0245 (Table 3.2) means that volatilization depletes $^{15}$N in $NH_3$ by 24.5‰ and enriches the remaining $NH_4^+$ with $^{15}$N.

Table 3.2 also shows that other N processes such as nitrification can result in a significant isotopic fractionation, while ammonification and the diffusion of $NH_4^+$ and $NO_3^-$ generate limited or no isotopic fractionation. For example, during the nitrification of $NH_4^+$ to $NO_3^-$, the enrichment factor of 25–28.5‰ could induce lighter $NO_3^-$ product (i.e., $^{15}$N depletion by 25–28.5‰). In contrast, the ammonification, which involves the production of $NH_4^+$ from SOM decomposition, causes only a small fractionation (0–3‰).

The variability in $\delta\ ^{15}$N values in soils and plants is caused by not only isotopic fractionation but also by other factors as discussed below (Robinson 2001; Bedard-Haughn et al. 2003).

**TABLE 3.2**

**Isotopic Fractionation Values (Median Values) for Some Nitrogen Processes**

| Nitrogen Processes | Bedard-Haughn et al. (2003) | | Robinson (2001) | |
|---|---|---|---|---|
| | $\alpha$ | $\beta$ | $\alpha^a$ | $\varepsilon$ |
| Ammonification ( $NH_4^+$ production from organic matter decomposition) | | 1.0025 | 1.000–1.005 | 0–5 |
| $NH_4^+$ assimilation into organic N by plants | 1.0158 | 1.0050 | 1.009–1.018 | 9–18 |
| $NO_3^-$ assimilation into organic N by plants | | | 1.000–1.019 | 0–19 |
| $NH_4^+$ assimilation into organic N by microbes | | | 1.014–1.020 | 14–20 |
| $NO_3^-$ or organic N assimilation into organic N by microbes | 1.0142 | | 1.013 | 13 |
| $NO_2^-$ assimilation by microbes | 1.0210 | | | |
| Biological $N_2$ fixation | 1.0020 | 1.0013 | 1.000–1.006 | 0–6 |
| $N_2O$ and $N_2$ production during $NO_3^-$ reduction (denitrification) | 1.0305 | 1.0185 | 1.028–1.033 | 28–33 |
| $N_2O$ and $N_2$ production during $NH_4^+$ oxidation (nitrification) | 1.0600 | 1.0350 | 1.035–1.060 | 35–60 |
| $NO_3^-$ production during nitrification | 1.0285 | 1.0250 | 1.015–1.035 | 15–35 |
| $NH_4^+$ exchange | | 1.0014 | | |
| $NH_3$ volatilization | 1.0400 | 1.0245 | 1.040–1.060 | 40–60 |
| Organic N assimilation by animals (deamination and transamination) | | | 1.001–1.006 | 1–6 |

[a] Calculated as $\alpha = 1 + (\varepsilon/1000)$.

### 3.3.2.1 Influence of Topography, Riparian Zones, and Soil Ages

The review conducted by Bedard-Haughn et al. (2003) has reported that topographic features, which control the rate of various hydrological and biological processes, can result in significantly different $\delta^{15}N$ between lower and upper slope positions. Soils in low lying parts of the landscape tend to have higher $\delta^{15}N$ than soils in well-drained areas, possibly due to greater denitrification, which causes a significant isotopic fractionation (Table 3.2). Enhanced denitrification in low lying areas was attributed to anaerobic conditions resulting from fluctuating water tables (Bedard-Haughn et al. 2003). In riparian landscapes, several soil N transformation processes such as nitrification and denitrification may occur over a very short distance, giving rise to a complex spatial pattern for $\delta^{15}N$ at the boundary between the uplands and the riparian zone and within the riparian zone itself (Bedard-Haughn et al. 2003).

There are inconsistent effects of soil ages on soil and plant $\delta^{15}N$ values (Vitousek et al. 1989; Martinelli et al. 1999; Brenner et al. 2001; Hyodo and Wardle 2009; Menge et al. 2011), probably attributing to the net effect of isotopic fraction associated with changes with time in N transformation processes and N inputs and N outputs (Templer et al. 2007; Menge et al. 2011).

### 3.3.2.2 Influence of Nitrogen Inputs from Atmosphere and Fertilizers

Atmospheric N inputs via dry and wet deposition have $\delta^{15}N$ of approximately 0‰, while fertilizers can have a range of $\delta^{15}N$ values. Bateman and Kelly (2007) reported a lower $\delta^{15}N$ value in synthetic fertilizers than organic-based fertilizers. Synthetic N fertilizers such as ammonium sulfate, potassium nitrate, urea, and a range of N-P-K mixtures, have a mean $\delta^{15}N$ of −0.2‰ (ranging from −5.9 to 6.6‰). In contrast, organic fertilizers including farmyard manure, cattle slurry, compost, seaweed extracts, and the animal excreta from grazing animals have higher $\delta^{15}N$ values. For example, median $\delta^{15}N$ values for seaweed extracts, blood and bone, fishmeal, and manure/compost are 2.5,

5.9, 7.5, and 8.1‰, respectively. The low δ $^{15}$N values for synthetic N fertilizers, except for ammonium sulfate $(NH_4)_2SO_4$ with a high δ $^{15}$N value of 6.6‰, are attributed to the source of N in these fertilizers, which is derived from atmospheric $N_2$ with the δ $^{15}$N value of 0‰ ($^{15}$N and $^{14}$N account for 0.3663% and 99.6337% of the atmospheric N and hence the $^{15}$N/$^{14}$N ratio in the atmospheric standard is constant at 0.003676 and hence δ $^{15}$N in the standard is 0‰; Section 3.3). The higher δ $^{15}$N values for $(NH_4)_2SO_4$ are attributed to the manufacturing processes that potentially caused $(NH_4)_2SO_4$ to be isotopically heavier than that produced from a direct synthesis by the Haber–Bosch process (Bateman and Kelly 2007). Organic fertilizers such as cattle slurry or farmyard manure have high δ $^{15}$N values, due to isotopic fractionation during digestion of forage and manure storage (e.g., $NH_3$ volatilization during storage, which is accompanied by a strong fractionation against the heavier $^{15}$N; Högberg 1997; Bateman and Kelly 2007).

The degree of $^{15}$N enrichment in soils and plants depends on not only the type but also the amount of fertilizer used (Watzka et al., 2006; Kriszan et al. 2009; Yuan et al. 2012). In soils receiving organic fertilizers, δ $^{15}$N values in soils and plants are found to be higher than those in soils receiving synthetic fertilizers, attributed to a higher $^{15}$N enrichment in organic manure, compared to synthetic fertilizers which have δ $^{15}$N close to 0‰ as a result of their synthesis from atmospheric $N_2$ through the Haber–Bosch technology (Watzka et al. 2006; Kriszan et al. 2009; Yuan et al. 2012).

### 3.3.2.3 Influence of Land Use and Management Practices

Soil δ $^{15}$N values were found to vary with land uses and land management intensity (Stevenson et al. 2010) with the highest δ $^{15}$N of 6.2‰ for intensively managed cropping farms, followed by dairy farms (5.4‰), sheep/beef farms (3.8‰), forestry (2.8‰), and indigenous (native) forests. They attributed their findings to the increase in fertilizer N inputs and enhanced isotope fractionation associated with N processes such as $NH_3$ volatilization, mineralization, nitrification, denitrification, and $NO_3^-$ leaching (particularly volatilization and $NO_3^-$ leaching) when there is a change in land uses from native forests to sheep/beef or dairy farms and to cropping lands.

Other workers (e.g., Kleinebecker et al. 2014; Kriszan et al. 2014) also reported the effects of land uses and management intensity on the $^{15}$N signatures (i.e., δ $^{15}$N) in soils, plant biomass, and animal tissues. For example, Kriszan et al. (2014) found that the δ $^{15}$N values of soils and harvested forage from high N input grasslands were elevated by 2.8‰ relative to those of low N input grasslands. Based on the findings obtained, Kriszan et al. (2014) have proposed that the δ $^{15}$N values in these N pools are useful integrators of past fertilizer and nutrient management histories, regardless of high natural δ $^{15}$N variations in agricultural systems.

### 3.3.2.4 Influence of Manure and Excreta from Grazing Animals

Various workers (e.g., Aranibar et al. 2008; Kriszan et al. 2014) have reported the enrichment of $^{15}$N in animal feces and the depletion of $^{15}$N in urine, relative to the $^{15}$N values in diet (i.e., forage eaten). They have attributed these findings to the isotopic fractionation during metabolic and digestion processes in animals.

Cattle slurry and animal manure, which are derived mainly from animal feces, are enriched in δ $^{15}$N values (6–20‰) due to isotopic fractionation of $NH_3$ volatilization, nitrification, and denitrification process during the storage of manure (Bedard-Haughn et al. 2003; Lee et al. 2011; Kriszan et al. 2014). In the survey of dairy farms with low and high N inputs (50 kg N/ha and 450 kg N/ha), Kriszan et al. (2014) have shown that cow urine from both low and high N input systems (termed LNI and HNI, respectively) had depleted δ $^{15}$N values (–1.13 to –2.94‰ for LNI and –0.08 to –0.25‰ for HNI), while cow feces had enriched δ $^{15}$N values (ranging from 1.58–3.55‰ for LNI and 4.58–6.08‰ for HNI).

Applications of animal manure and cattle slurry to farmlands or the regular urine and dung deposition by livestock have been found to enrich soil and plant δ $^{15}$N values (Aranibar et al. 2008; Blüthgen et al. 2012; Kleinebecker et al. 2014; Kriszan et al. 2009, 2014). This effect is attributed to the enriched δ $^{15}$N values of the fecal component and a strong fractionation during $NH_3$ volatilization from urine and manure (Kleinebecker et al. 2014; Kriszan et al. 2014). The fractionation would enhance a loss of $^{15}N$-depleted N and induce the $^{15}N$ enrichment in the remaining soil N pool. In addition, grazing animals through the regular return of animal excreta and animal trampling can influence SON inputs, soil aggregate stability, and soil pugging (Nguyen et al. 1998). These in turn can influence soil nitrification, denitrification, and $NH_3$ volatilization processes (Menneer et al. 2005; Hoeft et al. 2012; Yan et al. 2016), which generate isotopic fractionation and lead to $^{15}N$ enrichment in soils (Table 3.2; Högberg 1997; Bedard-Haughn et al. 2003; Watzka et al. 2006; Kriszan et al., 2009). Furthermore, the removal of plant biomass, which is usually more depleted in $\delta^{15}N$, compared to the soil from which the plant biomass is grown and subsequently grazed, can also contribute to an enhanced enrichment of soil $^{15}N$ (Ometto et al. 2006; Watzka et al. 2006; Kriszan et al. 2009, 2014; Kleinebecker et al. 2014).

### 3.3.2.5  Influence of Plants

Assimilation of soil $NH_4^+$ and soil $NO_3^-$ by plants can induce a small fractionation with an isotopic enrichment factor (ε) between −3 and 1‰ (Yoneyama et al. 2001; Table 3.2). However, plant $\delta^{15}N$ values can be modified by multiple interacting factors such as (1) the inherent variation in soil $\delta^{15}N$ values, (2) the type of N sources, (3) mycorrhizal-plant associations, (4) fractionation mechanisms that occur during plant uptake, and (5) specific temporal and spatial N acquisition strategies (Bedard-Haughn et al. 2003; He et al. 2009; Hobbie and Högberg 2012). For example, plants can become $^{15}N$ depleted, compared to soil $\delta^{15}N$ values because of $^{15}N$ discrimination during soil ammonification and nitrification that yield isotopically light $NH_4^+$ and $NO_3^-$ relative to SON (Marshall et al. 2007).

Plant $\delta^{15}N$ values for legume (grain and forage legume) are likely to be similar to atmospheric $\delta^{15}N$ (0.0‰) since these plants have a direct access to atmospheric $N_2$ through biological nitrogen fixation (BNF) and there is a negligible isotopic fractionation during the BNF process (Table 3.2). In non-$N_2$-fixing plants, plant $\delta^{15}N$ values normally vary between 3‰ and 5‰ (Bedard-Haughn et al. 2003) and are likely to reflect the soil $\delta^{15}N$ in the plant available soil N pool (soil $NH_4^+$ and soil $NO_3^-$). Plant $^{15}N$ signatures are dependent on N inputs from synthetic or organic sources and isotopic fractionation associated with N processes that influence the plant-available soil N pool (Bedard-Haughn et al. 2003; Fraser et al. 2011).

Plant $\delta^{15}N$ values are approximately 6–7 times more variable than $\delta^{15}N$ values in soil total N (Sutherland et al. 1993) and are generally lower than those in soils (i.e., plants are depleted in $^{15}N$ relative to soil; Hobbie and Hobbie 2006, 2008; Wallander et al. 2009; Kriszan et al. 2009; Hobbie and Högberg 2012). The high variability of plant $\delta^{15}N$ values, compared to soil $\delta^{15}N$ values, is probably attributed to the accessibility of plants to the soil mineral N pool which is more isotopically variable than total soil N or SON (Mariotti et al. 1981; Robinson 2001). In addition, the observed $^{15}N$ depletion in plants is probably due to several reasons (Hobbie and Hobbie 2006, 2008; Wallander et al. 2009; Kriszan et al. 2009; Hobbie and Högberg 2012). These include: (1) plant uptake of soil $^{14}NH_4^+$ and soil $^{14}NO_3^-$ that have been produced as a result of $^{15}N$ discrimination during soil ammonification and soil nitrification, (2) the transfer of $^{15}N$-depleted soil $NO_3^-$ to plants by mycorrhizal fungi, and (3) the discrimination against the heavy $^{15}N$ over the light $^{14}N$ during plant N uptake. The plant $\delta^{15}N$ values are therefore expected to reflect the net effect of a range of biochemical and environmental processes that occur at studied sites. In some studies, plant $\delta^{15}N$ values were found to reflect the $^{15}N$ enrichment of the inorganic soil and can be used as valuable tools to study ecosystem N dynamics, both as tracers or as integrative signals of soil N transformation processes (Korontzi et al. 2000; Amundson et al. 2003; Pardo et al. 2006; Kleinebecker et al. 2014). For example, Houlton et al. (2007) showed that foliar $\delta^{15}N$ values that reflect the $\delta^{15}N$ signatures of the plant's specific N sources (e.g., $NO_3^-$ and $NH_4^+$) can be valuable in revealing information on plant N uptake patterns and soil N turnover. In some other studied ecosystems, not only plant $\delta^{15}N$ but also soil $\delta^{15}N$ values can act as indicators of soil N cycling and plant uptake patterns. For example, Templer et al. (2007) have reported that $\delta^{15}N$ values in both fine roots and SON are better indicators of relative rates of soil N cycling than are aboveground (foliage) $\delta$ $^{15}$N values in the forests of the Catskill Mountains, New York. Similarly, Kahmen et al. (2008) in their study across 18 different grassland ecosystems have reported that foliar $\Delta$ $\delta$ $^{15}$N values (i.e., difference between $\delta$ $^{15}$N leaf and $\delta$ $^{15}$N soil) are valuable tools to assess plant N uptake patterns and to characterize the soil N cycle.

### 3.3.2.6  Influence of Soil Depth

Soil $\delta$ $^{15}$N values tend to increase with soil depth (Bedard-Haughn et al. 2003). This increase is partly attributed to the discrimination against $^{15}$N associated with isotopic fractionation during denitrification and during SOM decomposition and humification (Table 3.2). This discrimination results in surface soils with lower $\delta$ $^{15}$N values than deeper soils (Bedard-Haughn et al. 2003; Billy et al. 2010). Another reason for a higher soil $\delta$ $^{15}$N values with depth is the exploitation of deeper soil layers by ectomycorrhizal (EM) fungi and a subsequent accumulation of $^{15}$N-enriched EM fungi microbial biomass in old SOM (Billings and Richter 2006; Wallander et al. 2009; Billy et al. 2010).

### 3.3.3  Summary

Both $^{15}$N dilution technique and $^{15}$N natural abundance (NA) approach are used at a range of scales to investigate nutrient cycling and the fate of N sources in agricultural ecosystems. Information on their uses will be elaborated on in the following section.

The $^{15}$N dilution technique involves the application of N-labeled materials which can be costly, particularly at a landscape scale. In addition, the addition of labeled N materials (e.g., fertilizers or solution containing $NH_4^+$ or $NO_3^-$) can potentially influence the normal functioning of N cycling processes in natural (undisturbed) soil-plant systems. In contrast, the NA approach does not involve any addition of $^{15}$N-labeled materials but relies on the natural difference in $^{15}$N signatures between different components of N sources and sinks to follow the fate of N sources in soil-plant-water components. The difference in $\delta$ $^{15}$N between different N components is inherently small and measured in terms of per mil or part per thousand (‰). Any factor such as isotopic fractionation that increases the variability of $\delta$ $^{15}$N of N sources can make it difficult to interpret the difference between $\delta$ $^{15}$N values in N sources and those in N sinks. Isotopic fractionation is the separation of $^{15}$N and $^{14}$N during biochemical-physical processes because of their mass differences. Isotopic fraction associated with nitrification, volatilization, and denitrification processes can cause a depletion of $^{15}$N in products (i.e., $^{15}$N-depleted $NH_3$, $N_2$, $N_2O$, and $NO_3^-$) that are lost to the environment through volatilization, denitrification, or leaching and an enrichment of $^{15}$N in the residual substrate (e.g., $NH_4^+$, $NO_3^-$ and organic N) in soils.

Plant $\delta$ $^{15}$N values are also influenced by fractionation mechanisms during uptake. In addition, plant $\delta$ $^{15}$N values are dependent on a range of $\delta$ $^{15}$N signatures in soil N pools from which plants take up N for their nutrition.

## 3.4  SPECIFIC APPLICATIONS OF $^{15}$N STABLE ISOTOPES TECHNIQUES IN AGRICULTURAL SYSTEMS

### 3.4.1  Applications of $^{15}$N Dilution Method

The $^{15}$N Dilution Method is used to investigate N transformations in soils and N use efficiency of fertilizers, manure, and crop residues and the fate of N from these N sources, as influenced by the following management practices:

1. The forms of applied N (e.g., manure, urea, or ammonium sulfate; Hauck and Bouldin 1961; Paul and Beauchamp 1995; Bronson et al. 1999; IAEA 2001; Macdonald et al. 2002; Muñoz et al. 2003; Douxchamps et al. 2011).
2. The time of N application (e.g., autumn vs. spring; Glendining et al. 1997; Cookson et al. 2000).

3. Methods of fertilizer applications (e.g., banding vs. surface topdressing or split vs. single application) or the use of slow-release N fertilizers or nitrification inhibitors (e.g., Hauck et al. 1994; Delgado and Mosier 1996; Shoji et al. 2001).
4. The influence of biochar or conservation tillage on the availability of N fertilizers (e.g., Dong et al. 2012; Hüppi et al. 2016).

$^{15}$N isotopic dilution technique has also been used to assess $NH_3$ volatilization and to apportion gaseous denitrification from soils into $N_2$ and $N_2O$ (Fillery and de Datta 1986; Mulvaney and Boast, 1986; Arah 1997; Stevens et al. 1997; Bronson et al. 1999).

Studies using $^{15}$N-labeled materials have been conducted under a range of conditions, including laboratory incubation (Griffin 2007), field plots (ranging from microplots of 1–1.5 m$^2$ to plots of 10 m$^2$; Zapata and Van Cleemput 1986; Bhogal et al. 1997; Macdonald et al. 2002; Fan et al. 2007; Douxchamps et al. 2011), lysimeters (diameter 60–80 cm and 60–120 cm depth; Di et al. 1999; Hüppi et al. 2016), intact soil cores (5–15 cm diameter; 15–20 cm depth; Luxhøi et al. 2004; Wu et al. 2011; Dong et al. 2012; Bedard-Haughn et al. 2013). Several reviews (Follett 2001; Murphy et al. 2003; IAEA 2008) have provided comprehensive information on how to enhance the uniform application of $^{15}$N-labeled materials and to minimize experimental errors that may affect N isotopic dilution and hence gross N fluxes between different soil N pools.

$^{15}$N-labeled urea, ammonium sulfate, and potassium nitrate are commercially available, while some other $^{15}$N-labeled materials such as manure, crop residues, or controlled release fertilizers or urea of unusual particle size are not readily available and/or costly to be labeled with $^{15}$N. For example, the production of $^{15}$N-labeled sheep feces by feeding sheep with $^{15}$N labeled ryegrass hay, which was produced from a field plot labeled with a $^{15}$N-labeled fertilizer (Douxchamps et al. 2011), is time consuming and very expensive. Nguyen and Goh (1994) used a similar approach to investigate the fate of sulfur (S) from urine patches in a perennial ryegrass (*Lolium perenne*) white clover (*Trifolium repens*) pasture. They obtained $^{35}$S-labeled urine by feeding sheep with $^{35}$S-labeled herbage, which had been harvested from a ryegrass-clover pasture receiving $^{35}$S-labeled potassium sulfate solution. An alternative to this direct approach for labeling manure or crop residues is the use of indirect $^{15}$N labeling technique, in which a $^{15}$N labeled solution (e.g., $^{15}(NH_4)_2SO_4$), instead of labeled materials is added to soil-plant systems. One treatment then receives an amendment of an unlabeled test material while another treatment, referred to as a control treatment, receives no amendment. Thus, any $^{15}$N dilution observed in the amended treatment is attributed to the release of N from the unlabeled amendment. The proportion of N derived from manure or crop residue (%Ndfm or %Ndfr) using the indirect $^{15}$N labeling technique is calculated as follows:

$$\%Ndfm \ (or \ \%Ndfr) = [1 - APE \ in \ amended \ treatment/APE \ in \ a \ control \ treatment)] \times 100 \quad (3.13)$$

where APE stands for atom% $^{15}$N excess.

Pool substitution can occur when $^{15}$N-labeled solution and non-labeled manure or residues are added simultaneously. Pool substitution is the process in which labeled N stands proxy for unlabeled inorganic soil N that would otherwise have been abstracted (immobilized) from a common N pool containing both labeled and unlabeled N (Hart et al. 1986). As a result, there may be a rapid decline in the $^{15}$N abundance in the plant-available soil N pool and there may be a differential N immobilization rate between the amended and non-amended soils (Hart et al. 1986; Hood et al., 1999; Murphy et al. 2003). To overcome the problem of pool substitution, the soil is left for a period of time (up to 6 months) after the addition of $^{15}$N solution (with or without added carbon source) until an equilibrium of $^{15}$N in the soil N pool is reached before manure or residues are added (Hood et al. 2000, 2001; Douxchamps et al. 2011).

As already stated, the isotopic dilution technique has been used to determine NUE and fate of N from fertilizers, manure, and crop residues. Information obtained has helped to develop management practices that can tighten N cycling in soil-plant systems. It is out of the scope of this review to list all the findings obtained. Thus, only a few findings are listed below to highlight practices, such as time of application, sources of N fertilizers, impacts of stocking rates, cropping rotation, and environmental conditions, that may influence the fate of N applied and N losses from agricultural systems. For example, Di et al. (1999) examined relative leaching losses from $^{15}$N-labeled dairy-shed effluent (DSE) and $^{15}$N-labeled $NH_4^+$ fertilizer applied twice annually in autumn (May) and late spring (November) in New Zealand, each at 200 kg N/ha to a pasture of perennial ryegrass-white clover. The autumn-applied N had a higher potential for leaching from both sources than N applied in spring, but the dairy-shed effluent (DSE) had lower leaching losses in the year following application. Between 4.5 and 8.1% of the $^{15}$N-labeled mineral N from the DSE was lost, in contrast to 15.1–18.8% of the fertilizer N. The lower leaching loss of $^{15}$N from the applied DSE was attributed to the stimulated microbial activities and increased immobilization rates following the application of DSE (Di et al.1999). Using urea enriched with 5.26% atom% $^{15}$N (i.e., 4.89 atom% $^{15}$N excess), applied at 120–180 kg N/ha to study N dynamics in a rice-wheat (R-W) rotation in Southwest China, Fan et al. (2007) concluded that the split application of urea increased its NUE. $^{15}$N-labeled urea was lost mainly during the flooding anaerobic rice growing period and this loss accounted for 67–81% of the loss over the whole R-W cropping system. This finding is similar to that reported by Liu et al. (2005) in the same region of China and in a similar R-W cropping rotation in India (Singh et al. 2001). $^{15}$N-labeled urea was also employed in these studies to investigate the fate of applied urea, highlighting the usefulness of the $^{15}$N dilution technique to assess FNUE in soils where other potential N sources for plant growth existed.

Gross N mineralization rates in soils which have received different sheep stocking rates (ungrazed, low, moderate, and heavy at 0, 2.35, 4.8, and 7.85 sheep/ha during June–September grazing season) have been investigated by Wu et al. (2011) used the $^{15}$N dilution technique. Labeling was done using $NH_4^+$ solution containing 60 atom% $^{15}$N enrichment. The amount of added N was approximately 0.9 mg $NH_4^+$-N/kg soil. Over a 2-day short-term incubation, 30–60% of added excess $^{15}$N was recovered in studied soil-plant systems. At the highest stocking rate, $^{15}$N recovery decreased by 50% over this incubation period. Most of the injected $^{15}$N was recovered in the $NO_3^-$, plant biomass and soil microbial biomass pools, while recovery of $^{15}$N in the soil $NH_4^+$ and dissolved organic N pools was of minor importance. Grazing decreased $^{15}$N recovery both in plant and microbial N pools, but strongly promoted $NO_3^-$ accumulation in the soil and thus negatively affected potential ecosystem N retention. This appeared to be closely related to the grazing induced decline in easily degradable soil C availability at increasing stocking rate. In another study, Bedard-Haughn et al. (2013) used ($^{15}NH_4$)$_2SO_4$ and K$^{15}NO_3$ (both containing 99 atom% $^{15}$N) to investigate gross N mineralization and nitrification over a 24-h incubation period in a continuous wheat arable system and in a wheat-pulse rotation at the Scott and Swift Current sites of the Northern Great Plains of Canada. Their results indicated that the average gross mineralization (ammonification) rate varied from 1.4 ± 3.9 to 2.0 ± 4.0 mg $NH_4^+$-N/kg soil. day. The only significant difference ($P$ = 0.1) in gross N mineralization (GNM) among rotations was at the Swift Current site, where the fertilized continuous wheat rotation had the highest GNM rates (2.3 mg $NH_4^+$-N/kg soil. day), compared to non-fertilized continuous wheat rotation (1.21 mg $NH_4^+$-N/kg soil. day) and a wheat-lentil rotation (0.98 mg $NH_4^+$-N/kg soil. day). Measured gross nitrification rates were higher when grain or forage legume was in the rotation (1.85 mg $NO_3^-$ N/kg soil. day), compared to the continuous wheat rotation (0.59 mg $NO_3^-$ N/kg soil. day).

The $^{15}$N dilution technique has been found to assist in quantifying total N losses as $N_2$ and $N_2O$ emission (Selbie et al. 2015; Friedl et al. 2016; Sgouridis et al. 2016). Information obtained can be useful in identifying management factors that may influence the partition of $N_2$/$N_2O$ losses from the denitrification process. Selbie et al. (2015) have reported that co-denitrification, as $N_2$ accounted for 95–97% of denitrification-derived gaseous N losses (while $N_2O$ accounted for only 3–5%) from a pastoral soil in Ireland over 123 days following a deposition of a highly $^{15}$N-enriched urine (labeled with 98% atom $^{15}$N urea). The authors attributed the predominance of $N_2$ emissions to a range of possible factors such as the form of metabolizable carbon substrates present and change in soil pH (> 6) in urine-affected soil that may favor $N_2O$ reductase activity and therefore promote $N_2$ formation. However, in a different environment (a pastoral soil in sub-tropical Australia), Friedl et al. (2016) using the $^{15}N$ dilution technique and $^{15}N$ gas flux method (detailed information given in Sgouridis et al. 2016) have shown that $N_2O$ is the main component (60%) of denitrification N losses from applied $^{15}N$ nitrate fertilizer during a 2-day incubation period. High $NO_3^-$ concentrations as observed at the end of the incubation shifted the $N_2/N_2O$ ratio toward $N_2O$ and the authors attributed this shift to the decrease in the $C/NO_3^-$ ratio and/or as a consequence of preferential $NO_3^-$ reduction. Although N losses as $N_2$ do not have environmental impacts in terms of greenhouse effects as $N_2O$ (300 times as potent as $CO_2$; IPCC 2001), they may represent a significant loss of N from applied fertilizers and urine patches.

### 3.4.2 APPLICATIONS OF $^{15}N$ NATURAL ABUNDANCE APPROACH

The natural abundance (NA) approach has been used in many studies to provide insight into the main N transformation processes such as mineralization, denitrification, volatilization, and nitrification in the N cycle of agricultural systems (Högberg 1997; Robinson 2001; Sebilo et al. 2003; Oelmann et al. 2005; Billy et al. 2010, 2013; Bu et al. 2017). For example, Billy et al. (2010) have shown that $\delta^{15}N$ of SON can be used as a semi-quantitative indicator of the intensity of denitrification integrated over century-long periods in a sloping drained agricultural plot and adjacent uncultivated riparian buffer strips. The $\delta^{15}N$ values of SON in the top 80 cm depth (5.5–6.0‰) were found to be significantly higher than the primary N sources from which they are derived, namely, synthetic fertilizers, atmospheric deposition, and symbiotic or non-symbiotic $N_2$ fixation (all with $\delta^{15}N$ close to 0‰). The $\delta^{15}N$ of SON was found to increase with depth, reaching up to 8‰ at 80 cm depth, particularly at downhill field sites or in riparian buffer strips that were frequently waterlogged. Using the $\delta^{15}N$ values obtained and a simple algorithm model developed, the authors showed that an increase of 1‰ of $\delta^{15}N$ in the SON above the level of the primary N sources was equated to an additional denitrification of 10 kg N/ha. yr.

The NA approach has also been used to provide semi-quantitative assessment of N cycling in ecosystems and of interactions between plants or plant communities and the environment. The $\delta^{15}N$ information obtained provides insight into the functional value of biodiversity for ecosystem functioning (Kahmen et al. 2008; Gubsch et al. 2011; Kleinebecker et al. 2014). For example, studies conducted by Kleinebecker et al. (2014) in 150 grassland sites in three regions of Germany have shown that an increase in plant diversity was consistently linked to a decrease in $\delta^{15}N$ isotopic signals in soil and aboveground community biomass. In contrast, there were elevated $\delta^{15}N$ values of soils and plants in intensively managed grasslands with high N fertilizer inputs and high N losses. The elevated $\delta^{15}N$ in soils and plants in these grasslands was attributed to the isotopic fractionation associated with N losses such as leaching, denitrification, and $NH_3$ volatilization (Ometto et al. 2006; Watzka et al. 2006). This isotopic fractionation was expected to lead to a loss of $^{14}N$ from soils and an overall enrichment of $^{15}N$ in ecosystems with an opened N cycle. A decrease in soil and plant $\delta^{15}N$ values indicates a more closed N system. Based on the results obtained, Kleinebecker et al. (2014) have proposed that (1) $\delta^{15}N$ in soils and plants can provide an indication of the openness of the N cycle or semi-quantitative assessment of N use efficiency of different sources, and (2) an increase in richness in plant species in grasslands could lead to a more efficient use of N resources and a more closed N cycle. The increase in plant diversity would potentially reduce the negative impact of landscape scale eutrophication caused by agricultural land use.

The $\delta^{15}N$ values of plants and soils (and also animal tissues such as cows' milk and hair in dairying systems) provide not only a semi-quantitative assessment of NUE of current farm management practices but also an integrator of the history of fertilizer and nutrient management practices (Watzka et al. 2006; Kriszan et al. 2009, 2014). For example, Kriszan et al. (2014) found that the $\delta^{15}N$ values of the top 5 cm soil depth and plant biomass were significantly correlated ($r^2 = 0.71 - 0.85$;

$P < 0.001$) with the overall N input-output balances, stocking rates, fertilizer inputs, and animal tissues. They also found that: (1) the higher the $\delta^{15}N$ values in soil and biomass, the lower was the overall farm's N balances, and (2) the longer the timeframe of $^{15}N$ accumulation in the system, the stronger is the $^{15}N$ isotope signature of plants and soils influenced by isotopic fractionation. Based on the above results, Kriszan et al. (2014) have concluded that the $\delta^{15}N$ values can provide valuable information on effects of long-term N inputs and farm management practices on N cycling and losses in agricultural ecosystems.

The NA approach has been used to trace the contribution of N sources (e.g., fertilizers and animal waste) to $NO_3^-$ in receiving waters and hence the overall fate of $NO_3^-$ in catchment studies (Kellman 2005; Kellman and Hillaire-Marcel 2003; Granger et al. 2008; Kohzu et al. 2008; Ohte et al. 2010; Xue et al. 2010; Di Lorenzo et al. 2012; Yuan et al. 2012; Billy et al. 2013; Ding et al. 2015). For example, Ding et al. (2015) in their 7-month study of the Taige River in the East Plain Region of China have found that the $\delta^{15}N$ values of $NO_3^-$ in water samples collected from the Taige River varied from 4.56‰ to +8.99‰. These were in the range of $\delta^{15}N$ values of $NO_3^-$ in manure and urban sewage (+4‰ to +25‰), indicating that animal manure and urban sewage effluent were the main sources of $NO_3^-$ pollution in the Taige River. Ding et al. (2015) also observed a decrease in $NH_4^+$-N concentrations in the downstream of the river and concluded that nitrification of N-containing organic materials played a more important role than the assimilation of $NH_4^+$ by aquatic algae for this decreasing $NH_4^+$ concentrations in the downstream. They have arrived at this conclusion since N isotopic enrichment factors ($\varepsilon$) during $NH_4^+$ utilization were found to range from −13.88‰ to −29.00‰, which were similar to the N isotopic enrichment factors ($\varepsilon$) of −12‰ to −29‰ during nitrification as reported by Kendall (1998). Similar findings of the main sources of N pollutants in receiving waters were reported for other catchments (e.g., Taihu Lake and Wehi River) in China (Townsend-Small et al. 2007 and Xue et al. 2016). For example, Xue et al. (2016) reported the high $\delta^{15}N$ values of $NO_3$ (ranging from +8.3‰ to +27.0‰) for high $NO_3^-$ N-containing water samples (water samples with $NO_3^-$ N contents above 45 mg/L) collected from the Weihe River, the largest tributary of the Yellow River, China (Xue et al. 2016). Based on these results, Xue et al. (2016) have suggested that domestic sewage and agricultural activities including animal waste with high $^{15}N$ enrichment are the main sources of $NO_3^-$ pollution in the Weihe River. Denitrification was assumed by Xue et al. (2016) to have no significant impact during the studied period, except at the two sites within the Weihe River catchment where there were high values of $\delta^{15}N$- $NO_3^-$ in the water samples, while $NO_3^-$ N and dissolved oxygen (DO) concentrations in these samples were less than 12.5 mg/L and 2 mg/L, respectively. The isotopic fractionation associated with the denitrification probably resulted in the subsequent enrichment of $^{15}N$ in water $NO_3^-$.

The use of $\delta^{15}N$ values has been successfully used as tracers in landscapes where: (1) variability associated with $\delta^{15}N$ can be accounted for and (2) isotopic fractionation associated with different N transformation processes does not have a large effect on the $^{15}N$ isotopic signatures of N sources to be traced (Bedard-Haughn et al. 2003). However, in ecosystems where there is a variability in $\delta^{15}N$ values of N sources because of isotopic fractionation and major fluxes of N inputs and outputs, Robinson (2001) and Bedard-Haughn et al. (2003) have proposed the use of $\delta^{15}N$ as an integrator of N cycle, instead of as a tracer, to provide semi-quantitative estimates of N flows. In such ecosystems, it is recommended by Robinson (2001) and Bedard-Haughn et al. (2003) to use both the NA approach and the $^{15}N$-dilution method to provide a complete, accurate picture of the flow and fate of N sources. Since the $^{15}N$-dilution technique involves the application of $^{15}N$-labeled materials (atom% $^{15}N$ excess of 0.04–0.4 or atom% $^{15}N$ of 0.4063–0.7664), the $^{15}N$ value in a tested source may be sufficiently distinct and hence the inferences drawn may not be limited by isotopic fractionation that may occur. The $^{15}N$ dilution technique can be successfully used as a tracer to test hypotheses and to quantify N cycling through the landscape, regardless of background variability in $\delta^{15}N$ (Bedard-Haughn et al. 2003). The $\delta^{15}N$ can provide first approximation of N sources and cycling at a given site and the $^{15}N$-dilution method can be used to quantitatively test N cycling hypotheses that have been generated from the $\delta^{15}N$ pattern analysis. A similar approach has been proposed by Peoples et al. (2015) to be used to provide a quantitative estimate of a transfer of biologically fixed N from the legume to the companion non-legume. They have recommended this approach to overcome the effects of many factors including isotopic fractionation on the variability of plant and soil $\delta^{15}N$ values.

Since $\delta^{15}N$ values may be unable to distinguish various N sources in some ecosystems due to their overlapping ranges and the isotopic fractionations associated with N transformation processes (Bedard-Haughn et al. 2003), the additional use of oxygen (O) isotope ratio ($^{18}O/^{16}O$), as expressed as $\delta^{18}O$ (i.e., the ratio $^{18}O/^{16}O$ in a sample, relative to that in Vienna Standard Mean Ocean Water (VSMOW); IAEA 2001; Nguyen et al. 2011) can be valuable in providing additional information on microbial N processes, such as denitrification and $N_2O$ production, as well as in distinguishing the sources of $NO_3^-$ in waters (Pardo et al. 2004; Xue et al. 2010; Mingzhu et al. 2014).

## 3.5 OVERALL SUMMARY AND CONCLUSIONS

Both isotopic dilution and natural abundance approaches are being used to study the impact of environmental change and farming practices on N transformation processes and N use efficiency in agricultural ecosystems. The $^{15}N$ dilution approach is costlier than the NA because of the use of $^{15}N$-labeled materials in the dilution approach. Although $\delta^{15}N$ values in soil and plant can provide insight into the main N transformation processes in the N cycle of agricultural farms, the $\delta^{15}N$ values should be used with caution because of many factors that influence the variability of these values, making it difficult to compare or interpret $\delta^{15}N$ values of different N sources. The isotopic composition ($^{15}N$ and $^{14}N$) of soil N is altered during soil N transformations and hence the changes in soil $\delta^{15}N$ is expected to influence the variability in plant $\delta^{15}N$. The change in soil $\delta^{15}N$ is attributed to the isotopic discrimination between $^{14}N$ and $^{15}N$ in biochemical-physical processes due to their mass differences. Because of this discrimination, the product of each of these transformation processes is depleted in $^{15}N$ relative to the substrate, except when the conversion from an N substrate to an N product is complete. Consequently, N losses from the soil system during $NH_3$ volatilization, denitrification, and leaching lead to a depletion of $^{15}N$ in products that are lost to the environment ($NH_3$, $N_2$, $N_2O$, $NO_3^-$) and an enrichment of $^{15}N$ in the residual substrate (soil $NH_4^+$ and SON).

Since many factors including isotopic fraction can influence the variability of soil and plant $\delta^{15}N$ values, these values should be used as semi-quantitative indicators of: (1) predominant N processes at a given site (e.g., a semi-quantitative indicator of the intensity of denitrification in riparian zones), (2) N use efficiency in agricultural ecosystems, (3) the history of fertilizer and farm management practices, (4) the interaction between plants/plant communities and the environment, and (5) the functional value of biodiversity for ecosystem functioning. Where N sources and $\delta^{15}N$ values are known, $\delta^{15}N$ values in soil, plant, and water can be used to trace the impact of N sources on $NO_3^-$ losses to receiving waters. To provide comprehensive and quantitative information on N transformation processes and cycling in ecosystems, the $^{15}N$-dilution technique should be used to complement the NA approach. In addition, the use of dual isotope approach (i.e., the use of both $\delta^{15}N$ and $\delta^{18}O$) can provide more reliable information for tracing sources of $NO_3^-$ in water and $NO_3^-$ transformation processes.

## ACRONYMS

| | |
|---|---|
| **BNF** | Biological nitrogen fixation |
| **CO$_2$** | Carbon dioxide |
| **FNUE** | Fertilizer nitrogen use efficiency |
| **GHG** | Greenhouse gases |
| **ha** | hectare |
| **NH$_3$** | Ammonia gas |
| **NH$_4^+$** | Ammonium |

NO$_3^-$      Nitrate
N$_2$         Dinitrogen gas
N$_2$O        Nitrous oxide
NUE          Nitrogen (N) use efficiency
SOC          Soil organic carbon
SOM          Soil organic matter
SON          Soil organic nitrogen
Tg           Teragram
yr           year

## REFERENCES

Amundson, R., Austin, A. T., Schuur, E. A. G., Yoo, K., , V., Kendall, C., Uebersax, A., Brenner, D., and W. T. Baisden. 2003. Global patterns of the isotopic composition of soil and plant nitrogen. *Global Biogeochemical Cycles* 17: 1031, doi:10.1029/2002GB001903.

Arah, J. R. M. 1997. Apportioning nitrous oxide fluxes between nitrification and denitrification using gas-phase mass spectrometry. *Soil Biology and Biochemistry* 29: 1295–1299.

Aranibar, J. N., Anderson, I. C., Epstein, H. E., Feral, C. J. W., Swap, R. J., Ramontsho, J., and S. A. Macko. 2008. Nitrogen isotope composition of soils, C-3 and C-4 plants along land use gradients in southern Africa. *Journal of Arid Environments* 72: 326–337.

Ball, R., Keeney, D. R., Theobald, P. W., and P. Nes. 1979. *Agronomy Journal* 71:309–314.

Barnard, R., Barthes, L., and P. W. Leadley 2006. Short-term uptake of 15N by a grass and soil micro-organisms after long-term exposure to elevated CO2. *Plant and Soil* 280: 91–99.

Barraclough, D. 1991. The use of mean pool abundances to interpret [15]N tracer experiments. *Plant and Soil* 131: 89–96.

Barraclough, D. 1995. N[15] isotope dilution techniques to study nitrogen transformations and plant uptake. *Fertilizer Research* 42: 185–192.

Barraclough, P. D., Howarth, J. R., Jones, J. et al. 2010. Nitrogen efficiency of wheat: Genotypic and environmental variation and prospects for improvement. *European Journal of Agronomy* 33: 1–11.

Bates, G., Quin, B. F., and P. Bishop. 2015. Low-cost detection and treatment of fresh cow urine patches. In *Nutrient Management for the Farm, Catchment and Community*, L. D. Currie and L. L. Burkitt, Eds.. Occasional Report No. 28 Fertilizer and Lime Research Centre, Massey University, Palmerston North, New Zealand. http://flrc.massey.ac.nz/publications.html.

Bateman, A. S., and S. D. Kelly. 2007. Fertilizer nitrogen isotope signatures. *Isotopes in Environmental and Health Studies* 43: 237–247.

Batjes, N. H. 1996. Total carbon and nitrogen in the soils of the world. *European Journal of Soil Science* 47: 151–163.

Bedard-Haughn, A., van Groenigen, J. W., and C. van Kessel 2003. Tracing [15]N through landscapes: Potential uses and precautions. *Journal of Hydrology* 272: 175–190.

Bedard-Haughn, A., Comeau, L.-P., and A. Sangster 2013. Gross nitrogen mineralization in pulse-crop rotations on the Northern Great Plains. *Nutrient Cycling in Agroecosystems* 95: 159–174.

Beeson, K. O. 1946. The mineral composition of forage crops. *Bot. Rev.* 12: 424–455.

Berg, B., and C. McClaugherty. 2007. *Plant litter: Decomposition, humus formation, carbon sequestration*, 2nd ed. Springer.

Bhogal, A., Young, S. D., and R. Sylvester-Bradley 1997. Fate of [15]N-labelled fertilizer in a long-term field trial at Ropsley, UK. *Journal of Agricultural Science, Cambridge* 129: 49–63.

Billings, S. and D. Richter. 2006. Changes in stable isotopic signatures of soil nitrogen and carbon during 40 years of forest development. *Oecologia* 148: 325–333.

Billy, C., Billen, G., Sebilo, M., Birgand, F., and J. Tournebize. 2010. Nitrogen isotopic composition of leached nitrate and soil organic matter as an indicator of denitrification in a sloping drained agricultural plot and adjacent uncultivated riparian buffer strips. *Soil Biology and Biochemistry* 42: 108–117.

Billy, C., Birgand, F., Ansart, P., Peschard, J., and J. Tournebize. 2013. Factors controlling nitrate concentrations in surface waters of an artificially drained agricultural watershed. *Landscape Ecology* DOI 10.1007/s10980-013-9872-2.

Blüthgen, N., Dormann, C. F., Prati, D., Klaus, V. H., Kleinebecker, T., Hölzel, N., Alt, F. et al. 2012. A quantitative index of land-use intensity in grasslands: Integrating mowing, grazing and fertilization. *Basic and Applied Ecology* 13: 207–220.

Boesch, D. F., Brinsfield, R. B., and R. E. Magnien. 2001. Chesapeake Bay eutrophication: Scientific under-standing, ecosystem restoration, and challenges for agriculture. *Journal of Environmental Quality* 30: 303–320.

Bu, X. L., Zhao, C. X., Han, F. Y., Xue, J. H., and Y. B. Wu. 2017. Nitrate reduction in groundwater and iso-topic investigation of denitrification in integrated tree-grass riparian buffers in Taihu Lake watershed, eastern China. *Journal of Soil and Water Conservation* 72: 45–54.

Brady, N. C., and R. R. Weil. 1999. *The nature and properties of soils*. Upper Saddle River, NJ: Prentice Hall, Inc.

Brenner, D. L., Amundson, R., Baisden, W. T., Kendall, C., and J. Harden. 2001. Soil N and $^{15}$N variation with time in a California annual grassland ecosystem. *Geochimica et Cosmochimica Acta* 65: 4171–4186.

Bronson, K. F., Sparling, G. P., and I. R. P. Fillery 1999. Short term N dynamics following application of $^{15}$N-labeled urine to a sandy soil in summer. *Soil Biology and Biochemistry* 31: 1049–1057.

Busari, M. A., Kukal, S. S., Kaur, A., Bhatt, R., and A. A. Dulazi. 2015. Conservation tillage impacts on soil, crop and the environment. *International Soil and Water Conservation Research* 3: 119–129.

Butterbach-Bahl, K., Baggs, E. M., Dannenmann, M., Kiese, R., and S. Zechmeister-Boltenstern. 2013. Nitrous oxide emissions from soils: How well do we understand the processes and their controls? *Philosophical Transactions Royal Society B-Biological Science* B368: 20130122. http://dx.doi .org/10.1098/rstb.2013.0122.

Cameron, K. C., Di, H. J., and J. L. Moir. 2013. Nitrogen losses from the soil/plant system: A review. *Annals of Applied Biology* 162: 145–173.

Canfield, D. E., Glazer, A. N., and P. G. Falkowski. 2010. The evolution and future of Earth's nitrogen cycle. *Science* 330: 192–196.

Castellanos, J. Z., Follett, R. F., Mora, M., Sosa, A., and P. Vargas. 1998. $^{15}$N use under tillage treatments in a Vertisol. *Agronomy Abstract* 90: 244.

Chapman, D. F., Edwards, G. R., and J. B. Pinxterhuis. 2014. Plants for dairy grazing systems operating under nitrate leaching limits. *Proceedings of the New Zealand Society of Animal Production* 74:102–107.

Chen, D., Sun, J., Bai, M., Dassanayake, K. B., Denmead, O. T., and J. Hill. 2015. *Scientific Reports* 5: article number 16689. DOI 10.1038/srep16689.

Christensen, B. T., Petersen, J., and M. Schacht (ed.). 2008. Long-term field experiments—A unique research platform. *Proceedings of NJF Seminar 407* Askov Experimental Station and Sandbjerg Estate, June 2008, Denmark. DJF Plant Science No. 137.

Clark, D. A., and S. L. Harris. 1996. White clover or nitrogen fertilizer for dairying? *Agronomy Society of New Zealand*. Special Publication No. 11/Grassland Research and Practice Series No. 6, 107–114.

Compant, S., Clenent, C., and A. Sessitsch. 2010. Plant growth-promoting bacteria in the rhizo- and endo-sphere of plants: Their role, colonization, mechanisms involved and prospects for utilization. *Soil Biology and Biochemistry* 42: 669–678.

Cookson, W. R., Rowarth, J. S., and K. C. Cameron. 2000. The effect of autumn applied $^{15}$N-labelled fertilizer on nitrate leaching in a cultivated soil during winter. *Nutrient Cycling in Agroecosystems* 56: 99–107.

Crews, T. E., and M. B. Peoples. 2004. Legume versus fertilizer sources of nitrogen: ecological tradeoffs and human needs. *Agriculture, Ecosystems and Environment* 102: 279–297.

Cui, Z., Chen, X., Ju, X., and F. Zhang 2014. Managing agricultural nutrients for food security in China: Past, present, and future. *Agronomy Journal* 106: 191–198.

Davidson, E. A., Hart, S. C., Shanks, C. A., and M. K. Firestone. 1991. Measuring gross nitrogen mineraliza-tion, immobilization and nitrification by $^{15}$N isotopic pool dilution in intact soil cores. *Journal of Soil Science* 42: 335–349.

Delgado, J. A., and A. R. Mosier. 1996. Mitigation alternatives to decrease nitrous oxide emissions and urea-nitrogen loss and their effect on methane flux. *Journal of Environmental Quality* 25:1105–1111.

Delgado, J. A., Mosier, A. R., Follett, R. H., Follett, R. F. Westfall, D. G., Klemdtsson, L. K., and J. Vermeulen. 1996. Effect of N management on $N_2O$ and $CH_4$ fluxes and $^{15}$N recovery in an irrigated mountain meadow. *Nutrient Cycling in Agroecosystems* 46: 127–134.

Di, H. J., Cameron, K. C., Moore, S., and N. P. Smith 1999. Contributions to nitrogen leaching and pasture uptake by autumn-applied dairy effluent and ammonium fertilizer labeled with 15N isotope. *Plant and Soil* 210: 189–198.

Di, H. J., Cameron, K. C., Shen, J. P., Winefield, C. S., O'Callaghan M., Bowatte, S., and J. Z. He. 2009. Nitrification driven by bacteria and not archaea in nitrogen-rich grassland soils. *Nature Geoscience* 2: 621–624.

Di Lorenzo, T., Brilli, M., Del Tosto, D., Galassi, D. M. P., and M. Petitta. 2012. Nitrate source and fate at the catchment scale of the Vibrata River and aquifer (central Italy): An analysis by integrating component approaches and nitrogen isotopes. *Environmental Earth Sciences* 67: 2383–2398.

Ding, J., Xi, B., Xu, Q., Su, J., Huo, S., Liu, H., Yu, Y., and Y. Zhang. 2015. Assessment of the sources and transformations of nitrogen in a plain river network region using a stable isotope approach. *Journal of Environmental Sciences* 30: 198–206.

Dise, N. B, Ashmore, M., Belyazid, S. et al. 2011. Nitrogen to a threat to European terrestrial biodiversity. In *The European nitrogen assessment,* M. A. Sutton, C. M. Howard, J. W. Erisman et al., Eds. pp. 463–494. Cambridge University Press.

Dong, W., Hu, C., Zhang, Y., and D. Wu 2012. Gross mineralization, nitrification and N2O emission under different tillage in the North China. *Nutrient Cycling in Agroecosystems* 94: 237–247.

Doran, J. W. 1980. Soil microbial and biochemical changes associated with reduced tillage. *Soil Science Society of America Journal* 44: 765–771.

Douxchamps, S., Frossard, E., Bernasconi, S. M., van der Hoek, R., Schmidt, A., Rao, I. M., and A. Oberson. 2011. Nitrogen recoveries from organic amendments in crop and soil assessed by isotope 1 techniques under tropical field conditions. *Plant and Soil* 341: 179–192.

Eckard, R. J., and D. R. Franks. 1998. Strategic nitrogen fertilizer use on perennial ryegrass and white clover in north-western Tasmania. *Australian Journal of Experimental Agriculture* 38:155–160.

Evans, R. D. 2007. Soil nitrogen isotopic composition. In *Stable isotopes in ecology and environmental science,* 2nd ed. R. Michener and K. Lajtha, Eds., pp. 83–98. Oxford, UK: Blackwell Publishing Ltd.

Fan, M., Lu, S., Jiang, R., Liu, X., Zeng, X., Goulding, K. W. T., and F. Zhang. 2007. Nitrogen input, $^{15}$N balance and mineral N dynamics in a rice-wheat rotation in southwest China. *Nutrient Cycling in Agroecosystems* 79: 255–265.

FAO (Food and Agriculture Organization of the United Nations). 2012. *Current world fertilizer trends and outlook to 2016.* FAO, Rome. ftp://ftp.fao.org/ag/agp/docs/cwfto16.pdf.

FAO. 2016. *World fertilizer trends and outlook to 2019—Summary Report.* FAO, Rome. http://www.fao.org/3/a-i5627e.pdf.

Field, R. R. O., and P. R. Ball. 1978. Tactical use of fertiliser nitrogen. *Proceedings of the New Zealand Agronomy Society* 8: 129–133.

Fillery, I. R. P., and S. K. de Datta. 1986. Ammonia volatilisation from nitrogen sources applied to rice fields. I. Methodology for ammonia fluxes and $^{15}$N loss. *Soil Science Society of America Journal* 50: 86–91.

Follett, R. F. 2001. Innovative $^{15}$N microplot research techniques to study nitrogen use efficiency under different ecosystems. *Communications in Soil Science and Plant Analysis* 32: 951–979

Fowler, D., Pyle, J. A., Raven, J. A., and M. A. Sutton. 2013. The global nitrogen cycle in 701 the twenty-first century, introduction. *Philosophical Transactions Royal Society B-Biological Science* 368. doi:10.1098/rstb.2013.0165.

Fowler, D., Coyle, M., Skiba, U., Sutton, M. A., Cape, J. N., Reis, S., Sheppard, L. J. et al. 2013. The global nitrogen cycle in the twenty-first century. *Philosophical Transactions Royal Society B-Biological Science* 368. 20130164. http://dx.doi.org/10.1098/rstb.2013.0164.

Fraser, R. A., Bogaard, A., Heaton, T., Charles, M., Jones, G., Christensen, B. T., Halstead, P. et al. 2011. Manuring and stable nitrogen isotope ratios in cereals and pulses: Towards a new archaeobotanical approach to the inference of land use and dietary practices. *Journal of Archaeological Science* 38: 2790–2804.

Friedl, J., Scheer, C., Trappe, J., Rowlings, D. W., and P. R. Grace. 2016. Nitrogen turnover and $N_2$:$N_2O$ partitioning from agricultural soils—a simplified incubation assay. In *Proceedings of the 2016 International Nitrogen Initiative Conference, "Solutions to improve nitrogen use efficiency for the world,"* December 4–8, 2016, Melbourne, Australia. www.ini2016.com.

Galloway, J. N., Aber, J. D., Erisman, J. W., Seitzinger, S. P., Howarth, R. W., Cowling, E. B., and B. J. Cosby. 2003. The nitrogen cascade. *Bioscience* 53: 341–356.

Galloway, J. N., Dentener, F. J., Capone, D. G. et al. 2004. Nitrogen cycles: Past, present and future. *Biogeochemistry* 70: 153–226.

Galloway, J. N., Townsend, A. R., Erisman, J. W., Bekunda, M., Cai, Z. C., Freney, J. R., Martinelli, L. A. et al. 2008. Transformation of the nitrogen cycle: Recent trends, questions, and potential solutions. *Science* 320: 889–892.

Gardiner, C. A., Clough T. H., Cameron, K. C., Di, H. J., Edwards, G. R., and A. A. M. de Klein 2016. Potential for forage diet manipulation in New Zealand pasture ecosystems to mitigate ruminant urine derived $N_2O$ emissions: A review. *New Zealand Journal of Agricultural Research* 59:301–317. http://dx.doi.org/10.1080/00288233.2016.1190386.

Ge, Y., Zhang, J.-B., Zhang, L.-M., Yang, M., and J.-Z. He. 2008. Long-term fertilization regimes affect bacterial community structure and diversity of an agricultural soil in northern China. *Journal of Soils and Sediments* 8: 43–50.

Glendining, M. J., Poulton, P. R., Powlson, D. S., and D. S. Jenkinson. 1997. Fate of [15]N-labelled fertilizer applied to spring barley grown on soils of contrasting nutrient status. *Plant and Soil* 195: 83–98.

Goh, K. M. 1982. The significance of the N[15]technique in nitrogen balance studies. In *Nitrogen balances in New Zealand Ecosystems: Proceedings of a Workshop, May 1980*. P. W. Gandar and D. S. Bertand, Eds. Palmerston North, New Zealand: Publ. Dept. of Scientific and Industrial Research NZ.

Granger, S. J., Heaton, T. H., Bol, R., Bilotta, G. S., Butler, P., Haygarth, P. M., and P. N. Owens. 2008. Using δ [15]N and δ [18]O to evaluate the sources and pathways of NO3- in rainfall event discharge from drained agricultural grassland lysimeters at high temporal resolutions. *Rapid Communications in Mass Spectrometry* 22: 1681–1689.

Griffin, T. S. 2007. Estimates of gross transformation rates of dairy manure N using [15]N pool dilution. *Communications in Soil Science and Plant Analysis* 38: 1451–1465.

Gu, B., Chang, J., Min Y., Ge, Y., Zhu, Q., Galloway, J. N., and C. Peng. 2013 The role of industrial nitrogen in the global nitrogen biogeochemical cycle. *Scientific Report* 3, 2579. DOI:10.1038/srep02579.

Gubsch, M., Roscher, C., Gleixner, G., Habekost, M., Lipowsky, A., Schmid, B., Schulze, E. D., Steinbeiss, S., and N. Buchmann. 2011. Foliar and soil δ [15]N values reveal increased nitrogen partitioning among species in diverse grassland communities. *Plant Cell and Environment* 34: 895–908.

Gundersen, P., and V. N. Bashkin. 1994. Nitrogen Cycling. In *Biogeochemistry of small catchments: A tool for environmental research*, B. Moldan and J. Cerny, Eds. SCOPE, John Wiley & Sons, Ltd.

Harris, S. L., Clark, D. A., and E. B. L. Jansen. 1997. Optimum white clover content for milk production. *Proceedings of the New Zealand Society of Animal Production* 57: 169–171.

Hart, P. B. S., Rayner, J. H., and D. S. Jenkinson 1986. Influence of pool substitution on the interpretation of fertilizer experiments with [15]N. *Journal of Soil Science* 37: 389–403.

Hart, S. C., Nason, G. E., Myrold, D. D., and D. A. Perry 1994. Dynamics of gross nitrogen transformations in an old-growth forest: The carbon connection. *Ecology* 75: 880–891.

Hauck, R. D., and D. R. Bouldin 1961. Distribution of isotopic nitrogen gas during denitrifcation. *Nature* (London) 1991: 871–872.

Hauck, R. D., and J. M. Bremner 1976. Use of tracers for soil and fertiliser nitrogen research. *Advances in Agronomy* 28: 219–266.

Hauck, R. D., Meisinger, J. J., and R. L. Mulvaney 1994. Practical considerations in the use of nitrogen tracers in agricultural and environmental research. In *Methods of soil analysis, Part 2. Microbiological and biochemical properties*, pp. 907–950. Madison, WI: Soil Science Society of America Book Series, no. 5. Soil Science Society of America.

Hawkesford, M. J. 2014. Reducing the reliance on nitrogen fertilizer for wheat production. *Journal of Cereal Science* 59: 276–283.

He, X., Xu, M., Qiu, G. Y., and J. Zhou. 2009. Use of [15]N stable isotope to quantify nitrogen transfer between mycorrhizal plants. *Journal of Plant Ecology* 2: 107–118.

Heffer, P., and M. Prudhomme. 2016. Global nitrogen fertilizer demand and supply: Trend, current level and outlook. *7th International Nitrogen Initiative Conference*, December 4–8, 2016, Melbourne, Australia. *Journal of Cereal Science* 59(3): 276–283. DOI 10.1016/jcs.2013.12.001.

Hobbie, J. E., and E. A. Hobbie. 2006. [15]N in symbiotic fungi and plants estimates nitrogen and carbon flux rates in Arctic tundra. *Ecology* 87: 816–822.

Hobbie, J. E., and E. A. Hobbie. 2008. Natural abundance of N-15 in nitrogen limited forests and tundra can estimate nitrogen cycling through mycorrhizal fungi: A review. *Ecosystems* 11: 815–830.

Hobbie, E. A., and P. Högberg. 2012. Nitrogen isotopes link mycorrhizal fungi and plants to nitrogen dynamics. *New Phytologist* 196: 367–382.

Hobbs, P. R., Sayre, K., and R. Gupta. 2008. The role of conservation agriculture in sustainable agriculture. *Philosophical Transactions Royal Society B-Biological Science B* 363: 543–555.

Hoeft, I., Steude, K., Wrage, N. and E. Veldkamp. 2012. Response of nitrogen oxide emissions to grazer species and plant species composition in temperate agricultural grassland. *Agriculture, Ecosystems and Environment* 151: 34–43.

Högberg, P. 1997. Tansley Review No. 95 [15]N natural abundance in soil-plant systems *New Phytologist* 137: 179–203.

Hood, R. 2001. Evaluation of a new approach to the nitrogen-15 isotope dilution technique, to estimate crop N uptake from organic residues in the field. *Biology and Fertility of Soils* 34: 156–161.

Hood, R. C., N'Goran, K., Aigner, M., and G. Hardarson 1999. A comparison of direct and indirect [15]N isotope techniques for estimating crop uptake from organic residues. *Plant and Soil* 208: 259–270.

Hood, R. C., Merckx, R., Jensen, E. S., Powlson, D., Matijevic, M., and G. Hardarson 2000. Estimating crop N uptake from organic residues using a new approach to the [15]N isotope dilution technique. *Plant and Soil* 223: 33–46.

Houlton, B. Z., Sigman, D. M., Schuur, E. A. G., and L. O. Hedin. 2007. A climate driven switch in plant nitrogen acquisition within tropical forest communities. *Proceedings of the National Academy of Sciences of the United States of America (PNAS)* 104: 8902–8906.

Howarth, R.W. 2008. Coastal nitrogen pollution: A review of sources and trends globally and regionally. *Harmful Algae* 8: 14–20.

Hristov, A. N., Hanigan, M., Cole, A. et al. 2011. Review: Ammonia emissions from dairy farms and beef feedlots. *Canadian Journal of Animal Science* 91: 1–35.

Hüppi, R., Neftel, A., Lehmann, M. F., Krauss, M., Six, J., and J. Leifeld 2016. N use efficiencies and N2O emissions in two contrasting, biochar amended soils under winter wheat-cover crop-sorghum rotation. *Environmental Research Letters* 11, doi:10.1088/1748-9326/11/8/084013.

Hyodo, F., and D. A. Wardle. 2009. Effect of ecosystem retrogression on stable nitrogen and carbon isotopes of plants, soils and consumer organisms in boreal forest islands. *Rapid Communications in Mass Spectrometry* 23: 1892–1898.

IAEA (International Atomic Energy Agency). 2001. *Use of Isotope and Radiation Methods in Soil and Water Management and Crop Nutrition.* IAEA Training Manual Series 14. IAEA, Vienna.

IAEA. 2008. Guidelines on Nitrogen Management in Agricultural Systems. Training Course Series Number 29. IAEA, Vienna.

IFIA (International Fertilizer Industry Association). 2007. *Sustainable management of the nitrogen cycle in agriculture and mitigation of reactive nitrogen side effects,* 1st ed. Paris, France: IFA.

IPCC (Intergovernmental Panel on Climate Change). 2001. *Climate change 2001: The scientific basis. Contribution of working group 1 to the third assessment report of the intergovernmental panel on climate change.* J. T. Houghton, Y. Ding, D. J. Griggs, M. Noguer, P. J. van der Linden, X. Dai, K. Maskell, and C. A. Johnson. Eds. Cambridge, UK: Cambridge University Press.

Jarvis, S. C. 2000. Progress in studies of nitrate leaching from grassland soils. *Soil Use and Management* 16: 152–156.

Jones, L., Provins, A., Holland, M., Mills, G., Hayes, F., Emmett, B., Hall, G. et al. 2014. A review and application of the evidence for nitrogen impacts on ecosystem services. *Ecosystem Services* 7: 76–88.

Kahmen, A., Wanek, W., and N. Buchmann. 2008. Foliar $\delta^{15}N$ values characterize soil N cycling and reflect nitrate or ammonium preference of plants along a temperate grassland gradient. *Oecologia* 156: 861–870.

Kellman, L. M. 2005. A study of tile drain nitrate-$\delta^{15}N$ values as a tool for assessing nitrate sources in an agricultural region. *Nutrient Cycling in Agroecosystems* 71: 131–137.

Kellman L. M., and C. Hillaire-Marcel. 2003. Evaluation of nitrogen isotopes as indicators of nitrate contamination sources in an agricultural watershed. *Agriculture, Ecosystems and Environment* 95: 87–102.

Kendall, C. 1998. Tracing nitrogen sources and cycling in catchments. In *Isotope tracers in catchment hydrology.* C. Kendall, and J. J. McDonnell, Eds. pp. 519–576. Amsterdam: Elsevier.

Kirkham, D., and W. V. Bartholomew. 1954. Equations for following nutrient transformations in soil, utilizing tracer data. *Soil Science Society of America Proceedings* 18: 33–34.

Kleinebecker, T., Hölzel, N., Prati, D., Schmitt, B., Fischer, M. and V. H. Klaus. 2014. Evidence from the real world: $^{15}N$ natural abundances reveal enhanced nitrogen use at high plant diversity in Central European grasslands. *Journal of Ecology* 102: 456–465.

Kohzu, A., Miyajima, T., Tayasu, I., Yoshimizu, C., Hyodo, F., Matsui, K., Nakano, T., Wada, E., Fujita, N., and T. Nagata. 2008. Use of stable nitrogen isotope signatures of riparian macrophytes as an indicator of anthropogenic N inputs to river ecosystems. *Environmental Science and Technology* 42: 7837–7841.

Kong, A. Y. Y, Fonte, S. J., van Kessel, C., and J. Six. 2007. Soil aggregates control N cycling efficiency in long-term conventional and alternative cropping systems. *Nutrient Cycling in Agroecosystems* 79: 45–58.

Korontzi, S., Macko, S. A., Anderson, I. C., and M. A. Poth. 2000. A stable isotopic study to determine carbon and nitrogen cycling in a disturbed southern Californian forest ecosystem. *Global Biogeochemical Cycles* 14: 177–188.

Kriszan, M., Amelung, W., Schellberg, J., Gebbing, T., and W. Kühbauch. 2009. Long-term changes of the 15N natural abundance of plants and soil in a temperate grassland. *Plant and Soil* 325: 157–169.

Kriszan, M., Schellberga, J., Amelung, W., Gebbing, T., Pötschd, E. M., and W. Kühbauch. 2014. Revealing N management intensity on grassland farms based on natural $^{15}N$ abundance. *Agriculture, Ecosystems and Environment* 184: 158–167.

Lal, R. 2000. Soil management in the developing countries. *Soil Science* 165: 57–72.

Lal, R. 2004. Soil carbon sequestration impacts on global climate change and food security. *Science* 304: 1623–1627.

Ledgard, S. F., Penno, J. W., and M. S. Sprosen. 1999. Nitrogen inputs and losses from clover/grass pastures grazed by dairy cows, as affected by nitrogen fertilizer application. *The Journal of Agricultural Science* 132: 215–225.

Lehmann, J., and M. Kleber. 2015. The contentious nature of soil organic matter. *Nature* 528: 60–68.

Lehmann, M., Sigman, M., and W. Berelson. 2004. Coupling the $^{15}N/^{14}N$ and $^{18}O/^{16}O$ of nitrate as a constraint on benthic nitrogen cycling. *Marine Chemistry* 88: 1–20.

Lerman, A., Mackenzie, F. T., and L. M. Ver. 2004. Coupling of the perturbed C-N-P cycles in industrial time. *Aquatic Geochemistry* 10: 3–32.

Lee, C., Hristov, A. N., Cassidy, T. and K. Heyler. 2011. Nitrogen isotope fractionation and origin of ammonia nitrogen volatilized from cattle manure in simulated storage. *Atmosphere* 2: 256–270.

Liu, X. J., Ai, Y. W., Zhang, F. S., Lu, S. H., Zeng, X. Z., and M. S. Fan. 2005. Crop production, nitrogen recovery and water use efficiency in rice–wheat rotation as affected by non-flooded mulching cultivation (NFMC). *Nutrient Cycling in Agroecosystems* 71: 289–299.

Liu, J., Huang, W., Zhou, G., Zhang, D., Liu, S., and Y. Li. 2013. Nitrogen to phosphorus ratios of tree species in response to elevated carbon dioxide and nitrogen addition in subtropical forests. *Global Change Biology* 19: 208–216.

Liu, J., Ma, K., Ciais, P., and S. Polasky. 2016. Reducing human nitrogen use for food production. *Scientific Report* 6: 30104. doi: 10.1038/srep30104.

Low, A. J. 1972. The effect of cultivation on the structure and other physical characteristics of grassland and arable soils. *Journal of Soil Science* 23: 363–380.

Ludwig, B., Geisseler, D., Michel, K. et al. 2010. Effects of fertilization and soil management on crop yields and carbon stabilization in soils. A review. *Agronomy for Sustainable Development* 31: 361–372.

Luxhøi, J., Nielsen, N. E., and L. S. Jensen. 2004. Effect of soil heterogeneity on gross nitrogen mineralization measured by $^{15}N$-pool dilution techniques. *Plant and Soil* 262: 263–275.

Macdonald, A. J., Poulton, P. R., Stockdale, E. A., Powlson, D. S., and D. S. Jenkinson 2002. The fate of residual $^{15}N$-labelled fertilizer in arable soils: Its availability to subsequent crops and retention in soil. *Plant and Soil* 246: 123–137.

Mariotti, A. 1983. Atmospheric nitrogen is a reliable standard for natural $^{15}N$ abundance measurements. *Nature* 303: 685–687.

Mariotti, A. 1984. Natural $^{15}N$ abundance measurements and atmospheric nitrogen standard calibration. *Nature* 311: 251–252.

Mariotti, A., Germon, J. C., Hubert, P., Kaiser, P., Tardieux, A., and P. Tardieux. 1981. Experimental determination of kinetic isotope fractionations: Some principals; illustration for denitrification and nitrification processes. *Plant and Soil* 62: 413–430.

Marshall, J. D., Brooks, J. N., and K, Lajtha. 2007. Sources of variation in the stable isotopic composition of plants. In *Stable isotopes in ecology and environmental Science,* 2nd ed. R. Michener and K. Lajtha, Eds., pp. 22–60. Oxford, UK: Blackwell Publishing Ltd.

Martinelli, L. A., Piccolo, M. C., Townsend, A. R., Vitousek, P. M., Cuevas, E., McDowell, W., Robertson, G. P. et al. 1999. Nitrogen stable isotopic composition of leaves and soil: Tropical versus temperate forests. *Biogeochemistry* 46: 45–65.

Menge, D. N. L., W. Troy Baisden, W. T., Richardson, S. J., Peltzer, Duane, A. and M. M. Barbour. 2011. Declining foliar and litter δ $^{15}N$ diverge from soil, epiphyte and input δ $^{15}N$ along a 120 000 yr temperate rainforest chronosequence. *New Phytologist* 190: 941–952.

Menneer, J. C., Ledgard, S., McLay, C., and W. Silvester. 2005. Animal treading stimulates denitrification in soil under pasture. *Soil Biology and Biochemistry* 37: 1625–1629.

Menyailo, O. V., and B. A. Hungate. 2006. Stable isotope discrimination during soil denitrification: Production and consumption of nitrous oxide. *Global Biogeochemistry Cycles* 20. GB3025, doi:10 .1029/2005GB002527.

Mingzhu, L., Seyf-Laye, A.-S. M., Ibrahim, T., Gbandi, D.-B., and C. Honghan. 2014. Tracking sources of groundwater nitrate contamination using nitrogen and oxygen stable isotopes at Beijing area. *Environmental Earth Science* 72: 707–715.

Misselbrook, T., del Prado, A., and D. Chadwick. 2013. Opportunities for reducing environmental emissions from forage-based dairy farms. *Agricultural and Food Science*. 22: 93–107.

Mitsch, W. J., Day, J. W., Gilliam, J. W. et al. 2001. Reducing nitrogen loading to the Gulf of Mexico from the Mississippi River Basin: Strategies to counter a persistent ecological problem. *Bioscience* 515: 373–388.

Mulvaney, R. L., and C. W. Boast 1986. Equations for determination of nitrogen-15 labeled dinitrogen and nitrous oxide by mass spectrometry. *Soil Science Society of America Journal* 50: 360–363.

Muñoz, G. R., Powell, J. M., and K. A. Kelling 2003. Nitrogen budget and soil N dynamics after multiple applications of unlabeled or nitrogen 15-enriched dairy manure. *Soil Science Society of America Journal* 67: 817–825.

Murphy, D. V., Recous, S., Stockdale, E. A., Fillery, I. R. P., Jensen, L. S., Hatch, D. J., and K. W. T. Goulding 2003. Gross nitrogen fluxes in soil: Theory, measurement and application of $^{15}$N pool dilution techniques. *Advances in Agronomy* 79: 69–118.

Nelson, M. B., Martiny, A. C., and J. B. H. Martiny. 2016. Global biogeography of microbial nitrogen-cycling traits in soil. *Proceedings of the National Academy of Sciences of the United States of America (PNAS)* 113: 8033–8040.

Nguyen, M. L., and K. M. Goh. 1994. Distribution, transformations and recovery of urinary sulphur and sources of plant-available soil sulphur in irrigated pasture soil-plant systems treated with sulphur-35 labelled urine. *Journal of Agricultural Science, Cambridge* 122, Part 1: 91–105.

Nguyen, M. L., Sheath, G. W., Smith, C. M., and A. B. Cooper. 1998. Impact of cattle treading on hill land. 2. Soil physical properties and contaminant runoff. *New Zealand Journal of Agricultural Research* 41: 279–290.

Nguyen, L., Zapata, F., Lal, R., and G. Dercon. 2011. Role of nuclear and isotopic techniques in sustainable land management. In *World soil resources and food security, advances in soil science*, R. Lal and B. A. Stewart, Eds., pp. 345–418. Boca Raton, FL: CRC Press.

Niu, S., Classen, A. T., Duke, J. S., Kardol, P., Liu, L., Luo, Y., Rustard, L. et al. 2016. Global patterns and substrate-based mechanisms of the terrestrial nitrogen cycle. *Ecology Letters* 19: 697–709.

Oades, J. M. 1984. Soil organic matter and structural stability: Mechanisms and implications for management. *Plant and Soil* 78: 319–337.

Oelmann, Y., Wilcke, W., and R. Bol. 2005. Nitrogen-15 in $NO_3^-$ characterises differently reactive soil organic N pools. *Rapid Communications in Mass Spectrometry* 19: 3177–3181.

Ohte, N., Tayasu, I., Kohzu, A., Yoshimizu, C., Osaka, K., Makabe, A., Koba, K., Yoshida, N., and T. Nagata. 2010. Spatial distribution of nitrate sources of rivers in the Lake Biwa watershed, Japan: Controlling factors revealed by nitrogen and oxygen isotope values. *Water Resources Research* 46: 1–16, W07505, doi:10.1029/2009WR007871.

Ometto, J. P. H. B., Ehleringer, J. R., Domingues, T. F., Berry, J. A., Ishida, F. Y., Mazzi, E., Higuchi, N. et al. 2006. The stable carbon and nitrogen isotopic composition of vegetation in tropical forests of the Amazon Basin, Brazil. *Biogeochemistry* 79: 251–274.

Ostrom, P. H., Pitt, A. J., Sutka, R. L., Ostrom, N. E., Fang, L., and H. Gandhi. 2004. Characterization of isotopomer fractionation during consumption of nitrous oxide in pure culture and soils. *Eos Transactions American Geophysical Union 85 (47)*, Fall Meeting Supplement, Abstract B21A-0852.

Pardo, L. H., Kendall, C., Pett-Ridge, J., and C. C. Y. Chang. 2004. Evaluating the source of streamwater nitrate using $\delta^{15}$N and $\delta^{18}$O in nitrate in two watersheds in New Hampshire, USA. *Hydrological Processes* 18: 2699–2712.

Pardo L., Templer, P., Goodale, C., Duke, S., Groffman, P., Adams, M. B., Boeckx, P., Boogs, J., Campbell, J. et al. 2006. Regional assessment of N saturation using foliar $\delta$ $^{15}$N. *Biogeochemistry* 80:143–171.

Parkin, T. B. 1987. Soil microsites as a source of denitrification variability. *Soil Science Society of America Journal* 51: 1194–1199.

Parfitt, R. L., Baisden, W. T., Schipper, L. A., and A. D. Mackay. 2008. Nitrogen inputs and outputs for New Zealand at national and regional scales: Past present and future scenarios. *Journal of the Royal Society of New Zealand* 38: 71–87.

Paul, J. W., and E. G. Beauchamp 1995. Availability of manure slurry ammonium for corn using $^{15}$N-labelled $(NH_4)_2SO_4$. *Canadian Journal of Soil Science* 75: 35–42.

Peoples, M. B., Boyer, E. W., Goulding, K. W. T. et al. 2004. Pathways of nitrogen loss and their impacts on human health and the environment. In *Agriculture and the nitrogen cycle*. A. R. Mosier, J. K. Syers, and J. R. Freney, Eds., pp. 53–69, SCOPE: Island Press.

Peoples, M. B., Chalk, P. M., Unkovich M. J., and R. M. Boddey. 2015. Can differences in $^{15}$N natural abundance be used to quantify the transfer of nitrogen from legumes to neighbouring non-legume plant species? *Soil Biology and Biochemistry* 87: 97–109. http://dx.doi.org/10.1016/j.soilbio.2015.04.010.

Peoples, M. B., Hauggaard-Nielsen, H., and E. S. Jensen. 2009. The potential environmental benefits and risks derived from legumes in rotations. In *Nitrogen fixation in crop production*. D. W. Emerich, and H. B. Krishnan, Eds., pp. 349–385. Madison, WI: American Society of Agronomy, Crop Science Society of America, Soil Science Society of America.

Quin, B. F. 1982. The influence of grazing animals on nitrogen balances. In *Nitrogen balances in New Zealand ecosystems: Proceedings of a workshop May 1980*. P. W. Gandar and D. S. Bertaud, Eds. Palmerston North, New Zealand: Publ. Dept. of Scientific and Industrial Research, NZ.

Raymond, J., Siefert, J. L., Staples, C. R., and R. E. Blankship. 2004. The natural history of nitrogen fixation. *Molecular Biology and Evolution* 21: 541–554.

Reid, D. 1983. The combined use of fertilizer nitrogen and white clover as nitrogen sources for herbage growth. *Journal of Agricultural Science, Cambridge* 100: 612–623.

Robertson, G. P., and P. M. Groffman. 2015. Nitrogen transformations. In *Soil microbiology, ecology and biochemistry*, 4th ed. pp. 421–446. Burlington, MA: Academic Press.

Robinson, D. 2001. Change in $^{15}$N as an integrator of the nitrogen cycle. *Trends in Ecology and Evolution* 16: 153–162.

Saggar, S., Luo, J., Giltrap, D. L., and M. Maddena. 2009. Nitrous oxide emission processes, measurements, modelling and mitigation. In *Nitrous oxide emissions research progress*. A. J. Sheldon, and E. P. Barnhart, Eds., pp. 1–66, New York, NY: Nova Science Publishers, Inc.

Schepers, J. S., and A. R. Mosier. 1991. Accounting for nitrogen in nonequilibrium soil-crop systems. In *Managing nitrogen for groundwater quality and farm profitability*. R. F. Follett, D. R. Keeney, and R. M. Cruse, Eds., pp. 125–138. Madison, WI: Soil Science Society of America.

Sebilo, M., Billen, G., Grably, M., and A. Mariotti. 2003. Isotopic composition of nitrate–nitrogen as a marker of riparian and benthic denitrification at the scale of the whole Seine river system. *Biogeochemistry* 63: 35–51.

Sebilo, M., Billen, G., Grably, M., and A. Mariotti. 2006. Assessing nitrification and denitrification in the Seine river and estuary using chemical and isotopic techniques. *Ecosystems* 9: 564–577.

Selbie, D. R., Lanigan, G. J., Laughlin, R. J., Di, H. J., Moir, J. L., Cameron, K. C., Clough, T. J. et al. 2015. Confirmation of co-denitrification in grazed grassland. *Scientific Reports* 5: 17361, doi 10.1038 /srep17361.

Sgouridis, F., Stott, A., and S. Ullah. 2016. Application of the $^{15}$N gas-flux method for measuring in situ N$_2$ and N$_2$O fluxes due to denitrification in natural and semi-natural terrestrial ecosystems and comparison with the acetylene inhibition technique. *Biogeosciences* 13: 1821–1835.

Shi, Y., Parker, D. B., Cole, N. A., Auvermann, B. W., and J. E. Mehlhorn. 2001. *Transactions of the American Society of Agricultural and Biological Engineers (ASAE)* 44: 677–682.

Shibata, H., Branquinho, C., McDowell, W. H., Mitchell, M. J., Monteith, D. T., Tang, J., Arvola, L. et al. 2015. Consequence of altered nitrogen cycles in the coupled human and ecological system under changing climate: The need for long-term and site-based research. *Ambio* 44: 178–193.

Shoji, S., Delgado, J., Mosier, A., and Y. Miura. 2001. Use of controlled release fertilizers and nitrification inhibitors to increase nitrogen use efficiency and to conserve air and water quality. *Communications in Soil Science and Plant Analysis* 32: 1051–1070.

Singh, B., Bronson K. F., Singh, Y., Khera T. S., and E. Pasuquin. 2001. Nitrogen-15 balance as affected by rice straw management in rice–wheat rotation in northwest India. *Nutrient Cycling in Agroecosystems* 59: 227–237.

Smil, V. 1999. Nitrogen in crop production: An account of global flows. *Global Biogeochemical Cycles* 13: 647–662.

Smil, V. 2011. Nitrogen cycle and world food production. *World Agriculture* 2: 9–13.

Sorai, M., Yoshida, N., and M. Ishikawa. 2007. Biogeochemical simulation of nitrous oxide based on the major nitrogen processes. *Journal of Geophysical Research* 112, doi 10.1029/2005JG000109.

Stevens, R. J., Laughlin, R. J., Burn, L. C., Arah, J. R. M., and R. C. Hood. 1997. Measuring the contributions of nitrification and denitrification to the flux of nitrous oxide from soil. *Soil Biology and Biochemistry* 29: 139–151.

Stevens, C. J., Lind, E. M., Hautier, Y., Harpole, W. S., Borer, E. T., Hobbie, S. et al. 2015. Anthropogenic nitrogen deposition predicts local grassland primary production worldwide. *Ecology* 96: 1459–1465.

Stevenson, B. A., Parfitt, R. L., Schipper, L. A., Baisdend, W. T., and P. Mudge. 2010. Relationship between soil $^{15}$N, C/N and N losses across land uses in New Zealand. *Agriculture, Ecosystems and Environment* 139: 736–741.

Suddick, E. C., Whitney, P., Townsend, A. R., and E. A. Davidson. 2013. The role of nitrogen in climate change and the impacts of nitrogen–climate interactions in the United States: Foreword to thematic issue. *Biogeochemistry* 114: 1–10.

Sutherland, R. A., van Kessel, C., Farrell, R. E., and D. J. Pennock. 1993. Landscape-scale variations in plant and soil nitrogen-15 natural abundance. *Soil Science Society of America Journal* 57: 169–178.

Sutton, M.A., Howard, C.M., Erisman, J.W., Billen, G., Bleeker, A., Grennfelt, P., van Grinsven, H., and B. Grizzetti (Eds.) 2011. *The European nitrogen assessment: Sources, effects and policy perspectives*. Cambridge: Cambridge University Press.

Templer, P. H., Arthur, M. A., Lovett, G. M., and K. C. Weathers. 2007. Plant and soil natural abundance $\delta^{15}N$: indicators of relative rates of nitrogen cycling in temperate forest ecosystems. *Oecologia* 153: 399–406.

Thomson, A. J., Giannopoulos, G., Pretty, J., Baggs, E. M., and L. Richardson 2012. Biological sources and sinks of nitrous oxide and strategies to mitigate emissions. *Philosophical Transactions Royal Society B-Biological Science B*367: 1157–1168.

Townsend-Small, A., McCarthy, M. J., Brandes, J. A., Yang, L. Y., Zhang, L., Gardner, W.S., 2007. Stable isotopic composition of nitrate in Lake Taihu, China, and major inflow rivers. *Hydrobiologia* 581: 135–140.

UNDESA (United Nations Department of Economic and Social Affairs). 2015. *World Population Prospects: The 2015 Revision.* http://www.un.org/en/development/desa/news/population/2015-report.html.

van Groenigen, J. W., Huygens, D., Boeckx, P., Kuyper, Th. W., Lubbers, M. T., and Groffman, P. M. 2015. The soil N cycle: new insights and key challenges. *Soil* 1: 235–256.

Van Meter, K. J., Basu, N. B., Veenstra, J. J., and C. L. Burras. 2016. *Environ. Res. Lett.* 11: 035014. doi 10.1088/1748-9326/11/3/035014.

Velthof, G., Barot, S., Bloem, J. et al. 2011. Nitrogen as a threat to European soil quality. In *The European nitrogen assessment.* M. A. Sutton, C. M. Howard, J. W. Erisman et al., Eds., pp. 495–510. Cambridge University Press.

Vieten, B., Blunier, T., Neftel, A., Alewell, C., and F. Conen. 2007. Fractionation factors for stable isotopes of N and O during $N_2O$ reduction in soil depend on reaction rate constant. *Rapid Communications in Mass Spectrometry* 21: 846–850.

Vitousek, P. M., Shearer, G., and D. H. Kohl. 1989. Foliar $^{15}N$ natural abundance in Hawaiian rainforest: patterns and possible mechanisms. *Oecologia* 78: 383–388.

Wallander, H., Mörth, C-M., and R. Giesler. 2009. Increasing abundance of soil fungi is a driver for 15N enrichment in soil profiles along a chronosequence undergoing isostatic rebound in northern Sweden. *Oecologia* 160: 87–96.

Ward, B. 2012. The global nitrogen cycle. In *Fundamentals of geobiology.* A. H. Knoll, D. E. Canfield, and K. O. Konhauser, Eds. Blackwell Publishing Ltd.

Watzka, M., Buchgraber, K., and W. Wanek. 2006. Natural $^{15}N$ abundance of plants and soils under different management practise in a montane grassland. *Soil Biology and Biochemistry* 38, 1564–1576.

Whitehead, D. C., and J. Tinsley. 1963. The biochemistry of humus formation. *Journal of the Science of Food and Agriculture* 14: 849–857.

Wienhold, B. J., and D. L. Tanaka. 2001. Soil property changes during conversion from perennial vegetation to annual cropping. *Publications from USDA-ARS/UNL Faculty.* Paper 1218. http://digital commons .unl.edu/usdaarsfacpub/1218.

Wu, H., Dannenmann, M., Fanselow, N., Wolf, B., Yao, Z., Wu, X., Brüggemann, N., Zheng, X., Han, X., Dittert, K., and K. Butterbach-Bahl. 2011. Feedback of grazing on gross rates of N mineralization and inorganic N partitioning in steppe soils of Inner Mongolia. *Plant and Soil* 340: 127–139.

Xue, D., Bottea, J., De Baets, B., Accoe, F., Nestler, A., Taylor, P., Van Cleemputa, O., Berglund, M., and P. Boeckx. 2010. Present limitations and future prospects of stable isotope methods for nitrate source identification in surface- and groundwater. *Water Research* 43: 1159–1170.

Xue, Y., Song, J., Zhang, Y., Kong, F., Wen, M., and G. Zhang. 2016. Nitrate pollution and preliminary source identification of surface water in a semi-arid river basin, using isotopic and hydrochemical approaches. *Water* 8: 328; doi:10.3390/w8080328 (www.mdpi.com/journal/water).

Yan, R., Yang, G., Chen, B., Wang, X., Yan, Y., Xin, X., Li, L., Zhu, X., Bai, K., Rong, Y., and L. Hou. 2016. Effects of livestock grazing on soil nitrogen mineralization on Hulunber meadow steppe, China. *Plant, Soil and Environment* 62: 202–209.

Yoneyama, T., Matsumaru, T., Usui, K., and W. M. H. G. Engelaar. 2001. Discrimination of nitrogen isotopes during absorption of ammonium and nitrate at different nitrogen concentrations by rice (*Oryza sativa* L.) plants. *Plant, Cell and Environment* 24: 133–139.

Yuan, L., Pang, Z, and T. Huang. 2012. Integrated assessment on groundwater nitrate by unsaturated zone probing and aquifer sampling with environmental tracers. *Environmental Pollution* 171: 226–233.

Zaman, M., Nguyen, M. L., Blennerhassett, J. D., and B. F. Quin. 2008. Reducing $NH_3$, $N_2O$ and $NO_3^-$-N losses from a pasture soil with urease or nitrification inhibitors and elemental S-amended nitrogenous fertilizers. *Biology and Fertility of Soils* 44:693–705.

Zapata, F., and O. Van Cleemput, O. 1986 Recovery of $^{15}N$-labelled fertilizer by sugarbeet-spring wheat and winter rye-sugarbeet cropping sequences. *Fertilizer Research* 8: 269–278.

Zehr, J. P., Jenkins, B. D., Short, S. N., and G. F. Steward. 2003. Nitrogenase gene diversity and microbial community structure: A cross-system comparison. *Environmental Microbiology* 5: 539–554.

# 4 Nitrogen Loss in Snowmelt Runoff from Non-Point Agricultural Sources on the Canadian Prairies

*Kimberley D. Schneider, Arumugam Thiagarajan,
Barbara J. Cade-Menun, Brian G. McConkey,
and Henry F. Wilson*

## CONTENTS

## 4.1 INTRODUCTION

Nitrogen (N) is an essential element for crop growth, and many agricultural systems require N fertilizer inputs for optimal crop production. However, only a portion of the N applied as chemical fertilizer or manure is used by plants or retained in soil (Malhi et al. 2001). Nitrogen can be lost from agricultural landscapes through volatilization (as ammonia, $NH_3$), denitrification (in the gaseous forms $N_2O$, $NO$, $N_2$), in leaching of nitrate ($NO_3$-N), and in rainfall or snowmelt runoff. Nitrogen is transported in runoff in a variety of forms, including soluble inorganic [$NO_3$-N, ammonium-N ($NH_4$-N)], and dissolved organic forms (DON), or particulate-bound organic and inorganic N (Delgado 2002). Nitrogen export to surface water in runoff from agricultural sources is a major global concern; in a worldwide study of 946 watersheds, Alvarez-Cobelas et al. (2008)

found that N losses in runoff were consistently higher per unit area from agricultural than from non-agricultural land. Non-point agricultural sources of nutrients are typically more challenging to manage than point sources because of broader distribution throughout impacted watersheds and because these sources are often highly variable in time due to the role of weather in determining the frequency and intensity of runoff (Carpenter 1998; Corriveau et al. 2013). To improve fertilizer use efficiency and minimize risks to downstream water quality, the pathways of N loss from agriculture must be identified, along with management practices that increase or decrease N losses.

Although phosphorus (P) availability has been shown to limit algal production in freshwater lakes and is recognized as a major contributor to eutrophication (Sharpley et al. 1987), increases in N loading can also lead to further eutrophication of surface waters (Havis and Alberts 1993; Carpenter et al. 1998; Eickhout et al. 2006; USEPA 2015), particularly in waterbodies that have received high P loads naturally or with runoff from fertilized soils on adjacent lands (Hall et al. 1999; Dodds and Smith 2016). Many lakes in the Canadian Prairies are naturally eutrophic and dense summer algal blooms are common due to high nutrient loadings from large, fertile catchments, shallow depths, and high solar energy (Barica 1993). Paleolimnological studies in this region show high concentrations of bioavailable non-apatite P in lake sediments dated to before European settlement (Hall et al. 1999), indicating that P was not a limiting nutrient historically. Nutrient loading increased with European settlement in the 1900s following construction of the national railroad, decreased during the drought years of the 1930s, and then steadily increased from ~1945, which is similar to trends also observed in the U.S. northern Great Plains (NGP) region (Hall et al. 1999; Dodds and Smith 2016). Many of these lake have N:P ratios below 10 (Barica 1993; Hall et al. 1999; Pham et al. 2008; Finlay et al. 2010), which favors N control of eutrophication, and mesocosm studies have confirmed that N additions increase algal blooms (Finlay et al. 2010; Donald et al. 2013). However, most of the catchments of these lakes are used for crop production that relies on N fertilization (Malhi et al. 2001).

Another unique feature of the Canadian Prairies is the distinct seasonality of runoff. The region is semi-arid to sub-humid with high rates of evapotranspiration in the summer, so rainfall runoff from agricultural land is generally low. Instead, the majority of runoff occurs in the spring as snowmelt runoff while the soil is still frozen and infiltration is restricted (Gray and Landine 1988; Pomeroy et al. 2005). As such, management practices that minimize nutrient transport in rainfall runoff may not be appropriate for this region. However, management practices to reduce N transport in snowmelt have been poorly studied, despite the importance of N in eutrophication of lakes in this region. The objective of this chapter is to review the literature regarding N losses in snowmelt on the Canadian Prairies. The scope of this work is limited to N losses during snowmelt from non-point agricultural sources in the region; results from relevant studies that collected and chemically analyzed snowmelt runoff at the edge-of-field scale will be summarized and integrated with related knowledge about N losses from other similar regions. The main sources and transport factors affecting N loss in snowmelt runoff will be discussed, as well as impacts of agricultural management practices on these losses. Knowledge gaps and areas requiring future research will also be identified.

## 4.2 AGRICULTURE AND FERTILIZATION PRACTICES ON THE CANADIAN PRAIRIES

The Canadian Prairies are located in the northernmost portion of the NGP (Figure 4.1). The region has a continental climate, with long, cold winters and short, warm summers (Padbury et al. 2002; McGinn 2010). Average winter and summer temperatures are $-10^{\circ}$C and $15^{\circ}$C, respectively, with >2400 hours of sunshine annually. Winds are strong and westerly, with average annual surface wind speeds between 14 and 22 km h$^{-1}$, although gusts as strong as 171 km h$^{-1}$ have been recorded (McGinn 2010). Precipitation is highly variable and unpredictable, averaging 454 mm, with ~30% as snowfall, and summer drought occurs regularly (Gray and Landine 1988; Fang et al. 2010; McGinn

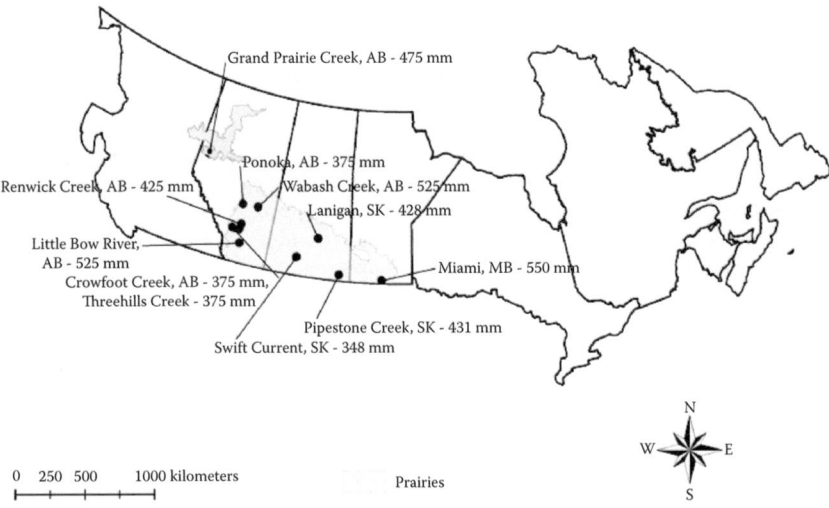

**FIGURE 4.1** Map of Canada showing the Prairie Ecozone in gray. The labeled sites are locations of prior snowmelt runoff studies, discussed in the text, with average annual precipitation for each site. AB, Alberta; MB, Manitoba; SK, Saskatchewan.

2010). The lowest precipitation is in the southwestern Prairies on the Alberta (AB)–Saskatchewan (SK) border (Figure 4.1) and precipitation increases eastward into Manitoba (MB) and northward (Figure 4.1). The soils in this region were primarily developed on medium-textured glacial till, with important cropland area developed on fine-textured sediments of glacial lakes. Soils are affected by the moisture gradient from east to west, from Black Chernozems (Udic Ustolls) in MB and eastern SK, to Dark Brown Chernozems (Typic Ustolls) and Brown Chernozems in SK and Dark Gray Luvisols (Mollic Typic Hapludalfs) and Gray Luvisols (Typic Hapludalfs) in western AB and eastern British Columbia (BC). Gleysols (Aquaols) and Vertisols (Cryerts, Aquerts) may also be present (Padbury et al. 2002). Prairie soils are generally deep, with good water holding capacity (Pomeroy et al. 2005).

Prior to European settlement, native grasslands covered most of the Canadian Prairie region. These were grazed by vast bison herds, and grass fires were common (Anderson and Cerkowniak 2010). In the early 1900s, much of the land was farmed as cropland, although stands of native grassland remained. To combat soil erosion during the severe drought of the 1930s, abandoned cropland was replanted with winter-hardy and drought-tolerant species, including crested wheatgrass [*Agropyron cristatum* (L.)] and alfalfa (*Medicago* spp.). Both native and planted (tame) grasslands are used to graze livestock, while the remaining arable land is used for crop production.

Globally, Canada is a significant agricultural producer, and the Canadian Prairies are extremely important for Canadian agriculture. Based on the 2011 Census of Agriculture (Table 4.1; Statistics Canada 2016) 82% of Canada's cropland area and 49% of Canadian farms are located in the Prairie Provinces. In addition, 90% of Canadian land in tame pasture (55% of Canadian farms with tame pasture) is located in the Prairie Provinces, as is 89% of natural land used for pasture (59% of farms). Farms in these provinces create 99% of Canadian non-soybean oilseed production [e.g., canola (*Brassica napus* L.), flax (*Linum usitatissiumum* L.)], 93% of pulse production [including pea (*Pisum sativum* L.), lentils (*Lens culinaris* L.)], 84% of wheat production [spring wheat (*Triticum aestivum* L.), durum wheat (*Triticum turgidum* var. *durum* L.)], 73% of other cereal production [e.g., barley (*Hordeum vulgare* L.), oats (*Avena sativa* L.)], 63% of beef ranching, 53% of hay production, 43% of horse production, and 24% of potato (*Solanum tuberosum* L.) farming (Table 4.2). Prairie farms represent 85% of the area in Canada with application of commercial fertilizers and 43% of

## TABLE 4.1
## Agricultural Land Use, Chemical Fertilizer Inputs Manure Production and Application

|  |  |  | Canada | MB[a] | SK | AB | BC-P[b] | Prairie Total (%)[c] |
|---|---|---|---|---|---|---|---|---|
| Land use | Land in crops | Farms[d] | 174343 | 13797 | 34185 | 36662 | 1378 | 866022 (49) |
|  |  | '000 ha[e] | 35350 | 4349 | 14729 | 9754 | 271 | 29102 (82) |
|  | Summerfallow land | Farms | 20221 | 1961 | 10378 | 5336 | 316 | 17991 (89) |
|  |  | '000 ha | 2085 | 96 | 1446 | 511 | 16 | 2069 (99) |
|  | Tame or seeded | Farms | 64949 | 4475 | 11759 | 19036 | 760 | 36060 (55) |
|  | pasture | '000 ha | 5533 | 415 | 2058 | 2396 | 100 | 497 (90) |
|  | Natural land for | Farms | 82865 | 8132 | 16372 | 23855 | 898 | 49257 (59) |
|  | pastures | '000 ha | 14703 | 1467 | 4817 | 6436 | 341 | 13061 (89) |
| Fertilizers | Use of commercial | Farms | 99599 | 8546 | 20679 | 19730 | 409 | 49364 (50) |
|  | fertilizers | '000 ha | 24918 | 3452 | 10489 | 7111 | 129 | 21181 (85) |
| Manure | Manure applied on | Farm | 99573 | 7453 | 13324 | 21494 | 740 | 43011 (43) |
|  | the farm |  |  |  |  |  |  |  |

*Source:* Statistics Canada. 2016. Results of the 2011 Canadian Census of Agriculture. http://www5.statcan.gc.ca/cansim
        /a03?lang=eng&pattern=004-0200..004-0242&p2=31.
[a] MB, Manitoba; SK, Saskatchewan; AB, Alberta; BC, British Columbia.
[b] Data are for the Peace River region of BC only.
[c] Total number for MB, SK, AB, BC-P (percent of total Canadian farms).
[d] Total farms reporting in the Census of Agriculture.
[e] Total hectares (× 1000) as reported by farmers completing the Census of Agriculture.

## TABLE 4.2
## Agricultural Farm Classes of the Canadian Prairies, Classified by the North American Industry Classification (NAICS) System

| NAICS Classification | Canada | MB[a] | SK | AB | BC-P[b] | Prairie Total (%)[c] |
|---|---|---|---|---|---|---|
| Total number of farms | 205730 | 15877 | 36952 | 43234 | 1560 | 97623 (47) |
| Cattle ranching and farming | 49613 | 4485 | 7455 | 12507 | 255 | 24702 (50) |
| Beef cattle ranching and farming[d] | 37406 | 4152 | 7314 | 12022 | 249 | 23737 (63) |
| Dairy cattle and milk production | 12207 | 333 | 141 | 485 | 6 | 965 (8) |
| Hog and pig farming | 3470 | 318 | 66 | 193 | 2 | 579 (17) |
| Poultry and egg production | 4484 | 253 | 115 | 339 | 6 | 713 (16) |
| Horse and other equine production | 13655 | 671 | 968 | 3995 | 182 | 5816 (43) |
| Soybean farming | 6471 | 322 | 16 | 4 | 0 | 342 (5) |
| Oilseed (not soybean) farming | 16508 | 2497 | 8592 | 5243 | 80 | 16412 (99) |
| Dry bean and bean (pulse) farming | 1617 | 57 | 1274 | 162 | 3 | 1496 (93) |
| Wheat farming | 8206 | 768 | 4017 | 2083 | 19 | 6887 (84) |
| Corn farming | 6160 | 48 | 6 | 9 | 0 | 63 (1) |
| Other grain farming | 22730 | 2926 | 8290 | 5191 | 76 | 16483 (73) |
| Potato farming | 1323 | 99 | 61 | 149 | 2 | 311 (24) |
| Hay farming | 24844 | 1638 | 3109 | 7799 | 668 | 13214 (53) |

*Source:* Statistics Canada. 2016. Results of the 2011 Canadian Census of Agriculture. http://www5.statcan.gc.ca/cansim
        /a03?lang=eng&pattern=004-0200..004-0242&p2=31.
[a] MB, Manitoba; SK, Saskatchewan; AB, Alberta; BC, British Columbia.
[b] Data are for the Peace River region of BC only.
[c] Total number of farms in MB, SK, AB, BC-P (percent of total Canadian farms).
[d] Includes feedlots.

farms where manure is applied (Table 4.1). Crop production is generally dryland; irrigation is rare (Padbury et al. 2002).

Early agriculture relied on summerfallow (no planted crops, weeds controlled by tillage) every second or third year to conserve moisture and release organic N (Padbury et al. 2002). However, there has been a shift over time from conventional (CT) to reduced tillage (RT) and zero tillage (NT), which have been widely adopted to improve water use efficiency, reduce soil erosion and degradation, and increase cropping diversity and intensity (Malhi et al. 2001). Nitrogen is the nutrient that is most frequently limiting to crops on the Canadian Prairies, and as a result N is applied at much greater rates than P, K, or S (Grant et al. 2002). In the past, fall N applications were common (Malhi et al. 2001; Grant et al. 2002), but there has been a shift to spring banding of coated or uncoated urea ($CO(NH_2)_2$) with seeding (Khakbazan et al. 2013). Although chemical fertilizers are more common, manure is used by some producers as a fertilizer or soil amendment for forage, with increased application in areas of concentrated livestock production including feedlots (Beaulieu 2004). Hog production systems are more common in MB than the western Canadian Prairie Provinces, and produce liquid manure that is generally injected or incorporated into the soil (Beaulieu 2004). Solid manures are usually spread on top of the soil, and are incorporated less frequently than liquid manure (Padbury et al. 2002). About one third of manure application in the Canadian Prairies is during the summer, with the remainder applied in the fall; application to frozen soils in winter is discouraged (Beaulieu 2004). However, it has become common in recent years to feed cattle in fields during the winter, rather than confining them in corrals. This practice will be discussed in more detail below.

## 4.3 TOPOGRAPHY, HYDROLOGY, AND NUTRIENT TRANSPORT IN SNOWMELT

The topography of the Canadian Prairies is gently rolling to hummocky terrain, with numerous un-drained depressions ("potholes") and larger glacial lake plains (Padbury et al. 2002; Pomeroy et al. 2005). This has resulted in poorly developed, discontinuous, and sparse drainage systems (Pomeroy et al. 2005). Under natural conditions, wetlands and potholes collect and store runoff, spilling only in very wet years. As such, many parts of the landscape do not frequently contribute runoff to the main rivers and streams (non-contributing areas). However, intensive modification of surface drainage for agriculture in some areas has channelized flow from potholes and wetlands to streams and rivers, increasing transport of water and nutrients (Pomeroy et al. 2005; Corriveau et al. 2013). Major spring flooding events during snowmelt and extreme precipitation events enhance hydrological connectivity and increase contributing areas within watershed of the region, even where connectivity may be limited for the rest of the year (Corriveau et al. 2013).

The hydrology of the Canadian Prairies is dominated by cold region processes. Winter typically lasts from late October until late March, during which soils are typically frozen to depths of ~1 m (Coles et al. 2017). There is a generally continuous snow cover, which varies with the moisture gradient from east to west, and which will be affected by sublimation and wind distribution (Pomeroy et al. 2005; Fang et al. 2010). Spring snowmelt is governed by the internal energy of the snowpack, with meltwater released when the snowpack becomes isothermal at 0°C (Gray and Landine 1988). The rate of snowmelt is influenced by factors controlling the influx of energy to the snowpack, including weather, vegetation, and topography (Dunne 1983). Water migration through the snowpack is dominated by vertical unsaturated flow to the frozen soil surface, which restricts infiltration, resulting in Hortonian overland flow (Dunne 1983). The amount of runoff during snowmelt can be explained by the relative infiltration of snowmelt into frozen soil, expressed as the runoff ratio (ratio of total runoff amount to entire winter snowfall amount or amount of snow cover water equivalent; Coles et al. 2017). The runoff ratio will increase with total snowfall, pre-melt snowpack, melt rate, and fall soil surface water content, all related to conditions that reduce relative infiltration into the soil. Management practices can influence many of these factors. For example, standing stubble from no-till can trap snow

(Nicholaichuk and Gray 1986; Elliott et al. 2001), while chisel ploughing or subsoiling can increase infiltration (Nicholaichuk and Gray 1986; Zuzel and Pikul 1987; Casson et al. 2008).

Most rainfall on the Canadian Prairies occurs in the spring and summer, and is rapidly consumed by evaporation or transpiration. Only very intense summer storms will generate runoff (Pomeroy et al. 2005). As such, the majority of runoff occurs during spring snowmelt (Nicholaichuk 1967; Granger et al. 1984; Casson et al. 2008; Tiessen et al. 2010). In watersheds with naturally high connectivity or enhanced connectivity through drainage, snowmelt will also contribute significantly to streamflow; for example, snowmelt runoff typically comprised 56–95% of total annual streamflow in three watersheds studied in southern MB (Glozier et al. 2006). As the major source of annual runoff, snowmelt contributes significantly to annual nutrient loads (Glozier et al. 2006; Tiessen et al. 2010). The transfer of nutrients to snowmelt runoff is less understood than for rainfall runoff. There are significant differences in the kinetics of the interaction of snowmelt water with soil and crop residues; typically, it is a slower process involving longer contact times. Particle detachment from frozen soils is limited (Panuska and Karthikeyan 2010), minimizing particulate nutrient transfer. However, this may be increased in years with high snowfall, in areas where high volumes of snow accumulate, or with snow with high moisture content (Tanasienko et al. 2011). As such, dissolved nutrients rather than particulate forms have been observed to be the dominant form of nutrients present in snowmelt runoff (Glozier et al. 2006). This contrasts with patterns observed in agricultural regions with annual runoff dominated by rainfall, where particulate N (PN) loads tend to be greater than dissolved N loads in runoff (Sharpley et al. 1987). Total N loads are closely associated with snowmelt volume (Liu et al. 2013), and seasonal climate patterns that lead to abrupt hydrological changes (e.g., flooding) will alter runoff volume and thus nutrient export (Gomi et al. 2002; Rattan et al. 2017). In runoff observed from fields in the region, drier years with less snow tend to result in an increase in flow-weighted nutrient concentrations on a daily basis, but will decrease the duration of runoff and thus the total load; conversely, wetter years have lower flow-weighted nutrient concentrations but higher total loads from longer runoff events (Casson et al. 2008; Liu et al. 2013; Cade-Menun unpublished data).

## 4.4   NON-POINT AGRICULTURAL SOURCES OF N LOSS

To the authors' best knowledge, only a handful of studies have been published that have investigated N loss in snowmelt runoff on the Canadian Prairies. Table 4.3 gives information about those studies, including reported fall soil $NO_3$-N concentrations, while Table 4.4 includes fertilization information and concentrations (annual flow-weighted mean concentrations, FWMC) and loads, where available, for $NO_3$-N (including $NO_2$-N for some studies), $NH_4$-N (and $NH_3$-N for some studies), total dissolved N (TDN), and total N (TN), as it was available in the publications. In MB, details are included from four publications from the South Tobacco Creek watershed project near Miami, MB. Tiessen et al. (2010) and Liu et al. (2014a) used the same paired watersheds to compare CT with NT and rotational tillage (RT; tillage in alternate years). Li et al. (2011) compared a control subwatershed with annual cropping to a treatment subwatershed in which five best management practices (BMPs) were implemented: a holding pond downstream of a cattle overwintering feedlot; riparian zone and grassed waterway management; grazing restriction; forage conversion; and nutrient management. Liu et al. (2014b) compared annual cropland to forage, using two sub-watersheds that started as annual cropland and were converted to forage, and two that started as forage and were converted to annual cropland. In eastern SK, Cade-Menun et al. (2013) compared annual cropland versus summer-grazed pasture during the snowmelt event in spring 2010, using five replicate subwatersheds (0.5–1.9 ha) for each land use type. Photos of snowmelt runoff collection at the site used for Cade-Menun et al. (2013) are shown in Figure 4.2. In central SK, Smith et al. (2011) compared the effects of in-field winter bale-grazing to control sites (two each) during a runoff event in spring 2009. In southwest SK, Nicholaichuk and Read (1978) compared six years of runoff data from a wheat-summerfallow rotation, collecting data from wheat stubble, summerfallow, and fertilized

**TABLE 4.3**

**Site Location, Soil Type, Annual Precipitation, and Tillage Details of the Studies Investigating N Losses in Snowmelt Runoff from Croplands and Pastures on the Canadian Prairies**

| References | Site Location[a] (lat, long) | Soil Type[b] | Annual Precipitation (Snow)[c] (mm) | Study Years | Tillage[d] | Study Area (ha) | Treatment Duration (y) | Cropping System/ Rotation Details[e] | Soil $NO_3$-N Mean[f] (range) (depth) (mg kg$^{-1}$) |
|---|---|---|---|---|---|---|---|---|---|
| Nicholaichuk et al. 1978 | Swift Current, SK (50°15'N, 107°43'W) | Brown Chernozem | 348 (96) | 1970–1976 | CT | ~8 (2 plots) | 6 | Spring wheat/summerfallow rotation; stubble | 7.6 ± 4.3 (0–15 cm) |
| | | | | | | ~4 | 6 | Spring wheat/summerfallow rotation; summerfallow | 18.6 ± 4.8 (0–15 cm) |
| | | | | | | ~4 | 2 | Spring wheat/summerfallow rotation; fertilized summerfallow | 34.2 ± 9.2 (0–15 cm) |
| Casson et al. 2008 | Crowfoot Creek (51°07'N, 113°19'W) | Dark brown Chernozem | 375 (120) | 2002–2005 | NT | 248 | | 2002: barley, wheat; 2003: canola, barley; 2004: barley, wheat; 2005: wheat | 13.6 (3.5–50.2) (0–15 cm) |
| | Grande Prairie Creek (55°24'N, 118°31'W) | Dark grey Chernozem | 445 (154) | 2002–2005 | CT | 62 | | 2002: barley, oat; 2003: barley, barley; 2004: canola, wheat; 2005: canola, oat | 15.4 (0.5–57.3) (0–15 cm) |
| | Renwick Creek (51°42'N, 113°31'W) | Black Chernozem | 417 (84) | 2002–2005 | RT | 26 | | 2002: canola, wheat; 2003: pea, wheat; 2004: rye, wheat; 2005: barley, canola | 9.2 (1.8–71.3) (0–15 cm) |
| | Threehills Creek (52°02'N, 113°37'W) | Black Chernozem | 417 (84) | 2002–2005 | NT | 51 | | 2002: canola; 2003: wheat, wheat; 2004: wheat, wheat; 2005: pea | 15.2 (1.1–59.9) (0–15 cm) |
| | Wabash Creek (54°03'N, 113°58'W) | Dark grey Chernozem | 461 (103) | 2002–2005 | CT | 33 | | 2002: barley; 2003: canola, wheat; 2004: canola, wheat; 2005: barley, wheat | 11.7 (3.5–66.6) (0–15 cm) |
| | Lower Little Bow (49°59'N, 112°33'W) | Dark brown Chernozem | 375 (112) | 2002–2005 | CT | 88 | | 2002–2005: corn silage; irrigated; Manured every 2 years | 33.4 (5.6–121.7) (0–15 cm) |
| | Ponoka (52°47'N, 113°38'W) | Black Chernozem | 525 (135) | 2002–2005 | CT | 30 | | 2002 & 2003: corn silage; 2004: canola; 2005: barley silage | 124.8 (31.2– 261.8) (0–15 cm) |

*(Continued)*

**TABLE 4.3 (CONTINUED)**
**Site Location, Soil Type, Annual Precipitation, and Tillage Details of the Studies Investigating N Losses in Snowmelt Runoff from Croplands and Pastures on the Canadian Prairies**

| References | Site Location[a] (lat, long) | Soil Type[b] | Annual Precipitation (Snow)[c] (mm) | Study Years | Tillage[d] | Study Area (ha) | Treatment Duration (y) | Cropping System/ Rotation Details[e] | Soil NO₃-N Mean[f] (range) (depth) (mg kg⁻¹) |
|---|---|---|---|---|---|---|---|---|---|
| Smith et al. 2011 | Lanigan, SK (51°51′N, 105°02′W) | Black Chernozem | 428 (100) | 2009 | None | 6 | 1 | Russian rye grass winter feeding area with a stocking density of 2218 cow-days ha⁻¹ for 87 days; no fertilizer | 3.8 (0.4–8.6) (0-10 cm) |
| | | | | | | 3 | 1 | Unfertilized Russian rye grass pasture, control | 4.3 (0.03–7.8) (0–10 cm) |
| Tiessen et al. 2010 | Miami, MB (49°20′N, 98°22′W) | Dark gray Chernozem | 550 (140) | 1993–2007 | CT | 4.2 | 4 | Years 1, 3: canola; year 2: barley; year 4: wheat; mineral fertilizer | 4.2 (2.6–5.5) (0-15 cm) |
| | | | | 1997–2007 | NT | 5.1 | 4 | | 3.3 (2.2–3.8) (0–15 cm) |
| Li et al. 2011 | Miami, MB (49°20′N, 98°22′W) | Dark gray Chernozem | 550 (140) | 2005–2010 | | 205 | | Treatment watershed, 5 BMPs: (a) Holding pond downstream of a beef cattle overwintering feedlot, 2 ha; (b) riparian zone, grassed waterway (43 ha); (c) grazing restriction (205 ha); (d) perennial forage conversion (26 ha); (e) nutrient management (121 ha) | N/A[g] |
| | | | | | | 207 | | Control watershed; NT | N/A |
| Cade-Menun et al. 2013 | Pipestone Creek, SK (50°00′N, 102°20′W) | Black Chernozem | 431 (116) | 2010 | None | 0.5–1.9 | >5 | Tame pasture (alfalfa + smooth brome grass); summer-grazed, hayed; no fertilizer | 2.0 ± 1.2 (0–15 cm) |

*(Continued)*

**TABLE 4.3 (CONTINUED)**

**Site Location, Soil Type, Annual Precipitation, and Tillage Details of the Studies Investigating N Losses in Snowmelt Runoff from Croplands and Pastures on the Canadian Prairies**

| References | Site Location[a] (lat, long) | Soil Type[b] | Annual Precipitation (Snow)[c] (mm) | Study Years | Tillage[d] | Study Area (ha) | Treatment Duration (y) | Cropping System/ Rotation Details[e] | Soil $NO_3$-N Mean[f] (range) (depth) (mg $kg^{-1}$) |
|---|---|---|---|---|---|---|---|---|---|
| Liu et el. 2014a | Miami, MB (49°20'N, 98°22'W) | Dark gray Chernozem | 550 (140) | 2004–2012, treatment years were 2008–2012 | NT | 0.6–1.1 | >10 | Wheat–Canola rotation; NPK fertilizer | 7.2 ± 3.3 (0–15 cm) |
| | | | | | CT | 4.2 | 9 | Canola, spring wheat; mineral fertilizer | 8 (4–15) (0–15 cm) |
| | | | | 2004–2007 | NT | 5.1 | 4 | Canola, barley, spring wheat | 8 (4–15) (0–15 cm) |
| | | | | 2008–2012 | RT | 5.2 | 5 | Canola, spring wheat; mineral fertilizer | 9 (7–15) (0–15 cm) |
| Liu et al. 2014b | Miami, MB (49°20'N, 98°22'W) | Dark gray Chernozem | 550 (140) | 2005–2012 | CT in crop years | 2.45–12.7 | 8 | 2005–2009: Cropland (flax, wheat, canola, oat); 2010–2012: forage (alfalfa, grass); mineral fertilizer | 9.5 (4–10)–9 (0–15 cm) |
| | | | | 2005–2012 | CT in crop years | 2.45–12.7 | 17 | 2005–2009: forage (alfalfa, grass); 2010–2012: Cropland (flax, wheat, canola, oat); mineral fertilizer | 4.6 (1–5) (0–15 cm) |

a Latitude and longitude.
b Canadian soil classification system.
c Total precipitation (snow).
d CT = conventional tillage, NT = no-till, RT = rotational tillage; None: pasture with no tillage history.
e Multiple sites were involved in the study.
f Mean ± standard deviation.
g N/A: Not available based on information in the publication.

**TABLE 4.4**

**Details on the Treatments, Fall N Application, and Flow-Weighted Concentrations of Nitrate-N ($NO_3$-N), Ammonium-N ($NH_4$-N), Total Dissolved N (TDN), and Total N Concentrations in Snowmelt Runoff from Studies in Canadian Prairie Region**

| References | Treatments/Details | Nutrient Management (kg N ha⁻¹y⁻¹) | | Snowmelt Runoff N Concentrations | | | |
|---|---|---|---|---|---|---|---|
| | | | | $NO_3$-N | $NH_4$-N | TDN | Total N |
| Nicholaichuk et al. 1978 | Stubble | | AFWMCᵃ (range), mg L⁻¹ | 0.2 ± 0.2 | N/Aᵇ | 1.4 ± 0.6 | N/A |
| | Summerfallow | 50 | AFWMC (range), mg L⁻¹ | 1.0 ± 0.8 | N/A | 4.7 ± 1.6 | N/A |
| | Fall fertilized summerfallow | | AFWMC (range), mg L⁻¹ | 2.0 ± 1.0 | N/A | 11.4 ± 4.5 | N/A |
| Casson et al. 2008 | Crowfoot Creek | 2002: barley (57), wheat (53) 2003: canola (46), barley (54) 2004: barley (0), wheat (6), wheat (50) 2005: wheat (50 &111) | AFWMC (range), mg L⁻¹ | 0.69 (0.00–3.28) | 0.21ᶜ (0.00–1.47) | N/A | 2.87 (0.47–8.54) |
| | Grande Prairie Creek | 2002: barley (57), oat (53) 2003: barley (0), barley (84) 2004: canola (78), wheat (78) 2005: canola (78), oat (78) | AFWMC (range), mg L⁻¹ | 1.53 (0.00–12.3) | 0.17 (0.00–0.99) | N/A | 3.56 (0.15–17.0) |
| | Renwich Creek | 2002: wheat/canola (34–95) 2003: pea (0), wheat (84) 2004: rye (165), wheat (0) 2005: barley (0), canola (0) | AFWMC (range), mg L⁻¹ | 0.64 (0.10–2.04) | 0.68 (0.02–2.9) | N/A | 5.11 (1.95–57.5) |
| | Threehills Creek | 2002: canola (158) 2003: wheat (157) 2004: wheat (165) 2005: pea (0) | AFWMC (range), mg L⁻¹ | 1.49 (0.31–9.59) | 0.61 (0.03–4.14) | N/A | 5.69 (2.38–18.8) |

(*Continued*)

**TABLE 4.4 (CONTINUED)**
**Details on the Treatments, Fall N Application, and Flow-Weighted Concentrations of Nitrate-N (NO$_3$-N), Ammonium-N (NH$_4$-N), Total Dissolved N (TDN), and Total N Concentrations in Snowmelt Runoff from Studies in Canadian Prairie Region**

| References | Treatments/Details | Nutrient Management (kg N ha$^{-1}$ y$^{-1}$) | | Snowmelt Runoff N Concentrations | | | |
|---|---|---|---|---|---|---|---|
| | | | | NO$_3$-N | NH$_4$-N | TDN | Total N |
| | Wabash Creek | 2002: barley (110) 2003: canola (110), wheat (121) 2004: canola (38), wheat (115) 2005: barley (7), wheat (7) | AFWMC (range), mg L$^{-1}$ | 5.05 (0.18–15.0) | 0.33 (0.03–2.76) | N/A | 8.17 (2.50–18.8) |
| | Ponoka | 2002: 100 Mg/ha cattle manure 2003: 112 kg N/ha fertilizer 2004: 100 Mg/ha cattle manure 2005: None applied | AFWMC (range), mg L$^{-1}$ | 1.12 (0.00–2.96) | 26.0 (1.9–109) | N/A | 39.3 (0.0–178.1) |
| | Little Bow River | 2002: None applied 2003: None applied 2004: cattle manure 2005: None applied | AFWMC (range), mg L$^{-1}$ | 6.55 (0.00–34.7) | 1.12 (0–11.1) | N/A | 6.8 (0.0–43.9) |
| Smith et al. 2011 | Bale-grazed pasture | Manure, urine, bales | Concentration (range) mg L$^{-1}$ | 0.25 (0–0.6) | 40.2 (10–80) | N/A | N/A |
| | Control pasture | None | Concentration (range) mg L$^{-1}$ | 0.19 (0–0.7) | 0.33 (0–0.5) | N/A | N/A |
| Tiessen et al. 2010 | Conventional tillage | Year 1: 90 - broadcast; Years 2-4: 67–78 banded; Year 4: 11 with seed | AFWMC (range), mg L$^{-1}$ | 5.45 (5.16–5.81) | 0.37 (0.25–0.44) | 7.08 (6.37–7.57) | 7.59 (7.37–7.96) |
| | | | Mean (range) of loads, kg ha$^{-1}$ | 3.90 (2.53–6.14) | 0.26 (0.16–0.48) | 5.06 (3.36–8.13) | 6.15 (3.50–8.76) |
| | No-till | | AFWMC (range), mg L$^{-1}$ | 0.84 (0.53–1.27) | 0.76 (0.23–2.12) | 2.72 (1.30–5.33) | 8.25 (1.81–5.33) |
| | | | Mean (range) of loads, kg ha$^{-1}$ | 0.58 (0.37–0.76) | 0.48 (0.12–1.28) | 1.83 (1.21–3.21) | 1.68 (1.50–3.21) |

*(Continued)*

## TABLE 4.4 (CONTINUED)

Details on the Treatments, Fall N Application, and Flow-Weighted Concentrations of Nitrate-N ($NO_3$-N), Ammonium-N ($NH_4$-N), Total Dissolved N (TDN), and Total N Concentrations in Snowmelt Runoff from Studies in Canadian Prairie Region

| References | Treatments/Details | Nutrient Management (kg N ha⁻¹y⁻¹) | | Snowmelt Runoff N Concentrations | | | |
|---|---|---|---|---|---|---|---|
| | | | | NO₃-N | NH₄-N | TDN | Total N |
| Li et al. 2011 | BMPs (see Table 4.3 and text) | N/A | AFWMC, mg L⁻¹ (load, kg ha⁻¹) | 1.1 (0.037) | 0.07 (0.031) | 1.8 (0.051) | 2.0 (0.055) |
| | Control watershed | N/A | AFWMC, mg L⁻¹ (load, kg ha⁻¹) | 2.7 (0.154) | 0.8 (0.0045) | 3.2 (0.173) | 3.4 (0.055) |
| Cade-Menun et al. 2013 | Summer grazed pasture | None | AFWMC (range), mg L⁻¹ | 0.19 (0.13–0.25) | 0.53 (0.26–0.70) | 2.61 (2.11–3.02) | 3.49 (2.51–6.26) |
| | No till cropland | 92 with seed | AFWMC (range), mg L⁻¹ | 0.17 (0.03–0.74) | 0.002 (0.00–0.01) | 2.38 (1.99–2.84) | 3.49 (2.25–3.53) |
| Liu et al. 2014a | Annual cropland, CT | 13–185 | AFWMC, mg L⁻¹ (load, kg ha⁻¹) | N/A | N/A | 7 (5) | 7 (6) |
| | Annual cropland, NT | 67–185 | AFWMC, mg L⁻¹ (load, kg ha⁻¹) | N/A | N/A | 3 (2) | 3 (2) |
| | Annual cropland, RT | 13–112 | AFWMC, mg L⁻¹ (load, kg ha⁻¹)[l] | N/A | N/A | 8 | 8 |
| Liu et al. 2014b | Annual cropland, all sites | 54–123 (spring, anhydrous NH₃) | AFWMC, mg L⁻¹ (load, kg ha⁻¹) | 9.08 (2.08) | 0.36 (0.12) | 9.73 (2.56) | 10.4 (2.89) |
| | Perennial forage, all sites | 0 | AFWMC, mg L⁻¹ (load, kg ha⁻¹) | 1.46 (0.96) | 1.25 (0.75) | 4.86 (2.73) | 5.73 (3.11) |

*Note:* Loads are included if available.

[a] AFWMC: Average of flow weighted means and range of concentrations.

[b] N/A - Not available.

[c] Concentrations for NH₄-N were not flow-weighted.

**FIGURE 4.2    (See color insert.)** Sample collection during snowmelt runoff in Saskatchewan. (a) The view looking up the hill from the flume. (b) The set-up for sample collection. The boards are berms to direct water to the flume, and the green box holds the auto-sampler for water collection, which is powered by the solar panel. (c) A close-up of the flume with water flowing through from top to bottom. The * indicates the tube through which the auto-sampler takes in water from the flume. All pictures were taken on the same day in March 2012, which was a year with little snow. The sites are the same as those used for Cade-Menun et al. (2013). Photo source: B. Cade-Menun.

summerfallow. Casson et al. (2008) summarized data from a four-year project at eight locations throughout AB; data from the seven sites with the most snowmelt runoff in the three years of runoff events are included in the summary tables. All seven sites were cropped; five received chemical fertilizer and two manure.

The summarized studies indicate that N inputs to snowmelt runoff come from soil, fertilizers, manure plus urine and plant material. In the following section, these sources of N are reviewed.

### 4.4.1  Soil

There was no strong relationship between soil N concentrations and N loss in snowmelt across the studies in Tables 4.3 and 4.4. Casson et al. (2008) found significant relationships between soil $NO_3$-N and $NO_3$-N and TN concentrations in runoff. Tiessen et al. (2010) reported greater $NO_3$-N concentrations in runoff from soils with greater fall soil $NO_3$-N concentrations, due to increased N mineralization from tillage. Summerfallow also increased fall soil $NO_3$-N concentrations from increased mineralization of organic matter and organic N (Campbell et al. 1994), which increased $NO_3$-N in snowmelt runoff the following spring (Nicholaichuk and Read 1978; McConkey et al. unpublished data). However, Cade-Menun et al. (2013) did not detect significant differences in $NO_3$-N concentrations in snowmelt runoff from cropland and pasture, although cropland fall soil $NO_3$-N concentrations were double those of pasture, and Liu et al (2014a) had no significant differences in soil $NO_3$-N, but found differences in runoff TDN and TN concentrations. Soil N status can also directly affect runoff N from soil loss due to erosion. Although this is generally lower in snowmelt runoff than rainfall runoff (Glozier et al. 2006), it can be increased by bare soil (Nicholaichuk and Read 1978), by repeated freeze-thaw cycles that reduce soil resistance to particle detachment, or by erosion from extremely high volumes of snowmelt runoff (Tanasienko et al. 2011).

### 4.4.2 Chemical Fertilizers

Fertilization is linked with soil $NO_3$ concentrations; thus, fertilizers can be seen as an indirect link to increased N loss by increasing the soil N that is available for loss. However, there is also a direct link to runoff loss from the fertilizers themselves. Fertilizers that are broadcast (surface applied) in the fall represent the greatest risk to N loss in snowmelt runoff, as there would be a concentrated source of soluble N at the soil surface. However, broadcast fertilizer applied to summerfallow also increased N loss (Nicholaichuk and Read 1978). In examining patterns of N loss over multiple years at an edge-of-field site in southern MB, Liu et al. (2013) identified a positive correlation between fall N application rate and FWMC of total N in snowmelt runoff, and suggested reducing N application rates to decrease runoff TN concentrations. Injection or incorporation of fertilizers in the spring, after snowmelt, is also recommended (Nicholaichuk and Read 1978; Tiessen et al. 2010).

### 4.4.3 Manure, Urine, and Dung

Losses of N in runoff from urine, dung, and manure will depend on the source and timing of application. Dung and urine can be deposited directly onto soil when cattle graze in fields from spring to fall, or during in-field winter bale-grazing. Manure that has been stock-piled or composted may also be applied by the producer, either as a fertilizer or as a soil amendment. Urea comprises 60–94% of total N in urine in cattle; the remainder is metabolic products from both the animal and from rumen microbes, such as allantoin, creatine, and amino acids, the majority of which degrade to $NH_4$-N. Specific concentrations of N compounds in urine will vary with dietary N and water intake (Bristow et al. 1992). The N concentration of dung will also vary with diet; however, fresh manure can have a concentration of 16.1 g N kg$^{-1}$ dung (Larney et al. 2006), composed primarily of organic N. A mature beef cow can produce 28 kg of feces and 9 kg of urine per day, that are 0.4 and 1.1% N, respectively (Bierman et al. 1999; Kelln et al. 2012). Urine may also lead to high losses of DON, from both the high concentration of DON in urine and the solubilizing effect it can have on soil organic matter (van Kessel et al. 2009). Nitrogen from dung and urine deposited during summer grazing can be lost through volatilization of $NH_4$-N, converted to $NO_3$-N by nitrification, immobilized by microbes or taken up by plants. However, dung and urine deposited in pastures during in-field winter bale-grazing will be retained, and more prone to loss in runoff. Composted or stockpiled manure will generally be lower in N than fresh manure, and much of the $NH_4$-N will have been volatilized or converted to $NO_3$-N (Larney et al. 2006).

Increases in total N concentration in snowmelt runoff have been observed following manure application (Casson et al. 2008), in summer-grazed pasture (Cade-Menun et al. 2013) and after in-field winter bale-grazing (Smith et al. 2011). However, the biggest difference compared to run-off from soils receiving chemical fertilizers is in the form of N in snowmelt runoff, with much higher concentrations of $NH_4$-N than $NO_3$-N from fields with manure and animals. Concentrations of $NH_4$-N were 200 times greater than $NO_3$-N in winter bale-grazing sites (Smith et al. 2011), and $NH_4$-N was 52% of total N in runoff from heavily manured sites, but 21% or less from sites receiving only chemical fertilizers (Casson et al. 2008). Loss of N was greatest in snowmelt runoff from one site to which manure was applied with no incorporation close to the date of soil freezing the previous fall, which increased both $NH_4$-N and $NO_3$-N in runoff (Casson et al. 2008). Although dung and urine can be sources of DON, Cade-Menun et al. (2013) reported more DON in runoff from cropland than pasture, especially early in the runoff event.

### 4.4.4 Vegetation and Plant Material

The leaching of N from vegetation and crop residues on the soil surface can represent an important source of soluble N loss with snowmelt (Timmons et al. 1970; Elliott 2013). This may be pronounced under conditions where the crop or vegetative material has been subject to freezing (Timmons et al.

1970; Miller et al. 1994; Elliott 2013) and the soil surface is frozen, preventing infiltration (Miller et al. 1994), as would be expected during snowmelt runoff on the Canadian Prairies.

The amount of N lost from over-wintering vegetation is affected by the tissue N concentration of the vegetation, the amount of biomass, the moisture content of the vegetation (influenced by growth stage), and plant species (Miller et al. 1994; Cermak et al. 2004; Elliott 2013). Vegetation that is still growing when temperatures drop and snow falls (e.g., perennial forage, riparian vegetation, newly planted winter crops, and cover crops) can contribute more water-soluble N than crop residues (i.e., stubble that has senesced), because these materials have higher tissue N concentrations, and their higher moisture contents increase cell lysing during freezing and subsequent thawing in spring (Timmons et al. 1970; Elliott 2013).

Liu et al. (2014a,b) investigated potential N loss by collecting crop residues in the fall, freezing them with water, and then thawing and analyzing the leachate. As expected, concentrations of total N were greater in leachate from perennial forage [alfalfa and timothy (*Phleum pretense* L.)] than from cropland (wheat and canola) stubble, and were higher in cropland stubble under NT than CT.

## 4.5  MANAGEMENT PRACTICES INFLUENCING N LOSS IN SNOWMELT RUNOFF

This section will review the available information about the effects of various agricultural management practices from the published studies of N loss in snowmelt runoff that are listed in Tables 4.3 and 4.4.

### 4.5.1  TILLAGE

Conservation tillage, including NT, has been widely adopted on the Canadian Prairies. Benefits include increased soil organic matter, decreased soil erosion due to residue cover, and improved water use efficiency (Carter 1994; Larney et al. 1994; Lal, 2000; Malhi et al. 2001).

Several studies have investigated the effects of tillage on N losses under snowmelt conditions in Canada. In MB, Tiessen et al. (2010) reported that NT reduced the flow weighted mean concentrations (FWMC) of TN, PN, TDN, and $NO_3 + NO_2$-N by 40, 20, 44, and 82%, respectively, when compared to more intensive CT. Corresponding annual nutrient loads experienced larger reductions and were 68, 63, 50, and 80%. In subsequent research on the same sites, an RT system with fall tillage every other year increased TDN loads in snowmelt runoff by 190% when compared with NT (Liu et al. 2014a). Liu et al. (2013) found that the number of tillage passes was linked to increases in TN FWMC. In AB, although soil $NO_3$-N concentrations were similar, $NO_3$-N FWMC were higher in CT than NT or RT (Casson et al. 2008). Sediment losses did not change significantly between RT and CT systems in the study by Liu et al. (2014a), but Tiessen et al. (2010) found annual sediment losses in runoff from NT were 61% of those in CT.

Tillage is expected to break soil aggregates and accelerate organic matter mineralization, increasing soil $NO_3$-N concentrations (Campbell et al. 2008). This effect was observed in the study by Tiessen et al. (2010), but not in that of Liu et al. (2014a). Deep fall tillage can also increase snowmelt infiltration, reducing snowmelt volumes and total nutrients in runoff (Casson et al. 2008). Reduced tillage and NT will leave residue on the soil surface, which may contribute N to runoff (Elliott 2013). This residue can trap snow on the soil surface, increasing nutrient export by increasing the length of the runoff event and the contact time of residues with runoff water (Casson et al. 2008; Cade-Menun et al. 2013; Liu et al. 2013, 2014a).

It is important to note that management practices can affect different nutrients in different ways. In the study by Liu et al. (2014a), an RT system (fall tillage every other year) was found to increase TDN losses (loads) in snowmelt runoff by 190% over a CT system; however, total P and total dissolved P losses were decreased by 42 and 56%, respectively. Thus, management practices that reduce N losses in snowmelt runoff must be balanced with those reducing P losses in snowmelt runoff, and vice versa.

### 4.5.2 Perennial Versus Annual Crops

Cover crops and riparian buffer zones are frequently used in other parts of Canada and elsewhere to reduce nutrient loss in cropland from rainfall runoff, especially particulate losses (Aronsson et al. 2016; Chase et al. 2016). Riparian buffer zones have been promoted as a BMP for the Canadian Prairies (Stewart et al. 2011). However, they are not widely used, and little is known about their effectiveness to control nutrient loss. In MB, research suggests that they can act as sinks or sources for P (Sheppard et al. 2006); however, this study did not include results for N. There have been no published studies to date directly evaluating the effectiveness of either cover crops or riparian buffer zones for N loss in snowmelt runoff from the Canadian Prairies. Erosion losses are less common with snowmelt than rainfall runoff, and there is increased risk of N loss from vegetation (Elliott 2013). The research comparing N loss in snowmelt from annual cropping with that from permanent pastures suggests that the effects can be mixed (Cade-Menun et al. 2013; Liu et al. 2014b). In SK, losses of $NH_4$-N and particulate N were significantly higher from summer grazed pastures than cropland, but losses of DON were higher from cropland than pasture (Cade-Menun et al. 2013). In MB, concentrations and loads of $NH_4$-N, and $NO_3$-N and concentrations of TDN and TN were lower in unfertilized perennial forage than cropland. Research from SK also suggests that permanent grasses may enhance infiltration compared to annual cropland, including when soil is frozen, likely due to the development of macropores (van der Kamp et al. 2003), but this study did not measure nutrient loss in snowmelt.

Further research is needed to determine the benefits and risks from using cover crops and riparian vegetation on the Canadian Prairies.

### 4.5.3 Timing of Fertilizer Applications/Manure Spreading

Losses of N increased with both chemical fertilizers and manure that were spread on soil in fall and not fully incorporated (Nicholaichuk and Read 1978; Casson et al. 2008), while fertilizers that were banded or injected in the spring after snowmelt runoff did not increase N in snowmelt runoff beyond their enrichment of soil $NO_3$-N (Tiessen et al. 2010). As such, injection or banding with seeding would minimize loss of chemical fertilizers. For manure, further research is needed to determine the optimal application time for soils where manure is not incorporated, although it appears that it would be best to avoid late fall applications. Additionally, N losses in snowmelt runoff could be reduced by applying only the N that will be used by the crop within the growing season, to reduce build-up of soil N concentrations (Casson et al. 2008; Li et al. 2011; Liu et al. 2013). The use of slow-release coated urea in the spring or fall may also reduce N losses (Kahkbazan et al. 2013) but this requires further research.

### 4.5.4 In-Field Winter Bale-Grazing

Recently, it has become common practice on the Canadian Prairies to feed cattle in fields during the winter, rather than confining them in corrals, with nearly half of all summer-pastured cows spending at least some portion of the winter out in pastures (Sheppard et al. 2016). With no need to use fuel to haul manure, this practice reduces production costs, and may retain nutrients in soil (Jungnitsch et al. 2011; Kelln et al. 2012). However, it has the potential to increase nutrient losses in snowmelt by depositing high concentrations of manure, dung, and feed (straw, hay) onto the surface of frozen soil prior to snowmelt (Smith et al. 2011; Jungnitsch et al. 2011; Kelln et al. 2012). Many studies across the Canadian Prairie provinces have investigated the effects of this practice on nutrient loss in snowmelt; to date, only the Smith et al. (2011) paper has been published. However, the results of studies from MB (Chen and Wilson unpublished data), southwest SK (Cade-Menun unpublished data), and AB (S. Reedyk, pers. comm.) are consistent with those of Smith et al. (2011) from central SK, showing that in-field winter bale-grazing significantly increases the concentration of $NH_4$-N in runoff compared to summer grazed pastures. Smith et al. (2011) used round bales, which can produce highly concentrated "hot spots" on the landscape of nutrients from urine, feces, and feed. An alternative practice is swath grazing, in which the feed, and thus feces and feed, is more evenly distributed across the pasture, reducing nutrient hot spots (Kelln et al. 2012). Research is needed to determine the effects of swath grazing on nutrient loss in snowmelt runoff. The alternative to extensive winter feeding in fields is more intensive, confined feeding. This does produce high concentrations in the corrals (Li et al. 2011). However, these can be managed as point sources of nutrients, such as with holding ponds for runoff water (Li et al. 2011). Extensive on-pasture feeding in winter shifts nutrient loss from a point source to a non-point source, which is harder to control.

### 4.5.5 Wetland Drainage and Restoration

As previously noted, the natural gently rolling terrain of the Canadian Prairies has a discontinuous and sparse drainage system in which numerous wetlands and potholes collect and store snowmelt runoff water, not contributing flow to the main rivers and streams in most years (Pomeroy et al. 2005). However, intensive modification of surface drainage for agriculture in some areas has channelized flow from potholes and wetlands to streams and rivers, increasing the transport of water and nutrients (Pomeroy et al. 2005; Corriveau et al. 2013). Considering this, reducing flow by restoring potholes and wetlands may reduce N loss in snowmelt runoff. None of the studies listed in Tables 4.3 and 4.4 tested wetland restoration. However, Li et al. (2011) included a holding pond below a beef cattle overwintering site in the suite of BMPs that were tested together, and reported that nutrient loss reductions from this BMP far exceeded the average reductions of all other treatment BMPs. Badiou et al. (2011) surveyed wetlands at 22 locations across the Canadian Prairies, including restored wetlands of various ages and reference wetlands that were never drained for agriculture. In the first year after restoration but not the second, water TN increased with wetland age (reference > long-term restored > newly restored). However, $N_2O$ fluxes were also higher over wetlands than surrounding upland soils. Thus, while wetland restoration may reduce transport of N across landscapes during snowmelt runoff, the net effect on greenhouse gas fluxes ($N_2O$ + $CH_4$ + $CO_2$) is complex (Badiou et al. 2011; Tangen et al. 2015) and further research is needed.

## 4.6 RECOMMENDATIONS AND FUTURE RESEARCH NEEDS

Nutrient loss in runoff from the Canadian Prairies is dominated by snowmelt runoff. As such, BMPs that work well in regions where runoff occurs primarily from rainfall may not be suitable (Tiessen et al. 2010). This is due to the unique characteristics of snowmelt runoff in that it primarily transports soluble nutrients, has low erosion potential, and may cause higher losses from over-winter vegetation due to freeze-thaw processes and low infiltration rates. Additional research is needed to develop BMPs for reducing N losses in snowmelt specific to the Canadian Prairie region. However, based on available literature, the following concepts deserve particular attention:

1. The number of studies focusing on N losses from snowmelt runoff that have been conducted in the Canadian Prairie Provinces is limited. Further research is needed, at plot, field, and watershed scales, to fully evaluate the factors controlling N loss from agriculture in this region. These studies should be conducted for multiple years, to cover a range of weather conditions.
2. Decreasing soluble $NO_3$-N concentrations left in the soil at the end of the growing season is likely to reduce losses with snowmelt the following spring. Balancing fertilizer and manure application rates and timing with crop removal will prevent accumulation in the soil. Reduced soluble $NO_3$-N can likely be accomplished by avoiding summer fallow seasons where increases in soil $NO_3$-N associated with mineralization peak in the fall,

leaving more N susceptible to loss in snowmelt runoff. If application rates in a given year are anticipated to exceed N removal with the crop, then fall-planted cover crops may offer an opportunity to retain soluble N; however, the potential for increases in nutrient losses from vegetation of cover crops with freeze-thaw is a topic deserving further research.

3. Avoid broadcasting N fertilizers and manures in the fall season without incorporation. Spring applications below the surface or with incorporation will reduce N losses in snowmelt, although incorporation with tillage may increase N loss in erosion in heavy summer rain events.

4. Coated urea is a slow-release fertilizer compared with uncoated urea. Research to date on the Canadian Prairies has focused on its effectiveness agronomically and economically (Kahkbazan et al. 2013), but not on its potential to reduce N loss in snowmelt runoff with fall application. Banding fall-applied $NH_4$-N fertilizers into the soil when soil temperature is $<7°C$ will minimize nitrification and thereby reduce soil $NO_3$-N concentrations (Tiessen et al. 2008), which should reduce N losses in runoff. Further research is needed with respect to timing, rate, and source of N fertilizers to reduce losses in snowmelt runoff.

5. In-field winter cattle feeding needs to be carefully conducted to minimize N losses. Care should be taken with respect to location of fields, to avoid direct losses to waterbodies. More research is needed about stocking rates, placement of feed (e.g., bale versus swath grazing), and appropriate siting to reduce runoff N losses.

6. Nitrogen loss from vegetation into snowmelt is still not fully understood. More research is needed at the field scale to determine management practices that limit loss, including the timing of haying and differences among perennial forage species. Information is also needed about potential N loss from below-ground plant material, especially from perennial plants.

7. Research is needed into the effectiveness of trapping snowmelt nutrients in restored wetlands or in holding ponds in combination with other management practices.

8. More research is needed about the effects of no-till and rotational tillage on N loss, balanced with the effects of this practice on P loss from agriculture.

It is clear that managers and producers must consider the benefits and potential consequences of each specific management practice when selecting strategies to reduce N loss from snowmelt runoff. It is clear that BMPs are not "one size fits all," but location-specific, and properties at each site should be considered when making recommendations. For example, the site-specific risk of erosion, soil characteristics, and balancing effects of practices on N and P losses, especially if contradictory, should be considered.

When considering the risk of nutrient loss to adjacent surface waters, risk assessment index approaches have been commonly used, although more frequently for P than N (Drewry et al. 2011). To the best of the authors' knowledge, an N index for snowmelt runoff on the Canadian Prairies has not been published, but would include a more tailored list of source and transport factors.

## 4.7 CONCLUSIONS

The Canadian Prairies are an important region for agriculture, both in Canada and globally. Many lakes in this region are limited by N as much or more than P; as such, movement of N from agricultural land to water is of concern, and management practices that minimize nutrient losses are essential. However, the Canadian Prairies differ from other regions in Canada because N losses are higher in snowmelt runoff than rainfall runoff. This means that management practices that are effective at minimizing N losses in other regions may not be appropriate on the Canadian Prairies. Given the limited number of studies reporting N losses in snowmelt runoff, more research is needed to fully understand the effects of agricultural management practices on N losses and forms of N in snowmelt runoff and to validate recommended best management practices.

## ABBREVIATIONS

| | |
|---|---|
| **AB** | Alberta |
| **BMPs** | best management practices |
| **BC** | British Columbia |
| **CT** | conventional tillage |
| **DON** | dissolved organic nitrogen |
| **FWMC** | flow weighted mean concentration |
| **K** | potassium |
| **MB** | Manitoba |
| **N** | nitrogen |
| **$N_2$** | nitrogen gas or dinitrogen |
| **$N_2O$** | nitrous oxide gas |
| **NGP** | northern Great Plains |
| **$NH_3$** | ammonia |
| **$NH_4$-N** | ammonium-nitrogen |
| **NO** | nitric oxide gas |
| **$NO_2$-N** | nitrite-nitrogen |
| **$NO_3$-N** | nitrate-nitrogen |
| **NT** | no-till or zero-tillage |
| **P** | phosphorus |
| **PN** | particulate nitrogen |
| **RT** | reduced tillage |
| **S** | sulfur |
| **SK** | Saskatchewan |
| **TDN** | total dissolved nitrogen |
| **TN** | total nitrogen |

## ACKNOWLEDGMENTS

We gratefully thank Sharon Reedyk and Dr. Barry Olsen for contributing information for this chapter.

## REFERENCES

Alvarez-Cobelas, M., D. G. Angeler, and S. Sánchez-Carrillo. 2008. Export of nitrogen from catchments: A worldwide analysis. *Environ. Pollut.* 156:261–269.

Anderson, D., and D. Cerkowniak. 2010. Soil formation in the Canadian Prairie region. *Prairie Soils Crops* 3:57–64.

Aronsson, H., E. M. Hansen, I. K. Thomsen, J. Liu, A. F. Øgaard, H. Känkänen, and B. Ulén. 2016. The ability of cover crops to reduce nitrogen and phosphorus losses from arable land in southern Scandinavia and Finland. *J. Soil Water Conserv.* 71:41–55.

Badiou, P., R. McDougal, D. Pennock, and B. Clark. 2011. Greenhouse gas emissions and carbon sequestration potential in restored wetlands of the Canadian prairie pothole region. *Wetl. Ecol. Manage.* 19:237–256.

Barica, J. 1993. Boundaries of ecological sustainability of prairie lakes and reservoirs. *Can. Water Resour. J.* 18:291–297.

Beaulieu, M. S. 2004. *Manure Management in Canada.* Vol. 1, No. 2. Catalogue #21-021-MIE-No.002. Minister of Industry, Government of Canada. 52 pp.

Bierman, S., J. E. Erickson, T. J. Klopfenstein, R. A. Stock, and D. H. Shain. 1999. Evaluation of nitrogen and organic matter balance in the feedlot as affected by level and source of dietary fibre. *J. Anim. Sci.* 77:1645–1653.

Bristow, A. W., D. C. Whitehead, and J. E. Cockburn. 1992. Nitrogenous constituents in the urine of cattle, sheep and goats. *J. Food Agr.* 59:387–394.

Cade-Menun, B. J., G. Bell, S. Baker-Ismail, Y. Fouli, K. Hodder, D. W. McMartin, C. Perez-Valdivia, and K. Wu. 2013. Nutrient loss from Saskatchewan cropland and pasture in spring snowmelt runoff. *Can. J. Soil Sci.* 93:445–458.

Campbell, C. A., G. P. Lafond, R. P. Zenter, and Y. W. Jame. 1994. Nitrate leaching in a Udic Haploboroll as influenced by fertilization and legumes. *J. Environ. Qual.* 23:195–201.

Campbell, C. A., R. P. Zentner, P. Basnyat, R. De Jong, R. Lemke, and R. Desjardins. 2008. Nitrogen mineralization under summer fallow and continuous wheat in the semiarid Canadian Prairie. *Can. J. Soil Sci.* 88:681–696.

Carpenter, S. R., N. F. Caraco, D. L. Correll, R. W. Howarth, A. N. Sharpley, and V. H. Smith. 1998. Nonpoint pollution of surface waters with phosphorus and nitrogen. *Ecol. Appl.* 8:559–568.

Carter, M. R. 1994. A review of conservation tillage strategies for humid temperate regions. *Soil Till Res* 31:289–301.

Casson, J. P., B. M. Olson, J. L. Little, and S. C. Nolan. 2008. Assessment of environmental sustainability in Alberta's Agricultural Watersheds Project Volume 4: Nitrogen loss in surface runoff. Alberta Agriculture and Rural Development. Lethbridge, AB, Canada.

Cermak, J. D., J. E. Gilley, B. Eghball, and B. J. Wienhold. 2004. Leaching and sorption of nitrogen and phosphorus by crop residue. *Trans. ASAE* 47:113–118.

Chase, J. W., G. A. Benoy, S. W. R. Hann, and J. M. Culp. 2016. Small differences in riparian vegetation significantly reduce land use impacts on stream flow and water quality in small agricultural watersheds. *J. Soil Water Conserv.* 71:194–205.

Coles, A. E., W. M. Appels, B. G. McConkey, and J. J. McDonnell. 2017. Climate change impacts on hillslope runoff on the northern Great Plains, 1962–2013. *J. Hydrol.* 550:538–548.

Corriveau, J., P. A. Chambers, and J. M. Culp. 2013. Seasonal variation in nutrient export along streams in the Northern Great Plains. *Water Air Soil Poll.* 224:1594–1610.

Delgado, J. A. 2002. Quantifying the loss mechanisms of nitrogen. *J. Soil Water Conserv.* 57:389–398.

Dodds, W. K., and V. H. Smith. 2016. Nitrogen, phosphorus and eutrophication in streams. *Inland Waters* 6:155–164.

Donald, D. B., M. J. Bogard, K. Finlay, L. Bunting, and P. R. Leavitt. 2013. Phytoplankton-specific response to enrichment of phosphorus-rich surface waters with ammonium, nitrate and urea. *PLoS ONE* 8:e53277.

Drewry, J. J., L. T. H. Newham, and R. S. B. Greene. 2011. Index models to evaluate the risk of phosphorus and nitrogen loss at catchment scales. *J Environ. Manage.* 92:639–649.

Dunne, T. 1983. Relation of field studies and modeling in the prediction of storm runoff. *J. Hydrol.* 65: 25–48.

Eickhout, B., A. F. Bouwman, and H. van Zeijts. 2006. The role of nitrogen in world food production and environmental sustainability. *Agr. Ecosys. Environ.* 116:4–14.

Elliott, J. 2013. Evaluating the potential contribution of vegetation as a nutrient source in snowmelt runoff. 2013. *Can. J. Soil Sci.* 93:435–443.

Elliott, J. A., A. J. Cessna, and C. R. Hilliard. 2001. Influence of tillage system on water quality and quantity in prairie pothole wetlands. *Can. Water Resour. J.* 26:165–181.

Fang, X., J. W. Pomeroy, C. J. Westbrook, X. Guo, A. G. Minke, and T. Brown. 2010. Prediction of snowmelt derived streamflow in a wetland dominated prairie basin, *Hydrol. Earth Syst. Sci.* 14:991–1006.

Finlay, K., A. Patoine, D. B. Donald, M.J. Bogard, and P. R. Leavitt. 2010. Experimental evidence that pollution with urea can degrade water quality in phosphorus-rich lakes of the Northern Great Plains. *Limnol. Oceanogr.* 55:1213–1230.

Glozier, N. E., J. A. Elliott, B. Holliday, J. Yarotski and B. Harker. 2006. Water quality characteristics and trends in a small agricultural watershed: South Tobacco Creek, Manitoba, 1992–2001. Environment Canada, Saskatoon, SK, Canada.

Gomi, T. R. C. Sidle, and J. S. Richardson. 2002. Understanding processes and downstream linkages of headwater systems. *BioScience* 52:905–916.

Granger, R. J., D. M. Gray, and G. E. Dyck. 1984. Snowmelt infiltration to frozen prairie soils. *Can. J. Earth Sci.* 21:669–677.

Grant, C. A., G. A. Peterson, and C. A. Campbell. 2002. Nutrient considerations for diversified cropping systems in the northern Great Plains. *Agron. J.* 94:186–198.

Gray, D. M., and P. G. Landine. 1988. An energy budget snowmelt model for the Canadian prairies. *Can. J. Earth Sci.* 25:1292–1303.

Hall, R. I., P. R. Leavitt, R. Quinlan, A. S. Dixit, and J. P. Smol. 1999. Effects of agriculture, urbanization and climate on water quality in the northern Great Plains. *Limnol. Oceanogr.* 44:739–756.

Havis, R. N., and E. E. Alberts. 1993. Nutrient leaching from field-decomposed corn and soybean residue under simulated rainfall. *Soil Sci. Soc. Am. J.* 57:211–218.

Jungnitsch, P. F., J. J. Schoenau, H. A. Lardner, and P. G. Jefferson. 2011. Winter feeding beef cattle on the western Canadian prairies: Impacts on soil nitrogen and phosphorus cycling and forage growth. *Agr. Ecosys. Environ.* 141:143–152.

Kelln, B., H. Lardner, J. Schoenau, and T. King. 2012. Effects of beef cow winter feeding systems, pen manure and compost on soil nitrogen and phosphorus amounts and distributions, soil density and crop biomass. *Nutr. Cycl. Agroecosys.* 92:183–194.

Khakbazan, M., C. A. Grant, G. Finlay, R. Wu, S. S. Malhi, F. Selles, G. W. Clayton, N. Z. Lupwayi, Y. K. Soon, and K. N. Harker. 2013. An economic study of controlled release urea and slip applications of nitrogen as compared with non-coated urea under conventional and reduced tillage management. *Can. J. Plant Sci.* 93:523–534.

Lal, R. 2000. Soil management in the developing countries. *Soil. Sci.* 165:57–72.

Larney, F. J., K. E. Buckley, X. Hao, and W. P. McCaughey. 2006. Fresh, stockpiled and composted beef cattle feedlot manure: Nutrient levels and mass balance estimates in Alberta and Manitoba. *J. Environ. Qual.* 35:1844–1854.

Larney, F. J., C. W. Lindwall, R. C. Lzaurralde, and A. P. Moulin. 1994. Tillage systems for soil and water conservation on the semi-arid Canadian Prairies. In *Conservation tillage in temperate agroecosystems,* ed. M.R. Carter, pp. 305–328. Boca Raton, FL: CRC Press.

Li, S., J. A. Elliott, K. H. D. Tiessen, J. Yarotski, D. A. Lobb, and D. N. Flaten. 2011. The effect of multiple beneficial management practices on hydrology and nutrient losses in a small watershed in the Canadian Prairies. *J. Environ. Qual.* 40:1627–1642.

Liu, K., J. A. Elliott, D. A. Lobb, D. N. Flaten, and J. Yarotski. 2013. Critical factors affecting field-scale losses of nitrogen and phosphorus in spring snowmelt runoff in the Canadian Prairies. *J. Environ. Hydrol.* 42:484–496.

Liu, K., J. A. Elliott, D. A. Lobb, D. N. Flaten, and J. Yarotski. 2014a. Conversion of conservation tillage to rotational tillage to reduce phosphorus losses during snowmelt runoff in the Canadian Prairies. *J. Environ. Qual.* 43:1679–1689.

Liu, K., J. A. Elliott, D. A. Lobb, D. N. Flaten, and J. Yarotski. 2014b. Nutrient and sediment losses in snowmelt runoff from perennial forage and annual cropland in the Canadian Prairies. *J. Environ. Qual.* 43:1644–1655.

Malhi, S. S., C. A. Grant, A. M. Johnston and K. S. Gill. 2001. Nitrogen fertilization management for no-till cereal production in the Canadian Great Plains: A review. *Soil Till. Res.* 60:101–122.

McGinn, S. M. 2010. Weather and climate patterns in Canada's prairie grasslands. *Arthropods of Canadian Grasslands* 1:105–119

Miller, M. H., E. G. Beauchamp, and J. D. Lauzon. 1994. Leaching of nitrogen and phosphorus from the biomass of three cover crop species. *J. Environ. Qual.* 23:267–272.

Nicholaichuk, W. 1967. Comparative watershed studies in southern Saskatchewan. *Trans. ASAE* 10:502–504.

Nicholaichuk, W., and D. Gray. 1986. Snow trapping and moisture infiltration enhancement. *Proceedings of the Moisture Management in Crop Production Conference.* Calgary Alberta.

Nicholaichuk, W., and D. W. L. Read. 1978. Nutrient runoff from fertilized and unfertilized fields in western Canada. *J. Environ. Qual.* 7:542–544.

Padbury, G., S. Waltman, J. Caprio, G. Coen, S. McGinn, D. Mortensen, G. Nielsen, and R. Sinclair. 2002. Agroecosystems and land resources of the Northern Great Plains. *Agron. J.* 94:251–261.

Panuska, J. C., and K. G. Karthikeyan. 2010. Phosphorus and organic matter enrichment in snowmelt and rainfall-runoff from three corn management systems. *Geoderma* 154:253–260.

Pham, S. V., P. R. Leavitt, S. McGowan, and P. Peres-Neto. 2008. Spatial variability of climate and land-use effects on lakes of the northern Great Plains. *Limnol. Oceanogr.* 53:728–742.

Pomeroy, J. W., D. de Boer, and L. W. Matz. 2005. Hydrology and water resources of Saskatchewan. Centre for Hydrology Report 1. University of Saskatchewan, Saskatoon. 25 pp.

Rattan, K. J., J. C. Corriveau, R. B. Brua, J. M. Culp, A. G. Yates, and P. A. Chambers. 2017. Quantifying seasonal variation in total phosphorus and nitrogen from prairie streams in the Red River Basin, Manitoba Canada. *Sci. Total Environ.* 575:649–659.

Sharpley, A. N., S. J. Smith, and J. W. Naney. 1987. Environmental impact of agricultural nitrogen and phosphorus use. *J. Agr. Food Chem.* 35:812–817.

Sheppard, S. C., M. I. Sheppard, J. Long, B. Sanipelli, and J. Tait. 2006. Runoff phosphorus retention in vegetated field margins on flat landscapes. *Can. J. Soil Sci.* 86:871–884.

Sheppard, S. C., S. Bittman, D. Macdonald, B. D. Amiro, and K. H. Ominski. 2016. Changes in land, feed and manure management practices on beef operations in Canada between 2005 and 2011. *Can. J. Anim. Sci.* 96:252–265.

Smith, A., J. Schoenau, H. A. Lardner, and J. Elliott. 2011. Nutrient export in run-off from an in-field cattle overwintering site in East-Central Saskatchewan. *Water Sci. Technol.* 64:1790–1795.

Statistics Canada. 2016. Results of the 2011 Canadian Census of Agriculture. http://www5.statcan.gc.ca /cansim/a03?lang=eng&pattern=004-0200..004-0242&p2=31.

Stewart, A., S. Reedyk, B. Franz, K. Fomradas, C. Hilliard, and S. Hall. 2011. A planning tool for design and location of vegetative buffers on watercourses in the Canadian prairies. *J. Soil Water Conserv.* 66 (4):97A–103A.

Tanasienko, A. A., O. P. Yakutina, and A. S. Chumbaev. 2011. Effect of snow among on runoff, soil loss and suspended sediments during periods of snowmelt on southern West Siberia. *Catena* 87:445–451.

Tangen, B. A., R. G. Finocchiaro, and R. A. Gleason. 2015. Effects of land use on greenhouse gas fluxes and soil properties of wetland catchments in the Prairie Pothole Region of North America. *Sci. Total Environ.* 533:391–409.

Tiessen, K. H. D., D. N. Flaten, P. R. Bullock, C. A. Grant, R. E. Karamanos, D. L. Burton, and M. H. Entz. 2008. Interactive effects of landscape position and time of application on the response of spring wheat to fall-banded urea. *Agron. J.* 100:557–563.

Tiessen, K. H. D., J. A. Elliot, J. Yarotski, D. A. Lobb, D. N. Flaten, and N. E. Glozier. 2010. Conventional and conservation tillage: Influence on seasonal runoff, sediment, and nutrient losses in the Canadian Prairies. *J. Environ. Qual.* 39:964–980.

Timmons, D. R., R. F. Holt, and J. J. Latterell. 1970. Leaching of crop residues as a source of nutrients in surface runoff water. *Water Resour. Res.* 6:1367–1375.

United States Environmental Protection Agency (USEPA). 2015. Preventing eutrophication: Scientific support for dual nutrient criteria. EPA 820-S015-001. https://www.epa.gov/sites/production/files/documents /nandpfactsheet.pdf.

van der Kamp, G., M. Hayashi, and D. Gallén. 2003. Comparing the hydrology of grassed and cultivated catchments in the semi-arid Canadian prairies. *Hydrol. Process.* 15:559–575.

van Kessel, C., T. Clough, and J. W. van Groenigen. 2009. Dissolved organic nitrogen: An overlooked pathway of nitrogen loss from agricultural systems? *J. Environ. Qual.* 38:393–401.

Zuzel, J. F., and J. L. Pikul Jr. 1987. Infiltration into a seasonally frozen agricultural soil. *J. Soil Water Conserv.* 42:447–450.

# 5 Denitrification in Soil

*Mark S. Coyne*

## CONTENTS

## 5.1    WHAT IS DENITRIFICATION?

Denitrification loosely refers to gaseous nitrogen loss from soil. In its strictest sense, denitrification is "biological reduction of nitrogen oxides (usually nitrate and nitrite) to molecular nitrogen or nitrogen oxides with a lower oxidation state of nitrogen by bacterial activity. The nitrogen oxides are used by the bacteria as terminal electron acceptors in place of oxygen during anaerobic or microaerophilic respiratory metabolism" (Soil Science Glossary Terms Committee 2008). Denitrification's defining characteristics are transforming soluble inorganic nitrogen to gaseous inorganic nitrogen and using this transformation to generate energy. We know it is not an exclusively prokaryotic process. We also know it is not exclusively anaerobic or microaerophilic, although these are the conditions in which denitrification most often occurs.

Denitrification places a central role in the nitrogen cycle (Figure 5.1) closing the gap between the $N_2$ fixation to organic N, organic N mineralization to ammonium $\left(NH_4^+\right)$, and $NH_4^+$ oxidation to nitrite $\left(NO_2^-\right)$ and ultimately nitrate $\left(NO_3^-\right)$. Without denitrification, the atmosphere would be stripped of $N_2$ over geologic time. It presents a "wicked problem" for agriculture and the environment (Doering 2017). It depletes plant available $NO_3^-$ from soil and thereby decreases fertilizer use efficiency. But it also removes $NO_3^-$ from surface and groundwater thereby protecting human health, improving water quality, and minimizing eutrophication. However, this often produces radiatively significant gases like nitric oxide (NO) and nitrous oxide ($N_2O$), which also deplete stratospheric ozone.

### 5.1.1    GASEOUS NITROGEN LOSS OTHER THAN BY DENITRIFICATION

It is worth briefly noting biological and abiological reactions that also contribute to gaseous N loss from soil because in many cases these processes coexist with denitrification. In much prior research,

**The nitrogen cycle**

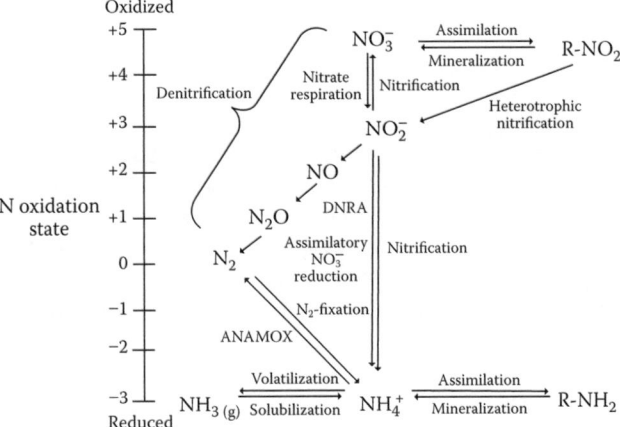

**FIGURE 5.1** The nitrogen cycle and denitrification's role in it. In terrestrial environments denitrification closes the gap between the product of $NH_4^+$ oxidation (nitrification) and $NH_4^+$ formation ($N_2$ fixation). Nitrogen oxides are sequentially reduced, typically beginning with $NO_3^-$, to lower oxidation states, which ultimately produces $N_2$ gas.

their activity would not have been distinguished from denitrification—but they are not denitrifying processes.

#### 5.1.1.1 Anammox (Anaerobic Ammonium Oxidation)

Anammox is a strictly anaerobic microbial process performed by members of the order *Planctomycetales* in which $NH_4^+$ oxidation can occur because $NO_2^-$ is used as an electron sink (Mulder et al. 1995). The final product is $N_2$ (van de Graaf et al. 1995) (Equation 5.1).

$$5NH_4^+ + 3NO_3^- \rightarrow 4N_2 + 9H_2O + 2H^+ \tag{5.1}$$

$$\Delta G^{\circ\prime} = -1483.5 \text{ kJ per reaction}$$

In this process $NO_2^-$ is not used as a terminal electron acceptor for respiratory metabolism. The Planctomycetes generate their energy from $NH_4^+$ oxidation through other mechanisms (van de Graaf et al. 1997).

Anammox occurs in marine environments (e.g., Kuypers et al. 2003) and estuaries (e.g., Crowe et al. 2012) where it competes with denitrification for $NO_3^-$. It was originally observed in wastewater treatment facilities (Schmid et al. 2005). Here, anammox is regarded as a more efficient means to remove residual inorganic N from treated water than conventional nitrification/denitrification treatments (e.g., Vlaeminck et al. 2010). It is likely that the organisms performing anammox can be found in soil, but because they are extremely slow growing few isolates have been cultured to taxonomic purity (e.g., *Kuenenia stuttgartiensis*). Consequently, their identification has almost exclusively been by molecular approaches (Schmid et al. 2005). This shows that a diverse group of Planctomycetes related to known anammox organisms occur in soil environments (Buckley et al. 2006; Humbert et al. 2010). It is not known if these bacteria actually perform anammox *in situ*. Due to environmental constraints, it is likely that anammox in terrestrial soils is inconsequential, but this supposition requires testing (Humbert et al. 2010).

#### 5.1.1.2 Chemodenitrification

Chemodenitrification refers to "nonbiological processes that produce gaseous nitrogen (molecular nitrogen or an oxide of nitrogen) or by chemical reactions involving nitrite" (Soil Science Glossary

Terms Committee 2008). Chemodenitrification is classically associated with $NO_2^-$ reduction by metals in acidic environments and gaseous N production by the Van Slyke reaction (Equation 5.2).

$$HNO_2 + RNH_2 \rightarrow N_2 + ROH + H_2O \tag{5.2}$$

where R-NH$_2$ represents any of a number of organic N compounds.

There are N transformations that also occur in circum-neutral soils. For example, Fe(II) can react abiologically with $NO_2^-$ to produce Fe(III) and $N_2O$ (Moraghan and Buresh 1977). Fe-containing minerals such as siderite (Rakshit et al. 2008) and magnetite (Dhakhal et al. 2013) can also reduce $NO_2^-$. Rakshit et al. (2016) found that Fe(II) associated with kaolinite reduced $NO_2^-$ with almost 80% recovery as $N_2O$ during a 50-hour incubation, but the $NO_2^-$ reduction rate was much slower than by denitrification.

### 5.1.1.3 Co-Denitrification

Co-denitrification refers to a process in which at least one nitrogen of the gaseous products of $NO_2^-$ reduction comes from a compound other than $NO_2^-$ (e.g., $N_3^-$ or $NH_2OH$). It is thought that dissimilatory $NO_2^-$ reductase catalyzes a nitrosyl transfer (nitrosation) from $NO_2^-$ to N-nucleophiles (Tanimoto et al. 1992) producing $N_2O$ and ultimately $N_2$. The N-donating compounds that are "denitrified" by such a system are utilized by $NO_3^-$ or $NO_2^-$ reducing systems; they are not themselves able to induce denitrification. The process occurs in bacteria (Garber and Hollocher 1982; Kim and Hollocher 1984) and fungi (Tanimoto et al. 1992) and can occur simultaneously with respiratory denitrification as determined by $^{15}$N and $^{18}$O labeling studies, and using inhibitors such as actinomycin and rotenone that prevent respiratory electron transfer (Garber and Hollocher 1982). *In vitro* and *in vivo* co-denitrification appears to account for only a few percent of the bacterial denitrification flux (Kim and Hollocher 1984) but in fungi it represents a much greater potential fate for $NO_2^-$ (Tanimoto et al. 1992). Laughlin and Stevens (2002) concluded that 92% of labeled $N_2$ in a grassland system was due to co-denitrification by fungi.

### 5.1.1.4 Other Losses

Nitrous oxide production can occur because of $NO_3^-$ respiration (Kaspar 1982; Smith 1982, 1983) and because of dissimilatory $NO_3^-$ reduction to $NH_4^+$ (DNRA) (Kaspar and Tiedje 1981). When $NO_2^-$ is added to these systems slight increases in $N_2O$ evolution occur. Although the mechanisms are not fully understood, it appears to be due to $NO_2^-$ reduction by $NO_3^-$ reductase (Smith 1983) in which the reaction serves as a detoxification mechanism, the bacteria being sensitive to $NO_2^-$ (Kaspar 1982). The $N_2O$ is not further reduced and the bacteria appear to gain little or no energy from $NO_2^-$ reduction.

## 5.2 HISTORICAL CONTEXT OF DENITRIFICATION AND DENITRIFICATION RESEARCH IN SOIL

That inorganic nitrogen, including $NO_3^-$, was lost from soil was well understood by early agricultural scientists even if they did not fully understand the mechanisms. Between 1867 and 1920 much of what we know about denitrification in soil was investigated. Payne (1981) provides an excellent review of this history. Between 1867 and 1875 $NO_3^-$ disappearance and gas liberation from agricultural waters was systematically investigated. The loss was associated with biological activity. The term "denitrification" was coined by Gayon and Dupetit in 1882 and its biological basis confirmed. By 1886 the first pure denitrifying cultures had been isolated, its respiratory nature identified, and the release of other gaseous N products noted. By 1888 denitrification was being used as a taxonomic property. While useful at the time, it turns out to be a fundamentally flawed approach based on molecular phylogeny. Regulation of denitrification by carbon and oxygen was well understood by 1895 and the first methods of its inhibition tested. By 1920 the pathway and substrates of denitrification were clear, there were an expanding variety of known denitrifiers, and their ubiquity in soils was appreciated.

Lack of good, simple analytical techniques to evaluate gaseous intermediates slowed denitrification research from 1920 to 1970. Mass balance studies seemed to indicate that denitrification was a major source of N loss in soil. Nõmmik (1956) conducted pioneering studies to determine soil factors controlling denitrification. There was also progress on pathways based on mutagenesis studies.

The subsequent 50-year period has been thoroughly dynamic. A major breakthrough in denitrification research occurred with the discovery (Federova et al. 1973) and application (Balderston and Payne 1976; Yoshinari and Knowles 1976) of the acetylene block technique to investigate field and lab denitrification. In combination with photoacoustic detection and gas chromatographic detection by thermal conductivity (TCD) and electron capture (ECD) (Kaspar and Tiedje 1980) this enabled observation of denitrification in nitrifiers, $N_2$-fixers, and fungi, in diverse terrestrial environments.

Detection methods for process were soon followed in the 1980s by immunological studies, molecular cloning, and physical mapping of denitrification genes to enable definitive establishment of the enzymes and pathway of denitrification. Shortly thereafter, molecular probes, particular for two types of dissimilatory $NO_2^-$ reductases, were developed to detect and evaluate the presence and diversity of microbes with the capacity to denitrify. Current research is increasingly recognizing the contribution nitrifiers and fungi (Mothapo et al. 2013) make to trace gas evolution from terrestrial soils via denitrification, and the most recent advances in isotopomer analysis (e.g., Sutka et al. 2006) are helping to distinguish the actual sources of those gases.

## 5.3  BIOCHEMISTRY AND GENETICS OF DENITRIFICATION

To fully appreciate denitrification in complex soil environments—and its control—one ought to be familiar with the biochemistry, enzymology, and genetic regulation of the process. These concepts will be briefly addressed in this section. For a more detailed review, see Zumft (1997).

### 5.3.1  PATHWAY

The pathway of denitrification is simple. It has four reduction steps. Each step is catalyzed by a distinct enzyme (Equation 5.3):

$$\overset{\text{Nar}}{NO_3^-} \rightarrow \overset{\text{Nir}}{NO_2^-} \rightarrow \overset{\text{Nor}}{NO} \rightarrow \overset{\text{Nos}}{N_2O} \rightarrow N_2 \tag{5.3}$$

The pathway can be truncated in a denitrifying organism, meaning it cannot perform all steps. The most common truncation occurs because organisms cannot further reduce $N_2O$ to $N_2$. Denitrifying fungi, for example, are generally unable to reduce $N_2O$ to $N_2$. Figure 5.2 shows a comparison of the pathways and enzymes involved.

#### 5.3.1.1  Denitrification Enzymes

##### 5.3.1.1.1  Nitrate Reductase (Nar)

Dissimilatory $NO_3^-$ reductases in bacteria are membrane-bound molybdoproteins. Some bacteria also have a periplasmic $NO_3^-$ reductase (Nap) (Philippot et al. 2007). They are distinct from assimilatory $NO_3^-$ reductases in being regulated by $O_2$ rather than $NH_4^+$ (Tiedje 1988). In anaerobic conditions and the absence of $NH_4^+$ assimilatory $NO_3^-$ reductases in bacteria can initiate denitrification if dissimilatory enzymes are absent or inactivated. The $NO_3^-$ reductase of archaea is analogous to its bacterial counterparts, although it appears to have a slightly different position in the membrane (Zumft and Kroneck 2007). Some fungi (e.g., *Fusarium*) have a relatively unique dissimilatory

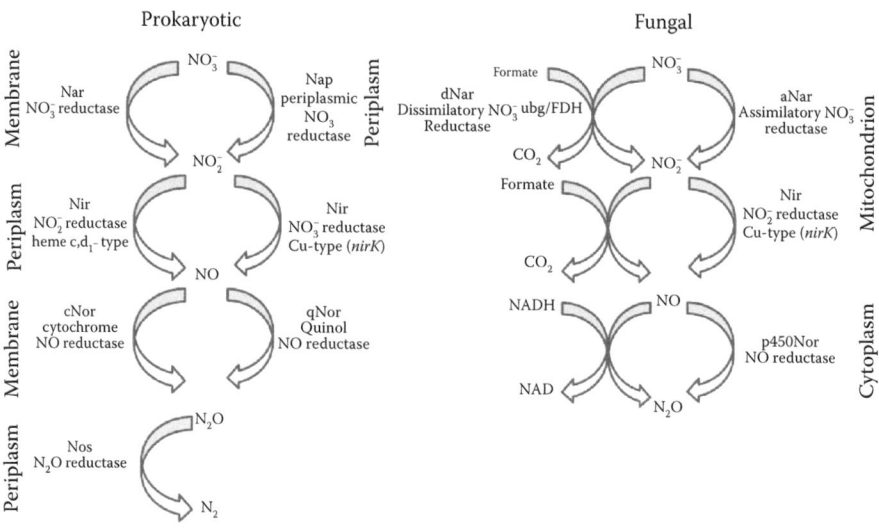

**FIGURE 5.2**   Spatial distribution of denitrifying enzymes in prokaryotic and fungal cells (adapted from Ye and Tiedje 1994; Philippot et al. 2007; Ma et al. 2008; Mothapo et al. 2015). N-oxides cycle through denitrifying enzymes bound to or associated with the cell membrane and periplasm in prokaryotes. Intermediate gases in the periplasmic space can be readily stripped by mass flow if the N-oxide flux through the enzyme system is impeded. Nar and Nap refer to a membrane bound and periplasmic $NO_3^-$ reductase, respectively. Nir refers to two distinct types of $NO_2^-$ reductase. cNor and qNor refer to two distinct types of NO reductase that use cytochromes (c) or quinol (q) as the direct electron donors to the enzyme. N-oxides cycle through denitrifying enzymes bound to or associated with the mitochondria and active during hypoxic conditions in fungi. aNar refers to assimilatory $NO_3^-$ reductase while dNar refers to a dissimilatory reductase linked to a ubiquinone formate dehydrogenase (FDH). NirK refers to Cu-type $NO_2^-$ reductase. p450Nor refers to NO reductase. (Not to scale.)

$NO_3^-$ reductase (dNar) that employs a ubiquinol-formate dehydrogenase couple to transfer electrons to $NO_3^-$ reductase (Ma et al. 2008; Mothapo et al. 2015). Most fungi also have an assimilatory $NO_3^-$ reductase (aNAr) that forms $NO_2^-$ to initiate the denitrification process (Mothapo et al. 2015). In fungi, these enzymes are associated with the mitochondria.

### 5.3.1.1.2   *Nitrite Reductase (Nir)*

Nitrite reduction is the defining step in denitrification and produces the gaseous intermediate NO. Nitrite reductase is located on the periplasmic side of the cell membrane. The reduction is catalyzed by two evolutionarily distinct $NO_2^-$ reductases; one a heme $c,d_1$- type (NirS) and the other a metallo protein with Cu at the reaction center (NirK). The Cu-type appears to be more prevalent in soil. Bacteria have one or the other of these $NO_2^-$ reductases, as do archaea (Zumft and Kroneck 2007). Fungal Nir is of the Cu-type (NirK) and associated with the mitochondria.

### 5.3.1.1.3   *Nitric Oxide Reductase (Nor)*

Bacterial NO reductases are also of two types. Both are heme Cu oxidases but receive their electrons from different sources. cNor receives its electrons from cytochrome $c$ or pseudoazurin (Philippot et al. 2007). qNor receives its electrons from quinol. Archaeal NO reductases appear to be of the qNor type (Zumft and Kroneck 2007). Both enzymes are membrane-bound.

   Fungal NO reductases are quite different from their prokaryotic counterparts. They are members of the cytochrome *p450* super family associated with detoxification mechanisms (Shoun and Tanimoto 1991). In fungi they are either loosely associated with the mitochondria or cytoplasmic. Relative to the NO reductase in bacteria and archaea, the p450Nor in fungi is five times faster, making the $NO_3^-/NO_2^-$ reduction steps the rate-limiting process (Mothapo et al. 2015).

*5.3.1.1.4   Nitrous Oxide Reductase (Nos)*

Bacteria and archaea have a Cu-based metallo-protein, $N_2O$ reductase (Nos), that reduces $N_2O$ to $N_2$—the final product of denitrification. This enzyme is also periplasmic. Some of the archaea appear to have $N_2O$ reductases sufficiently distinct from bacterial versions that they are not apparent by homology search (Zumft and Kroneck 2007). Fungi appear to lack a distinct $N_2O$ reductase and accumulate $N_2O$ as the terminal product of denitrification. If fungi produce $N_2$ it is believed to be the result of co-denitrification, a non-respiratory process.

### 5.3.1.2   Inhibitors

Each of the individual enzymes involved in denitrification has potential inhibitors. Tungsten (W), for example, can replace the molybdenum (Mo) in $NO_3^-$ reductase and render the enzyme inoperative. Cyanide (CN) is one of many inhibitors of electron transfer by cytochromes and is effective against NirS while the Cu-chelator, Diethyldithiocarbamate (DDC), is a potent inhibitor of NirK (Shapleigh and Payne 1985; Coyne and Tiedje 1990a). Cyanide (Heiss et al. 1989) and antimycin A (Urata and Satoh 1985) can specifically inhibit NO reductase as does DDC (Heiss et al. 1989). The most widely used and effective inhibitor of $N_2O$ reductase is acetylene ($C_2H_2$), which can completely inhibit $N_2O$ reductase in concentrations as low as 0.1 kPa (Knowles 1990).

One of the features that makes denitrification a "wicked problem" in soil is the lack of specific inhibitors for the processes that target denitrifiers and denitrifiers alone. There are no known commercial inhibitors of denitrification and the current inhibitors are practically useless for field application; not that one would want to do so. Inhibiting $NO_3^-$ reductase allows $NO_3^-$ accumulation and promotes leaching of a soluble, mobile, groundwater pollutant. Inhibiting $NO_2^-$ reductase causes $NO_2^-$ to accumulate, which is toxic to plants and people (Pearce et al. 1998). Nitric oxide is already chemically reactive and its accumulation would only lead to $NO_2^-$ formation (if reoxidized by soil) or release into the atmosphere where it catalyzes ozone ($O_3$) destruction (Conrad 1990; Haden et al. 2016). Inhibiting $N_2O$ reductase would cause further release of $N_2O$ into the atmosphere. Nitrous oxide is a potent greenhouse gas with $O_3$ destroying features and 300 times the contribution to global warming as an equivalent amount of $CO_2$ (Lassey and Harvey 2007). Management of denitrification in soil must therefore focus less on its inhibition than on driving the process to completion and the production of $N_2$ as the terminal product.

### 5.3.1.3   Enzyme Induction and Persistence

Denitrification enzymes in prokaryotes are not constitutive. Too little is known about fungal denitrification at present to make the same assessment. For isolates growing aerobically, there is normally a lag period after transition to denitrifying growth, which represents the synthesis of new enzymes (Smith and Tiedje 1979a). Within hours, however, the complete denitrification pathway can be fully induced (Sexstone et al. 1985a). Hypoxia or anaerobiosis, and to a lesser extent the presence of N-oxides, primarily triggers full induction of denitrification enzymes. Once synthesized, the denitrification enzymes persist for long periods (Smith and Parsons 1985) subject only to inherent cellular catabolism and anabolism. This persistence in soil, even in dry aerobic conditions, means the capacity for denitrification to commence immediately upon rewetting is ever present (Firestone et al. 1980; Sexstone et al. 1985a).

Three explanations have been offered for this persistence (Tiedje 1988):

1. The indigenous denitrifiers are physically associated with the soil matrix and protected during drying events in contrast to newly added (allochthonous) denitrifiers.
2. There is a second denitrifier population in soil that is slow growing, not easily isolated or culturable, and persists in a desiccating environment.
3. Denitrification enzymes in aged indigenous organism or cell fragments are stabilized and remain active.

Of these explanations, the first is most plausible. The second has some plausibility by analogy to the numerical dominance and activity of slow growing ammonia-oxidizing archaea in soil. The third explanation is doubtful. Unlike hydrolytic extracellular enzymes such as urease, phosphatase, and sulfatase, denitrifying enzymes require a means of electron transfer through membrane components or other carriers that is lost when cells are disrupted.

Likewise, despite the propensity of dissimilatory $NO_2^-$ reduction to $NH_4^+$ (DNRA) to be more competitive than denitrification in persistently anaerobic environments, high denitrification rates can be observed. Obligate anaerobes (e.g., *Shewanella denitrificans*) are known. So, this phenomenon has been associated with the potential existence of as-yet uncultured denitrifiers contributing to this activity (Tiedje 1988).

## 5.3.2 GENETICS AND MOLECULAR BIOLOGY

### 5.3.2.1 Denitrification Genes

At least 60 genes are involved in denitrification ranging from the genes for structural proteins to those affecting regulation (i.e., electron flow) and other accessory components (i.e., insertion in membranes, assembly of metals in active sites) (Zumft 1997). These genes may be found chromosomally or on plasmids. They may be tightly linked, as in the case of pseudomonads, or scattered (Viebrock and Zumft 1987). The principle genes and their role are identified in Table 5.1.

### 5.3.2.2 Use of Denitrification Genes as Molecular Probes

One of the great values of a thorough knowledge of denitrification genetics is the ability to establish pathway relationships. Viebrock and Zumft (1987) used transposon Tn5 mutagenesis to create Nos⁻ mutants and identify the relevance of $N_2O$ reductase in the denitrification pathway. Later, Zumft et al. (1988) used transposon Tn5 mutagenesis of cytochrome $cd_1$-$NO_2^-$ reductase to demonstrate NO reductase activity was preserved, and therefore represented a distinct enzymatic step in the denitrification pathway. Another value of genetic information is its use to specifically detect and identify denitrifiers in soil habitats, which has enormous value for studying community diversity.

Because the characteristic enzyme of denitrification is $NO_2^-$ reductase, molecular probes for the *nirK* and *nirS* gene have been used most often to characterize denitrifying populations. This has revealed that, although denitrifiers with NirS are quite frequently isolated from soil (Gamble et al. 1977; Coyne et al. 1989) NirK is probably more abundant in denitrifiers (Braker et al. 2000). Any attempt to enumerate denitrifiers in the soil by molecular techniques would need to use probes

**TABLE 5.1**
**Significant Genes in Denitrification**

| Gene | Role | Reference |
|---|---|---|
| narG | Large subunit of Nar | Zumft 1997 |
| narH | Small subunit of Nar | Zumft 1997 |
| napA | Periplasmic Nar | Philippot et al. 2007 |
| nirK | Cu-type Nir | Long et al. 2015, Wei et al. 2015 |
| nirS | Heme-type Nir | Braker et al. 2000; 2001 |
| cnorB | c-type Nor | Philippot et al. 2007 |
| qnorZ | q-type Nor of Archaea | Zumft and Kroneck 2007 |
| x | Plasmid-bound Nor (*Ralstonia eutropha*) | Zumft 1997 |
| p450nor | Fungal Nor | Higgins et al. 2016, Novinscak et al. 2016, |
| nosZ | Nos | Zumft 1997 |
| nosDFY | Accessory genes | Zumft and Kroneck 2007 |

for both genes. The *nirK* gene proved useful in identifying denitrification capacity in soil fungi (Mothapo et al. 2015). But efforts to distinctly identify fungal denitrifiers based on *nirK* have been hampered by similarity to bacterial counterparts. Consequently, more recent studies examining fungal denitrification have employed more fungal specific *nirK* probes (Long et al. 2015; Wei et al. 2015). In addition, the distinct *p450nor* of fungi has been used as the basis for a discriminating gene probe (Higgins et al. 2016; Novinscak et al. 2016).

Molecular probes are used less often to quantify denitrifiers than to characterize their diversity in various environments, although both approaches have been used. As with most molecular probe approaches, it has revealed that the true population of denitrifiers is woefully underrepresented (numerically and phylogenetically) by culturable isolates.

## 5.4 DENITRIFYING ORGANISMS AND PHYLOGENY

What makes a microorganism a denitrifier? Tiedje (1988; 1994) proposed the following criteria:

1. Growth yield is enhanced proportionally to the amount of N-oxide present; add more $NO_3^-$ and you get more cells.
2. Yield increase is greater than if the N-oxide was simply used as an electron sink.
3. 80% or more of the $NO_3^-$ or $NO_2^-$ is converted to $N_2O$ or $N_2$.
4. Nearly stoichiometric accumulation of $N_2O$ from $NO_3^-$ or $NO_2^-$ in the presence of acetylene ($C_2H_2$).
5. A rapid rate of $NO_3^-$ reduction (e.g. > 10 nmol N gas mg$^{-1}$ protein min$^{-1}$).
6. Presence of one of the two types of dissimilatory $NO_2^-$ reductases (i.e., the cytochrome $cd_1$ or Cu type) or one of the other denitrification specific enzymes or genes.

Enzyme extraction and immunological characterization are not obsolete techniques for studying denitrifiers. But relative to molecular methodology they are far more cumbersome and tedious (and in the case of antibodies—less humane).

These criteria were developed for bacteria. Their application to archaea is problematic given the difficulty in cultivating many archaea. Likewise, their application to fungal denitrifiers may have to be modified to reflect the capacity for fermentative growth, and relatively low rates of gas production despite the presence of denitrification specific genes.

### 5.4.1 BACTERIA

Denitrifying bacteria are distributed among a wide variety of physiological and morphological groups (Table 5.2).

#### 5.4.1.1 Numerical Estimates

The variability of physiological characteristics means that cultural enumeration of denitrifiers inherently underestimates the true population (Martin et al. 1988). In addition to denitrifiers that will not grow in the culture media (viable but not active) there are denitrifiers that cannot grow in the culture media. Because the most common method of enumerating denitrifiers—the Most Probable Number (MPN) technique—relies on dilution to extinction and subsequent growth of the remaining population to bring about a discernible change in media, even if the cells are present and active, if they cannot grow they are unlikely to produce a measurable response in the time allowed for incubation. This would be particularly true of lithotrophic, carboxidotrophic, or phototropic denitrifiers present in soils where enumeration with nutrient broth (a C-rich media specific for heterotrophs) is used.

Quantification of denitrifiers by qPCR avoids the need for culture, but suffers from its own biases that include: efficiency of DNA extraction, selectivity of primer sets, inclusiveness of molecular

**TABLE 5.2**
**Short List of Genera Containing Denitrifying Bacteria (Some genera fit in more than one category)**

| Organotrophs | | Phototrophs |
|---|---|---|
| Aerobic heterotrophs | Fermentative | |
| (nonspecific) | *Chromobacterium* | *Rhodobacter* |
| *Achromobacter* | *Empedobacter* | *Rhodoplanes* |
| *Acidovorax* | *Janthinobacterium* | *Rhodopseudomonas* |
| *Acinetobacter* | *Wolinella* | *Roseobacter* |
| *Agrobacterium* | | |
| *Alcaligenes* | | |
| *Alteromonas* | Gliding | Lithotrophs |
| *Bacillus* | *Cytophaga* | $H_2$ oxidizing |
| *Brachymonas* | *Flexibacter* | *Paracoccus* |
| *Campylobacter* | | *Ralstonia* |
| *Comamonas* | Magnetotactic | |
| *Corynebacterium* | *Aquaspirillum* | S oxidizing |
| *Eikenella* | | *Beggiatoa* |
| *Flavobacterium* | | *Thiobacillus* |
| *Gluconobacter* | $N_2$ fixing | *Thiomicrospora* |
| *Herbaspirillum* | *Aquaspirillum* | *Thioploca* |
| *Jonesia* | *Azoarcus* | *Thiosphaera* |
| *Kingella* | *Azospirillum* | |
| *Moraxella* | *Bradyrhizobium* | |
| *Morococcus* | *Rhizobium* | $NH_3$ oxidizing |
| *Neisseria* | *Sinorhizobium* | *Nitrosomonas* |
| *Ochrobactrum* | | *Nitrosospira* |
| *Oligella* | Spore forming | |
| *Pseudomona* | *Bacillus* | |
| *Thauera* | | |
| *Sphingobacterium* | Thermophiles | |
| *Tsukamurella* | *Aquifex* | |
| *Xanthomonas* | *Thermothrix* | |

Budding
  *Blastobacter*
  *Hyphomicrobium*

Carboxidotrophic
  *Zavarzinia*

*Source:* Adapted from Tiedje, J.M. 1988. Ecology of denitrification and dissimilatory nitrate reduction to ammonium, pp. 179–243. In A.J.B. Zehnder (Ed.). *Biology of anaerobic organisms*. John Wiley & Sons, Inc., New York; Zumft, W. 1997. Cell biology and molecular basis of denitrification. *Molecular Biology Reviews* 61:533–616; and Coyne, M.S. 2008. Biological denitrification. pp. 197–249. In J.S. Schepers and W. Raun (Eds.), *Nitrogen in agricultural systems*. Agronomy Monograph 49. ASA-CSSA-SSSA, Madison, WI.

probes, and efficiency of amplification (Wallenstein et al. 2006). Nevertheless, this is increasingly becoming the method of choice to enumerate soil denitrifiers (Philippot et al. 2007).

By far most studies that have enumerated denitrifiers have focused on the heterotrophic population. Heterotrophic denitrifiers in soil, as enumerated by MPN procedures, range from $10^4$ to $10^8$ g$^{-1}$ soil depending on the soil type, management, treatment, and media of enumeration (Coyne 2008). In a typical survey of geographically diverse soils, for example, Gamble et al. (1977) recorded MPN values ranging from $1.2 \times 10^4$ to $7 \times 10^6$ denitrifiers per g. In one extreme example (a poultry waste fermenter) they observed $3.5 \times 10^{10}$ denitrifiers per mL. Davidson et al. (1985) in various sites from a hardwood forest in North Carolina, found between $1.2 \times 10^1$ and $4 \times 10^4$ MPN denitrifiers. These numbers represent about 10–20% of the total $NO_3^-$ respiring heterotrophs (Staley and Griffin 1981) and 15–20% of the total culturable population (Payne 1981). The actual ratio of denitrifiers to nitrate reducers to total facultative anaerobes in Gamble et al.'s study was 0.2:0.8:1.0.

Culture-free methods of enumeration, in contrast, give denitrifier numbers about 10-fold higher than are determined by MPN. For example, Henry et al. (2006) used the *nosZ* and *nirK* genes to observe copy numbers in various soils ranging from $10^5$ to $10^7$ per g of soil. 16S rRNA copy numbers ranged from $10^8$ to $10^9$ per g soil indicating that denitrifiers as a general physiological class represented 5–6% of the total prokaryotic population.

### 5.4.1.2 Population Dynamics

#### 5.4.1.2.1 Competition with Other Soil Organisms

Denitrification presumably makes denitrifiers more competitive than other heterotrophs in environments that promote growth of facultative anaerobes (Tiedje et al. 1982); but this has been difficult to demonstrate. Based on various environments in which denitrifiers have been enumerated, they consistently make up a minority of the microbial population. An exception may be in subsurface environments in which stratification of water and carbon occurs (Section 5.4.5.1). This also appears to be true for denitrifiers in the root rhizosphere; with the capacity to denitrify they are more competitive with other soil heterotrophs than without it (Bothe 1992; Clays-Josserand et al. 1995; Philippot et al. 1995; Ghiglione et al. 2000).

Smith and Tiedje (1980) evaluated the competitive success of two denitrifying *Pseudomonas* species in aerobic and anaerobic soils (with $NO_3^-$); one competed better as a denitrifier and the other competed better as an aerobic heterotroph. Denitrifiers struggle to compete with other $NO_3^-$ reducing organisms in some environments, particularly when excess C is present. For example, on a theoretical basis denitrifying growth should allow microorganisms to outcompete $NH_4^+$ dissimilators when the ratio of electron donors (i.e., C) to electron acceptors (i.e., $NO_3^-$) is low (Tiedje 1988), which is characteristic of aerobic soils. The opposite is true when that ratio is reversed. Chemostat studies in which *Pseudomonas stutzeri* (a denitrifier) competed with *Klebsiella* (an $NH_4^+$ dissimilator) supported that hypothesis; the *Klebsiella* outcompeted the *Pseudomonas* at higher C/$NO_3^-$ ratios (Rehr and Klemme 1988). In the ultimate organic chemostat, the animal rumen, which represents a persistently anaerobic C-rich environment with continuous flow, denitrification activity was completely absent as the denitrifiers were outcompeted (Kaspar and Tiedje 1981).

#### 5.4.1.2.2 Community Dynamics

Culture-free techniques are the method of choice to evaluate denitrifier community dynamics. Their population dynamics are just beginning to be understood but their structure in the long term seems to be most affected by C availability, pH, and moisture and temperature regimes rather than $NO_3^-$ (Wallenstein et al. 2006).

Soil type does and doesn't make a difference. In a forested upland versus an adjacent wetland marsh there was very little overlap in *nirK* sequence; the *nirK* sequences were more diverse in the marsh soil (Prieme et al. 2002). Likewise, in two soils differing in $N_2O/N_2$ production ratios, *nosZ* genes showed divergent communities (Cheneby et al. 1998). In contrast, an adjacent forest and meadow shared most of the *nosZ* genes, although there were enough differences in abundance and habitat specific variants to distinguish the two soils (Rich et al. 2003). Similarly, in a soil transect across several soil habitats the *nosZ* genes were in common and the gene profiles and activity did not vary the same across habitat (Rich and Myrold 2004). Wallenstein et al. (2006) propose that these studies suggest *nosZ* is a more "cosmopolitan" molecular marker and less affected by distal (e.g., environmental) controls. The implications are clear, however, for trying to describe variability among denitrifier communities in different soil types if that variability is influenced by the molecular marker chosen for analysis—a cosmopolitan marker will imply less diversity than more specific markers.

Denitrifier communities experience seasonal and annual shifts. Based on *nirS* and *nirK*, denitrifiers were less abundant in summer than other seasons, which is consistent with soil populations in general (Mergel et al. 2001). There was also inter-annual variation in forest and meadow soils each fall over a 5-year period in a study by Boyle et al. (2006).

Overall N deposition seems to affect denitrifier communities even if $NO_3^-$ does not. *nirK* and *nirS* abundance were less in fertilized than unfertilized forest soils (Wallenstein 2004). Fertilized plots had a different community structure than unfertilized plots (Wolsing and Prieme 2004). There were differences in enzyme kinetics of $NO_3^-$ and $NO_2^-$ reduction in three organic soils differing in N content (Holtan-Hartwig et al. 2000). Likewise, Cavigelli and Robertson (2000) observed differences in enzyme activity, microbial species composition, and enzyme sensitivity to pH and $O_2$ in two soils differing in plant community and disturbance.

Plant community appears to be a driver for denitrifier community differences but not in a clear-cut manner. Phylogenetically, *Agrobacterium*-related denitrifiers were enriched in planted versus unplanted soils (Cheneby et al. 2004). Shannon diversity index based on *nirK* and *nirS* meta-analysis (Dell et al. 2010) ranged from 0.8 to 4.1 for environments ranging from a mixed deciduous forest in Michigan to a wheat field in Moselle, France.

Rhizosphere and soil type are two factors interacting simultaneously to affect the composition of the indigenous soil community, and presumably the denitrifiers therein (Philippot et al. 2007). Surprisingly, although root C is an important driver of activity, it was not an important driver of denitrifier community diversity (Mounier et al. 2004). Boyle et al. (2006), in a reciprocal transfer experiment, saw the denitrification enzyme activity but not the population profile change.

Age of the plant community can influence denitrifier composition. Dell et al. (2010) evaluated denitrifier diversity in turfgrass systems ranging from new (1 year) to old (95 years). Relative to an adjacent forest, the turf systems had a more diverse denitrifier population, driven in large part by pH. Among different turf systems there was a significant difference in denitrifier population by age, with older systems (25 and 95 years old) having greater diversity than younger systems.

## 5.4.2 ARCHAEA

There are few reports of denitrifying archaea, which is related to the difficulty in their isolation and cultivation. *Halobacterium marismortui* (Werber and Mevarech 1978) and *Halobacterium denitrificans* (Tomlinson et al. 1986) are two denitrifying extremophiles isolated from the Dead Sea. Other extremophilic denitrifying archaea are *Haloarcula marismortui* (Yoshimatsu et al. 2000) and *Pyrobaculum aerophilum* (Völkl et al. 1993). Treusch et al. (2005) used molecular probes for $NO_2^-$ reductase to identify denitrifying potential in $NH_3$-oxidizing mesophyllic crenarcheota. The archaea appear to resemble denitrifying bacteria in terms of the enzymes utilized for denitrification and the presence of truncation in the denitrification pathway (Zumft and Kroneck 2007).

### 5.4.3   Fungi

#### 5.4.3.1   Evidence for Fungal Denitrification

One of the most significant recent advances in soil denitrification studies has been an appreciation of the role fungi play. Mothapo et al. (2015) recently reviewed this topic. Since Bollag and Tung (1972) first observed $N_2O$ production by *Fusarium oxysporum* and *F. solani*, an increasing number of studies have reported isolates with this property and the presence of genes for denitrifying enzymes that vary in similarity to their prokaryotic counterparts (Table 5.3). The dominant feature of fungal denitrification is the absence of functional $N_2O$ reductase and relaxed $O_2$ regulation of $NO_2^-$ reductase, which enables fungi to denitrify at much higher $O_2$ contents than prokaryotic denitrifers.

Isotope studies indicate $N_2$ can be one of the products of denitrifying fungi, but this is thought to be more likely evidence of co-denitrificaton (Shoun et al. 1992). Potential rates of fungal denitrification are orders of magnitude lower than prokaryotes in pure culture studies (Chen et al. 2015b; Mothapo et al. 2015). But the dominance of fungal biomass in soil and their ability to denitrify in much drier soil environments than prokaryotes means their contribution to denitrification, particularly to $N_2O$ evolution from soil, can be substantial (Herold et al. 2012).

#### 5.4.3.2   Mechanism of Fungal Denitrification

Fungal denitrification employs a pathway that is metabolically similar to—but not quite identical with—prokaryotic denitrification. Some of these differences are illustrated in Figure 5.2. Nitrite reductase is located on the mitochondria. Unlike prokaryotic denitrification, a member of the cytochrome *p450* superfamily was identified as a key player in the further reduction of NO to $N_2O$ (Shoun and Tanimoto 1991). *p450* acts as an NO-reductase (p450Nor) in the cytosol of fungi utilizing NADH as a direct electron ($e^-$) donor (Ma et al. 2008). The entire pathway is linked to formate oxidation in hypoxic conditions via formate dehydrogenase (FDH), which utilizes $NO_3^-$ and $NO_2^-$ as electron acceptors to generate ATP.

#### 5.4.3.3   Numbers of Fungal Denitrifiers

Unlike prokaryotes, few attempts have been made to quantify fungal denitrifiers. Most studies have confined themselves to observation of the percentage of culturable fungal isolates that denitrify. Hence, 68 of 151 isolates produced $N_2O$ (Mothapo et al. 2013) representing 16 genera and 35 species, with production rates ranging from < 0.01 to > 10,000 nmol $N_2O$-N $d^{-1}$. Of 214 isolates representing 15 distinct morphological groups, 151 denitrified (Higgins et al. 2016). Of 492 fungal isolates among 10 genera, 27 had *p450nor* or *nirK* (Novinscak et al. 2016); prevalence did not follow a soil ecosystem distribution. In a compost, 10 of 25 fungal isolates denitrified and were dominated by *Byssochlamys nivea*. *Aspergillus, Fusarium, Penicillium*, and *Trichoderma* account for about 50% of isolates with *Fusarium* the only genus that is consistently isolated (Mothapo et al. 2015).

#### 5.4.3.4   Contributions of Fungal Denitrification to $N_2O$ Evolution

Robertson and Tiedje (1987) provided tantalizing evidence for the significance of fungal denitrification when they observed that neither abiotic process, nitrification, nor denitrification could fully account for the $N_2O$ evolved from recirculating gas cores. The principal means by which investigators have tried to distinguish between fungal and prokaryotic denitrification is the use of organism-specific biocides: cyclohexamide for fungi and streptomycin for prokaryotes (e.g., Laughlin and Stevens 2002; Chen et al. 2015a). Depending on the environment and soil moisture content (%WFPS) adding cyclohexamide has a significantly greater effect on $N_2O$ evolution than streptomycin, suggesting a significantly greater role of fungi in $N_2O$ evolution. As aeration status increases, so does the apparent contribution of fungi to $N_2O$ evolution (Herold et al. 2012).

For example, in a desert grassland cyclohexamide decreased $N_2O$ evolution by 85% relative to a 53% reduction by streptomycin (Crenshaw et al. 2008). Similarly, in an arable peat soil in Indonesia,

**TABLE 5.3**

**Short List of Cultured Denitrifying Fungi**

| Isolate | Reference |
|---|---|
| *Alternaria alternate* | Mothapo et al. 2015 |
| *Aspergillus flavus* | Mothapo et al. 2015 |
| *A. fumigatus* | Mothapo et al. 2015 |
| *A. niger* | Wei et al. 2015 |
| *A. oryzae* | Mothapo et al. 2013 |
| *A. terreus* | Mothapo et al. 2013 |
| *A. versicolor* | Mothapo et al. 2013 |
| *Bionectria ochroleuca* | Mothapo et al. 2013 |
| *Botrytis cinerea* | Mothapo et al. 2015 |
| *Byssochlamys nivea* | Novinscak et al. 2016 |
| *Chaetomium spp.* | Shoun et al. 1992 |
| *Ch. fumicola* | Shoun et al. 1992 |
| *Ch. globosum* | Long et al. 2015 |
| *Chloridium spp.* | Novinscak et al. 2016 |
| *Cylindrocarpon tonkinense* | Shoun et al. 1992 |
| *Fusarium avenaceum* | Wei et al. 2015 |
| *F. equiseti* | Wei et al. 2015 |
| *F. lini* | Shoun et al. 1992 |
| *F. oxysporum* | Bollag and Tung 1972 |
| *F. solani* | Bollag and Tung 1972 |
| *Gibberella fujikori* | Shoun et al. 1992 |
| *Hypocrea koningii* | Mothapo et al. 2013 |
| *Metarhizium pingshaense* | Mothapo et al. 2013 |
| *Mucor fragilis* | Mothapo et al. 2013 |
| *Neocosmospora vasinfecta* | Mothapo et al. 2013 |
| *Ophiocordyceps heteropoda* | Mothapo et al. 2015 |
| *Penicillium chrysogenum* | Long et al. 2015 |
| *P. pinophilum* | Mothapo et al. 2013 |
| *P. janthinellum* | Mothapo et al. 2013 |
| *P. purpurogenum* | Wei et al. 2015 |
| *P. restrictum* | Mothapo et al. 2013 |
| *P. verruculosum* | Mothapo et al. 2013 |
| *Rhizomucor sp.* | Wei et al. 2015 |
| *Talaromyces flavus* | Shoun et al. 1992 |
| *Trichocladium spp.* | Novinscak et al. 2016 |
| *Trichoderma hamatum* | Shoun et al. 1992 |
| *T. piluliferum* | Mothapo et al. 2013 |
| *T. spirali* | Mothapo et al. 2013 |
| *Tylospora fibrillose* | Mothapo et al. 2015 |
| *Volutella ciliata* | Novinscak et al. 2016 |

cyclohexamide reduced $N_2O$ evolution 81% relative to a 31% reduction when streptomycin was used (Yanai et al. 2007) (Table 5.4). Prokaryotes contributed 54% and fungi 18% to potential denitrification rates across a range of pH (4.5–7.5) and tillage (Herold et al. 2012).

Between a redox potential of 250 to 400 mV Eh, fungi represented 35% of the total denitrification in a sediment system while bacteria represented approximately 2%. In contrast, at –200 mV Eh fungi represented only 18% of the denitrification while bacteria represented 65% (Seo and Delaune

**TABLE 5.4**

**Contribution of Fungi to N$_2$O Evolution from Soil Based on Antimicrobial Inhibition Studies**

| % Reduction with Fungal Inhibition | System | Reference |
|---|---|---|
| 89% | Perennial ryegrass | Laughlin and Stevens 2002 |
| 81% | Arable peat soil | Yanai et al. 2007 |
| 65% | Prairie | Ma et al. 2008 |
| 63% | Semiarid grassland | Crenshaw et al. 2008 |
| 40–51% | Cropland | Chen et al. 2015b |
| 18% | Grass field | Herold et al. 2012 |

2010). Across numerous studies evaluating agricultural systems ranging from grasslands to organic farms fungi had a median contribution of 62% to N$_2$O flux while the median bacterial contribution was about 42% (Mothapo et al. 2015). It is worth noting an apparently inverse relationship between the lability of organic matter and water content with the contribution of fungi to N$_2$O flux.

One caution with these studies is they are typically performed in relatively aerated soils and the absence of an inhibitor of N$_2$O reductase, such as C$_2$H$_2$. Thus, they underestimate the actual amount of denitrification occurring when cyclohexamide alone is used because they do not account for the full denitrifying potential of the prokaryotic population. Nitrous oxide evolution from soil is not synonymous with denitrification in soil—only a component of it.

### 5.4.4 PHYLOGENY

Some 60 genera of prokaryotes denitrify (Zumft 1997; Philippot et al. 2007). Based on the distribution of denitrification among various prokaryotic physiological groups (see Table 5.2) the phylogeny of denitrification among the prokaryotes can best be described as eccentric, with dubious value as a defining characteristic for taxonomy. Nitrite reductase (*nirK* or *nir S*) is typically used as the gene of choice for phylogenetic studies although *nosZ* for N$_2$O reductase has also been used. *nirK* appears to dominate in soil, and when primers for this gene have been used to amplify denitrifying populations the trait appears in five distinct clades: two well-separated bacterial clades, archaea, NH$_3$-oxidizing bacteria, and fungi (Long et al. 2015). The diversity of the *nirK* gene among these clades follows the order (from greatest to least): bacterial clade 1 > bacterial clade 2 > archaea > NH$_3$-oxidizing bacteria > fungi (Long et al. 2015).

From the perspective of phylogeny, denitrifiers are found among all three microbial domains (Table 5.5). Among the bacteria, the proteobacteria are most often represented by cultured isolates.

Based on 16s rRNA phylogeny of isolated prokaryotic species Betlach (1982) proposed that denitrifiers evolved from a common ancestor within the purple photosynthetic group and distinctly from other NO$_3^-$ respiring prokaryotes. This would place its evolution at a similar time to the evolution of aerobic respiration. Based on homology to archael denitrifiers, Zumft and Kroneck (2007) place the evolution at a far earlier date, in some common ancestor prior to the period in which bacterial and archaeal domains separated. Although convergent evolution occurs, and lateral gene transfer is a well-known phenomenon, that denitrification is present in all three domains of life argues that it is a very old biological process indeed.

The phylogeny of denitrifying fungi is primarily based on sequence comparison of the *nirK* gene for NO$_2^-$ reductase (e.g., Wei et al. 2015) following amplification by PCR. Because there is significant homology between the *nirK* of prokaryotes and fungi, much recent work has been devoted to developing more fungi-specific *nirK* primers (e.g., Long et al. 2015; Chen et al. 2016) and developing primers for *p450nor*, which is uniquely fungal (e.g., Higgins et al. 2016; Novinscak et al. 2016).

**TABLE 5.5**

**Short List of Cultured Denitrifying Organisms**

| Domain | Phylum or Class | Order | Genus | Species |
|---|---|---|---|---|
| Bacteria | Firmicutes | | *Bacillus* | *azotoformans* |
| | | | *Paenibacillus* | *terrae* |
| | Actinomycetes | | *Corynebacterium* | *nephridii* |
| | | | *Streptomyces* | *thioluteus* |
| | Bacteroides | | *Flavobacterium* | *denitrificans* |
| | | | *Flexibacter* | *cadiensis* |
| | Aquifaceae | | *Hydrogenobacter* | *thermophilus* |
| | Alpha proteobacteria | | *Azospirillum* | *lipoferum* |
| | | | *Bradyrhizobium* | *japonicum* |
| | | | *Hyphomicrobium spp.* | |
| | | | *Paracoccus* | *pantotrophus* |
| | | | *Rhodobacter* | *sphaeroides* |
| | | | *Rhodopseudomonas* | *salustris* |
| | | | *Sinorhizobium* | *meliloti* |
| | Beta proteobacteria | | *Alcaligenes* | *faecalis* |
| | | | *Aquaspirillum* | *magnetotacticum* |
| | | | *Burkholderia spp.* | |
| | | | *Kingella* | *denitrificans* |
| | | | *Neisseria spp.* | |
| | | | *Nitrosomonas* | *europeae* |
| | | | *Thiobacillus* | *denitrificans* |
| | Gamma proteobacteria | | *Halomonas* | *desiderata* |
| | | | *Pseudomonas* | *fluorescens* |
| | | | *Shewanella* | *denitrificans* |
| | | | *Thioalkavibrio* | *denitrificans* |
| | Epsilon proteobacteria | | *Nitratifactor* | *salsuginis* |
| | | | *Thiomicrospira* | *denitrificans* |
| Archaea | | | *Haloarcula* | *marismortui* |
| | | | *Halobacterium* | *denitrificans* |
| | | | *Pyrobaculum* | *aerophilum* |
| Eucarya | Eumycota (Dikarya) | | | |
| | | Hypocreales | *Fusarium* | *oxysporum* |

*Source:*  Adapted from Philippot, L., S. Hallin, and M. Schloter. 2007. Ecology of denitrifying prokaryotes in agricultural soil. *Advances in Agronomy* 96:249–305.

Of 119 $N_2O$-producing fungi in 60 genera, 90% appear among the Ascomycota (e.g., *Alternaria, Aspergillus, Botrytis, Fusarium, Penicillium, Trichoderma*), 7% among the Basidiomycota, and 3% among the Zygomycota (Mothapo et al. 2015). Within the Ascomycota, denitrifying fungi are found in the orders Eurotiales, Hypocreales, and Sordorialis (Wei et al. 2015) as well as among the Saccharomycetes and Dothideomycetes. Denitrifying fungi have not been systematically evaluated in lichens or mycorrhizal fungi, although an ectomycorrhizal fungus (*Paxillus involutus*) appears to denitrify (Mothapo et al. 2015).

## 5.4.5 DISTRIBUTION

If you are looking for denitrifiers in soil, where will you find them? The diversity of physiological groups suggests that denitrifiers are ubiquitous in soils and sediments, whether these environments are persistently anaerobic, aerobic, or somewhere in-between. As a general rule, the driving factors of C, $O_2$, and $NO_3^-$ (see Section 5.5) are likely to indicate the potential for the highest denitrifier populations. Where it is wet or potentially wet, there is a lot of C, and at least the prospect for N input to the soil by fertilization, nitrification, and $N_2$-fixation, there lays a prime spot for high denitrifier population.

### 5.4.5.1 Spatial Distribution

There has been surprisingly little work performed on the spatial distribution of denitrifiers across landscapes. Where such work occurs it typically evaluates transects across various soil types or uses based on composites of random samples. In part this is due to the extreme variability of denitrification rates that are measured on a spatial basis, which implies equally high variability in denitrifier numbers (see Section 5.6.1.4).

### 5.4.5.2 Depth Distribution

Like most heterotrophs, denitrifiers stratify in those soil environments with the greatest amount of C (typically at the soil surface) and rapidly decrease with soil, although they can be found at depth. Parkin and Meisinger (1989) found denitrifier numbers to decline exponentially from $10^5$ to $10^6$ cells $g^{-1}$ at the soil surface in cropland to undetectable levels at 180 cm depth. At a watershed stream site dominated by herbs and shrubs in North Carolina Davidson et al. (1985) measured MPN denitrifiers of $4 \times 10^4$ $g^{-1}$ at 0–10 cm, $6.3 \times 10^3$ $g^{-1}$ at 10–20 cm, and $3.2 \times 10^3$ $g^{-1}$ at 20–30 cm depth. Stratification of denitrifiers at the soil surface is probably truer for denitrifying fungi because of their obligately heterotrophic lifestyle than it is for bacteria. However, denitrifying fungal distribution has not been adequately examined. *Fusarium oxysporum* is ubiquitous in soil samples that have been examined for fungal denitrification (Mothapo et al. 2013) but population numbers for fungal denitrifiers as a whole are not known.

It is certainly possible to find denitrifiers at depth in the soil environment, usually associated with boundary layers (e.g., buried horizons, impermeable layers) in the soil environment (Rice and Rogers 1993). These environments can potentially select for denitrifiers if they promote hypoxic and/or C-rich environments. In a cropped Zanesville silt loam with a fragipan (an impermeable layer) at approximately 65 cm depth, the population of denitrifiers increased, and their ratio with respect to other heterotrophs increased 64% immediately above the fragipan (Figure 5.3) (Fairchild et al. 1997, 1999).

Likewise, in an unamended Zanesville soil in pasture in which fragipans were found at three distinct depths, there was evidence of soil conditions favoring denitrifiers immediately above the fragipans (presumably periodic hypoxic conditions due to perching water tables) (Figure 5.4) (Wu 2008).

### 5.4.5.3 Rhizosphere

The rhizosphere is an environment in which horizontal stratification of denitrifiers can occur as a consequence of distance from the C-rich rhizosphere environment. Nitrate respiring, denitrifying, and $NH_4^+$ dissimilating bacteria varied in a ratio of 36:45:12 in rhizosphere soil to 43:22:18 in bulk soil surrounding cattails (*Typha angustifolia* L.) (Brunel et al. 1992). Nitrate dissimilating organisms, including denitrifiers, were present in greater numbers in the rhizosphere of flax and tomatoes compared to bulk soil (Clays-Josserand et al. 1995). The proportion of denitrifiers as part of the $NO_3^-$ dissimilating population in tomatoes decreased as one moved from root tissue (94%) to rhizoplane (75%) to rhizosphere (68%) to bulk soil (44%). The high proportion of denitrifiers in root tissue is almost certainly associated with selective enrichment of microbes in plant tissue that happen to be denitrifiers rather than active selection of the capacity for denitrification itself.

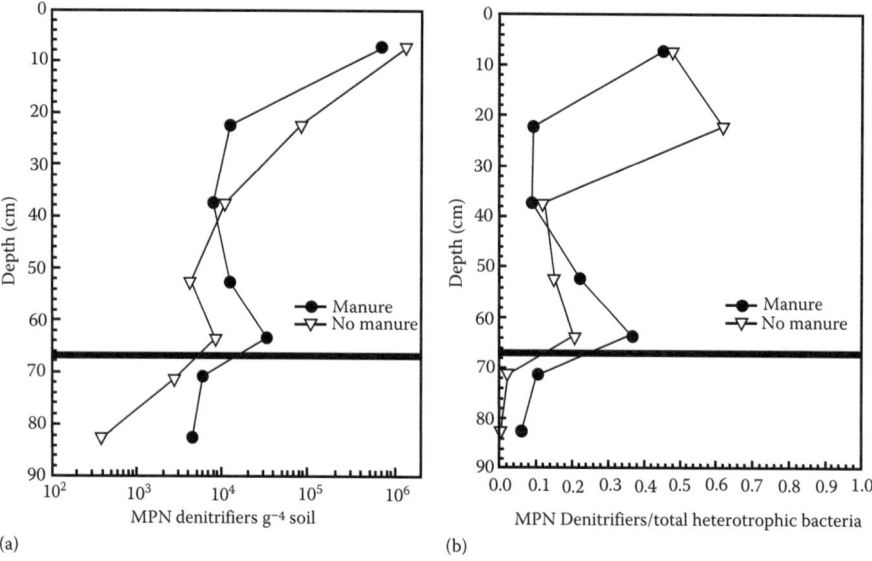

**FIGURE 5.3** Stratification of denitrifier MPN in a fragipan soil. In the presence and absence of manure, denitrifier MPN declined exponentially with depth until reaching a fragipan at approximately 65 cm whereupon populations increased (a). Relative to overall heterotrophs in the same soil depths, the ratio of denitrifiers to total facultatively anaerobic heterotrophs also increased at the fragipan (b). (From Fairchild, M.A. 1997. Denitrifier ecology in fragipan soils of Kentucky. PhD Dissertation, University of Kentucky, Lexington, KY.)

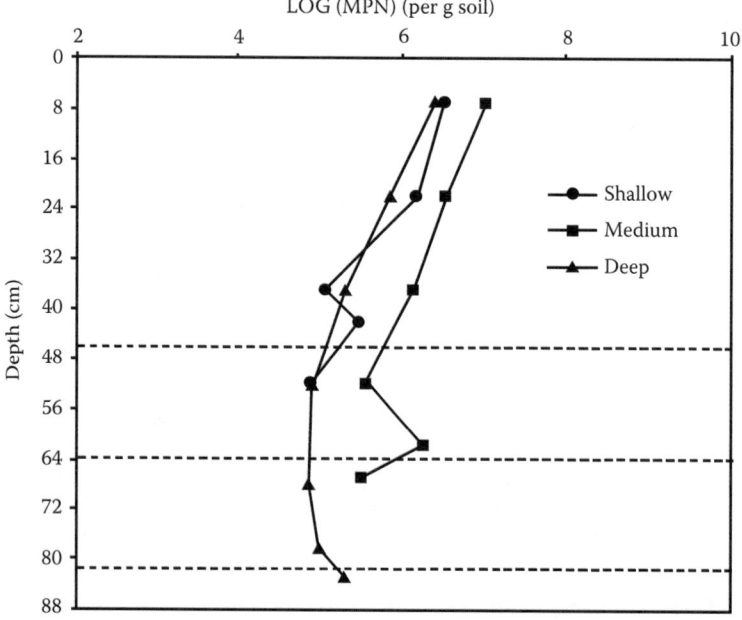

**FIGURE 5.4** Stratification of denitrifier MPN above three depths of fragipans (47, 64, and 82 cm) in a Zanesville silt loam in Kentucky. In each case, the denitrifier MPN increased above the fragipan. (From Wu, Tingting. 2008. Denitrifier ecology in a fragipan soil of Kentucky. M.S. Thesis, University of Kentucky, Lexington, KY.)

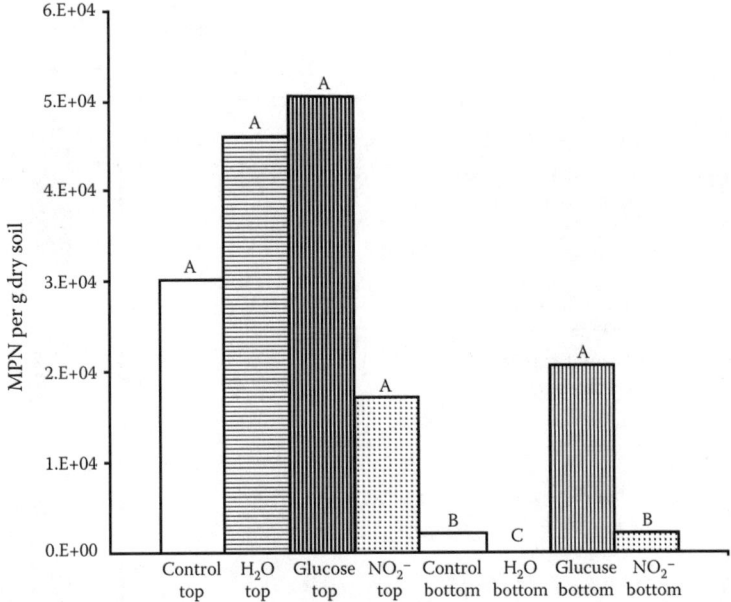

**FIGURE 5.5** MPN denitrifiers in intact soil columns of Zanesville silt loam soil from Kentucky with a fragipan at 55–75 cm depth. Soils were incubated with a perched water table of 10 cm above the fragipan for 12 weeks. Top refers to the upper 15 cm of soil; bottom refers to the 15 cm of soil immediately above the fragipan. Columns with similar letters are not significantly different. While various treatments (water, $NO_3^-$, glucose) had no significant effect on MPN in the upper soil profile, when glucose or $NO_3^-$ were added to the soil immediately above the fragipan they significantly increased the denitifier MPN relative to water alone. (From Wu, Tingting. 2008. Denitrifier ecology in a fragipan soil of Kentucky. M.S. Thesis, University of Kentucky, Lexington, KY.)

### 5.4.5.4  Denitrifying Hotspots

"Hotspots," zones of intense denitrifying activity, can be naturally occurring (Parkin 1987). In artificially created hotspots to mimic burial of crop residue, Murray et al. (1995) showed that increased denitrification activity was coupled with a 1–3 log increase in MPN at the site of elevated activity. Similarly, Wu (2008) showed that when soil above a fragipan was allowed to perch water, addition of C and $NO_3^-$ stimulated denitrifier MPN relative to a water control (Figure 5.5).

## 5.5  REGULATORY CONTROL OF DENITRIFICATION IN SOIL

Although individual denitrifying organisms may respond differently to their environment, overall, they share common responses to environmental regulators of the process itself. Considering that all the steps in the pathway produce free intermediates, denitrification can be viewed as a series of interconnected pipes. At full capacity and flux the volume of the pipes is analogous to the total denitrification capacity of the soil. The rate of flux is determined by the capacity of the smallest pipe, or better yet, by the connections between each pipe, which are analogous to the denitrification enzymes in each step. The slowest or least efficient enzyme represents the flux limiting step. If intermediates meet this rate limiting step, they can be lost to the environment. This is the "Hole-in-the-Pipe" model of denitrification (Firestone and Davidson 1989) and is illustrated in Figure 5.6.

The classic sequence of denitrification derives from work by Cooper and Smith (1963) which illustrated the disappearance and appearance of various N-oxides in wet soils (5% greater than field capacity) under a He atmosphere during a 90-hour incubation. In all cases multiple intermediates formed regardless of whether the immediate precursors were still forming or decreasing in

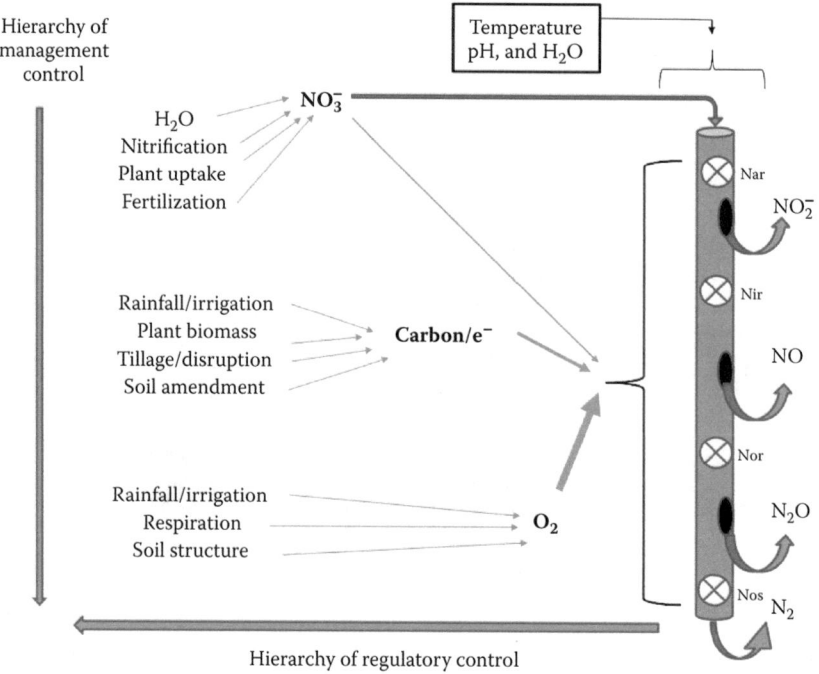

**FIGURE 5.6** Hierarchical control of denitrification in soils by regulation and management. Temperature, pH, and water content are global environmental regulators of the size and activity of microbial populations in soil. Once a critical mass of viable organisms is present, their activity is controlled by whether $NO_3^-$ is present. N-oxide flux through the denitrification pathway is controlled by the rate at which it passes through each of the respective enzymes, which act as checkpoints, and are independently regulated by $O_2$, carbon (or $e^-$) availability, and $NO_3^-$ and $NO_2^-$. If flux is impeded at any point in the pathway, it allows intermediates to be more readily released into the environment (the hole-in-the-pipe model). Each of the primary regulatory features controlling denitrification can be managed to greater or lesser extents by field practice. (Adapted from Robertson, G.P. 1989. Nitrification and denitrification in humid tropical ecosystems: Potential controls on nitrogen retention, pp. 55–69. In J. Procter (Ed.), *Mineral nutrients in tropical forest and savanna ecosystems.* Blackwell Science Publishes, Oxford; Davidson, E.A. 1991. Fluxes of nitrous oxide and nitric oxide from terrestrial ecosystems, pp. 219–235. In J.E. Rogers and W.B. Whitman (Eds.), *Microbial production and consumption of greenhouse gases: Methane, nitrogen oxide, and halomethanes.* American Society for Microbiology, Washington, DC; and Coyne, M.S. 2008. Biological denitrification. pp, 197–249. In J.S. Schepers and W. Raun (Eds.), *Nitrogen in agricultural systems.* Agronomy Monograph 49. ASA-CSSA-SSSA, Madison, WI.)

concentration. On an enzymatic basis this can be represented by the individual kinetics of each process (Figure 5.7). Betlach and Tiedje (1981) demonstrated that transient $NO_2^-$, NO, and $N_2O$ accumulation in denitrifying heterotrophs could be explained as the simple relationship between the rates of $NO_3^-$ and $NO_2^-$ reductase in the cell, as well as subsequent reductases.

### 5.5.1 Oxygen

Oxygen, carbon, and $NO_3^-$ are the dominant regulators of denitrification in the environment. Nitrate (and other N-oxides) have a direct effect as substrates and possibly inducers. Without them denitrification would not occur. In the presence or absence of $NO_3^-$, however, $O_2$ primarily influences the synthesis and activity of denitrification enzymes. Denitrification normally does not occur in well-aerated soils and evidence that it does (e.g., $N_2O$ production) is usually attributed to denitrifying

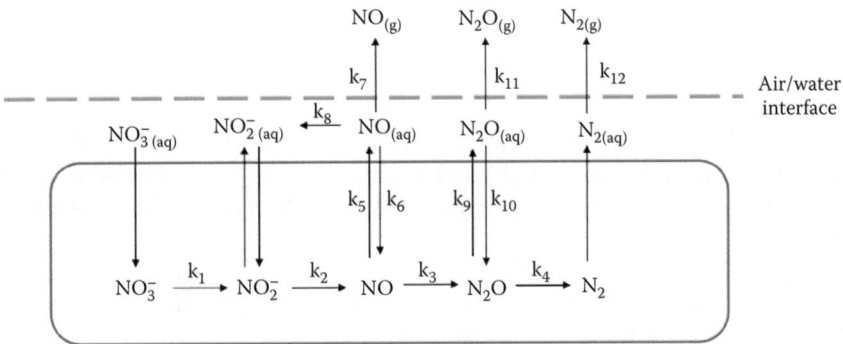

**FIGURE 5.7** Influence of production and consumption rates (k) on the release of denitrification intermediates into the soil environment. The denitrification rate and the likelihood that trace gas intermediates will be lost to the soil atmosphere reflect the summation of various production and consumption rates. Individual enzyme activity rates ($k_1$ to $k_4$) control N-oxide flux through the pathway; the slowest rate determines the overall rate. Where slower rates follow faster rates, intermediates can be lost to the aqueous environment ($k_5$, $k_9$) and gaseous intermediates subsequently lost to the soil atmosphere at rates depending on diffusion gradients through water films ($k_7$, $k_{11}$, $k_{12}$) or react with metals and organic matter in the soil environment ($k_8$). At sufficiently high enzymatic rates some of the gaseous intermediates have measurable uptake rates ($k_6$, $k_{10}$). $N_2$ is not taken up by denitrifiers, except for those organisms (e.g., *Bradyrhizobium japonicum*) that also have the potential to fix $N_2$ (Coyne and Focht 1987). (Modified from Firestone, M.K. and E.A. Davidson. 1989. Microbiological basis of NO and $N_2O$ production and consumption in soil, pp. 7–21. In M.O. Andreae and D.S. Schimel (Eds.). *Exchange of trace gases between terrestrial ecosystems and the environment.* John Wiley & Sons, Ltd., New York.)

microsites. At soil atmosphere concentrations < 6% $O_2$ (by volume) one can start to see slight increases in denitrification activity, but this does not become pronounced until the soil atmosphere has <1% $O_2$ (Tiedje 1988). At <0.2% atmospheric $O_2$ the denitrification activity rapidly accelerates. In a typical soil, this is not difficult to achieve given water content and diffusional gradients.

Thresholds for $O_2$ concentrations that permit denitrification vary widely, <3 to >10 μmol L$^{-1}$ (Tiedje 1988). The key feature for an individual organism is whether the $O_2$ is saturating the terminal cytochrome oxidase (usually at 2–4 μmol L$^{-1}$) and whether the organism can still denitrify, as in *Thiosphaera pantotropha* (Robertson and Kuenen 1990).

Denitrification enzymes are synthesized in a stepwise fashion as a function of the $O_2$ content in their immediate environment (Coyne and Tiedje 1990b) but not necessarily in the same order in which the enzymes become active. In *Pseudomonas stutzeri*, for example, $N_2O$ reductase can be synthesized at 50% $O_2$ saturation prior to the synthesis of $NO_3^-$ reductase at 25% $O_2$ saturation (Korner and Zumft 1989). Once synthesized, the repression of denitrification enzymes does appear to be a stepwise process in the typical denitrifier, with $N_2O$ reductase, NO reductase, $NO_2^-$ reductase, and finally $NO_3^-$ reductase activity being suppressed as the dissolved $O_2$ concentration increases (Hochstein et al. 1984).

Because of the differential expression and repression of denitrification enzymes in soil, there is potential for considerable overlap between organisms producing $N_2O$ because of metabolic activity (i.e., nitrifiers) as aerobic soils becoming increasingly water saturated. Thus, fungal denitrifiers contribute less to $N_2O$ evolution from soil when it is relatively dry and relatively wet because the former does not have the hypoxic conditions necessary for activity and the latter better favors prokaryotic denitrifiers. Fungal denitrifiers also have considerable variation of the response to $O_2$ concentration. The highest producers experience a decline as $O_2$ increases from 0 to 20%. The lowest producers have the highest $N_2O$ production at low $O_2$ and 20% $O_2$. Some, like *Gibberella*, are largely unaffected (Mothapo et al. 2013).

### 5.5.2   Carbon

#### 5.5.2.1   Effect on Rates

As a general rule, denitrification is C-limited in soil and when C is added to potential denitrification assays (typically as glucose) the rates dramatically increase (Bowman and Focht 1974). This has been demonstrated consistently. Consequently, bioreactors designed to remove groundwater $NO_3^-$ at field edge typically involved some type of C-rich media through which the water flows (see Section 5.6.2.2).

Soil C is typically more complex than simple subunits and this affects denitrification rates. The rhizosphere is seemingly C-rich but only part of this C is readily available to denitrifiers for use (Christensen et al. 1990) and the rhizosphere effect, though stimulating, is of very narrow influence in soil (< 5 mm) (Smith and Tiedje 1979b). Counterintuitively, a meta-analysis of experiments examining denitrification enzyme activity in soils from studies with $CO_2$ enrichment showed less rather than more activity (Barnard et al. 2005). This was attributed to lower $NO_3^-$ concentrations at the elevated $CO_2$ although unlimited $NO_3^-$ in potential assays should have mitigated that effect.

Wetting and drying, freezing and thawing, and physical disruption through tillage and bioturbation are all processes leading to increased C availability in soil environments and therefore increased denitrification. Rather than differential induction and suppression of denitrification enzymes, Groffman and Tiedje (1988) attributed hysteresis in denitrification during wetting and drying cycles to greater soil respiration and N mineralization as soils passed through a wetting phase relative to a drying phase. Repeated cycles of C release, however, can deplete C and reduce denitrification capacity even in the presence of adequate amounts of $NO_3^-$ (Sexstone et al. 1985a).

Cover crops are intended to both deplete residual inorganic N and provide additional C to soil. An added benefit also appears to be that root exudates persist long enough in soil in the absence of active plant roots to facilitate denitrification. Preliminary studies indicate that the extent to which residual $NO_3^-$ is removed from anaerobically incubated soils is related to soil depth and the amount of permanganate oxidizable carbon ($POX_c$) present in the soil (Coyne unpublished).

#### 5.5.2.2   Effect on Product Formation

Carbon availability to heterotrophic denitrifiers also affects the distribution of final products. The mole fraction of $N_2O/N_2$ increases as C becomes more limited; as C becomes more abundant the $N_2O/N_2$ ratio decreases (Nommik 1956; Smith and Tiedje 1979b). From a physiological perspective this is evidence that without sufficient C there are insufficient electrons to completely reduce $N_2$ or support the active growth and activity of denitrifiers in denitrifying conditions. When there is abundant C to $NO_3^-$ in soil the dominant product is $N_2$. Thus, in the "Hole-in-the-Pipe" model (Figure 5.6) lack of C restricts the process. If C is available, $N_2$ becomes the dominant product; if C is limiting ultimately $N_2O$ accumulates (Firestone and Davidson 1989). This can be demonstrated in chemostat studies in which C is used as the limiting growth factor (Coyne unpublished). This is one of the clearest arguments for the centrality of increasing soil organic matter content in agricultural soils as a simple means of combatting agriculture's contribution to trace gas production and global warming.

#### 5.5.2.3   Effects on Relative Contributions of Bacteria and Fungi to $N_2O$ Production

Substrate complexity influences the relative contribution of prokaryotes and fungi to denitrification. Prokaryotic contribution to $N_2O$ production (as a surrogate measure for soil denitrification) in suboxic soils (80% WFPS) is higher for glucose-amended than cellulose-amended soils compared to fungi; so is their contribution to $N_2O$ production in soils amended with winter pea residue compared to switchgrass residue (Chen et al. 2015a). This is partly explained by the population response of prokaryotic denitrifiers to readily available substrates, because their population density increases relative to fungi when readily available substrates are present (Chen et al. 2015a).

**FIGURE 2.1** Nodules on faba bean. (With permission from Dr. M. Unkovich; School of Agriculture, Food and Wine, The University of Adelaide, Australia; http://www.soilquality.org.au/factsheets/legumes -and-nitrogen-fixation-south-australia.)

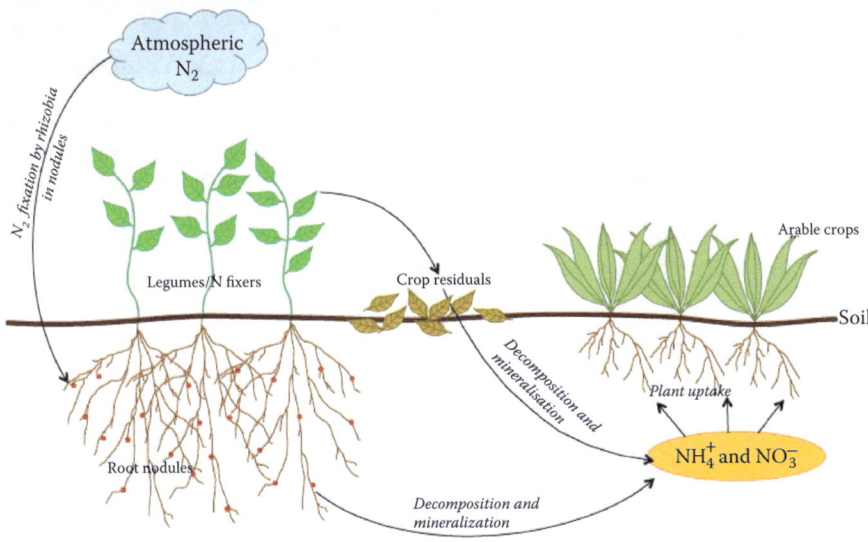

**FIGURE 2.2** Inputs of biological nitrogen fixation from grain (pulses) and forage legumes to arable crops via decomposition and subsequent mineralization of dead nodules and crop residues.

**FIGURE 4.2** Sample collection during snowmelt runoff in Saskatchewan. (a) The view looking up the hill from the flume. (b) The set-up for sample collection. The boards are berms to direct water to the flume, and the green box holds the auto-sampler for water collection, which is powered by the solar panel. (c) A close-up of the flume with water flowing through from top to bottom. The * indicates the tube through which the auto-sampler takes in water from the flume. All pictures were taken on the same day in March 2012, which was a year with little snow. The sites are the same as those used for Cade-Menun et al. (2013). Photo source: B. Cade-Menun.

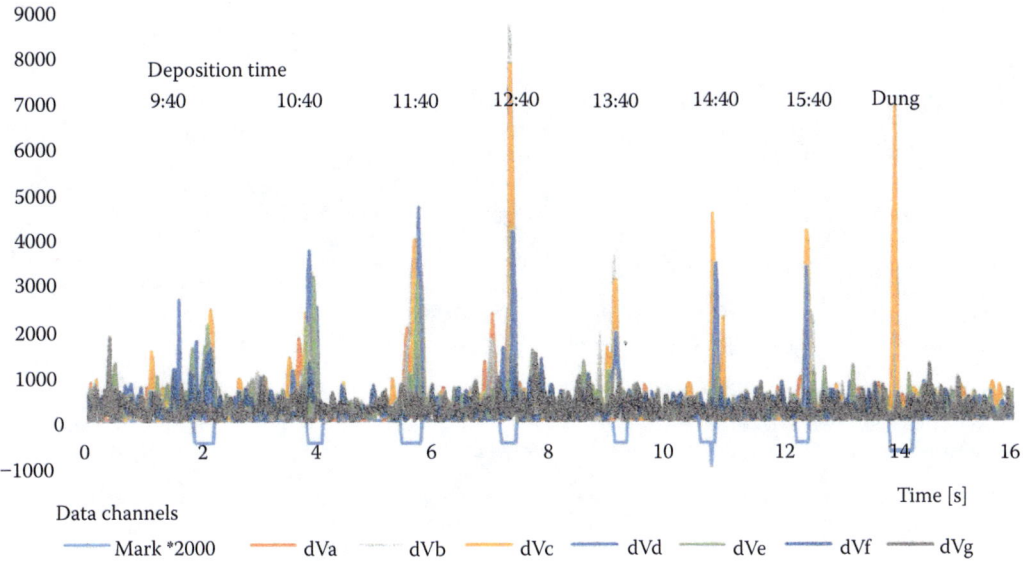

**FIGURE 7.4** Detection of several manually placed urine patches (position indicated by bracket below axis), with dung patch placed at the end. Detection data taken at 5 pm on the same day as deposition. Note the single channel detection of the dung patch, compared to the multichannel detection of urine patches.

**PHOTO 7.2** Beef grazing an electric fence—controlled forage crop in the UK.

**PHOTO 7.3**   A herd coming in for milking on an IGPF in New Zealand.

**PHOTO 7.4**   A typical rotary milking platform on an IGPF in New Zealand.

**PHOTO 7.5**   A large beef feedlot (CAFO) in Texas.

**PHOTO 7.6**   The first Spikey-8 fresh urine patch detection and treatment device in use on an IGPF in New Zealand, September 2017.

**PHOTO 10.1**    Intercropping of groundnut (left) with maize and soybean (right) with rice in Umiam, Meghalaya.

**FIGURE 13.7**    Polymer coated and uncoated commercial urea.

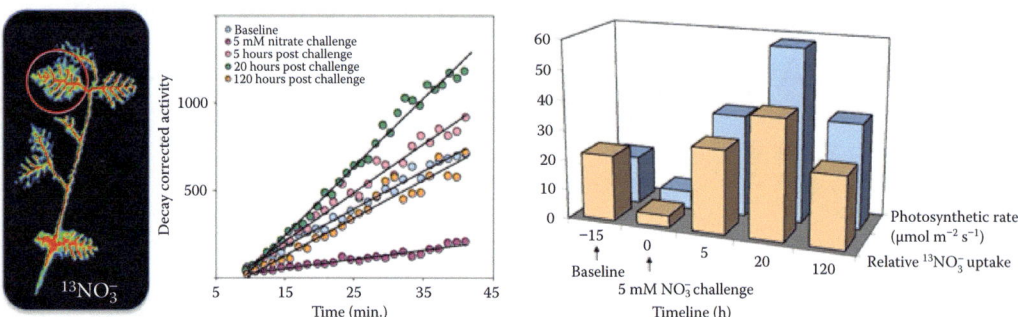

**FIGURE 14.5** Radiotracer imaging using $^{13}$N-nitrate coupled with positron emission tomography shows changes in nitrate acquisition of a young poplar sapling before and after a nitrate challenge. The kinetic uptake curves reflect the amount of $^{13}$N accumulation in the red circled mature leaf. The short half-life of $^{13}$N ($t_{1/2}$ 9.97 min) also affords opportunity to reimage the sapling over time after repeated administration of new doses of tracer to the roots. At administration, nitrate acquisition is initially shut down, but quickly increases in the subsequent hours. By 120 hours, nitrate acquisition is seen to return to baseline response. An infrared gas exchange cell placed on the same mature leaf maps changes in leaf photosynthetic capacity ($\mu$mol mm$^{-1}$ s$^{-1}$). Carbon input to the sapling, reflected by changes in leaf photosynthetic capacity, is seen to correlate with changes in nitrate acquisition.

### 5.5.3 Nitrogen

In the absence of inorganic N, denitrification does not occur, even if the capacity is present. Increasing the amount of N in soil does not appear to affect the denitrification rate, but it does affect the distribution of final products. Firestone et al. (1980) showed that as the concentration of either $NO_3^-$ or $NO_2^-$ increased from 0 to 20 mg kg$^{-1}$, the proportion of $N_2O$ produced during denitrification increased; up to 20% in the case of $NO_3^-$ and over 80% in the case of $NO_2^-$. There was a very interesting interaction with pH. With no added $NO_3$. the amount of $N_2O$ produced at pH 4.9 or 6.5 in a Brookston soil was minimal. When 10 mg kg$^{-1}$ $NO_3^-$ was added, however, the amount of $N_2O$ produced at pH 6.5 rose to nearly 20% and to over 60% at pH 4.9. One explanation for this phenomenon is that $N_2O$ reductase is most sensitive to pH among the denitrification enzymes.

### 5.5.4 pH

For a thorough review of pH effects on denitrification, see Ŝimek and Cooper (2002). Denitrification has been observed in soils with pH as low as 3.5 and as high as 11 with an optimum between pH 6–8 (Coyne 2008). Denitrification rates in acid soils generally increase if the pH is raised; similar studies do not appear to have been conducted with alkaline soils artificially acidified. Denitrifiers are cultured with difficulty (if at all) from acid soils, but if these soils are alkalinized and incubated, culturable denitrifiers can represent up to 15% of the total population (Blosl and Conrad 1992). This does not mean the denitrifiers in such soils have adapted to the new pH environment. Although Herold et al. (2012) suggested bacteria were adapting to increasing pH along a pH gradient from 4.5 to 7.5, the adaptation was maximum at pH 5.9. Parkin et al. (1985b) demonstrated that the pH optimum of denitrification from an acid soil was approximately 4.1 while that from a neutral soil was approximately 6 and denitrification rates in both soils declined above or below those optima. The overall denitrification rate in the neutral soil was approximately three times higher than that in the acid soil (6000 ng $N_2O$-N g$^{-1}$ d$^{-1}$ vs. 1600 ng $N_2O$-N g$^{-1}$ d$^{-1}$). That Parkin et al. (1985b) were unable to isolate a denitrifying culture from acid soils and maintain acid tolerance after initial enrichment and subculture strongly argues for a subset of non-culturable denitrifiers specifically adapted to acid conditions and amenable to discovery by molecular techniques.

Chen et al. (2015b) found that maximum denitrification occurred at 55–90% WFPS and pH 7–8 and fungi contributed more to denitrification suboxically and in acidic conditions. As pH increased from 4.5 to 7.5 Herold et al. (2012) found that fungal contribution to $N_2O$ evolution decreased, as did Mothapo et al. (2013) in a separate study.

Appropriate liming practices to raise soil pH to neutral or slightly alkaline conditions would appear to be another management practice based on denitrifier physiology that could reduce the contribution of fungi to aerobic denitrification and better ensure prokaryotic denitrification proceeds completely to $N_2$ because the mole fraction of $N_2O$ in denitrification increases as soils become more acidic. However, Chen et al. (2013) argue that despite decreasing $N_2O$ mole fraction, raising pH in soil will ultimately increase $N_2O$ evolution, particularly in C-amended environments, because the overall denitrification rate increases.

### 5.5.5 Temperature

Denitrification has been measured at temperatures ranging from 4 to 60°C (Coyne 2008) with at least one record of loss at a temperature of 1°C (Jacobsen and Alexander 1980). The $Q_{10}$ (rate change per 10°C increase or decrease in temperature) ranges from 1.5 to 3.0 between 10 and 35°C. Cooper and Smith (1963), for example, recorded a $Q_{10}$ of 2 for denitrification in this temperature range. Below 15°C denitrification rates decline significantly, as they do above 45°C. Temperatures vary diurnally and seasonally, and consequently, so do denitrification rates.

Denitrifier populations adapt to a given environmental temperature and there may or may not be adaptability to temperature change beyond that niche. For denitrifiers isolated from a temperate soil (<20°C), Gamble et al. (1977) found that 68% grew at 4°C, 22% grew at room temperature (ca. 15–20°C) and only 10% grew at 41°C. Of denitrifiers isolated from a tropical soil at >41°C, none grew at 4°C, 33% grew only at room temperature, and 67% grew at 41°C. It should be noted that these isolates were derived from cultures originally obtained from the soils in question at room temperature.

Herbert and Nedwell (1990) also found optima associated with denitrifiers from various environments; winter-sampled sediment had psychrophilic temperature optima while summer-sampled sediment had mesophilic temperature optima. In contrast to Parkin et al. (1985b), they could establish new temperature optima in their sediment samples within four weeks by incubating them at warmer or colder temperatures as appropriate. It remains to be demonstrated if this is an adaptation of individual organisms or a change in dominance by an existing diverse population, which could be revealed by genomic analysis.

Temperature effects on denitrification products have been observed. Lensi and Chalamet (1982) observed that the $N_2O/N_2$ ratio increased as temperatures declined below 20°C. Bailey (1976) found that NO was the dominant product at 6–8°C, $N_2O$ was the dominant product at 10°C, and $N_2$ the dominant product above 15°C. From this perspective it appears that warmer soils (within temperature ranges typical of plant production systems) appear to promote more complete denitrification.

## 5.6 SOIL MANAGEMENT AND DENITRIFICATION

It is important to remember that many studies evaluating denitrification from soils do so solely on the basis of $N_2O$ evolution. While denitrification may be a major source of $N_2O$, it is not the only source, and in some environments, it may be a minor source. In the absence of specific treatments partitioning denitrification from other processes, $N_2O$ reflects the integration of prokaryotic denitrification, fungal denitrification, nitrifier denitrification, and potential abiotic sources (Robertson and Tiedje 1987).

### 5.6.1 TILLAGE

Pioneering work on the consequences of tillage and no-tillage for denitrification was performed at the University of Kentucky by Rice and Smith (1982). At least initially, $N_2O$ flux from intact soil cores utilizing the acetylene block method was substantially greater than similar cores from plow tillage soil. After about a day, the rates were similar, and after 3 days rates from plow tillage cores were somewhat greater. This argues that the soil environment in no-tillage soils is primed for denitrification relative to the tilled soil, but after time available C or $NO_3^-$ becomes limiting. Compared to prairie soil, an adjacent cultivated soil (with ½ to ¼ the organic C content) had significantly lower denitrification rates in intact cores and 1/10 the denitrification enzyme activity of unburned or burned prairie (Groffman et al. 1993a).

#### 5.6.1.1 Soil Structure Effects

One of the principle reasons for the difference in denitrification rates between tilled and untilled soils, in addition to the higher C levels in untilled soils, is the persistence of larger soil aggregates in untilled soils. Sexstone et al. (1985b) found that the smallest aggregates having an anaerobic center in an otherwise aerobic environment was 4 mm in diameter. $O_2$ concentrations fall steeply from the surface of aggregates to the interior. Nitrate is potentially diffusion limited in aggregates larger than 2 mm in an anaerobic soil (Myrold and Tiedje 1985).

No studies having been conducted to partition $N_2O$ flux between nitrification and denitrification as a function of aeration can now be considered complete if they did not also include tests to partition fungi from the process. Although $C_2H_2$ inhibits nitrification, fungal denitrification stops at $N_2O$ and earlier studies would not have fully accounted for it. No-tillage soils have more C and greater fungal biomass than tilled soils, leading to more capacity for fungal denitrification. No fungal isolates producing more than 1000 mmol $N_2O$ $d^{-1}$ were isolated from non-agricultural sites (Mothapo et al. 2013). This suggests selection for either particular genera or particular populations able to compete effectively for readily available resources in cultivated environments.

## 5.6.2 IRRIGATION AND DRAINAGE

Irrigation and rainfall stimulate denitrification, and significant bursts of denitrification follow wetting events. In Sexstone et al.'s study (1985a) the burst was faster in a sandy loam than clay loam soil, but persisted longer in the clay loam soil, which ultimately had higher denitrification rates and greater denitrification losses. In rare instances, denitrification was not increased with a wetting event, and this was attributed to either $NO_3^-$ or C limitation. Irrigation and rainfall will cause greater $NO_3^-$ leaching in no-tillage than plow tillage soils due to greater macroporosity, and this will remove some $NO_3^-$ that could be denitrified within the plant root zone (Stoddard et al. 2005).

### 5.6.2.1 $NO_3^-$ Leaching

Locally, $NO_3^-$ leaching is detrimental to nitrogen use efficiency. Regionally, it can have severe environmental and health consequences.

Nitrate losses from cropland are subject to the vagaries of management practice and weather. Tile drainage, for example, essentially brings the field into the stream. In a tile drain study, Kladivko et al. (1999) found $NO_3^-$ losses in drainage ranging from 14 to 105 kg $ha^{-1}$, primarily in the non-growing season. The greatest lost was at 5-m spacing relative to 20-m spacing. In a study from the U.S. Midwest, 56% of the area in five study sites contributing agricultural drainage to the St. Joseph River basin in Indiana was artificially drained. Flow weighted discharge over a 5-year period from these sites ranged from 0.08 to 6.55 mg $NO_3^-$ - N $L^{-1}$ (annual averages ranged from 1.1 to 5.5 mg $NO_3^-$ -N) (Smith et al. 2008). These flows amounted to 0.05 to 19 kg $ha^{-1}y^{-1}$ $NO_3^-N$.

Using organic N sources does not provide immunity from $NO_3^-$ leaching. In organic wheat production in Italy, 25–33 kg N $ha^{-1}$ as $NO_3^-$ were lost annually over a 2-year study (Tosti et al. 2016). Drainage solution concentrations could reach as high as 30 mg $L^{-1}$ $NO_3^-$ -N below 90 cm depth. The greatest $NO_3^-$ loss occurred at the onset of drainage prior to crop growth. A persistent problem with relict feedlots is the long-term contribution of the animal wastes to $NO_3^-$ leaching. Peña-Yewtukhiw et al. (2009) studied a site in Kentucky in which an abandoned feedlot served as a point source of soluble $NO_3^-$ contaminating a domestic well (to 45 mg $L^{-1}$ $NO_3^-N$). Nitrate concentrations at depth were consistently lower in soils mapped to Loring soils (Oxyaquic fragiudalf) suggesting that perched water tables above the fragipan helped mitigate the $NO_3^-$ concentrations by denitrification (see Section 5.4.5.2). Because the bulk of soil with elevated organic N was within the first 30 cm of soil, the solution to this problem was excavation and distribution of the most affected soil over surround agricultural fields (J. Grove, personal communication).

Drought contributes to $NO_3^-$ persistence in soil, which can subsequently leach in later years. For 100 wadeable streams in the U.S. Midwest, 17% had unusually high $NO_3^-$ concentrations (19.8 vs. 5.31 mg $L^{-1}$ $NO_3^-N$ (Van Metre et al. 2016)). These sites corresponded with higher residual $NO_3^-$ levels. Overall stream $NO_3^-$ concentrations ranged from < 0.04 to 41.8 mg $L^{-1}$ $NO_3^-$ N. Likewise, a high $NO_3^-$ concentration was observed in discharges to the Illinois River in 2013 after a 2012 drought (McIsaac et al. 2016). There was a significant correlation between flow-weights $NO_3^-N$ concentration and cumulative residual agricultural inputs to the watershed in a 6-year window, although, overall, better N management seemed to lead to a downward trend in river $NO_3^-$ N concentration from 1990–2014.

Irrigated agriculture relies on substantial leaching to remove salt accumulation. This has the effect of removing soluble $NO_3^-$ below the root zone and into groundwater. Irrigated agriculture in California illustrates the adverse consequences of this practice. While surface waters generally contain <1 mg $L^{-1}$ $NO_3^-$ -N, groundwater $NO_3^-$ typically has a background concentration of

2 mg L$^{-1}$ NO$_3^-$ -N (Rosenstock et al. 2016). From the 1950s to the present, well-water NO$_3^-$ concentrations have been increasing 0.08 mg L$^{-1}$ NO$_3^-$ N annually. By 2010 over 60% of monitored wells were above natural background levels, 40% were above 5 mg L$^{-1}$ NO$_3^-$ -N, and 20% were above the MCL of 10 mg L$^{-1}$ NO$_3^-$ -N, which is regarded as a concentration at which chronic exposure (especially for infants) becomes a health issue.

One of the important limitations to mitigating high NO$_3^-$ concentrations in groundwater is the diffuse nature of this water source. A second limitation is that there can be a 5- to 50-year lag before leachate manifests itself in groundwater; remediation strategies applied today may therefore take decades to show effect (Rosenstock et al. 2016).

The Mississippi/Missouri river drainage basin in the central part of the United States illustrates another adverse regional consequence of NO$_3^-$ escaping potential denitrification in soil by leaching or drainage. Nitrate, in addition to other nutrients, is regarded as a major contributor to hypoxia in the Gulf of Mexico. The long-term (since 1985) area of the hypoxic zone is slightly less than 15,000 km$^2$ (USEPA 2015). Nitrate discharge from the basin in the 2011–2014 sampling period was approximately 130,000 metric tons of NO$_3^-$ -N. States contributing to nutrient discharge are tasked with significant reductions in loss by 2035 to reduce the hypoxic zone to <5000 km$^2$. The interim target for 2025 is a 20% reduction in discharge (USEPA 2015).

### 5.6.2.2  Point of Discharge Denitrification

Edge of field loss of NO$_3^-$ has been addressed by attempting to force surface runoff into the phreatic zone where reducing conditions can stimulate denitrification. This is most often accomplished by vegetated filter strips of either grass or other types of vegetation (Handayani et al. 2008). Barfield et al. (1998) demonstrated on silt loam soil in central Kentucky that NO$_3^-$ trapping by grass filter strips ranged from 95% with 4.6 m lengths to 98% with 9.1 m lengths.

Reactive biofilters are a relatively new approach to mitigating NO$_3^-$ loss in runoff and shallow groundwater. Bioreactors are designed to intercept flow at field edge and by means of enrichment as well as managed anaerobic conditions to facilitate denitrification. Bioreactors have been used since 1995 to treat groundwater and have been adapted to treat agricultural tile drainage waster and to polish onsite-waste-water. Three major versions of bioreactors are walls installed parallel to stream banks, beds, through which point sources of discharge (i.e., tile drain outlets), and columns.

Addy et al. (2016) performed a meta-analysis of studies using wood chip bioreactors of these three types. Mean NO$_3^-$ N removal ranged from 0.7 to 4.7 g m$^{-3}$ d$^{-1}$, with bed and lab column reactors performing significantly better than wall reactors. Key factors influencing successful mitigation were temperature (Q$_{10}$ of 2.15) and long hydraulic residence time (>6 hours). Reactors also tended to have their best performance the year of installation.

The idea has also been explored to use soil properties and sediments in drainage ditches to amplify their potential for NO$_3^-$ reduction at edge of field sites (Kröger et al. 2007, 2014). One of the main limitations to this application is the lack of labile C in some of these environments relative to true wetland environments used to mitigate NO$_3^-$ in runoff.

### 5.6.3  Liming and Fertilization

Liming generally benefits overall denitrification rates. Liming pastures, for example, increases denitrification capacity even though other soil parameters do not change (Isabella and Hopkins 1994). Because liming adversely influences the competitiveness of fungi relative to prokaryotes, it can be considered a potential means to select against denitrifying fungi whose contribution to aerobic N$_2$O loss from soil declines as soil pH rises.

Fertilization increases denitrification by removing at least one limiting factor—NO$_3^-$—regardless of whether one fertilizes with urea, NH$_4^+$, or NO$_3^-$. Groffman et al. (1993a) showed that burned and unburned prairies dramatically responded to NO$_3^-$ addition, increasing from less than 50 g ha$^{-1}$ d$^{-1}$ denitrification to >900 g ha d$^{-1}$ when NO$_3^-$ was added. In an unfertilized environment Myrold and

Tiedje (1985) suggest there is some potential for $NO_3^-$ diffusion to limit denitrification rates. The stimulation is hazardous to soil N loss if fertilization occurs in wet fields, Schnabel and Stout (1994) reported losses of 40% of the N applied (110 kg N ha$^{-1}$) in a wet field 2 weeks after fertilization relative to comparable losses of only 2.5 kg ha$^{-1}$ in a similarly fertilized well-drained soil.

Urea hydrolysis can lead to temporary $NO_2^-$ accumulation in soil and this has an indirect effect on denitrification because of the sensitivity of some denitrifiers to elevated $NO_2^-$ levels and the general toxicity of $NO_2^-$. Orthophosphate $\left(H_xPO_4^{y-}\right)$ was reported to inhibit $NO_2^-$ reduction in a pure culture of *Pseudomonas* with a $c,d_1$ type $NO_2^-$ reductase (Barak and van Rijn 2000). It is not known whether this is a generalized phenomenon or specific to the denitrifier used in the study.

### 5.6.3.1 Enhanced Efficiency Fertilizers

Enhanced efficiency fertilizers are one component of the "4R" strategy for soil nutrient management (right rate, right time, right location, right formulation). These products reduce nutrient loss by runoff, leaching, sorption, and volatilization relative to reference materials, and enhance plant uptake or at least accessibility to nutrients. Among the mechanisms of action are controlled or metered release (i.e., slow-release) and bioinhibition (Hauck 1985). Either mechanism limits denitrification by controlling the availability of $NO_3^-$ to denitrifying organisms. Controlled release usually includes: (1) coating soluble nutrients with relatively insoluble compounds such as elemental sulfur (e.g., sulfur coated urea); (2) using materials of limited solubility such as magnesium ammonium phosphate; and (3) using materials of reduced mineralizability (e.g., urea-aldehydes, melamine).

Coatings include materials with minute porosity through which dissolved materials diffuse; coatings that must be broken by physical abrasion, chemical, or biological action; and coatings that are semipermeable and allow water influx until in internal pressure potential ruptures containment. Various coatings have been used, such as asphalts, tars, gums, latexes, oils, paraffins, acrylic and epoxy resins, polyolefins, polyvinyl chloride, polystyrene, polyurethane, urea formaldehydes, vinyl acetate, and sulfur (Hauck, 1985). Bioinhibitors include urease inhibitors (e.g., NBPT, thiourea) or nitrification inhibitors (e.g., DCD, nitrapyrin,).

Mutiple studies have shown beneficial effects of enhanced efficiency fertilizers on both yield and reduced $N_2O$ loss. A meta-analysis for enhanced efficiency fertilizers used with major grain crops (maize, *Zea mays* L.; rice, *Oryza sativa* L.; wheat, *Triticum aestivum* L.) showed that controlled release fertilizer and bioinhibitors consistently reduced $N_2O$ loss relative to conventional fertilizers. Bioinhibitors rather than controlled release fertilizers increased plant yield (Chatterjee et al. 2016). In golf fairways, $N_2O$ emissions were 1.9 kg N ha$^{-1}$y$^{-1}$ with polymer-coated urea, 6.5 kg N with nitrification and urease inhibitors, and 7.6 kg N with a balanced methylene urea treatment (Gillette et al. 2016). In sandy loam soil, nitrapyrin, dicyandiamide (DCD) + nitrapyrin, and polymer-coated urea all significantly reduced $N_2O$ evolution (Awale and Chatterjee 2016). Pre-planted N-fertilizers treated with either nitrapyrin or polymer coated had 19% higher yields than untreated controls in temporarily flooded fields (Kaur et al. 2017). However, neither $NO_3^-$ leaching or yield were significantly influenced by polyacrylamide-coated biosolids relative to an uncoated control (Mailapalli and Thompson 2012).

### 5.6.4 Organic C Management

### 5.6.4.1 Plants

Plants have a complex relationship with denitrification that can be both positive and negative (Coyne 2008):

1. They exude C, which serves as an electron donor for denitrification and stimulates growth in the heterotrophic population.
2. They consume $O_2$, which can create anaerobic microsites as well as stimulate respiration and $O_2$ consumption in heterotrophs.

3. They stimulate $NH_4^+$ production and subsequent nitrification.
4. They can aerate flooded soils through aerenchyma.
5. They can create drier soils by evapotranspiration.
6. They can remove $NH_4^+$ and $NO_3^-$ by uptake.

Overall the response seems positive, and denitrification is higher in cropped than uncropped fields. Plants also affect the distribution of final denitrification products. In organic-poor crop and pasture soils $N_2O$ accumulated but did not do so when living plant roots were present (Stephanson 1972). Basche et al. (2014) conducted a meta-analysis of cover crops and $N_2O$ flux and reported that in 40% of cases cover crops decreased flux.

### 5.6.4.2 Organic Amendments

A common theme in denitrification since 1895 has been that adding organic matter, particularly labile organic matter, will stimulate denitrification activity and rates (Payne 1981). When organic wastes are injected they create artificial "hotspots" in soil that mimic naturally occurring organic debris. Rice et al. (1988) showed that readily available organic C wastes can have a huge influence on the potential for denitrification in two ways. First, rapid decomposition releases $NH_4^+$ that can diffuse into aerobic sites, nitrify, and diffuse into anaerobic zones. Anaerobic zones surrounding wastes develop because of rapid $O_2$ depletion by actively growing heterotrophs. This results in upwards of 50% of the N in the organic amendments being lost through denitrification.

### 5.6.5 PESTICIDES AND HERBICIDES

Philippot et al. (2007) reviewed the effects of various herbicides, fungicides, and insecticides on denitrification activity. These compounds are generally nonspecific as they target whole populations and can both increase and decrease potential activity. An exception will be nitrifier denitrification because nitrifiers seem to be inherently more susceptible to pesticides. Increases in activity have been attributed to release of labile C from dead cells or use of the pesticide as an electron donor. Negative effects have been attributed to cell death as well as nonspecific stress. An interesting observation was that fungicides had a more negative influence on potential denitrification than either herbicides or insecticides. Fungicides will obviously have an adverse effect on fungal denitrifiers globally, but specific studies investigating this effect have not been done.

## 5.7 ECOSYSTEM SCALE DENITRIFICATION STUDIES

### 5.7.1 SPATIAL VARIABILITY

Denitrification is a tremendously variable process in the soil environment as a consequence of the unique confluences of substrate, electron donor, and hypoxia necessary for the process to occur. Robertson et al. (1988) employed a geostatistical approach to identify locations promoting denitrification in an abandoned agricultural field. The denitrification rates were log normally distributed (a common feature) with a spatial range of 11.5 m. Soil respiration, potential nitrification, and moisture content were the three environmental factors most associated with the denitrification rate.

Parkin (1990) conducted a similar study in an agricultural field over an inference space of just $1.2 \times 2$ m. Adjacent locations could have denitrification rates differing by as much as 2–3 orders of magnitude. Potential activity in these locations sometimes, but not always, followed the same trend. In general, denitrification rates measured in a spatial context tend to have a great many low rates and a few abnormally high rates.

Whereas Robertson et al. (1988) found N availability and process was highly patterned in relation to plant community, Groffman and Tiedje (1989) found that soil texture and drainage were the main factors driving denitrification in forest soils. The patterning of activity would therefore seem to be driven by landscape scale features that differ from ecosystem to ecosystem.

### 5.7.2 TOPOGRAPHIC INFLUENCES

Topography does and doesn't matter depending on the ecosystem in which denitrification is measured. Groffman et al. (1993a) found no relation between topographic position (summit, backslope, toeslope) in a tallgrass prairie regardless of management (Konza Prairie, Kansas). In contrast, Pennock et al. (1992) found very specific relationships in an irrigated field between slope position and slope curvature (convex vs. concave), with those positions receiving water having greater evidence for denitrification. Young and Briggs (2005) found the most consistent reductions in $NO_3$-N in groundwater occurring in those transition zones where water moved from well-drained cropland to poorly drained riparian buffer soils.

### 5.7.3 SEASONAL AND ANNUAL VARIATION

Van Kessel et al.'s work (1993) illustrates a common trend in cultivated soils. Prior to tillage and planting, maximum denitrification is low. It increases in the first half of the growing season and decreases thereafter as C, $NO_3^-$, and water are depleted in soil. By fall it is at low ebb. Riparian soils can also show a seasonal shift. In all years, potential denitrification was greater in spring and summer than fall (Merrill and Benning 2006). As previously noted (Section 5.4.1.2.2) over a 5-year period the population dynamics of denitrifiers changed, which most likely has an effect on overall rates. Merrill and Bennning (2006) found that from year-to-year and season to season, potential denitrification rates could fluctuate.

### 5.7.4 GLOBAL AND REGIONAL GAS FLUXES

The major environmental concern with denitrification is its contribution to greenhouse gas emission and global warming when full reduction of $NO_3^-$ does not occur and $N_2O$ is evolved from soil. Multiple studies indicate that two key factors lead to exponential increases in $N_2O$ evolution from soil: water in excess of 70% WHC capacity and excess N fertilization (e.g., Hatfield 2017; Cai et al., 2016). Estimates from the IPCC range from 0.003 to 0.03 kg $N_2O$-N per kg N fertilizer applied (IPCC 2006) suggesting that $N_2O$ loss is highly site and practice specific.

Nitrous oxide has almost 300 times the global warming potential as an equivalent amount of $CO_2$ and an atmospheric residence time of over 100 years. It contributes about 7% to global radiative forcing, and it is increasing in the atmosphere at an estimated rate of 0.26% $y^{-1}$ (Tian et al. 2015). The global production range of $N_2O$ is estimated from 6.7 to 36.6 Tg N $y^{-1}$ (IPSS 2001). The Dynamic Land Ecosystem Model (DLEM) used by Tian et al. (2015) suggested that from 1998 to 2010 global $N_2O$ production was 12.52 ± 0.74 Tg N $y^{-1}$ with the greatest increases occurring in Northern and Southern tropical areas (e.g., Southeast Asia and South America). On a global basis, $N_2O$ production (in Tg $N_2O$-N $y^{-1}$) for various biomes ranked as follows: forests (4.28), cropland (3.36), shrubland (2.82), wetlands (0.97), and grassland (0.82) (Tian et al. 2015).

Comprehensive N-assessment data from California indicates that N-fertilizers are the primary cause (80%) of recent increases in $N_2O$ flux (Rosenstock et al. 2016). Soils represented 63% of the $N_2O$ flux in California (approximately 24 Gg N $y^{-1}$ in 2005) (Haden et al. 2016). The amount of $N_2O$ produced represents about 1.4% of the N applied; but this is highly site and management specific. Beyond 300 kg N fertilizer $ha^{-1}$ $N_2O$ production increases exponentially. For example, in rice production, where flooding occurs as part of the production practice, $N_2O$ loss is estimated to represent only 0.3% of the N applied (Rosenstock et al. 2016). Sixty percent of total agricultural $N_2O$ emissions in California come from producing just five crops: cotton (18%), almonds (11%), wheat (11%), processing tomatoes (8%), and lettuce (7%) (Rosenstock et al. 2016).

## 5.8  MEASUREMENT AND ASSESSMENT

How does one measure denitrification? You can look at the organisms themselves (enumeration and characterization) or you can look at the process. In looking at the process you can examine the disappearance of its substrates ($NO_3^-$, $NO_2^-$ NO, $N_2O$) or the appearance of its products ($NO_2^-$ NO, $N_2O$, $N_2$). In looking at the process you can also assess how rapidly it occurs in optimal conditions (potential assays) as well as in field conditions.

### 5.8.1  POPULATION MEASUREMENTS

Most studies enumerating soil denitrifiers use variations of the most probable number (MPN) procedure in either test tubes or microtiter plates (Rowe et al. 1977). Anaerobic growth on solid media is not definitive because many organisms can grow by fermentation or $NO_3^-$ respiration. Nutrient broth is one of the preferred media and $NO_3^-$ disappearance as indicated by a spot test is regarded as the criterion for presumptive evidence of denitrifiers (Focht and Joseph 1973; Tiedje 1994). Sealed tubes can also be incubated with acetylene so that $N_2O$ production in the headspace also becomes a criterion for a positive response. Additional modifications include using inverted Durham tubes (saccharimeter tubes) that collect gas bubbles (presumably $N_2$, which is much less soluble than either $CO_2$ or $N_2O$) or completely filling tubes sealed with rubber septa, and evaluating presence or absence of gas bubbles.

The selectivity of the media and the requirement for active growth to produce a measurable response are the greatest limitations for the MPN technique relative to enumeration by molecular probes. As a general rule, MPN enumeration will give values for denitrifiers orders of magnitude less than quantitative PCR. A major reason is media selectivity. For example, in unpublished work by El-Galil et al., which enumerated denitrifiers by MPN in certified organic soils, a tremendous difference in population occurred depending on whether tryptic soy broth (TSB) or nutrient broth (NB) was used (Table 5.6). Increasing the pH of the incubating media from acidic to neutral increased the estimated MPN in samples from cropland or forest (Davidson et al. 1985) but did not otherwise change the relative relationship of MPN between sites.

---

**TABLE 5.6**

**MPN of Nitrate Reducers in Tryptic Soy Broth (TSB) and Nutrient Broth (NB) (Values are MPN $g^{-1}$ oven dry soil)**

| Site | Field | TSB | NB |
|---|---|---|---|
| Kentucky State | Vetch | $7.51 \times 10^6$ | $6.36 \times 10^2$ |
| | Vetch/Rye | $3.22 \times 10^6$ | $1.89 \times 10^3$ |
| | Sod | $2.95 \times 10^6$ | $1.47 \times 10^4$ |
| Farm 1 | Sod | $4.69 \times 10^6$ | $2.95 \times 10^6$ |
| | F2 | $7.24 \times 10^7$ | $1.88 \times 10^6$ |
| | F3 | $9.38 \times 10^6$ | $6.57 \times 10^5$ |
| | F5 | $1.00 \times 10^8$ | $9.30 \times 10^2$ |
| Farm 2 | F1 | $2.28 \times 10^5$ | $3.08 \times 10^1$ |
| | F2 | $4.69 \times 10^5$ | $3.08 \times 10^3$ |
| | F3 | $1.74 \times 10^6$ | $1.74 \times 10^3$ |
| | F4 | $1.47 \times 10^6$ | $2.41 \times 10^4$ |

---

It is becoming more common to use molecular probes to evaluate copy numbers of specific denitrification enzymes (see Section 5.4.1.1). While this is more sensitive to detecting lower numbers of denitrifiers ($10^1$ to $10^2$ target molecules per reaction; Henry et al. 2006), it has several embedded assumptions:

1. The probes nonselectively amplify all denitrifiers.
2. The probe target represents a gene that is present in the denitrifier (no truncation of genes).
3. The probe targets all versions of the gene (*nirS* and *nirK*, for example).
4. The probes recognize the targeted denitrification gene.

Any or all of these assumptions are likely to be violated, particularly the last one because molecular probes are typically based on sequences in culturable organisms. As genetic data bases grow larger and more inclusive, however, this will become less of an issue.

## 5.8.2 POTENTIAL MEASUREMENTS

Potential measurements integrate all denitrification enzymes into a single process in which conditions for activity are optimized (temperature, pH, anaerobiosis, $NO_3^-$ supply, and C availability). Usually these are performed as soil slurries with anaerobic headspace in a buffered solution. The standard method (denitrification enzyme assay, DEA) was developed by Smith and Tiedje (1979a) to look at the phases of denitrification as soils were placed in denitrifying conditions. Acetylene is used to inhibit $N_2O$ reduction and facilitate detection. Generally, $N_2O$ production commences immediately and is linear for approximately four hours, thereafter increasing. This is taken as an indication that microbial growth is occurring and new denitrification enzymes are being produced. Consequently, chloramphenicol is often added to the potential assay to prevent new enzyme synthesis. The linear rate of $N_2O$ production is therefore a reflection of the denitrification capacity of the existing enzymes in the system when substrates are unlimited and repression by $O_2$ is removed. Specific activity can be inferred from the potential rates if the microbial population is known. But Martin et al. (1988) determined that this was of dubious value given that MPN counts are unlikely to reflect the true population size. Rather, the DEA is an extremely useful comparative tool for the potential of denitrification in different soils and an index of the denitrifying biomass (Martin et al. 1988).

Potential measurements vary widely depending on source and treatment (Table 5.7) with rates ranging from < 0.01 to 8.75 µg N $g^{-1}$ $h^{-1}$.

## 5.8.3 ACETYLENE INHIBITION

Measuring $N_2$ production to determine the rate of denitrification is technically feasible but operationally difficult in terms of measuring small changes in the total $N_2$ content of air or even an alternate gas (because of contamination from ambient air). Utilizing $^{15}N$-labeled substrates gets around this problem by measuring $^{15}N_2$ as the final product but adds the additional expense of a mass spectrometer. Measuring denitrification is facilitated by adding acetylene ($C_2H_2$), a specific inhibitor of $N_2O$ reduction to $N_2$ by $N_2O$ reductase. Concentrations as low as 10 kPa are effective (Klemedtsson et al. 1990). Because $N_2O$ is present in the atmosphere at naturally low concentrations (<0.33 ppm by vol.) and it can be detected with great sensitivity by gas chromatography, its production and accumulation make for a convenient measure of the denitrification rate, especially because in optimum conditions $N_2O$ reduction is not the rate limiting step in denitrification. Techniques for utilizing acetylene as an inhibitor were reviewed by Tiedje et al. (1989).

**TABLE 5.7**

**Potential Denitrification Enzyme Activity in Various Soils (Units Converted Where Necessary for Comparison)**

| Soil | Treatment | Rate (N$_2$O–N) | Reference |
|------|-----------|-----------------|-----------|
| **Agricultural** | | | |
| Brookston loam | no additions | 0.18–0.47 µg N g$^{-1}$ h$^{-1}$ | Smith and Tiedje 1979a |
| Brookston loam | no additions | 0.95–1.43 µg N g$^{-1}$ h$^{-1}$ | Firestone and Tiedje 1979 |
| Conover loam | no additions | 0.22–0.71 µg N g$^{-1}$ h$^{-1}$ | Firestone and Tiedje 1979 |
| Brookston loam | 0.15% glucose | 0.06 µg N g$^{-1}$ h$^{-1}$ | Smith and Tiedje 1979a |
| Brookston loam | 0.10% glucose | 2.37 µg N g$^{-1}$ h$^{-1}$ | Smith and Tiedje 1979a |
| Lanton | glucose + NO$_3^-$ | 1.08 µg N g$^{-1}$ h$^{-1}$ | Martin et al. 1988 |
| Matapeake silt loam, plowed | glucose + NO$_3^-$ | 0.10 µg N g$^{-1}$ h$^{-1}$ | Parkin and Meisinger 1989 |
| Matapeake silt loam, no-till | glucose + NO$_3^-$ | 0.40 µg N g$^{-1}$ h$^{-1}$ | Parkin and Meisinger 1989 |
| Spinks loamy sand (pre-incubated at 6% water content) | | 0.05 µg N g$^{-1}$ h$^{-1}$ | Smith and Tiedje 1979a |
| Spinks loamy sand (pre-incubated at 6% water content) | | 0.13 µg N g$^{-1}$ h$^{-1}$ | Smith and Tiedje 1979a |
| **Forest** | | | |
| Conifer/*Alnus* sp. | glucose + NO$_3^-$ | <0.01–1.90 µg N g$^{-1}$ h$^{-1}$ | Vermes and Myrold 1992 |
| *Oxalis* deciduous | no additions | 0.83–1.25 µg N g$^{-1}$ h$^{-1}$ | Müller et al. 1980 |
| Pine | no additions | 0.54–0.67 µg N g$^{-1}$ h$^{-1}$ | Müller et al. 1980 |
| Poplar (*Liriodendron* sp.)/Alder (*Alnus* sp.) | glucose + NO$_3^-$ | 0.43 ug N g$^{-1}$ h$^{-1}$ | Klingensmith and Van-Cleve 1993 |
| Spruce | no additions | 0.33–0.92 µg N g$^{-1}$ h$^{-1}$ | Müller et al. 1980 |
| *Alnus* sp. | no additions | 2.50–8.75 ug N g$^{-1}$ h$^{-1}$ | Müller et al. 1980 |
| Sedge pine (*Pinus* sp.) | no additions | 13.3–22 ug N g$^{-1}$ h$^{-1}$ | Müller et al. 1980 |
| **Riparian** | | | |
| Undefined | glucose + NO$_3^-$ | 0.30–1.20 µg N g$^{-1}$ h$^{-1}$ | Lowrance 1992 |
| Well drained | glucose + NO$_3^-$ | <0.01–0.09 µg Ng$^{-1}$ h$^{-1}$ | Groffman et al. 1992 |
| Somewhat poorly drained | glucose + NO$_3^-$ | 0.01–0.10 g$^{-1}$ h$^{-1}$ | Groffman et al. 1992 |
| Poorly drained | glucose + NO$_3^-$ | 0.10–1.08 µg N g$^{-1}$ h$^{-1}$ | Groffman et al. 1992 |
| Very poorly drained | glucose + NO$_3^-$ | 0.05–1.06 µg N g$^{-1}$ h$^{-1}$ | Groffman et al. 1992 |
| Montane | succinate + NO$_3^-$ | 0.78–1.35 mg N g$^{-1}$ h$^{-1}$ | Merrill and Benning 2006 |
| **Prairie** | | | |
| Dwight | unburned | 0.33 µg N g$^{-1}$ h$^{-1}$ | Groffman et al. 1993a |
| Dwight | burned | 0.31 µg N g$^{-1}$ h$^{-1}$ | Groffman et al. 1993a |
| Benfield | burned/grazed | 0.23 µg N g$^{-1}$ h$^{-1}$ | Groffman et al. 1993a |
| **Wetlands/Sediments** | | | |
| Natural wetland | glucose + NO$_3^-$ | <0.1–0.9 µg N g$^{-1}$ h$^{-1}$ | Duncan and Groffman 1994 |
| Constructed wetland | glucose + NO$_3^-$ | 0.17–0.36 µg N g$^{-1}$ h$^{-1}$ | Duncan and Groffman 1994 |
| Undefined | glucose + NO$_3^-$ | <0.01–6.44 µg N g$^{-1}$ h$^{-1}$ | Tiedje et al. 1982 |
| Pine (*Pinus* sp.) bog | no addition | 0.83 µg N g$^{-1}$ h$^{-1}$ | Müller et al. 1980 |
| *Sphagnum* bog | no addition | 1.25 µg N g$^{-1}$ h$^{-1}$ | Müller et al. 1980 |

*Source:* Modified from Coyne, M.S. 2008. Biological denitrification. pp. 197–249. In J.S. Schepers and W. Raun (Eds.), *Nitrogen in agricultural systems*. Agronomy Monograph 49. ASA-CSSA-SSSA, Madison, WI.

The acetylene inhibition technique is not without its problems. In addition to the operational issues with delivering a gas into a soil environment and the diffusion problems that exist therein, the following biological issues have been identified with its use:

1. Lack of, or reduced, inhibitory effect at low $NO_3^-$ concentrations.
2. Metabolism in soil after long-term exposure; *Rhodococcus rhodochrous* and *Pelobacter acetylenicus* can both utilize $C_2H_2$ (Knowles 1990).
3. Reduced effectiveness in the presence of $H_2S$.
4. Toxic impurities in industrial supplies of acetylene.

The ideal inhibitor should have no nontarget effects but studies have shown the following adverse effects (Knowles 1990):

1. $N_2$-fixation ($C_2H_2$ is a preferential substrate for nitrogenase).
2. Nitrification ($C_2H_2$ is a specific inhibitor of ammonia monooxygenase).
3. Methanogenesis ($C_2H_2$ inhibits methane monooxygenase).
4. Growth of sulfate reducing bacteria (e.g., *Desulfovibrio desulfuricans*).

One approach to estimating *quasi* undisturbed denitrification rates is removal of intact cores from a site and incubation in a sealed chamber with a 10% acetylene atmosphere, perfusion of the intact cores with acetylene-saturated water (Montgomery et al. 1997), or vigorous sparging of small cores with acetylene. Examples of published rates by this method are in Table 5.8. Typically, soil cores are only retrieved from the upper 0–30 cm of soil. Relative to measuring denitrification rates based on $^{15}N$ evolution, use of acetylene was not significantly different (Parkin et al. 1985a).

Core size matters. Parkin (1990) showed that as core size increased from a diameter of 1.75 to 5.40 cm the median ($0.82–4.05$ ng-N $g^{-1}d^{-1}$), mean ($3.14–26.4$ ng-N $g^{-1}d^{-1}$), and CV (269–338%) of denitrification all increased. This is probably a reflection that the larger cores encompass a greater range of microsite variability.

### 5.8.4 *In Situ* Measurements

Potentially more environmentally relevant denitrification rates are obtained by static or dynamic soil covers. Static covers measure the accumulation rate with times in a sealed or vented system. Dynamic covers flush a steady stream of gas over the covered area. In both cases, the covers are slightly inserted into the soil to form a tight seal. While soil covers have most often been used by removing samples and storing them in gas-tight syringes, increasingly automation using photo-acoustic and infrared detectors are being used directly for in-field measurements (Kreba 2013). Table 5.9 provides examples of some published rates.

### 5.8.5 Isotope Analysis

Isotopes have a long history of use in denitrification studies. Tiedje et al. pioneered the use of the radioactive isotope, $^{13}N$, to conduct short-term denitrification studies (Tiedje et al. 1979). The disadvantage of $^{13}N$ is its extremely short half-life (13 minutes) and therefore the necessity for close proximity of a $^{13}N$ source (i.e., a cyclotron). Using $^{15}N$ labeled substrates to generate $^{15}N_2$ is a long-standing approach to denitrification measurements—sensitive, albeit expensive. Isotope pairing studies can be done to determine the source and rates of $N_2$ production based on relative ratios of $^{28}N_2$, $^{29}N_2$, and $^{30}N_2$ (Steingruber et al. 2001).

Of particular concern is the source of $N_2O$ from soil, which simple mass spectrometry does not adequately distinguish. It is well known that biological reactions in soil preferentially use lighter

**TABLE 5.8**
**Denitrification Activity of Intact Soil Cores (Units Converted from the Original Data to Facilitate Comparison)**

| Ecosystem | Treatment | Rate | Reference |
|---|---|---|---|
| **Cultivated** | | | |
| | Intact + $C_2H_2$ | 0.06–0.11 µg N $g^{-1}$ $d^{-1}$ | Christensen et al. 1990a |
| | Repacked + $C_2H_2$ | 1.4–1.9 µg N $g^{-1}$ $d^{-1}$ | Aulakh et al. 1991a |
| | Repacked + $C_2H_2$ + Vetch | 5.1–10.0 µg N $g^{-1}$ $d^{-1}$ | Aulakh et al. 1991a |
| | $^{15}N$ gas loss + residues | 2.0–2.9 µg N $g^{-1}$ $d^{-1}$ | Aulakh et al. 1991b |
| | Intact + $C_2H_2$ | 9.5–29.5 µg N $g^{-1}$ $d^{-1}$ | Heaney et al. 1992 |
| Maize | | | |
| (plowed) | Intact + $C_2H_2$ | 0.33–104 mg N $m^{-2}d^{-1}$ | Rice and Smith 1982 |
| (no-tillage) | Intact + $C_2H_2$ | 2.7–212 mg N $m^{-2}d^{-1}$ | Rice and Smith 1982 |
| **Successional Pasture** | Intact + $C_2H_2$ | 47.3 µg $m^{-2}$ $d^{-1}$ | Robertson et al. 1988 |
| **Bare soil** | Intact + $C_2H_2$ | 23–49 mg N $m^{-2}$ $h^{-1}$ | Rice et al. 1988 |
| | Waste-amended | | |
| | Intact $C_2H_2$ | <0.5 mg N $m^{-2}$ $h^{-1}$ | Groffman et al. 1993b |
| **Forest** | | | |
| Sugar Maple/Red Oak | Intact + $C_2H_2$ | 0.01 mg N $m^{-2}d^{-1}$ | Merrill and Zak 1992 |
| Sugar Maple/Basswood | Intact + $C_2H_2$ | 0.02 mg N $m^{-2}d^{-1}$ | Merrill and Zak 1992 |
| Silver Maple/Red Maple | Intact + $C_2H_2$ | 0.04 mg N $m^{-2}d^{-1}$ | Merrill and Zak 1992 |
| Poplar/Alder | Intact + $C_2H_2$ | 0.26 µg N $g^{-1}$ $d^{-1}$ | Klingensmith and Van-Cleve 1993 |
| Beech/Sugar Maple | Intact + $C_2H_2$ | 6–17 mg N $m^{-2}d^{-1}$ | Groffman et al. 1993b |
| Conifer/*Alnus* sp. | Intact + $C_2H_2$ | 3.4–19.3 kg N $ha^{-1}$ $yr^{-1}$ | Ruz-Jerez et al. 1994 |
| Red Oak/SugarMaple/Oak | Intact + $C_2H_2$ | 120 mg N $m^{-2}$ $yr^{-1}$ (<5 mg N $m^{-2}d^{-1}$) | Holmes and Zak 1999 |
| Sugar Maple/Basswood | Intact + $C_2H_2$ | 350 mg N $m^{-2}$ $yr^{-1}$ (<5 mg N $m^{-2}d^{-1}$) | Holmes and Zak 1999, Holmes and Zak 1999 |
| **Desert** | | | |
| Intact + $C_2H_2$ | 7.22 kg N $ha^{-1}$ $yr^{-1}$ | Peterjohn and Schlesinger 1991 | |
| **Prairie** | | | |
| Perennial tallgrass | | | |
| Burned | Intact + $C_2H_2$ | <0.5–1.5 mg N $m^{-2}d^{-1}$ | Groffman et al. 1993a |
| Burned/grazed | Intact + $C_2H_2$ | <0.5–2.0 mg N $m^{-2}d^{-1}$ | Groffman et al. 1993a |
| Unburned | Intact + $C_2H_2$ | 1.5–5.0 mg N $m^{-2}$ $d^{-1}$ | Groffman et al. 1993a |
| **Sod** | | | |
| | Intact with $^{15}N$ | 0.06–0.11 µg N $g^{-1}$ $d^{-1}$ | Christensen et al. 1990a |
| | Intact + $C_2H_2$ | 9–16 kg N $ha^{-1}$ $d^{-1}$ | Groffman et al. 1991 |
| **Riparian** | | | |
| | Intact + $C_2H_2$ | 0.4–1.4 kg N $ha^{-1}$ $d^{-1}$ | Groffman et al. 1991 |
| | Intact + $C_2H_2$ | 8.8–21.9 mg N $m^{-2}$ $d^{-1}$ | Ambus and Christensen 1993 |
| | Intact + $C_2H_2$ | 120 mg N $m^{-2}$ $d^{-1}$ | Schipper et al. 1993 |
| | Intact + $C_2H_2$ | <5–40 kg N $ha^{-1}$ $yr^{-1}$ | Hanson et al. 1994 |

*Source:* Coyne, M.S. 2008. Biological denitrification, pp. 197–249. In J.S. Schepers and W. Raun (Eds.), *Nitrogen in agricultural systems.* Agronomy Monograph 49. ASA-CSSA-SSSA, Madison, WI.

## TABLE 5.9
## Denitrification Activity in Various Soils (Units Converted from the Original Data to Facilitate Comparison)

| Ecosystem | Treatment | Rate | Reference |
|---|---|---|---|
| **Cultivated** | | | |
| Rice (flooded) | Static cover + $C_2H_2$ | 0.12 kg N ha$^{-1}$ d$^{-1}$ | Lindau et al. 1990 |
| | | | |
| **Grass filter** | Static cover | 0.76 mg N m$^{-2}$ h$^{-1}$ | Coyne et al. 1994 |
| **Forest** | | | |
| Willow | Static cover + $C_2H_2$ | 0.34 mg N m$^{-2}$ h$^-$ | Klingensmith and Van-Cleve, 1993 |
| Pine | Static cover + $C_2H_2$ | 1.68 mg N m$^{-2}$ d$^{-1}$ | Hixson et al. 1990 |
| Conifer/*Alnus* sp. | Static cover + $C_2H_2$ | <0.01–0.04 kg N ha$^{-1}$ d$^{-1}$ | Vermes and Myrold 1992 |
| Beech | Static cover + $C_2H_2$ | 0.6–4.5 kg N ha$^{-1}$ yr$^{-1}$ | Struwe and Kjoller 1994 |
| | | | |
| **Sod** | | | |
| | Static cover + $C_2H_2$ | 21.6 mg N m$^{-2}$ d$^{-1}$ | Lensi and Chalamet 1982 |
| | Static cover + $^{15}$N | 20.7 mg N m$^{-2}$ d$^{-1}$ | Christensen et al. 1990a |
| **Riparian** | | | |
| | Static cover + $C_2H_2$ | 2.5 mg N m$^{-2}$ d$^{-1}$ | Hixson et al. 1990 |

*Source:* Coyne, M.S. 2008. Biological denitrification, pp. 197–249. In J.S. Schepers and W. Raun (Eds.), *Nitrogen in agricultural systems.* Agronomy Monograph 49. ASA-CSSA-SSSA, Madison, WI.

isotopes. The $N_2O$ produced during nitrification and denitrification (and presumably fungal denitrification) is similarly depleted in $^{15}$N as lighter isotopes are preferentially used (Pérez 2005). The extent of depletion in nitrification ranges from –26 to –66.5‰ while in denitrification it ranges from –13 to –42‰. The reason that $^{15}$N depletion in denitrification is not quite as great is that further reduction of $N_2O$ to $N_2$ in denitrification results in a slight enrichment of the remaining $N_2O$. The extent of $^{15}$N depletion may also be a feature of the organisms present. Consequently, isotopic comparisons of $^{15}N_2O$ from various soil environments (e.g., temperate grasslands in England to tropical rain forests in Costa Rica) are extreme, ranging from –12 to –71‰.

The $N_2O$ molecule is linear ($N^\beta N^\alpha O$) and it has recently been observed that different processes differentially label each N in the linear molecule. The terminal N is designated the $\beta$ site and the central N the $\alpha$ site. Variation in labeling of these sites leads to different "isotopomers"—$^{15}N^{14}NO$ and $^{14}N^{15}NO$—having the same mass. The significance of this phenomenon is that, depending on the process, the central N atom in $N_2O$ will be differentially labeled (Pérez 2005).

Sutka et al. (2006) found, on average, that the site preference of nitrification was 33‰ (meaning the $\alpha$ site had preferentially more $^{15}$N than the $\beta$ site, while the site preference was approximately 0‰ for denitrification—regardless of whether that $N_2O$ came from heterotrophic denitrification or nitrifier denitrification. Thus, isotopomer analysis utilizing the natural abundance of $^{15}$N in soils appears to be a very promising means of teasing out relative contributions of specific populations and processes, such as denitrification, to $N_2O$ evolution from soil environments without relying on fertilization with stable isotopes. For example, Maeda et al. (2010) used isotopomer analysis of $N_2O$ evolution from composting manure to show that denitrification was the main source immediately after turning.

## 5.9   CONCLUDING REMARKS

The old saying, "there is nothing new under the sun," hardly applies to soil denitrification. True, much of the work on pathways was identified in the late nineteenth century. But today we are investigating the genetic underpinning of those pathways. True, much of the work on the major regulating factors of soil denitrification was worked out in the mid-twentieth century. But today we are taking advantage of new techniques in isotope technology and gene regulation to better know when those regulating factors matter, and how they influence the competitiveness of denitrifiers with other organisms. Since the start of the twenty-first century, the advances in molecular microbiology have allowed us to look at the dynamics of denitrifier populations, including those that are non-culturable. It was convenient to believe that only bacteria were denitrifiers, but new techniques have forced us to broaden our horizons about which organisms (such as fungi and archaea) actually constitute the denitrifying population. The potential for aerobic denitrification makes its contribution to global nitrogen cycling all that more interesting to study.

Soil denitrification and its control, as well as soil denitrification and its management, are as pertinent today as ever. With increased concern about agriculture's contribution to global warming and climate change, and on-going concerns about preservation of water resources, there remains the need to better understand soil denitrification and how to use it advantageously. The questions remain the same, but the tools of investigation are different. Like as not, new tools will raise new questions that could not or were not answered previously. The answers are waiting.

## ACKNOWLEDGMENTS

This work was supported by the USDA National Institute of Food and Agriculture Hatch project KY007090 with additional support from a Natural Resources Conservation Service Conservation Innovation Grant, and a USDA-ARS specific cooperative agreement. This paper (16-06-111) is in connection with a project of the Kentucky Agricultural Experiment Station and is published with approval of the Director. Mention of trade names is for information only and does not imply endorsement by the Kentucky Agricultural Experiment Station.

## REFERENCES

Addy, K., A.J. Gold, L.E. Christianson, M.B. David. L.A. Schippers, and N.A. Ratigan. 2016. Denitrifying bioreactors for nitrate-removal: A meta-analysis. *Journal of Environmental Quality* 45:873–881.
Ambus, P., and S. Christensen. 1993. Denitrification variability and control in a riparian fen irrigated with agricultural drainage water. *Soil Biology & Biochemistry* 25:915–923.
Aulakh, M.S., J.W. Doran, D.T. Walters, and J.F. Power. 1991a. Legume residue and soil water effects on denitrification in soils of different textures. *Soil Biology & Biochemistry* 23:1161–1167.
Aulakh, M.S., J.W. Doran, D.T. Walters, A.R. Mosier, and D.D. Francis. 1991b. Crop residue type and placement effects on denitrification and mineralization. *Soil Science Society of America Journal* 55:1020–1025.
Awale, R. and A. Chatterjee. 2016. Enhanced efficiency nitrogen products influence ammonia volatilization and nitrous oxide emission from two contrasting soils. *Agronomy Journal* 109:47–57.
Bailey, L.D. 1976. Effect of temperature and roots on denitrification in soil. *Canadian Journal of Soil Science* 56:79–87.
Balderston, W., B. Sherr, and W. Payne. 1976. Blockage by acetylene of nitrous oxide reduction in *Pseudomonas perfectomarinus*. *Applied and Environmental Microbiology* 31:504–508.
Barak, Y., and J. van Rijn. 2000. Relationship between nitrite reduction and active phosphate uptake in the phosphate-accumulating denitrifier *Pseudomonas* sp. strain JR 12. *Applied and Environmental Microbiology* 66:5236–5240.
Barfield, B.J., R.L. Blevins, A.W. Fogle, C. E. Madison, S. Inamdar, D.I. Carey, and V.P. Evangelou. 1998. Water quality impacts on natural filter strips in karst areas. *Transactions of the ASAE* 41:371–381.
Barnard, R., P.W. Leadley, and B.S. Hungate. 2005. Global change, nitrification, and denitrification: A review. *Global Biogeochemical Cycles* 19: GB1007, doi:10.029/2004GB002282.

Basche, A.D., F.E. Miguez, T.C. Kaspar, and M.J. Castellano. 2014. Do cover crops increase or decrease nitrous oxide emissions? A meta-analysis. *Journal of Soil and Water Conservation* 69:471–482.

Betlach, M.R. 1982. Evolution of bacterial denitrification and denitrifier diversity. *Antonie van Leeuwenhoek* 48: 585–607.

Betlach, M.R. and J.M. Tiedje. 1981. Kinetic explanation for accumulation of nitrite, nitric oxide, and nitrous oxide during bacterial denitrification. *Applied and Environmental Microbiology* 42:1074–1084.

Blosl, M. and R. Conrad. 1992. Influence of an increased pH on the composition of the nitrate- reducing microbial populations in an anaerobically incubated acidic forest soil. *Systematic and Applied Microbiology* 15:624–627.

Bollag, J.-M. and G. Tung. 1972. Nitrous oxide release by soil fungi. *Soil Biology and Biochemistry* 4:271–276.

Bothe, H., G. Jost, M. Schloter, B. Ward, and K. Wirtzel. 2000. Molecular analysis of ammonia oxidation and denitrification in natural environments. *FEMS Microbiology Reviews* 24:673–690.

Bowman, R.A., and D.D. Focht. 1974. The influence of glucose and nitrate concentrations upon denitrification rates in sandy soils. *Soil Biology & Biochemistry* 6:297–301.

Boyle, S.A., J.J. Rich, P.J. Botttomley, J.K. Cromack, and D.D. Myrold. 2006. Reciprocal transfer effects on denitrifying community composition and activity at forest and meadow sites in the Cascade Mountains of Oregon. *Soil Biology & Biochemistry* 38:870–878.

Braker, G., J. Zhou, L. Wu, A. Devol, and J.M. Tiedje. 2000. Nitrite reductase genes (*nirK* and *nirS*) as functional markers to investigate diversity of denitrifying bacteria in Pacific northwest marine sediment communities. *Applied and Environmental Microbiology* 66:2096–2104.

Braker, G., H.L. Ayala-del-Rio, A.H. Devol, A. Fesefeldt, and J.M. Tiedje. 2001. Community structure of denitrifiers, *Bacteria* and Archaea along redox gradients in Pacific northwest marine sediments by terminal restriction *fragment* length polymorphism analysis of amplified nitrite reductase (*nirS*) and 16s RNA genes. *Applied and Environmental Microbiology* 67:1893–1901.

Brunel, B., J.D. Janse, H.J. Laanbroek, and J.W. Woldendorp. 1992. Effect of transient oxic conditions on the composition of the nitrate-reducing community from the rhizosphere of *Typha angustifolia*. *Microbial Ecology* 24:51–61.

Buckley, D.H., V. Huangyutitham, T.A. Nelson, A. Rumberger, and J.E. Thies. 2006. Diversity of Planctomycetes in soil in relation to soil history and environmental heterogeneity. *Applied and Environmental Microbiology* 72:4522–4531.

Cai, Z., S.G.A. Hendratna, Y.D.M. Xu, and B.D. Hanson. 2016. Key factors, soil nitrogen processes, and nitrite accumulation affecting nitrous oxide emissions. *Soil Science Society of America Journal* 80:1560–1571.

Cavigelli, M.A. and G.P. Robertson. 2000. The functional significance of denitrifier community composition in a terrestrial ecosystem. *Ecology* 81:1402–1414.

Chatterjee, R.T.A., R. Awale, D.A. McGranahan, and A. Daigh. 2016. Effect of enhanced efficiency fertilizers on nitrous oxide emissions and crop yields: A meta-analysis. *Soil Science Society of America Journal* 80:1121–1134.

Chen, H.H., X. Li, F. Hu, and W. Shi. 2013. Soil nitrous oxide emissions following crop residue addition: A meta-analysis. *Global Change Biology* doi:10.1111/gcb.12274.

Chen, H., F. Yu, and W. Shi. 2016. Detection of $N_2O$-producing fungi in environment using nitrite reductase gene (*nirK*)-targeting primers. *Fungal Biology* http://dx.doi.org/10.1016/j.funbio.2016.07.012 (in press)

Chen, H., N.V. Mothapo, and W. Shi. 2015a. Fungal and bacterial $N_2O$ production regulated by soil amendments of simple and complex substrates. *Soil Biology & Biochemistry* 84:116–126.

Chen, H., N. Mothapo, and W. Shi. 2015b. Soil moisture and pH control relative contributions of fungi and bacteria to $N_2O$ production. *Microbial Ecology* 69:180–191.

Cheneby, D., A. Hartmann, C. Henault, E. Topp, and J.C. Germon. 1998. Diversity of denitrifying microflora and ability to reduce $N_2O$ in two soils. *Biology and Fertility of Soils* 28:19–26.

Cheneby, D., S. Perrez, C. Devroe, S. Hallet, Y. Couton, F. Bizouard, G. Iuretig, J.C. Germon, and L. Philippot. 2004. Denitrifying bacteria in bulk and maize-rhiziospheric soil: Diversity and $N_2O$-reducing abilities. *Canadian Journal of Microbiology* 50:469–474.

Christensen, S., P. Groffman, A. Mosier, and D.R. Zak. 1990. Rhizosphere denitrification: A minor process but indicator of decomposition activity, pp. 199–211. In N.P. Revsbech and J. Sørensen (Eds.), *Denitrification in soils and sediments*. Plenum Press, New York.

Clays-Josserand, A., P. Lemanceau, L. Philippot, and R. Lensi. 1995. Influence of two plant species (flax and tomato) on the distribution of nitrogen dissimilative abilities within fluorescent *Pseudomonas* spp. *Applied and Environmental Microbiology* 61:1745–1749.

Conrad, R. 1990. Flux of NOx between soil and atmosphere: Importance and soil microbial metabolism, pp. 105–128. In N.P. Revsbech and J. Sørensen (Eds.), *Denitrification in soil and sediment*. Plenum Press, New York.

Cooper, G.S., and R.L. Smith. 1963. Sequence of products formed during denitrification in some diverse western soils. *Soil Science Society of America Proceedings* 27:659–662.

Coyne, M.S. 2008. Biological denitrification. pp. 197–249. In J.S. Schepers and W. Raun (Eds.), *Nitrogen in agricultural systems*. Agronomy Monograph 49. ASA-CSSA-SSSA, Madison, WI.

Coyne, M.S., A. Arunakumari, B. Averill, and J.M. Tiedje. 1989. Immunological identification and distribution of dissimilatory heme $cd_1$ and nonheme Cu nitrite reductases in denitrifying bacteria. *Applied and Environmental Microbiology* 55:2924–2931.

Coyne, M.S., and D.D. Focht. 1987. Nitrous oxide reduction in nodules: Denitrification or nitrogen fixation? *Applied and Environmental Microbiology* 53:1168–1170.

Coyne, M.S., R.A. Gilfillen, and R.L. Blevins. 1994. Nitrous oxide flux from poultry-manured erosion plots and grass filters after simulated rain. *Journal of Environmental Quality* 23:831–834.

Coyne, M.S., and J.M. Tiedje. 1990a. Distribution and diversity of dissimilatory $NO_3^-$ reductases in denitrifying bacteria, pp. 21–35. In N.P. Revsbech and J. Sørensen (Eds.), *Denitrification in soil and sediment*. Plenum Press, New York.

Coyne, M.S., and J.M. Tiedje. 1990b. Induction of denitrifying enzymes in oxygen-limited *Achromobacter cycloclastes* continuous culture. *FEMS Microbiology Ecology* 73:263–270.

Crenshaw, C.L., C. Luber, R.L. Sinsabaugh, and L.K. Stavely. 2008. Fungal control of nitrous oxide production in semiarid grassland. *Biogeochemistry* 87:17–27.

Crowe, S., D. Canfield, A. Mucci, B. Sundby, and R. Maranger. 2012. Anammox, denitrification and fixed-nitrogen removal in sediments from the Lower St. Lawrence estuary. *Biogeosciences* 9:4309–4321.

Davidson, E.A. 1991. Fluxes of nitrous oxide and nitric oxide from terrestrial ecosystems, pp. 219–235. In J.E. Rogers and W.B. Whitman (Eds.), *Microbial production and consumption of greenhouse gases: Methane, nitrogen oxide, and halomethanes*. American Society for Microbiology, Washington, DC.

Davidson, E.A., M.K. Strand, and L.F. Galloway. 1985. Evaluation of the most probable number method for enumerating denitrifying bacteria. *Soil Science Society of America Journal* 49:642–645.

Dell, E.A., D. Bowman, T. Rufty, and W. Shi. 2010. The community composition of soil-denitrifying bacteria from a turfgrass environment. *Research in Microbiology* 161:315–325.

Dhakal, P., C. Matocha, F. Huggins, and M. Vandiviere. 2013. Nitrite reactivity with magnetite. *Environmental Science and Technology* 47:6206–6213.

Doering, O.C. 2017. Economic and policy issues of phosphorus management in agroecosystems, pp. 133–150. In R. Lal and B.A. Stewart (Eds.), *Soil phosphorus*. CRC Press, Boca Raton, FL.

Duncan, C.P., and P.M. Groffman. 1994. Comparing microbial parameters in natural and constructed wetlands. *Journal of Environmental Quality* 23:298–305.

Fairchild, M.A. 1997. Denitrifier ecology in fragipan soils of Kentucky. PhD Dissertation, University of Kentucky, Lexington, KY.

Fairchild, M.A., M.S. Coyne, J.H. Grove, and W.O. Thom. 1999. Denitrifying bacteria stratify above fragipans. *Soil Science* 164:190–196.

Federova, R.I., E.I. Mileklina, and N.I. Ilyukhina. 1973. Possibility of using the "gas exchange" method to detect extraterrestrial life: Identification of nitrogen-fixing organisms. *Akad.Nauk SSR Izvestia Ser. Biol.* 6:797–806.

Firestone, M.K, and E.A. Davidson. 1989. Microbiological basis of NO and $N_2O$ production and consumption in soil, pp. 7–21. In M.O. Andreae and D.S. Schimel (Eds.), *Exchange of trace gases between terrestrial ecosystems and the environment*. John Wiley & Sons, Ltd., New York.

Firestone, M.K., R.B. Firestone, and J.M. Tiedje. 1980. Nitrous oxide from soil denitrification: Factors controlling its biological production. *Science* 208:749–751.

Firestone, M.K., and J.M. Tiedje. 1979. Temporal change in nitrous oxide and dinitrogen from denitrification following onset of anaerobiosis. *Applied and Environmental Microbiology* 38:673–679.

Focht, D.D., and H. Joseph. 1973. An improved method for the determination of denitrifying bacteria. *Soil Science Society of America Proceedings* 37:698–699.

Gamble, T.N., M.R. Betlach, and J.M. Tiedje. 1977. Numerically dominant denitrifying bacteria from world soils. *Applied and Environmental Microbiology* 33:926–939.

Garber, E.A., and T.C. Hollocher. 1992. 15N, 18O tracer studies on the activation of nitrite by denitrifying bacteria. *Journal of Biological Chemistry* 257:8091–8097.

Ghiglione, J.-F., G. Gourbierè, P. Potier, L. Philippot, and R. Lensi. 2000. Role of respiratory nitrate reductase in ability of *Pseudomonas fluorescens* YT101 to colonize the rhizosphere of maize. *Applied and Environmental Microbiology* 66:4012–4016.

Gillette, K.L., Y. Qian, R.F. Follett, and S. Del Grosso. 2016. Nitrous oxide emissions from a golf course fairway and rough after application of different nitrogen fertilizer. *Journal of Environmental Quality* 45:1788–1795.

Groffman, P.M., E.A. Axelrod, J.L. Lemunyon, and W.M. Sullivan. 1991. Denitrification in grass and forest vegetated filter strips. *Journal of Environmental Quality* 20:671–674.

Groffman, P.M., A.J. Gold, and R.C. Simmons. 1992. Nitrate dynamics in riparian forests: Microbial studies. *Journal of Environmental Quality* 21:666–671.

Groffman, P.M., C.W. Rice, and J.M. Tiedje. 1993a. Denitrification in a tallgrass prairie landscape. *Ecology* 74:855–862.

Groffman, P.M., and J.M. Tiedje. 1988. Denitrification hysteresis during wetting and drying cycles in soil. *Soil Science Society of America Journal* 52:1626–1629.

Groffman, P.M., and J.M. Tiedje. 1989. Denitrification in north temperate forests soils: Spatial and temporal patterns and the landscape and seasonal scale. *Soil Biology & Biochemistry* 21:613–620.

Groffman, P.M., D.R. Zak, S. Christensen, A. Mosier, and J.M. Tiedje. 1993b. Early spring nitrogen dynamics in a temperate forest landscape. *Ecology* 74:1579–1585.

Haden, V.R., D. Liptzin et al. 2016. Ecosystem services and human well-being, pp. 113–193. In T.P. Tomich, S.B. Brodt, R.A. Dahlgre, and K.M. Scow (Eds.), *The California nitrogen assessment: Challenges and solutions for people, agriculture, and the environment.* Regents of the University of California, Berkeley, CA.

Handayani, I.P., M.S. Coyne, C. Barton, and S. Workman. 2008. Soil carbon pools and aggregation following stream restoration in a riparian corridor: Bernheim Forest, Kentucky. *Journal of Environmental Monitoring and Restoration* 4:11–28.

Hanson, G.C., P.M. Groffman, and A.J. Gold. 1994. Denitrification in riparian wetlands receiving high and low groundwater nitrate inputs. *Journal of Environmental Quality* 23:917–922.

Hatfield, J.L. 2017. Soil and nitrogen management to reduce nitrous oxide emissions, pp. 91–108. In A. Chatterjee and D. Clay (Eds.), *Soil fertility management in agroecosytems.* ASA-CSSA-SSSA, Inc. Madison, WI.

Hauck, R.D. 1985. Slow-release and bioinhibitor-amended nitrogen fertilizers, pp. 293- 322. In R.A. Olsen (Ed.), *Fertilizer technology and use,* 3rd ed. Soil Science Society of America, Inc. Madison, WI.

Heaney, D.J., M. Nyborg, E.D. Solberg, S.S. Malhi, and J. Ashworth. 1992. Overwinter nitrate loss and denitrification potential of cultivated soils in Alberta. *Soil Biology & Biochemistry* 24:877–884.

Heis, B., K. Frunzke, and W.G. Zumft. 1989. Formation of the N-N bond from nitric oxide by a membrane-bound cytochrome *bc* complex of nitrate-respiring (denitrifying) *Pseudomonas stutzeri. Journal of Bacteriology* 171:3288–3297.

Henry, S., D. Bru, B. Stres, S. Hallet, and L. Philippot. 2006. Quantitative detection of *nosZ* gene, encoding nitrous oxide reductase, and comparison of the abundances of 16S rRNA, *narG*, *nirK*, and *nosZ* genes in soils. *Applied and Environmental Microbiology* 72:5181–5189.

Herbert, R.A., and D.B. Nedwell. 1990. Role of environmental factors in regulating nitrate respiration in intertidal sediments, pp. 77–90. In N.P. Revsbech and J. Sørensen (Eds.), *Denitrification in soil and sediment.* Plenum Press, New York.

Herold, M.B., E.M. Baggs, and T.J. Daniell. 2012. Fungal and bacterial denitrification are differently affected by long-term pH amendment and cultivation of arable soil. *Soil Biology & Biochemistry* 54:35–35.

Higgins, S.A., A. Welsh, L.H. Orellana, K. Konstantinidis, J.C. Chee-Sanford, R. Sanford, C.W. Schadt, and F.E. Löffler. 2016. Detection and diversity of fungal nitric oxide reductase genes (*p450nor*) in agricultural soils. *Applied and Environmental Microbiology* 82:2919–2928.

Hixson, S.E., R.F. Walker, and C.M. Skau. 1990. Soil denitrification rates in four subalpine plant communities of the Sierra Nevada. *Journal of Environmental Quality* 19:617–620.

Hochstein, L.I., M. Betlach, and G. Kritikos. 1984. The effect of oxygen on denitrification during steady-state growth of *Paracoccus halodenitrificans. Archives of Microbiology* 137:74–78.

Holmes, W.E., and D.R. Zak. 1999. Soil microbial control of nitrogen loss following clear-cut harvest in northern hardwood ecosystems. *Ecological Applications* 9:202–215.

Holtan-Hartwig, L., P. Dorsch, and L.R. Bakken. 2000. Comparison of denitrifying communities in organic soils: Kinetics of $NO_3^-$ and $N_2O$ reduction. *Soil Biology & Biochemistry* 32:833–843.

Humbert, S.S. Tarnawski, N. Fromin, M.-P. Mallet, M. Aragno, and J. Zopfi. 2010. Molecular detection of anammox bacteria in terrestrial ecosystems: Distribution and diversity. *The ISME Journal* 4:450–454.

IPCC. 2001. Climate change 2001: The scientific basis. Contributions of Working Group I to the Third Assessment Report of the Intergovernmental Panel on Climate Change. Cambridge University Press, New York.

IPCC. 2006. 2006 IPCC guidelines for national greenhouse gas inventories, prepared by the National Greenhouse Gas Inventories Program, IGES, Kanagawa, Japan.

Isabella, B.L., and D.W. Hopkins. 1994. Nitrogen transformations in a peaty soil improved for pastoral agriculture. *Soil Use and Management* 10:107–111.

Jacobsen, S.N., and M. Alexander. 1980. Nitrate loss from soil in relation to temperature, carbon source, and denitrifier populations. *Soil Biology & Biochemistry* 12:501–505.

Kaspar, H. 1982. Nitrite reduction to nitrous oxide by propionibacteria: Detoxification mechanism. *Archives of Microbiology* 133:126–130.

Kaspar, H., and J. Tiedje. 1980. Response of electron-capture detector to hydrogen, oxygen, nitrogen, carbon dioxide, nitric oxide, and nitrous oxide. *Journal of Chromatography* 193:142–147.

Kaspar, H., and J. Tiedje. 1981. Dissimilatory reduction of nitrate and nitrite in the bovine rumen: Nitrous oxide production and effect of acetylene. *Applied and Environmental Microbiology* 41:705–709.

Kaur, G., B.A. Zurweller, K.A. Nelson, p. P. Motavalli, and C.J. Dudenhoffer. 2017. Soil waterlogging and nitrogen fertilizer management effects on corn and soybean yields. *Agronomy Journal* 109:97–106.

Kim, C.-H., and T.C. Hollocher. 1984. Catalysis of nitrosyl transfer reactions by a dissimilatory nitrite reductase (cytochrome $c,d_1$). *Journal of Biological Chemistry* 259:2092–2099.

Kladivko, E.J., J. Grochulska, R.F. Turco, G.E. Van Scyoc, and J.D. Engel. 1999. Pesticide and nitrate transport into subsurface tile drains of different spacings. *Journal of Environmental Quality* 28:997–1004.

Klemeddtsson, L., G. Hansson, and A. Mosier. The use of acetylene for the quantification of $N_2$ and $N_2O$ production from biological process in soil, pp. 167–180. In N.P. Revsbech and J. Sørensen (Eds.), *Denitrification in soil and sediment*. Plenum Press, New York.

Klingensmith, K.M., and K. Van-Cleve. 1993. Denitrification and nitrogen fixation in floodplain successional soils along the Tanana River, interior Alaska. *Canadian Journal of Forestry Research* 23:956–963.

Knowles, R. 1990. Acetylene inhibition technique: Development, advantages, and potential problems, pp. 151–166. In N.P. Revsbech and J. Sørensen (Eds.), *Denitrification in soil and sediment*. Plenum Press, New York.

Korner, H., and W.G. Zumft. 1989. Expression of denitrification enzymes in response to the dissolved oxygen level and respiratory substrate in continuous culture of *Pseudomonas stutzeri*. *Applied and Environmental Microbiology* 55:167–1676.

Kreba, S. 2013. Land use impact on soil gas and soil water transport properties. PhD Dissertation, University of Kentucky, Lexington, Kentucky.

Kröger, R., M.M. Holland, M.T. Moore, and C.M. Cooper. 2007. Hydrological variability and agricultural drainage ditch inorganic nitrogen reduction capacity. *Journal of Environmental Quality* 36:1646–1652.

Kröger, R., J.T. Scott, and J.M.P. Czarbecki. 2014. Denitrification potential of low-grade weirs and agricultural drainage ditch sediments in the lower Mississippi Alluvial Valley. *Ecological Engineering* 73:168–175.

Kuypers, M., M. Sliekers, G. Lavik, M. Schmid, B. Jørgensen, J. Kuenen, J. Damasté, M. Strous, and M. Jetten. 2003. Anaerobic ammonium oxidation by anammox bacteria in the Black Sea. *Nature* 422:608–611.

Lassey, K., and M. Harvey. 2007. Nitrous oxide: The serious side of laughing gas. *Water & Atmosphere* 15:10–11.

Laughlin, R.J., and R.J. Stevens. 2002. Evidence for fungal dominance of denitrification and codenitrification in a grassland soil. *Soil Science Society of America Journal* 66:1540–1548.

Lensi, R., and A. Chalamet. 1982. Denitrification in waterlogged soils: *In situ* temperature dependent variations. *Soil Biology & Biochemistry* 14:51–55.

Lindau, C.W., R.D. DeLaune, W.H. Patrick, and P.K. Bollich. 1990. Fertilizer effects on dinitrogen, nitrous oxide, and methane emissions from lowland rice. *Soil Science Society of America Journal* 54:1789–1794.

Long, A., B. Song, K. Fridey, and A. Silva. 2015. Detection and diversity of copper containing nitrite reductase genes (*nirK*) in prokaryotic and fungal communities in agricultural soils. *FEMS Microbiology Ecology* 91:1–9.

Lowrance, R. 1992. Groundwater nitrate and denitrification in a coastal plain riparian forest. *Journal of Environmental Quality* 21:401–405.

Ma, W.K., R.E. Farrell, and S.D. Siciliano. 2008. Soil formate regulates the fungal nitrous oxide emission pathway. *Applied and Environmental Microbiology* 74:6690–6696.

Maeda, K., S. Toyoda, R. Shimojima, T. Osada, D. Hanajima, R. Morioka, and N. Yoshida. Source of nitrous oxide emissions during the cow manure composting process as revealed by isotopomer analysis of and *amo* abundance in betaproteobacterial ammonia-oxidizing bacteria. *Applied and Environmental Microbiology* 76:1555–1562.

Mailapalli, D.R., and A.M. Thompson. 2012. Nitrogen leaching from Saybrook soil amended with biosolid and polyacrylamide. *Journal of Water Resources and Protection* 4:968–979.

Martin, K., L.L. Parsons, R.E. Murray, and M. Scott Smith. 1988. Dynamics of soil denitrifier populations: Relationships between enzyme activity, most-probable-number counts, and actual N gas loss. *Applied and Environmental Microbiology* 54: 2711–2716.

McIsaac, G.F., M.B. David, and G.Z. Gertner. 2016. Illinois River nitrate-nitrogen concentrations and loads: Long-term variation and association with watershed nitrogen inputs. *Journal of Environmental Quality* 45:1268–1275.

Mergel A., K. Kloos, and H. Bothe. 2001. Seasonal fluctuations in the population of denitrifying and $N_2$-fixing bacteria in an acid soil of a Norway spruce forest. *Plant and Soil* 230:145–160.

Merrill, A.G., and T.L. Benning. 2006. Ecosystem type difference in nitrogen process rates and controls in the riparian zone of a montane landscape. *Forest Ecology and Management* 222:145–161.

Montgomery, E., M.S. Coyne, and G.W. Thomas. 1997. Denitrification can cause variable $NO_3^-$ concentrations in shallow groundwater. *Soil Science* 162:1–10.

Mothapo, N., H. Chen, M.A. Cubeta, J.M. Grossman, F. Fuller, and W. Shi. 2015. Phylogenetic, taxonomic and functional diversity of fungal denitrifiers and associated $N_2O$ production efficacy. *Soil Biology & Biochemistry* 83:160–175.

Mothapo, N., H. Chen, M. Cubeta, and W. Shi. 2013. Nitrous oxide producing activity of diverse fungi from distinct agroecosystems. *Soil Biology & Biochemistry* 66:94–101.

Moraghan, J., and R. Buresh. 1977. Chemical reduction of nitrite and nitrous oxide by ferrous iron. *Soil Science Society of America Journal* 41:47–49.

Mounier, E., S. Hallet, D. Chenby, E. Benizri, Y. Gruet, C. Nguyen, S. Piutti et al. 2004. Influence of maize mucilage on the diversity and activity of the denitrifying community. *Environmental Microbiology* 6:301–312.

Mulder, A., A. van de Graaf, L. Robertson, and J. Kuenen. 1995. Anaerobic ammonium oxidation discovered in a denitrifying fluidized bed reactor. *FEMS Microbiology Ecology* 16:177–184.

Müller, M., V. Sundman, and J. Skujins. 1980. Denitrification in low pH spodosols and peats determined with the acetylene inhibition method. *Applied and Environmental Microbiology* 40:235–239.

Murray, R.E., Y. Feig, and J.M. Tiedje. 1995. Spatial heterogeneity in the distribution of denitrifying bacteria associated with denitrification activity zones. *Applied and Environmental Microbiology* 61:2791–2793.

Myrold, D.D., and J.M. Tiedje. 1985. Diffusional constraints on denitrification in soil. *Soil Science Society of America Journal* 49:651–657.

Nõmmik, H. 1956. Investigations on denitrification in soil. *Acta Agriculturae Scandanivica* 6:197–228.

Novinscak, A., C. Goyer, B.J. Zebarth, D.L. Burton, M.H. Chantigny, and M. Filion. 2016. Novel *P450nor* gene detection assay used to characterize the prevalence and diversity of soil fungal denitrifiers. *Applied and Environmental Microbiology* 82:4560–4569.

Parkin, T.B. 1987. Soil microsites as a source of denitrification variability. *Soil Science Society of America Journal* 51:1194–1199.

Parkin, T.B. 1990. Characterizing the variability of soil denitrification, pp. 213–228. In N.P. Revsbech and J. Sørensen (Eds.), *Denitrification in soil and sediment.* Plenum Press, New York.

Parkin, T.B., and J.J. Meisinger. 1989. Denitrification below the crop rooting zone as influenced by surface tillage. *Journal of Environmental Quality* 18:12–16.

Parkin, T.B., A.J. Sexstone, and J.M. Tiedje. 1985a. Comparison of field denitrification rates determined by acetylene-based soil core and nitrogen-15 methods. *Soil Science Society of America Journal* 49:94–99.

Parkin, T., A.J. Sexstone, and J.M. Tiedje. 1985b. Adaptation of denitrifying populations to low soil pH. *Applied and Environmental Microbiology* 49:1053–1056.

Payne, W. 1981. *Denitrification.* John Wiley & Sons, New York.

Pearce, R.C., Y. Zhan, and M.S. Coyne. 1998. Nitrogen transformations in the tobacco float system. *Tobacco Science* 42:82–88.

Peña-Yewtukhiw, E.M., J.H. Grove, E.G. Beck, and J.S. Dinger. 2009. Effect of soil and absence/presence of an abandoned feedlot on determining the area sourcing nitrate to a contaminated domestic well. *Soil Science* 174:56–64.

Pennock, D.J., C. Van-Kessel, R.E. Farrell, and R.A. Sutherland. 1992. Landscape-scale variations in denitrification. *Soil Science Society of America Journal* 56:770–776.

Pérez, T. 2005. Factors that control the isotopic composition of N₂O from soil emissions, pp. 69–84. In L.B. Flanagan, J.R. Ehleringer, and D.E. Pataki (Eds.), *Stable isotopes and biosphere-atmosphere interactions: Processes and biological controls*. Elsevier/Academic Press, New York.

Peterjohn, W.T., and W.H. Schlesinger. 1991. Factors controlling denitrification in a Chihuahuan Desert ecosystem. *Soil Science Society of America Journal* 55:1694–1701.

Philippot, L. 2002. Denitrifying genes in bacterial and Archaeal genomes. *Biochimica et Biophysica Acta—Gene Structure and Expression* 1577:355–376.

Philippot, L., A. Clays-Josserand, and R. Lensi. 1995. Use of Tn5 mutants to assess the role of the dissimilative nitrite reductase in the competitive abilities of two *Pseudomonas* strains in soil. *Applied and Environmental Microbiology* 61:1426–1430.

Philippot, L., S. Hallin, and M. Schloter. 2007. Ecology of denitrifying prokaryotes in agricultural soil. *Advances in Agronomy* 96:249–305.

Prieme, A., G. Braker, and J.M. Tiedje. 2002. Diversity of nitrite reductase (*nirK* and *nirS*) gene fragments in forested upland and wetland soils. *Applied and Environmental Microbiology* 68:1893–1924.

Rakshit, S., C. Matocha, and M. Coyne. 2008. Nitrite reduction by siderite. *Soil Science Society of America Journal* 72:1070–1077.

Rakshit, S., C. Matocha, M. Coyne, and D. Sarkar. 2016. Nitrite reduction by Fe(II) associated with kaolinite. *International Journal of Environmental Science and Technology* DOI 10.1007/s13762-016-0971-x.

Rehr, B., and J.-H. Klemme. 1988. Competition for nitrate between denitrifying *Pseudomonas stutzeri* and nitrate ammonifying enterobacteria. *FEMS Microbiology Ecology* 62:51–58.

Rice, C.W., and K.L. Rogers. 1993. Denitrification in subsurface environments: Potential source for atmospheric nitrous oxide, pp. 121–132. In D.E. Rolston et al. (Eds.), *Agricultural ecosystem effects on trace gases and global climate change*. American Society for Agronomy, Madison, WI.

Rice, C.W., P.E. Sierzega, J.M. Tiedje, and L.W. Jacobs. 1988. Stimulated denitrification in the microenvironment of a biodegradable organic waste injected into soil. *Soil Science Society of America Journal* 52:102–108.

Rice, C.W., and M.S. Smith. 1982. Denitrification in no-till and plowed soils. *Soil Science Society of America Journal* 46:1168–1173.

Rich, J., R.S. Heichen, P. Bottomley, K. Cromack, and D.D. Myrold. 2003. Community composition and functioning of denitrifying bacteria from adjacent meadow and forest soils. *Applied and Environmental Microbiology* 69:5974–5982.

Rich, J., and D.D. Myrold. 2004. Community composition and activities of denitrifying bacteria from adjacent agricultural soil, riparian soil, and creek sediment in Oregon, USA. *Soil Biology & Biochemistry* 36:1431–1441.

Robertson, G.P. 1989. Nitrification and denitrification in humid tropical ecosystems: Potential controls on nitrogen retention, pp. 55–69. In J. Procter (Ed.), *Mineral nutrients in tropical forest and savanna ecosystems*. Blackwell Science Publishers, Oxford.

Robertson, G.P., M.A. Huston, F.E. Evans, and J.M. Tiedje. 1988. Spatial variability in a successional plant community: Patterns of nitrogen availability. *Ecology* 69:1517–1524.

Robertson, G.P., and J.M. Tiedje. 1987. Nitrous oxide sources in aerobic soils: Nitrification, denitrification, and other biological processes. *Soil Biology & Biochemistry* 19:187–193.

Robertson, L.A., and J.G. Kuenen. 1990. Physiological and ecological aspects of aerobic denitrification: A link with heterotrophic nitrification?, pp. 91–104. In N.P. Revsbech and J. Sørensen (Eds.), *Denitrification in soil and sediment*. Plenum Press, New York.

Rosenstock, T.L. et al. 2016. Responses: Technologies and practice, pp. 211–230. In T.P. Tomich et al. (Eds.), *The California nitrogen assessment*. University of California Press, Oakland, CA.

Rowe, R., R. Todd, and J. Waide. 1977. Microtechnique for most probable number analysis. *Applied and Environmental Microbiology* 33:675–680.

Ruz-Jerez, B.E., R.E. White, and P.R. Ball. 1994. Long-term measurement of denitrification in three contrasting pastures grazed by sheep. *Soil Biology & Biochemistry* 26:29–39.

Schipper, L.A., A.B. Cooper, C.G. Harfoot, and W.J. Dyck. 1993. Regulators of denitrification in an organic riparian soil. *Soil Biology & Biochemistry* 25:925–933.

Schmid, M., B. Maas, A. Dapena, K. van de Pas-Schoonen, J. van de Vossenberg, B. Kartal, L. van Niftrik et al. 2005. Biomarkers for *in situ* detection of anaerobic ammonium-oxidizing (anammox) bacteria. *Applied and Environmental Microbiology* 71:1677–1684.

Schnabel, R.R., and W.L. Stout. 1994. Denitrification loss from two Pennsylvania floodplain soils. *Journal of Environmental Quality* 23:344–348.

Seo, D.C., and R.D. DeLaune. 2010. Fungal and bacterial mediated denitrification in wetlands: Influence of sediment redox condition. *Water Research* 44:2441–2450.

Sexstone, A.J., T.B. Parkin, and J.M. Tiedje. 1985a. Temporal response of denitrification rates to rainfall and irrigation. *Soil Science Society of America Journal* 49:99–103.

Sexstone, A.J., N.P. Revsbech, T.B. Parkin, and J.M. Tiedje. 1985b. Direct measurement of oxygen profiles and denitrification rates in soil aggregates. *Soil Science Society of America Journal* 49:645–651.

Shapleigh, J.P., and W.J. Payne. Differentiation of $c$, $d_1$ cytochrome and copper nitrite reductase production in denitrifiers. *FEMS Microbiology Letters* 26:275–279.

Shoun, H., D.-H. Kim, H. Uchiyama, and J. Sugiyama. 1992. Denitrification by fungi. *FEMS Microbiology and Ecology* 94:277–282.

Shoun, H., and T. Tanimoto. 1991. Denitrification by the fungus *Fusarium oxysporum* and involvement of cytochrome P-450 in the respiratory nitrite reduction. *The Journal of Biological Chemistry* 266:11078–11082.

Ŝimek, M., and J.E. Cooper. 2002. The influence of soil pH on denitrification: Progress towards the understanding of this interaction over the last 50 years. *European Journal of Soil Science* 53:345–354.

Smith, D.R., S.J. Livingston, B.W. Zuecher, M. Larose, G.C. Heathman, and C. Huang. 2008. Nutrient lossess from row crop agriculture in Indiana. *Journal of Soil and Water Conservation* 63:396–409.

Smith, M.S. 1982. Dissimilatory reduction of $NO_2^-$ to $NH_4^+$ and $N_2O$ by a soil *Citrobacter* sp. *Applied and Environmental Microbiology* 43:854–860.

Smith, M.S. 1983. Nitrous oxide production by *Escherichia coli* is correlated with nitrate reductase activity. *Applied and Environmental Microbiology* 45:1545–1547.

Smith, M.S., and L.L. Parsons. 1985. Persistence of denitrifying enzyme activity in dried soils. *Applied and Environmental Microbiology* 49:316–320.

Smith, M.S., and J.M. Tiedje. 1979a. Phases of denitrification following oxygen depletion in soil. *Soil Biology & Biochemistry* 11:261–267.

Smith, M.S., and J.M. Tiedje. 1979b. The effect of roots on soil denitrification. *Soil Science Society of America Journal* 43:951–955.

Smith, M.S., and J.M. Tiedje. 1980. Growth and survival of antibiotic-resistant denitrifier strains in soil. *Canadian Journal of Microbiology* 26:854–856.

Soil Science Glossary Terms Committee. 2008. Glossary of Soil Science Terms 2008. Soil Science Society of America, Madison WI.

Staley, T.E., and J.B. Griffin. 1981. Simultaneous enumeration of denitrifying and nitrate-reducing bacteria in soil by a microtiter most-probable-number procedure. *Soil Biology & Biochemistry* 13:385–388.

Steingruber, S.M., J. Friedrich, R. Gächter, and B. Wehrli. 2001. Measurement of denitrification in sediments with the 15N isotope pairing technique. *Applied and Environmental Microbiology* 67:3771–3778.

Stephanson, R.C. 1972. Soil denitrification in sealed soil-plant systems. I. Effect of plants, soil water content, and soil organic matter content. *Plant and Soil* 37:141–149.

Stoddard, C.S., J.H. Grove, M.S. Coyne, and W.O. Thom. 2005. Fertilizer, tillage, and dairy manure contributions to nitrate and herbicide leaching. *Journal of Environmental Quality* 34:1354–1362.

Struwe, S., and A. Kjoller. 1994. Potential for $N_2O$ production from beech (*Fagus silvaticus*) forest soils with varying pH. *Soil Biology & Biochemistry* 26:1003–1009.

Sutka, R., N. Ostrom, P. Ostrom, J. Breznak, H. Ganhi, J. Pitt, and F. Li. 2006. Distinguishing nitrous oxide production from nitrification and denitrification on the basis of isotopomer abundances. *Applied and Environmental Microbiology* 72:638–644.

Tanimoto, T., K.-I Hatano, D.-H. Kim, H. Uchiyama, and H. Shoun. 1992. Co-denitrification by the denitrifying system of the fungus *Fusarium oxysporum*. *FEMS Microbiology Letters* 93:177–180.

Tian, H., G. Chen, C. Lu, X. Xu, W. Ren, B. Zhang, K. Banger, B. Tao, S. Pan, M. Liu, C. Zhang, L. Bruhwiler, and S. Wofsy. 2015. Global methane and nitrous oxide emissions from terrestrial ecosystems due to multiple environmental changes. *Ecosystem Health and Sustainability* 1(1):4. http://dx.doi.org/10.1890/EHS14-0015.1.

Tiedje, J.M. 1988. Ecology of denitrification and dissimilatory nitrate reduction to ammonium, pp. 179–243. In A.J.B. Zehnder (Ed.), *Biology of anaerobic organisms*. John Wiley & Sons, Inc., New York.

Tiedje, J.M. 1994. Denitrifiers, pp. 245–267. In R.W. Weaver et al. (Eds.), *Methods of soil analysis, part 2: Microbiological and biochemical methods*. Soil Science Society of America, Inc., Madison, WI.

Tiedje, J.M., R.B. Firestone, M.K. Firestone, M.R. Betlach, M.S. Smith, and W.H. Caskey. 1979. Methods for the production and use of Nitrogen-13 in studies of denitrification. *Soil Science Society of America Journal* 43:709–715.

Tiedje, J.M., A.J. Sexstone, D.D. Myrold, and J.A. Robinson. 1982. Denitrification: Ecological niches, competition and survival. *Antonie van Leeuwenhoek* 48:569–583.

Tiedje, J.M., S. Simkins, and P.M. Groffman. 1989. Perspectives on measurement of denitrification in the field including recommended protocols for acetylene based methods, pp. 217–240. In M. Clarholm and L. Bergstrom (Eds.), *Ecology of arable land: Perspectives and challenges.* Kluwer Academic Publishers, Holland.

Tomlinson, G., L. Jahnke, and L. Hochstein. 1986. *Halobacterium denitrificans* sp-nov, an extremely halophilic denitrifying bacterium. *International Journal of Systematic Bacteriology* 36:66–70.

Tosti, G., M. Farneselli, P. Benincasa, and M. Guiducci. 2016. Nitrogen fertilization strategies for organic wheat production: Crop yield and nitrate leaching. *Agronomy Journal* 108:770–781.

Treush, A.H., S. Leininger, A. Kletzin, S.C. Schuster, H.-P. Klenk, and C. Schleper. 2005. Novel genes for nitrite reductase and Amo-related proteins indicate a role of uncultivated mesophilic crenarchaeota in nitrogen cycling. *Environmental Microbiology* 7:1985–1995.

Urata, K., and T. Satoh. 1985. Mechanism of nitrite reduction to nitrous oxide in a photodenitrifier, *Rhodopseudomonas sphaeroides* forma sp. *denitrificans. Biochimica et Biophysica Acta* 841:201–207.

USEPA. 2015. Mississippi River/Gulf of Mexico Watershed Nutrient Task Force 2015 Report to Congress. Office of Wetlands, Oceans, and Watersheds. U.S. Environmental Protection Agency. Washington, DC, pp. 1–98.

van de Graaf, A., P. de Bruijn, L. Robertson, M. Jetten, and J. Kuenen. 1997. Metabolic pathway of anaerobic ammonium oxidation on the basis of $^{15}$N studies in a fluidized bed reactor. *Microbiology* 143:2415–2431.

van de Graaf, A., A. Mulder, P. de Bruijn, M. Jetten, L. Robertson, and J. Kuenen. 1995. Anaerobic oxidation of ammonium is a biologically mediated process. *Applied and Environmental Microbiology* 61:1246–1251.

Van Kessel, C., D.J. Pennock, and R.E. Farrell. 1993. Seasonal variation in denitrification and nitrous oxide evolution at the landscape scale. *Soil Biology & Biochemistry* 57:988–995.

Van Metre, P.C., J.W. Frey, M. Musgrove, D. Nakagaki, S. Qi, B. Mahler, M.E. Wieczorek, and D.T. Button. 2016. High nitrate concentrations in some Midwest United States streams in 2013 after the 2012 drought. *Journal of Environmental Quality* 45:1696–1704.

Vermes, J.F., and D.D. Myrold. 1992. Denitrification in forest soils of Oregon. *Canadian Journal of Forestry Research* 22:504–512.

Viebrock, A., and W.G. Zumft. 1987. Physical mapping of transposon Tn5 insertions defines a gene cluster functional in nitrous oxide respiration by *Pseudomonas stutzeri. Journal of Bacteriology* 169:4577–4580.

Vlaeminck, S., A. Terada, B. Smets, H. De Clippeleir, T. Schaubroeck, S. Bolca, L. Demeestere, J. Mast, N. Boon, M. Carballa, and W. Verstraete. 2010. Aggregate size and architecture determine microbial activity balance for one-stage partial nitritation and anammox. *Applied and Environmental Microbiology* 76:900–909.

Völkl, P., R. Huber, E. Drobner, R. Rachel, S. Burggraf, A. Trincone, and K.O. Stetter. 1993. *Pyrobaculum aerophilum* sp.-nov., a novel nitrate-reducing hyperthermophilic archaeum. *Applied and Environmental Microbiology* 59:2918–2926.

Wallenstein, M.D. 2004. Effects of increased nitrogen deposition on forest soil nitrogen cycling and microbial community structure. Dissertation. Duke University, Durham, NC.

Wallenstein, M.D., D.D. Myrold, M. Firestone, and M. Voytek. 2006. Environmental controls on denitrifying communities and denitrification rates: Insights from molecular methods. *Ecological Applications* 16:2143–2152.

Wei, W., K. Isobe, Y. Shiratori, T. Nishizawa, N. Ohte, Y. Ise, S. Otsuka, and K. Senoo. 2015. Development of PCR primers targeting fungal *nirK* to study fungal denitrification in the environment. *Soil Biology & Biochemistry* 81:282–286.

Werber, M, and M. Mevarech. 1978. Induction of a dissimilatory reduction pathway of nitrate in *Halobacterium* of Dead Sea—possible role for 2 Fe-Ferredoxin isolated from this organism. *Archives of Biochemistry and Biophysics* 186:60–65.

Wolsing, M., and A. Prieme. 2004. Observation of high seasonal variation in community structure of denitrifying bacteria in arable soil receiving artificial fertilizer and cattle manure by determining T-RLFP of *nir* gene fragments. *FEMS Microbiology Ecology* 48:261–271.

Wu, Tingting. 2008. Denitrifier ecology in a fragipan soil of Kentucky. M.S. Thesis, University of Kentucky, Lexington, KY.

Yanai, Y., K. Toyota, T. Morishita, F. Takakai, R. Hatano, S.H. Limin, U. Darung, and S. Dohong. 2007. Fungal $N_2O$ production in an arable peat soil in central Kalimantan, Indonesia. *Soil Science and Plant Nutrition* 53:806–811.

Ye, R., B. Averill, and J.M. Tiedje. 1994. Denitrification: Production and consumption of nitric oxide. *Applied and Environmental Microbiology* 60:1053–1058.

Yoshimatsu, K.T., T. Sakurai, and T. Fujiwara. 2000. Purification and characterization of dissimilatory nitrate reductase from a denitrifying halophilic archaeon, *Haloarcula marismortui*. *FEBS Letters* 470:216–220.

Yoshinari, T., and R. Knowles. 1976. Acetylene inhibition of nitrous oxide reduction by denitrifying bacteria. *Biochemical and Biophysical Research Communications* 69:705–710.

Young, E.O., and R.D. Briggs. 2005. Shallow ground water nitrate-N and ammonium-N in cropland and riparian buffers. *Agriculture, Ecosystems, and Environment* 109:297–309.

Zumft, W. 1997. Cell biology and molecular basis of denitrification. *Molecular Biology Reviews* 61:533–616.

Zumft, W.G., K. Döhler, H. Körner, S. Löchelt, A. Viebrock, and K. Frunzke. 1988. Defects in cytochrome $cd_1$-dependent nitrite respiration of transposon Tn5-induced mutants from *Pseudomonas stutzeri*. *Archives of Microbiology* 149:492–498.

Zumft, W.G., and P.M.H. Kroneck. 2007. Respiratory transformation of nitrous oxide ($N_2O$) to dinitrogen by bacteria and archaea. *Advances in Microbial Physiology* 52:107–227.

# 6 Nitrogen Balances and Nitrogen Use Efficiency in the Nordic Countries

*Anne Falk Øgaard and Marianne Bechmann*

## CONTENTS

## 6.1 INTRODUCTION

Nitrates from agricultural sources cause pollution of both water and air. For surface water, eutrophication of marine waters caused by nitrate is of special concern. The rapid increase in hypoxia in coastal areas around the world is due to the excessive inputs of plant nutrients, such as nitrogen (N) and phosphorus (P) by human activities. The sources of these nutrients include agriculture, sewage, and atmospheric deposition of N containing compounds from the burning of fossil fuels. The nutrients stimulate the growth of algae causing problems with eutrophication. The algae sink to the bottom and use the oxygen when they decompose. If mixing of the bottom waters is slow, such that oxygen stocks are not renewed, hypoxia can occur. The oxygen-deficient areas may be referred to as dead-zones either in marine, brackish, or fresh-water bodies. Hypoxia has occurred in the Baltic Sea and been identified as one of the world's largest anthropogenic "dead zones" (low oxygen zone), leading to death of organisms that live on the bottom. The total area of bottom covered with hypoxic waters with oxygen concentrations less than 2 mg/l in the Baltic Sea has averaged 49,000 km$^2$ over the last 40 years (Conley et al. 2009).

Nitrogen balance, surplus or deficit, in agricultural production has been identified as an indicator for risk of N losses from land to water (OECD 2001). The N surplus in agriculture also adds to the problem of greenhouse gas (GHG) emissions by increased emissions of nitrous oxides, which contributes to climate change.

At the plot scale, the relationship between N application, N balance, and N concentrations in surface and subsurface runoff is well-documented in studies comparing plots under similar conditions regarding soil and weather (Goulding 2000; Korsaeth and Eltun 2000). Korsaeth and Eltun (2000) showed that N balances for arable cropping systems explained 86% of the variation in N loss. Also, on the field scale, such relationships have been proven (Jansons et al. 2003; Salo and Turtola 2006).

For three Latvian fields monitored for 6 years, Jansons et al. (2003) showed that 57% of the variation in N loss was explained by N balance.

On the other hand, N is usually the main nutrient affecting grain yields. It is therefore applied in large quantities to maintain optimal yields. Nitrogen application response curves show diminishing yield increments (Mengel and Kirkby 1987). Consequently, increasing N application results in reduced N use efficiency (NUE), calculated as the amount of harvested N divided by the amount of applied N. In other words, the N surplus increases with increasing N application, because the crop utilizes a diminishing part of applied N when the application rate increases. Therefore, aiming at economically optimal fertilization is often in conflict with the aim of minimizing N losses.

In the Nordic countries, there has been a strong focus on agricultural pollution over the last decades and various regulations have been in place to improve the NUE. However, still there is a significant N surplus in the agriculture.

In this chapter, we present trends in consumption of N fertilizers at the national scale for the Nordic countries. On catchment scale, we present N balances, corresponding NUE and N losses. Long-term time series on farming practices from 23 intensively monitored catchments in the Nordic countries were used as case study areas in the description. Data on farming practice in three small agricultural catchments in Norway were used to compare N application to cereals with obtained yields, calculate N balances and NUE, and study the degree to which farmers' fertilizer practice is in accordance with fertilizer recommendations. Finally, we discuss the different regulations on N application in the Nordic countries in relation to their effect on N balances.

## 6.2 CONSUMPTION OF N FERTILIZER IN NORDIC COUNTRIES

The average fertilizer application shows considerable difference between the Nordic countries (Figure 6.1). In 1990, the average N application per ha of arable land in Denmark was about 80% higher than that in Sweden. Since then, the N application in Denmark has shown an almost 50% reduction, mainly by improved use of manure for crop production, due to restrictions in the use of mineral fertilizer. This huge reduction is a result of several nutrient action plans, focusing on N, that have been implemented during the last 25 years to reduce nitrate leaching (Kronvang et al. 2008). Already in the late 1980s, Denmark formed legislation aimed at a 50% nutrient reduction to the sea, and a long list of mandatory measures targeted toward more efficient use of organic N in

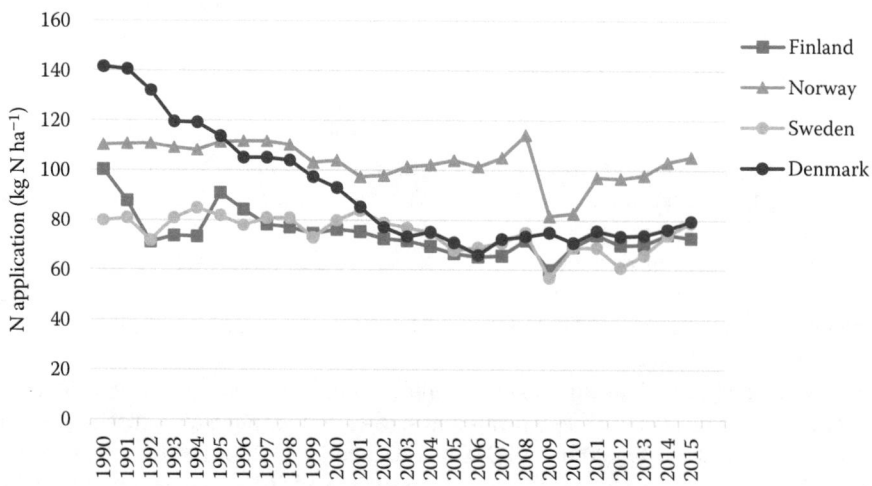

**FIGURE 6.1** Consumption of N fertilizer in the Nordic countries from 1990 to 2015, kg N ha$^{-1}$ agricultural land. Data from national statistics in the four countries.

manure was implemented (Andersen et al. 2014). Both pollution of drinking water with nitrate and eutrophication of fjords and coastal water have been drivers for the reduction in N application rates. Also in Sweden and Finland, where N application was much lower than in Denmark in the end of the 1980s, N application has decreased slightly since 1990. Average N application is now between 60 and 80 kg per ha of agricultural land for Denmark, Sweden, and Finland. In Norway, average N application was at a level between Denmark and Sweden/Finland in the end of 1980s, but the total change in fertilizer consumption has been small except for 2008–2010, resulting in the highest N consumption among the Nordic countries since 1996. Expected increased prices in Norway in 2009 resulted in more sale for the 2009 season in 2008 and less in 2009. Increased prices influenced the N fertilizer consumption in 2010. The reason for no trend in consumption of N in mineral fertilizer in Norway during the 1990s and 2000s is that most focus has been on reducing P losses from agriculture (Bechmann et al. 2014).

## 6.3 NITROGEN FIELD BALANCES IN THE NORDIC COUNTRIES

Average N balances for 2007–2011 were investigated for 23 catchments in the Nordic countries. The mass-balance of N at the field scale includes the following factors: N applied in manure and fertilizer, N removed in yield, N fixation by plants, and N deposition. The N balance includes changes in the soil N pool and N loss to water (Figure 6.2).

The N input and the calculated N balances showed a large variation between catchments (Figures 6.3 and 6.4). The N balances varied between a deficit of −12 and a surplus of 132 kg ha$^{-1}$yr$^{-1}$. Nitrogen balances from 89 kg N ha$^{-1}$yr$^{-1}$ and higher were found for 7 of the 23 catchments; namely two Norwegian catchments (Time and Kolstad), four of the Danish catchments (Odder, Horndrup, Lillebæk, and Bolbro), and one Swedish catchment (F26).

The variation in N balances is a result of variation in N application and N yield, in addition to N fixation and N deposition. The N balances in catchments with higher manure application rates tend to be higher than in catchments with lower rates (Figures 6.3 and 6.4). Higher N surplus when using manure is explained by higher N application, because of lower plant availability of manure N compared to N in mineral fertilizer and the ammonium volatilization while spreading manure.

The catchment Volbu, which showed a negative N balance on average for the monitoring period, is located in a mountain region in Norway with low intensity in agriculture. The soil N content may be decreasing within this catchment due to plant uptake of N being higher than the N application. The catchment, Time, with the highest N application and N surplus, on the other hand, is located in a region in Norway with an intensive livestock farming. These two catchments demonstrate the huge regional differences that may occur within one country.

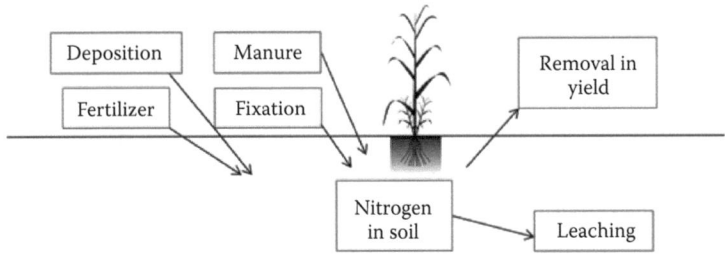

N balance = input − output of nitrogen =
fertilizer-N + manure-N + fixation-N + deposition-N − yield-N

**FIGURE 6.2** Schematic presentation of the N balance.

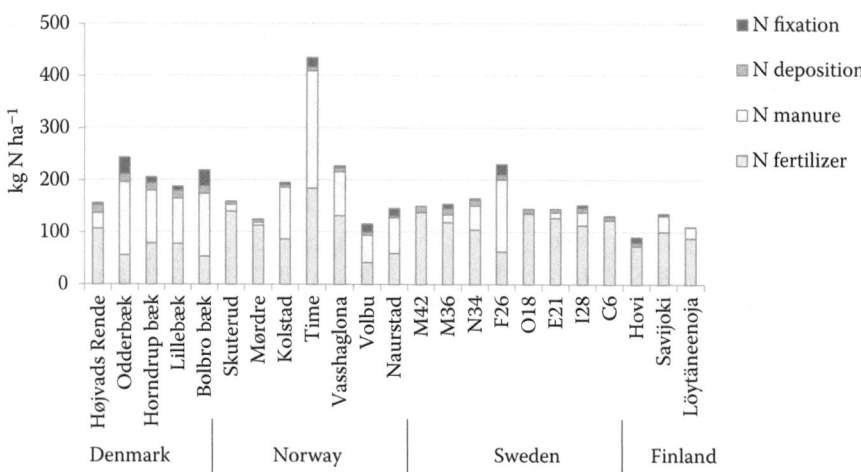

**FIGURE 6.3**   Average N inputs from fertilizer, manure, N deposition, and N fixation from 2007 to 2011 for agricultural land in 23 catchments in Denmark, Norway, Sweden, and Finland. (Data from Bechmann, M., Blicher-Mathiesen, G., Kyllmar, K., Iital, A., Lagzdins, A., and Salo, T. 2014. Nitrogen application, balances and effect on water quality in the Nordic-Baltic countries. *Agric. Ecosyst. Environ.* 198:104–113.)

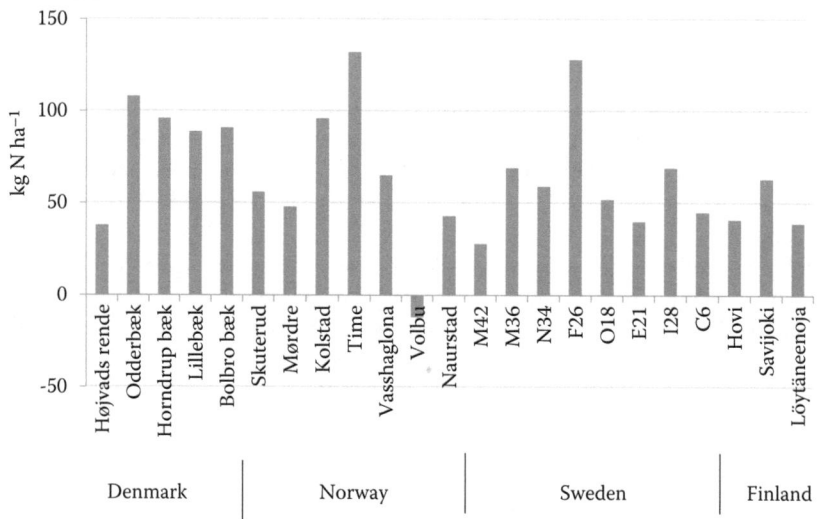

**FIGURE 6.4**   Average N balance from 2007 to 2011 for agricultural land in 23 catchments in Denmark, Norway, Sweden, and Finland. (Data from Bechmann, M., Blicher-Mathiesen, G., Kyllmar, K., Iital, A., Lagzdins, A., and Salo, T. 2014. Nitrogen application, balances and effect on water quality in the Nordic-Baltic countries. *Agric. Ecosyst. Environ.* 198:104–113.)

The decrease in use of mineral fertilizer in Denmark (Figure 6.1) is reflected in a decreasing trend in N balances in three of the five Danish catchments. The N balances of these three catchments decreased from 1991 to 2004, after which it stabilized at an N balance of around 100 kg N ha$^{-1}$yr$^{-1}$ (Figure 6.5). However, this is still a higher N surplus than for most of the other Nordic countries (Figure 6.4). Windolf et al. (2012) also showed a significant decreasing trend in N balances for 10 catchments in a larger scale study in Denmark.

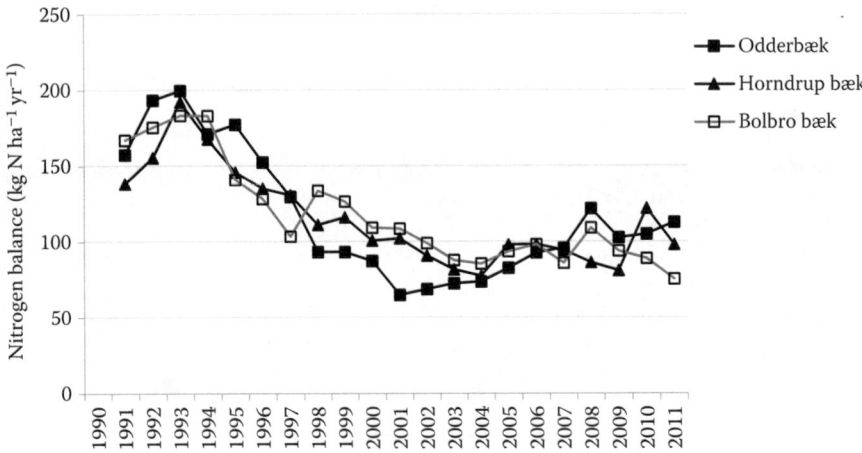

**FIGURE 6.5** Yearly average N balances for three Danish catchments in 1991–2011. (Data from Bechmann, M., Blicher-Mathiesen, G., Kyllmar, K., Iital, A., Lagzdins, A., and Salo, T. 2014. Nitrogen application, balances and effect on water quality in the Nordic-Baltic countries. *Agric. Ecosyst. Environ.* 198:104–113.)

Data at the national scale show for Sweden that the surplus in N balances has decreased from 58 in 1995 to 32 kg $ha^{-1}$ in 2009 (Bechmann et al. 2014), whereas for Finland, the N balances decreased from 1990 to 2005 and leveled at approximately 50 kg $ha^{-1}yr^{-1}$ after ammonia volatilization was accounted for (Salo et al. 2007).

## 6.4 NITROGEN LOSSES FROM AGRICULTURE IN THE NORDIC COUNTRIES

The overall range in annual long-term mean total N losses in 35 investigated Nordic agricultural catchments was 6–102 kg N $ha^{-1}$ (Stålnacke et al. 2014). Stålnacke et al. (2014) showed that nearly one-third of the catchments showed statistically significant downward trends in N losses or concentrations. They found that the most prominent decreasing trends occurred in the catchments in Denmark and Sweden. The decreasing use of mineral fertilizer in Denmark has led to decreasing N balances and may explain downward trends in N losses. In four catchments with long-term time series of decreasing N balances, a decrease in the N concentration in the streams was detected (Bechmann et al. 2014). However, Bechmann et al. (2014) concluded that a large and long-term decrease in N balance is required to detect decreased N concentrations in runoff at the catchment scale. Further, leaching responses to the field N balance are different for loamy and sandy soils and depend on farm type (Blicher-Mathiesen et al. 2014). In addition to reduced N surplus, measures that reduce N leaching during autumn and winter such as use of catch crops influence N losses (Aronsson et al. 2016). Further, climate change is expected to impact N losses. For the Nordic region increased precipitation is expected in the future to occur mainly outside the growing season, which results in increased runoff leading to increased N losses (Øygarden et al. 2014). Hence, climate change may counteract the positive effect of mitigation measures.

## 6.5 NITROGEN FERTILIZATION, N BALANCES, AND N USE EFFICIENCY FOR CEREALS

In Norway, the recommended N application rate is related to the expected yield, content of soil organic matter (SOM), and preceding crop. However, with large year-to-year variations in growth conditions, especially temperature and precipitation, there is high uncertainty regarding expected yields. In a meta-analysis of 61 field experiments in Finland, Valkama et al. (2013) found the average

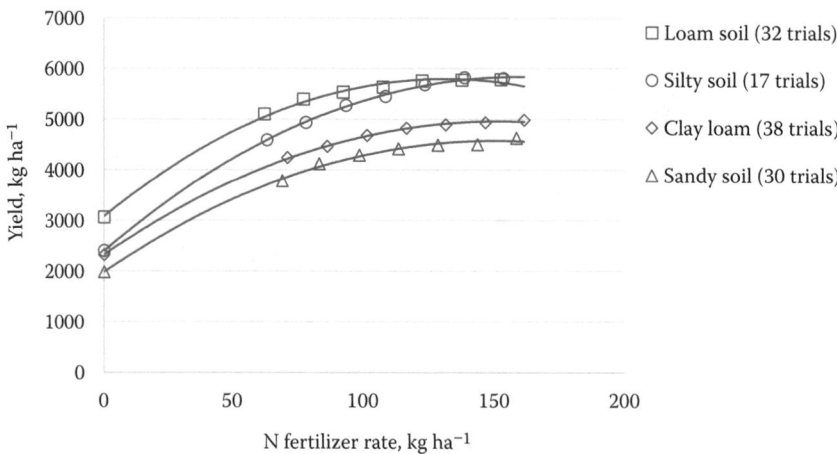

**FIGURE 6.6** Average yield response to N fertilization for barley on four soil textural groups. (Data from Riley, H., Hoel, B.O., & Kristoffersen, A.Ø. 2012. Economic and environmental opimization of nitrogen fertilizer recommendations for cereals in Norway. *Acta Acta Agric. Scand., Sect. B, Soil and Plant Sci.* 62:387–400.)

year-to-year variations in control yields to be 25%, 22%, and 15% for barley (*Hordeum vulgare*), spring wheat (*Triticum aestivum* L.), and oats (*Avena sativa*), respectively. Further, the grain yield response to N fertilization varies between sites due to variations in soil fertility (Riley et al. 2012; Figure 6.6). Figure 6.6 shows that average yield response to N fertilization for barley on clay and sandy soils was lower than on loam and silty soils. The lower yield response to N fertilization on clay and sandy soils resulted in higher N surplus by farmers' normal N fertilization on these soils than on loam and silty soils (about 40 kg N ha$^{-1}$ vs. 20–25 kg N ha$^{-1}$).

The Norwegian Agricultural Monitoring Programme (JOVA) provides a long-term time series on farming practices and yields in a number of small agricultural catchments in different regions of Norway (nibio.no/JOVA). Based on these data, N applications were related to yields, N balances, and N use efficiencies for different crops. Further, the degree to which farmers' practice is in accordance to recommended N application was studied. Here, N balances were limited to the farm management balance at the field level, and were calculated as applied N minus N removed with grain and straw. The NUE was calculated as the amount of harvested N divided by the amount of applied N.

## 6.5.1 NITROGEN APPLICATION COMPARED TO RECOMMENDED N APPLICATION

The average amount of applied N to cereals in three JOVA catchments dominated by cereals was higher, and sometimes considerably so, than the recommended N application for the obtained yield. The discrepancy between applied N and recommended N application for cereals varied between catchments. For two of the catchments average N surplus compared to recommended N fertilization for barley and oats was 22–24 kg N ha$^{-1}$, whereas in the third catchment the surplus was 8 kg N ha$^{-1}$, indicating regional differences in fertilization practice. One of the catchments had a higher discrepancy for wheat compared to barley and oats with an average surplus of 39 kg N ha$^{-1}$. In a Norwegian survey of yields and nutrient application on 250 fields in the cereal regions of Norway, Riley et al. (2002) found a discrepancy between applied N and recommended N application which were in line or lower than the values for the JOVA catchments. These authors found that the N application relative to the recommended N application was on average 10, 9, and 20 kg N ha$^{-1}$ higher for barley, oats, and wheat, respectively. The larger N surplus for wheat than for barley and oats was explained by the fact that farmers aim at a high protein concentration in wheat because this gives better prices for wheat.

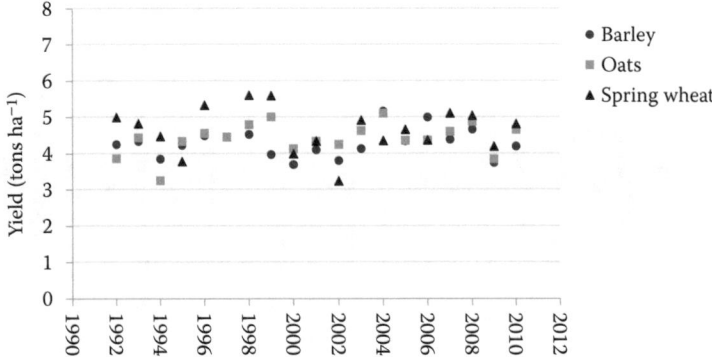

**FIGURE 6.7** Yearly average grain yields of barley, oats, and spring wheat at the JOVA catchment Mørdre. (Modified from Øgaard, A.F. 2014. Nitrogen balance and nitrogen use efficiency in cereal production in Norway. *Acta Agric. Scand., Sect. B, Soil and Plant Sci.* 63, Supplement 2. 146–155.)

As shown in Figure 6.7, an example for one of the catchments, there is a large variation in yield between years. It is recommended to plan for the average yield obtained in recent years, but the results presented above show higher N application than recommended for the obtained average yields. This indicates that farmers plan for a yield higher than the average yield, probably to be certain that there are sufficient nutrients to support a high yield. The ratio of N fertilizer cost to grain prices is low in Norway compared to other countries in Europe (Riley et al. 2012). In Norway, the ratio costs per kg N/price per kg grain often ranges from 4 to 7, whereas in Sweden, for instance, the ratio ranges from 7 to 9 (Albertsson 2011). The Norwegian ratio means that a yield increase of 4 to 7 kg grain is sufficient to cover the extra cost of applying one additional kilogram of N.

### 6.5.2 Nitrogen Balance

Higher N application than the recommended amount results in a high N surplus for cereals in the three catchments of the case study. The average N surplus was 40–51 kg N ha$^{-1}$ for barley, 22–38 kg N ha$^{-1}$ for oats, and 33–61 kg N ha$^{-1}$ for spring wheat. The observed N surpluses in these catchments were considerably higher than those found in the Swedish Monitoring Programme for Agriculture (Kyllmar et al. 2006). Here, the average N surplus for barley in 27 small agricultural catchments was 11 kg N ha$^{-1}$. The recommended N application rates for barley and oats are also lower in Sweden than in Norway, probably because of a higher ratio between N fertilizer costs and grain prices in Sweden (Albertsson 2011; www.bioforsk.no/gjodslingshandbok). For instance, in Sweden it is recommended to apply 11 kg less N ha$^{-1}$ at a yield level of 5 tonnes barley or oats than in Norway. The lower recommended N application rate may partly explain the lower N surplus in barley production in the monitored catchments in Sweden compared to that observed in Norway. Further, for Norwegian cereal production, it is shown that through economic optimization, N application may be considerably higher than the amount of N removed with the crop (Riley et al. 2012). The high N surplus in Norwegian cereals production is therefore probably a result of farmers' attempt to optimize their economy.

### 6.5.3 Nitrogen Use Efficiencies

Based on data from 23 agricultural dominated catchments in the Nordic countries, the NUE (calculated as N removed with yield as percent of applied N) varied from 50% to above 100%, with 17 of the 23 between 60 and 80% (Figure 6.8). The lowest NUE was obtained in F26, which is a Swedish catchment dominated by livestock production. There was, however, no general relationship between the percentage of N applied as manure and the NUE.

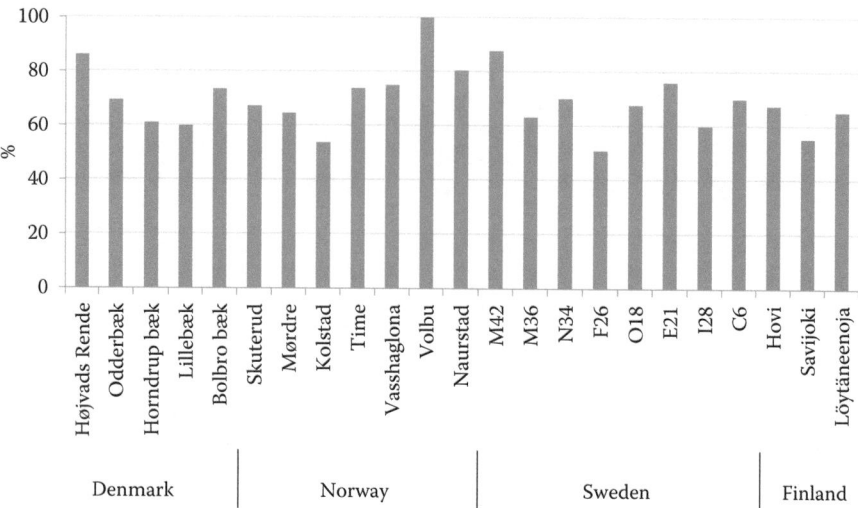

**FIGURE 6.8**  Average N use efficiency from 2007 to 2011 for agricultural land in 23 catchments in Denmark, Norway, Sweden, and Finland. (Data from Bechmann, M., Blicher-Mathiesen, G., Kyllmar, K., Iital, A., Lagzdins, A., and Salo, T. 2014. Nitrogen application, balances and effect on water quality in the Nordic-Baltic countries. *Agric. Ecosyst. Environ.* 198:104–113.)

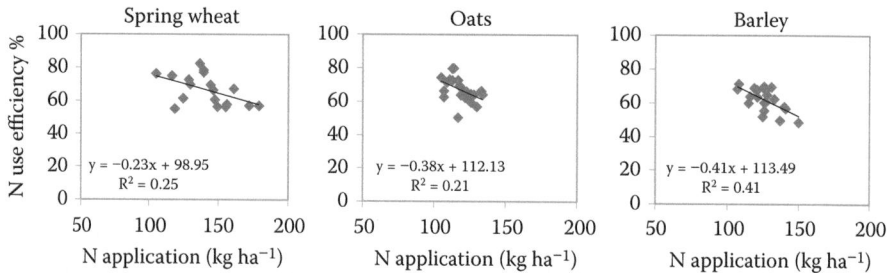

**FIGURE 6.9**  Relationship between N application and N use efficiency for spring wheat, oats, and barley at the JOVA catchment Mørdre. (Modified from Øgaard, A.F. 2014. Nitrogen balance and nitrogen use efficiency in cereal production in Norway. *Acta Agric. Scand., Sect. B, Soil and Plant Sci.* 63, Supplement 2. 146–155.)

In Norway, the average NUE for barley varied between 61% and 71% for the three Norwegian catchments dominated by cereals. The average NUE for oats and spring wheat was somewhat higher than for barley, with average values varying between 67% and 82% for oats and between 65% and 77% for spring wheat. As shown for one of the catchments in Figure 6.9, NUE was dependent on N application and declined with increased N application. Based on data from a Finnish farm published in Rankinen et al. (2007), the NUE was 71% at an average application of 110 kg N ha$^{-1}$ for cereals. This was at approximately the same level as that found in the Norwegian catchments. In the Swedish catchments (Kyllmar et al. 2006), the average N application to barley was low (90 kg N ha$^{-1}$), and consequently the NUE was high (87%) compared to the Norwegian and Finnish data.

## 6.6  REGULATIONS AND EFFECT ON NITROGEN CONSUMPTION

The Norwegian catchment data clearly demonstrate the importance of moderate N applications when aiming at reducing N surpluses in order to reduce N losses. Nitrogen application for higher yield levels than those obtained in reality results in low NUE and high N surplus in most years.

The conflict between aims of high yields and minimizing N losses has led to implementation of regulations for N application for protecting the water bodies and restricting GHG emissions. Various political strategies to reduce losses of N from agricultural land have been implemented. For example, the Nitrates Directive (91/676/EEC) aims to protect water quality across Europe by preventing nitrates from agricultural sources polluting ground and surface water and by promoting the use of good farming practices. Already in the late 1980s, Denmark and Sweden at the highest policy level, formed a legislation aimed at a 50% nutrient reduction to the sea. Both countries have measures targeted toward more efficient use of organic N in manure (Andersen et al. 2014). In Denmark, several nutrient action plans focusing on N have been implemented during the last 25 years (Kronvang et al. 2008). In Sweden, measures against N leaching have been implemented by a combination of subsidies for measures such as growing catch crops and spring ploughing, and regulations mainly related to manure application. In Finland, the Agri-Environmental Programme has set maximum N rates for most crops since 1995. In Norway, subsidies for agricultural management practices to reduce nutrient losses from agricultural land have been available since the beginning of the 1990s. These practices were, however, mainly focusing on P and not as much on N (Bechmann 2014). Regulations related to N comprised limits regarding livestock density, obligatory nutrient management plan, and subsidies for catch crops.

The differences in regulations targeted toward reduced N losses have clearly resulted in different trends regarding N application and N losses between the Nordic countries, as shown in Figure 6.1. Denmark with several nutrient action plans, focusing on N since the late 1980s has shown an almost 50% reduction in N application during the same period, whereas in Norway with low focus on reducing N application, there has been no trend in consumption of N in mineral fertilizer during the 1990s and 2000s.

## 6.7    CONCLUSION

There is a large surplus in the N balances for all the Nordic countries and a potential to decrease the surplus and increase the N use efficiency by a more precise N application rate. There are different trends in consumption of N fertilizers at the national scale for the Nordic countries, reflecting various political strategies to reduce N losses from agricultural land. For Norway with few regulations regarding N application, there is no decreasing trend in N application since 1990, which contrasts with Denmark with several nutrient action plans focusing on N and a considerable reduction in N application. The conflict between aims of high yields and minimizing N losses results in the need of regulations to obtain reduced N losses to air and water.

## REFERENCES

Albertsson, B. 2011. Riktlinjer för gödsling och kalkning 2012. Jordbruksverket, Jordbruksinformation 21-2011.90 pp. (In Swedish.)

Andersen, H.E., Blicher-Mathiesen, G., Bechmann, M., Povilaitis, A., Iital, A., Lagzdins, A., and Kyllmar, K. 2014. Mitigating diffuse nitrogen losses in the Nordic-Baltic countries. *Agric. Ecosyst. Environ.* 198:127–134.

Aronsson, H., Hansen, E.M., Thomsen, I.K., Liu, J., Øgaard A.F., Känkänen, H., and Ulén, B. 2016. The ability of cover crops to reduce nitrogen and phosphorus losses from arable land in southern Scandinavia and Finland—A review. *J. Soil and Water Conservation* 71:41–55.

Bechmann, M., Blicher-Mathiesen, G., Kyllmar, K., Iital, A., Lagzdins, A., and Salo, T. 2014. Nitrogen application, balances and effect on water quality in the Nordic-Baltic countries. *Agric. Ecosyst. Environ.* 198:104–113.

Bechmann, M. 2014. Long-term monitoring of nitrogen in surface and subsurface runoff from small agricultural dominated catchments in Norway. *Agric. Ecosyst. Environ.* 198:13–24.

Blicher-Mathiesen, G., Andersen H.E., and Larsen, S.E. 2014. Nitrogen field balances and suction cup-measured N leaching in Danish catchments. *Agric. Ecosyst. Environ.* 198:36–43.

Conley, D.J., Carstensen, J., Vaquer-Sunyer, R., Duarte, C.M. 2009. Ecosystem thresholds with hypoxia. *Hydrobiologia* 629:21–29.

Goulding, K.W.T. 2000. Nitrate leaching from arable and horticultural land. *Soil Use Manage* 16, 145–151.

Jansons, V., Busmanis, P., Dzalbe, I., and Kirsteina, D. 2003. Catchment and drainage field nitrogen balances and nitrogen loss in three agriculturally influenced Latvian watersheds. *Eur. J. Agron.* 20:173–179.

Korsaeth, A., and Eltun, R. 2000. Nitrogen mass balances in conventional, integrated and ecological cropping systems and the relationship between balance calculations and nitrogen run-off in an 8-year field experiment in Norway. *Agric. Ecosyst. Environ.* 79:199–214.

Kronvang, B., Andersen, H.E., Børgesen, C.D., Dalgaard, T., Larsen, S.E., Bødestrand, J., and Blicher-Mathiesen, G. 2008. Effects of policy measures implemented in Denmark on nitrogen pollution of the aquatic environment. *Environ. Sci. Pollut. Manage* 11:144–152.

Kyllmar, K., Carlsson, C., Gustafson, A., Ulén, B., and Johnsson, H. 2006. Nutrient discharge from small catchments in Sweden. Characterisation and trends. *Agric. Ecosyst. Environ.* 115:15–26.

Mengel, K., and Kirkby E.A. 1987. Principles of plant nutrition. International Potash Institute, Bern, Switzerland.

Nitrates Directive. 1991. EC Environment. Verified June 30, 2013.

OECD. 2001. Environmental Indicators for Agriculture, Methods and Results. Organization for Economic Co-operation and Development, Paris, France.

Rankinen, K., Salo, T., Granlund, K., and Rita, H. 2007. Simulated nitrogen leaching, nitrogen mass field balances and their correlation on four farms in south-western Finland during the period 2000–2005. *Agric. Food Sci.*16:387–406.

Riley, H., Hoel, B., Kristoffersen, A.Ø., and Tandsæther, H. 2002. N-gjødsling til korn: Anbefalinger og praksis. In: Abrahmsen, U. (Eds.). Jord- og plantekultur 2002. *Planteforsk Grønn forskning* 01/2002, pp. 75–80. (In Norwegian.)

Riley, H., Hoel, B.O., & Kristoffersen, A.Ø. 2012. Economic and environmental opimization of nitrogen fertilizer recommendations for cereals in Norway. *Acta Acta Agric. Scand., Sect. B, Soil and Plant Sci.* 62:387–400.

Salo, T., and Turtola, E. 2006. Nitrogen balance as an indicator of nitrogen leaching in Finland. *Agric. Ecosyst. Environ.* 113:98–107.

Salo, T., Lemola, R., and Esala, M. 2007. National and regional net nitrogen balances in Finland 1990–2005. *Agric. Food Sci.* 16:366–375.

Stålnacke, P., Bechmann, M., Blicher-Mathiesen, G., Iital, A., Jansons, V., Koskiaho, J., Kyllmar, K. et al. 2014. Temporal trends in nitrogen concentrations and losses from agricultural catchments in the Nordic and Baltic countries. *Agric. Ecosyst. Environ.* 198:94–103.

Valkama, E., Salo, T., Esala, M., and Turtola, E. 2013. Nitrogen balances and yields of spring cereals as affected by nitrogen fertilization in northern conditions: A meta-analysis. *Agric. Ecosyst. Environ.* 164:1–13.

Windolf, J., Blicher-Mathiesen, G., Carstensen, J., and Kronvang, B. 2012. Changes in nitrogen loads to estuaries following implementation of governmental action plans in Denmark: A paired catchment and estuary approach for analysing regional responses. *Environ. Sci. Policy* 24:24–33.

Øgaard, A.F. 2014. Nitrogen balance and nitrogen use efficiency in cereal production in Norway. *Acta Agric. Scand., Sect. B, Soil and Plant Sci.*63, Supplement 2. 146–155.

Øygarden, L., Deelstra, J., Lagzdins, A., Bechmann, M., Greipsland, I., Kyllmar, K., Povilaitis, A., and Iital, A. 2014. Climate change and the potential effects on hydrology and nitrogen losses in the Nordic-Baltic region. *Agric. Ecosyst. Environ.* 198: 114–126.

# 7 Intensive Dairy and Beef Systems

## N Loss Mitigations and Barriers to Their Adoption

*Bert F. Quin*

## CONTENTS

## 7.1    INTRODUCTION

This chapter discusses both existing and potential new environmental N loss mitigation options for intensive dairy and beef operations in cool-temperate and temperate climates, in the context of understanding the barriers to their adoption. These include physical practicality, cost-effectiveness, consumer attitudes, the attitudes of governments toward regulating the intensive dairy and beef industry, and the interlinked issue of the level of understanding and acceptance by society of the effects of N mitigation regulations on retail prices and the availability of dairy and milk products. Most N emissions are directly excreta-related. Depending on the basis of comparison (e.g., volume, N, phosphorus (P) or bacterial content such as *E. coli* (Rose et al. 2015)), each cattle beast excretes 30–100 times that of an adult human. The safe and productive recycling of this excreta is a key consideration if we wish to continue the trend of increasing consumption of beef meat and milk products.

### 7.1.1    Mixed Cropped and Grazed Farms (MCGFs)

Rising disposable incomes in western countries in the 1950s lead to increasing demand for dairy products and beef meat. In western Europe, the presence of a myriad of small rural population centers, small land holdings, cold winters, and extensive road and rail networks combined to restrain the emergence of large corporate farms. The average size of dairy and beef farms remains quite small, with most in the range of 20–100 ha. These farms, combining a mixture of cut and carry cropping and grazing (MCGFs) include full animal housing for the colder months, with cut and carry crops and cut pasture being fed, and to a lesser extent brought-in feeds (Photo 7.1). Animals are usually put out to graze pasture or forage crops (Photo 7.2) in the warmer months, although many dairy operations are becoming essentially fully housed. Intensification of production has taken place primarily through breeding more productive animals, crops and pastures, improved fertilizer and soil fertility management, and brought-in feed.

It is not the intention of this chapter to review the entire area of N management and loss mitigation in MCGF systems. In the European Union (EU), there have been three main directives

**PHOTO 7.1**    A large dairy house on a MCGF in Europe.

**PHOTO 7.2    (See color insert.)** Beef grazing an electric fence—controlled forage crop in the UK.

instigated by the European Commission (EC); the Nitrate Directive of 1991, the Water Framework Directive of 2000, and the National Ceilings Directive of 2001. These have been very influential in the development of a documented, common understanding in the EU and neighboring countries of both the N loss mechanisms and how they can be reduced, and of (to a somewhat slower extent) the implementation of regulations regarding farm management practices, and N inputs and outputs, designed to gradually reduce N losses to more environmentally sustainable levels. In particular, the risks of increased N losses to the environment associated, directly and indirectly, with use of fertilizer N and application of manure, are comprehensively understood and gradually being managed. Assessments of the MCGF system include those by Bussink and Oenema (1998);

Godfray et al. (2010); Pilgrim et al. (2010); Del Prado et al. (2013); and Misselbrook et al. (2013). Wright (2005) and Misselbrook et al. (2013) describe meat and milk production from different systems, concentrating on potential production limitations to MCGFs in different climates (Wright 2005), or on opportunities for reducing environmental N emissions from MCGFs in temperate climates (Misselbrook et al. 2013).

### 7.1.2 Intensively Grazed Pasture Farms (IGPFs)

Since the 1950s, increasing areas of extensive-grazing, forestry and cropping land in temperate climates with milder winters have been converted to pasture for intensive grazing (Photos 7.3 and 7.4). This involves the use of fencing (permanent and electric), increased fertilizer inputs, and the sowing

PHOTO 7.3    (See color insert.) A herd coming in for milking on an IGPF in New Zealand.

PHOTO 7.4    (See color insert.) A typical rotary milking platform on an IGPF in New Zealand.

of improved plant varieties, especially of ryegrass (*Lolium perenne* L.) and white clover (*Trifolium repens*). IGPFs have little animal housing (except for weaning replacements where not contracted out) and are still the norm in New Zealand and south-east Australia, and are making a comeback in other countries such as Ireland and in the mid-south United States. In IGPFs, environmental concerns predominantly relate to (1) increasing pollution of streams and rivers from milking shed effluent (Li et al. 2015), (2) nitrate leaching into streams, lakes, and groundwater, the majority of which comes from urine patches, especially on free-draining soils (Quin et al. 1977; Ball et al. 1979), and more recently (3) nitrous oxide ($N_2O$) emissions, again mainly from urine patches (Saggar et al. 2009). Clover/ryegrass pastures have a high N content (typically 3.5–4.0%), approximately double the animal's dietary N requirement. The excess N is voided in the urine, which typically contains 0.6–1.0% N (Saggar et al. 2009). Urine deposition creates 0.4 $m^2$ urine "patches" (Moir et al. 2011), in which N deposition rates are typically 600–1000 kg N/ha, well above the ability of the affected pasture to recover (Ball et al. 1979; Saggar et al. 2009).

IGPFs are loosely defined here by what might be called the "Triple 80" rule, namely, systems where the animals (1) receive at least 80% of their feed intake directly through grazing of pasture and fodder crops, (2) spend at least 80% of their productive lives outdoors, and (3) are rotated in mobs around permanently or electrically fenced paddocks or areas for at least 80% of the year, rather than being set-stocked. While production of up to 80% of that achievable to that under rotational grazing can be achieved with set-stocking, this practice leads to very uneven redistribution of excreta, with disproportionately large amounts being deposited in "campsites" under shade or near shelter, water troughs, and gates (Gillingham et al. 2007). This leads to increased nutrient loss to the environment from these sites, and to lower production from the remainder of the grazed area. Environmental and economic pressures are both accelerating the adoption of rotational grazing in IGPFs.

### 7.1.3 Confined Animal Feeding Operations (CAFOs)

Since the 1950s, improving large-scale crop husbandry knowledge, the development of more effective herbicides and pesticides, decreasing fuel costs, and more efficient harvesting equipment combined to reduce costs of production of grain crops in drier climates, particularly in the United States and Canada. It became increasingly more profitable to set up large, fully housed dairy houses or confined beef cattle operations, collectively called confined animal feeding operations or CAFOs, and bring grain (in combination with local cut-and-carry feeds) to the animals (Photo 7.5). With reducing transport costs, grain feed could be brought long distances, from other regions and even

**PHOTO 7.5**   (**See color insert.**) A large beef feedlot (CAFO) in Texas.

imported from other countries. It is now estimated that only 3% of cattle in the United States spend most or all of their lives grazing pasture. In CAFOs, the major environmental concerns are methane emissions belched from the animal's rumen, and ammonia and $N_2O$ emissions from excreta (Hutchinson et al. 1982). Much of the ammonia is redeposited as ammonium within a short distance; elevated levels of atmospheric ammonia have been recorded to a maximum of 800 m downwind of CAFOs (McGinn et al. 2008). However, gaseous ammonia reacts quickly with sulfur and nitrogen oxides to form the very fine (PM < 2.5 micron) particles responsible for serious respiratory illnesses (Hristov et al. 2011). Surface or groundwater quality deterioration, and gaseous N losses, occur because of manure, slurry, and effluent application, and also through redeposition of gaseous ammonia. Rotz (2004) summarized management options available to reduce N losses from both CAFOs and MCGFs, with only brief reference to IGPFs and the urine patch issue. Hristov et al. 2011 presented a very thorough review of ammonia emissions and mitigations from North American housed dairy and beef feedlot operations and European housed and mainly housed systems, but again with little reference to IGPFs.

This chapter discusses the existing and new environmental N loss mitigation options for intensive animal operations in cool-temperate and temperate climates, in the context of barriers to their adoption. These include issues of practicality and cost-effectiveness for the food producer, and perhaps conflicting societal expectations regarding food costs and environmental quality.

Not discussed here is the extensive grazing of low-fertility, unfertilized natural grassland with beef cattle which is still the main form of beef production in many climates and countries. Many of these systems, for example, in Siberia, Mongolia, and China, are believed to be slowly degrading due to increasing nutrient removal in meat and climatic change (Gilmanov 1995). In the small mixed-cropping farms prevalent in developing countries, particularly those with tropical climates such as India and Nigeria, a small number of dairy cows and animals grown for meat are kept on part of the farm, and receive all or most of their feed intake from cut and carry crops grown on the farm (Wright 2005). These farm systems still produce most of the world's milk, and perhaps 35% of its beef, but are not discussed further. Finally, the semi-controlled grazing of tropical pastures, for example in Brazil and Queensland, Australia, is a quite different system to temperate grassland and is not discussed here; it is noted, however, that ammonia volatilization from both urine (Vallis et al. 1982) and fertilizer N (Harper et al. 1983) are common concerns.

Countries with cattle populations (including dairy) of more than 10 m, and the numbers of milking cows, are listed in Table 7.1. The author's estimates of the percentages of total cattle confined in feedlots, and of dairy cows farmed on MCGFs, are also given. New Zealand has only one feedlot; Australia has only 0.87 m cattle (3% of its 26-m total) in CAFOs.

## 7.2   NITROGEN MANAGEMENT IN MCGFS: THE NETHERLANDS EXPERIENCE

Nitrogen management and the development of N loss mitigation strategies in MCGFs in Europe has been the subject of much research; implementation of mitigations is increasingly being legislated for, most commonly by minimizing allowable farm N surpluses (total N inputs minus N in produce and that measured in the soil). In the EU, agreed time-frames for reaching nitrate leaching and gaseous emission targets under the Nitrate, Water Framework and National Ceilings Directives vary quite widely between countries, largely reflecting the extent of the N surplus that existed prior to the Nitrate Directive of 1991. However, the goal is to have all countries meeting EU targets by 2027. Regarding the future management of IGPFs and CAFOs in other parts of the world, therefore, it is worth examining the success or otherwise of the implementation of N mitigation strategies in those EU counties with existing large N surpluses.

The Netherlands (along with parts of Belgium and northern Germany) has easily the largest N surpluses in the EU. This arises from the concentration of increasing numbers of stock (primarily dairy, beef, and pigs) on small farms; this has been made possible by improved breeding of animals

**TABLE 7.1**

**Beef (2016) and Dairy Cattle (2012) Inventories: Countries with at Least 10 m Cattle**

| Rank/Country | Climate | Tot. Cattle (m head) | Main System in Use | % in Feedlots (Estimated) | Total Dairy Cows (m Lactating)[f] | % in MCGFs (est) |
|---|---|---|---|---|---|---|
| 1 India | Ems | 303 | Small MCGFs | <5 | 44 | 95 |
| 2 Brazil | Ewd | 219 | CAPOs | 50 | 23 | 75 |
| 3 China | Thhs | 100 | Ext. grazing | 30 | 12 | 90 |
| 4 USA | Thhs | 92 | Large CAFOs | 97[b] | 9 | 20[c] |
| 5 EU[a] | Thhs | 89 | Med. MCGFs | <5 | 20 | 90 |
| 6 Argentina | Thhs | 53 | Ext. grazing | 25 | 3 | 60 |
| 7 Australia | Thhs | 27 | Ext. grazing | 3 | 2 | 10[d] |
| 8 Russia | Shhs | 19 | Ext. grazing | 5 | 9 | 65 |
| 9 Mexico | Ed | 17 | Feedlots | 60[b] | 2 | 90 |
| 10 Turkey | Thhs | 14 | Small MCGFs | <5% | 4 | 80 |
| 11 Uruguay | Eh | 12 | Ext. grazing | 20 | 1 | 90 |
| 12 Canada | Shcs | 12 | Feedlots | 90[b] | 1 | 20[c] |
| 13 New Zealand | Thws | 10 | IGPFs | <1 | 5 | <5[e] |
| Total of top 13 | | 967 | | | 133 | |
| Total world | | 998 | | | 264 | |

*Sources:* Total cattle numbers from USDA (2017); total lactating dairy cows from CIWF.org (2012).

*Notes:* Climate codes in main cattle areas: Ems equatorial, monsoonal; Ewd eq., winter dry; Thhs temperate, humid, hot dry summer; Shhs snow, humid, hot summer; Ed eq., dry; Eh eq., humid; Shcs snow, humid, cool summer; Thws temperate, humid, warm summer.

[a] Includes UK.
[b] At any one time, 20% may be grazing before returning to feedlot.
[c] 70% in fully housed operations (CAFOs).
[d] 80% on IGPFs.
[e] 95% in IGPFs.
[f] Note ranking for lactating dairy cows is quite different to that for total cattle.

and crops, the increased reliance on year-round housing, the increasing use of fertilizer N, and the increasing importation of feed (van Grinsven et al. 2010). The somewhat predictable outcome of all this is that the quantities of manure and slurry being produced are far more than the ability of the existing soil/plant system to recycle N in ways which avoid large losses in nitrate leaching, ammonia volatilization, and $N_2O$ emissions. Consequently, the Netherlands has had the worst surface and groundwater quality in the EU. Their food production enterprises have been profitable in purely financial terms, but far less so from an environmental or broad societal perspective (Van Grinsven et al. 2016). Data from the Karkendamm Experimental Farm in northern Germany and the De Marke Experimental farm in the Netherlands are shown in Figure 7.1, from Wachendorf and Golinski (2006). A linear relationship between nitrate leaching and N surplus per hectare was found, with each 100 kg N additional surplus resulting in a 20 kg N extra leaching (Lampe et al. 2004; Trott et al 2004; Wachendorf et al. 2004).

The De Marke dairy cow farming system for MCGFs, located on sandy soil in Gelderland province, was designed in 1993 to minimize external inputs of feed and fertilizer, and to maximize the use of feed and manure produced on-farm (Wachendorf and Golinski 2006). The farm is 55 ha, consisting of 44 ha of rotated grass and maize (3 years in grass followed by 3–5 years in maize) and 11 ha of permanent grassland. First year maize receives no fertilizer N to minimize nitrate loss from mineralization of soil organic N. About 75% of the slurry output is applied to the grassland fields

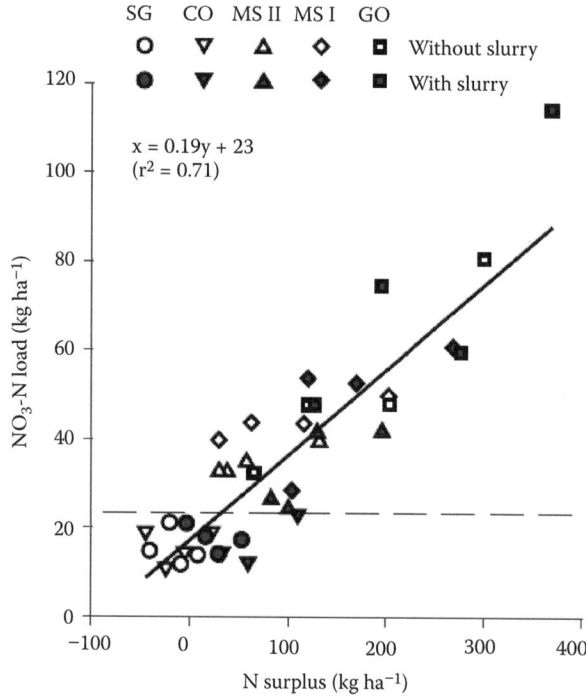

**FIGURE 7.1** Relationship between annual N surplus on grassland. GO = grazing only; MSI/II = mixed system with one cut and two cuts in spring, respectively, and subsequent grazing; CO = cutting only; SG = simulated grazing. (From Wachendorf, M., Buchter, M., Trott, H., and F. Taube 2004. Performance and environmental effects of forage production on sandy soils. II. Impact of defoliation system and nitrogen input on nitrate leaching leaching losses. *Grass and Forage Sci.* 59:56–68.).

by shallow injection to minimize ammonia volatilization; grassland now receives an additional 107 kg N ha[-1]. The farm has provided a long-term ongoing comparison with current farm "best practice," and has demonstrated that N surpluses can be greatly reduced by the adoption of easily implemented mitigations. Hilhorst et al. (2001) reported an annual N surplus for the mid-1990s of 146 kg N ha[-1], compared to 408 kg N ha[-1] for an average conventional farm. However, modeling by Rotz et al. (2005) indicated that a De Marke system farm would require an additional 9 ha in maize to attain the EU directive goals, which correspond to an N surplus of about 100 kg N ha[-1]. This additional land is of course simply not available; the clear implication is that stocking rate reductions of approximately 20% are required to meet EC directives. Their modeling did show, however, that conventional farms that have adopted newer technology have reduced their N surplus from 340 to 270 kg N ha[-1] (Table 7.2), the improvement coming mainly from the use of enclosed rather than bottom-loaded manure storage for 6 months, manure injection instead of surface application, and by restricting grazing to half days. This is still 120 kg N ha[-1] too high. Instead of forcing a direct reduction in stock numbers to meet the EC directives, and the mass farm bankruptcies that would follow, the Netherlands government has been working slowly toward them by regulating fertilizer N inputs and requiring dairy farmers to pay those cropping farmers who can take some of their manure; depending on the area, some of this needs to be composted or otherwise treated. This can result in N loss transfer; nitrate leaching may be reduced at the expense of increased ammonia emissions. Consequently, much manure (especially that of cattle, because of its higher N to P ratio than pig manure) has also been composted and exported for application on low fertility soils elsewhere in the EU such as in eastern Germany.

**TABLE 7.2**

**N Budgets for Mixed Cropping/Grazing Dairy Farms (MCGFs) in The Netherlands**

| | Actual Farm Data[a] (1990s) | | Simulated Farms[b] (Early 2000s) | | |
|---|---|---|---|---|---|
| | 'De Marke' Farm | Conv. Farm | 'De Marke' | Conv. Old | Conv. New |
| **Inputs (kg N ha$^{-1}$)** | | | | | |
| Fertilizer N | | | 64 | 242 | |
| Concentrates and feed | | | 94 | 145 | |
| Biological N fixation | | | 11 | 0 | |
| Deposition | | | 49 | 49 | |
| **Total inputs[c]** | **223** | **486** | **228** | **430** | **364** |
| **Outputs (kg N ha$^{-1}$)** | | | | | |
| Milk | 66 | 64 | 75 | 89 | 95 |
| Animals and misc.[d] | 11 | 14 | | | |
| **Environmental loss (surplus)** | **146** | **408** | **153** | **341** | **270** |
| Including: | | | | | |
| Ammonia | | | 27 | 167 | 62 |
| Denitrification | | | 38 | 45 | 27 |
| Leaching | | | 55 | 107 | 62 |

[a]  Modified presentation, from Hilhorst et al. 2001, Wachendorf and Golinski 2006.
[b]  Modified presentation, from Rotz et al. 2005, Wachendorf and Golinski 2006.
[c]  Total inputs only available for the simulated comparison.
[d]  Outputs in milk and animals combined for the measured comparison.

Overall, progress in reducing N (and P) losses to the environment in the Netherlands was substantial from 1991 until about 2003, but with very little progress since then (Figure 7.2, from Fraters et al. 2015 and van Grinsven et al. 2016). There are many contributing reasons for this, including (1) a complicated and even conflicting combination of requirements by different EC directives; (2) the impacts of a variety of government budgetary regulations to stimulate or restrict agricultural

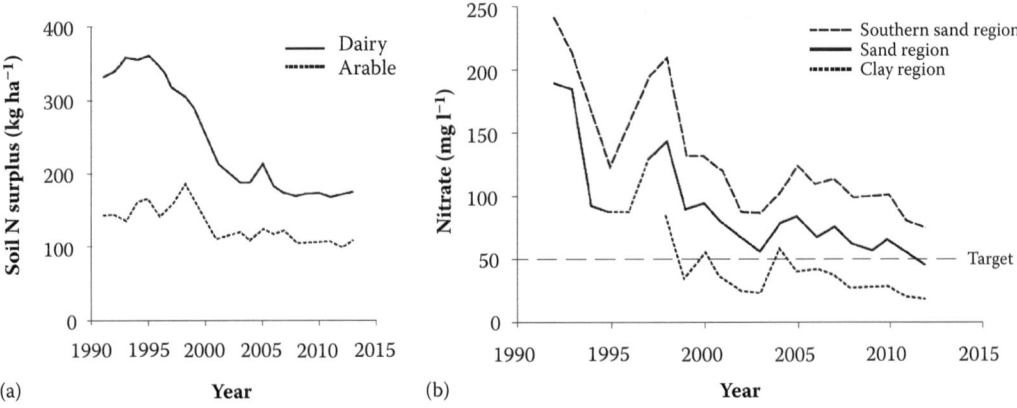

**FIGURE 7.2**  (a) Farm level soil N and P205 surpluses for the dairy sector. (b) Nitrate (mg l-1) in the upper groundwater in agricultural area for the total and south-eastern sand region and for the clay region. (From Fraters, B., van Leeuwen, T., Boumans, L., and J. Reijs 2015. Use of long-term monitoring data to derive a relationship between nitrogen surplus and nitrate leaching for grassland and arable land on well-drained sandy soils in The Netherlands. *Acta. Agr. Scand.* B-S P 65 (suppl. 2), pp. 144–154.)

production, and their expiry dates; (3) the introduction of manure-disposal regulations that, because of enforcement difficulties are open to avoidance or fraud, and (4) an underlying attitude of complacency now that average nitrate levels in groundwater are close to the target of 50 mg $l^{-1}$ nitrate (11.3 mg $NO_3$-N $l^{-1}$). This is despite little or no improvement in surface water quality, measured as both N or phosphorus (P), relative to goals set in the EC directives. The mandatory adoption of waterway fencing and riparian strips is proceeding very slowly. Taken altogether, it seems that further reductions in environmental N losses, and any improvement in surface water quality, will require a new round of far more coordinated regulations and food production-orientated government strategies. This is particularly true given that recent modeling of the effects of nitrate levels in drinking water on the incidence of cancer suggest that the drinking standard for nitrate in drinking water should ultimately be reduced to 25 mg $l^{-1}$ (Van Grinsven et al. 2010). Even to meet current nitrate and gaseous emission standards, it seems very clear that stocking rates in MCGFs in the Netherlands are going to have to be reduced by 20%, unless ways are quickly found to greatly improve the efficiency of N inputs in both feed and fertilizer. Much attention is now being paid to the former, but the great need for improvements in fertilizer nitrogen utilization efficiency has been largely ignored.

## 7.3  NITROGEN IN IGPFS: A HISTORICAL REVIEW

### 7.3.1  CLIMATE AND SOIL LIMITATIONS TO IGPFS

Pasture plants in drier, non-irrigated temperate climates need to be able to tolerate significant moisture stress (Crush et al. 2007). Intensive grazing requires soils that are well aggregated and reasonably free-draining. Even initially free-draining soils can become "pugged" due to treading pressure if they are not well aggregated and managed. Pugging greatly reduces root penetration and the establishment and growth of high growth-potential pasture plants (Menneer et al. 2005; Drewry et al. 2008).

Irrigation water where available has overcome the limitations to intensive grazing imposed by low rainfall, but can result in increased nitrate leaching. Poor drainage has been overcome by installation of physical soil drainage systems such as mole tiles, but these can lead to increased nitrate leaching (Houlbrooke et al. 2003). Reductions in N fixation are being overcome by the increased use of fertilizer N, largely in the form of urea produced in the Haber–Bosch industrial nitrogen fixing process. This process is now calculated to be producing over 100 Tg N (100 million tonnes) per annum of fertilizer N; nearly double that of current estimates of biological N fixation by agricultural legumes (Fowler et al. 2013).

Based on the parameter of unacceptable increases in nitrate leaching into surface waters and groundwater aquifers, intensively grazed pasture farming as currently practiced can already be categorized as unsustainable in many areas; increasing greenhouse gas losses reinforce this conclusion.

In New Zealand, increasing export opportunities for dairy products has resulted in large low-rainfall areas of Canterbury in the South Island, and Hawks Bay in the North Island, both comprising mainly shallow free-draining soils, converted from dryland sheep and beef farming to irrigated intensive dairying. Greatly increased fertilizer N use (200–400 kg N $ha^{-1}$ annually) and increased excreta return has resulted in widespread increases in nitrate levels in groundwater and bacterial and sediment deterioration in surface water quality. Regulations requiring increased irrigation efficiency have reduced leaching during the irrigation season, but large N losses occur (typically 30–80 kg N $ha^{-1}$) during autumn and winter. Intensification of existing North Island dairy and beef grazing farms has had broadly similar effects.

The dairy industry in Ireland has declared its intention of increasing dairy production by 50%, through intensification of grazing. Although the predominantly low infiltration rate of soils in Ireland means lower nitrate leaching in general (K. Richards, Teagasc, pers. comm.), the potential for increased denitrification losses of nitrous oxide ($N_2O$) is considerable. It is therefore timely to revisit the IGPF system to assess what needs to be done to regain true sustainability.

## 7.3.2   THE N CYCLE IN IGPFS: FROM CLOVER N FIXATION TO FERTILIZER N

The introduction of improved pasture species, particularly mixes of ryegrasses and clover, gave large increases in production under grazing (Walker 1956). For biological N fixation through the clover root nodule—*rhizobia* symbiosis to reach its potential—soils need to be brought up to an appropriate level of nutrient fertility for the clover component, especially with regard to P, potassium (K), sulfur (S), and the trace element molybdenum, a catalyst in the *nitrogenase* enzyme responsible for atmospheric N fixation by the *Rhizobia* bacteria in root nodules (Vitousek et al. 2013). Annual N fixation is typically in the range 150–300 kg N ha$^{-1}$ (Hoglund et al. 1979; Ledgard and Steele 1992; Thomas 1992).

By the late 1970s, it was known that the largest losses from the N cycle in IGPFs originated from the animal urine patch, principally as nitrate leaching and/or ammonia volatilization (Quin 1977; Ball et al. 1979; Field and Ball 1982; Eckard et al. 2001). This was largely a function of the high N content of the mixed herbage, typically 3.5–4% N. This is approximately double the dietary requirement of grazing animals; the excess N is voided in the urine, mainly as urea (75–90%), with lesser quantities of other N compounds (Bristow et al. 1992). Quin (1982) listed options for reducing nitrate leaching losses from urine patches (Table 7.3). Thirty-five years later, there is renewed investigation into the potential of most of these methods, and some newer ones (Table 7.3). These mitigation options are discussed in Section 7.4.

### TABLE 7.3
### Possible Methods for Increasing the Efficiency of Urine-N Recycling on Intensively Grazed Pastures

| Principle | Methods | Potential (1982) | Situation (2016) |
|---|---|---|---|
| Reduce opportunity for camping | Rotation grazing | Moderate | *Practiced throughout NZ* |
| Increase frequency of urination for more even distribution | (i) Insert catheter | Poor—high cost, risk of infection | *Investigated, not introduced* |
| | (ii) Breed animals with smaller bladders | Moderate | *Part of reason for increased interest in milking sheep* |
| Reduce N applcn rate per urination to incr recovery | (i) Breed low N-content pasture plants | Mod. to good, but feed quality may suffer | *Subject of large current research program* |
| | (ii) Increase thirst and urine volume with salt | Slight | *Trial found 15% leaching redn (Ledgard et al. 2016)* |
| Increase spreading of urine | (i) Breed animals that urinate while walking | Slight | *No known research* |
| | (ii) Attach device to cow to disperse urine flow and/or add inhibitors | Mod. to good—being investigated | *Several devices patented; production likely 2017* |
| **More recent options being investigated:** | | | |
| Reduce time on grass | Use stand-off pads—excreta collected | Poor—extra cost, ammonia volatn | *Reduces nitrate leaching; but is just 'N transfer'* |
| Reduce winter leaching | Follow winter fodder crops with catch crops | Good where too wet to allow pasture grazing | *Comparisons of alternative cash crops underway* |
| Detect and treat urine patches post-grazing | Spikey® EC detection and ORUN® treatment | Very high—extra grass growth makes economic | *Commercialization to start 2017 (Quin et al. 2016)* |

*Note:*   A flashback to Quin (1982) and the situation in 2016.

Increasing demand for competitively priced produce encouraged dairy farmers to use increasing quantities of fertilizer N. Initially, fertilizer N was essentially used as a tactical application, to bring feed on earlier in spring or continue its growth later in autumn (Field and Ball 1978, O'Connor 1982; Eckard and Franks 1998). The initial increase in pasture production in the 3–6 weeks after fertilizer N application in early spring was often followed by a similar period of lower production due to clover suppression (O'Connor 1982). However, researchers quickly confirmed what farmers had realized, namely, repeated applications of N avoided growth suppressions and increased annual production by 25–45% (Wilman and Wright 1983; Frame and Newbold 1984; Reid 1993; Harris et al. 1994; Clark and Harris 1996). This was shown to be accompanied by a large decline in the percentage of clover present, from a typical 25–30% in the absence of fertilizer N to 5–15% in pastures receiving more than 300 kg N/ha as fertilizer annually. The increase in total production made the use of fertilizer N economic, but has raised more environmental concerns. New high-growth potential ryegrasses required higher N contents to reach their full growth potential, meaning more urine-N deposited. Direct and indirect nitrate leaching from the fertilizer N also increased (Ledgard et al. 1999; Jarvis 2000). Increases in inputs and production saw more marginal soils being pushed closer to their sustainable limits. This was exacerbated by the increasing use of brought-in feeds when milk prices were high. Assessments by the author suggest that these may have stabilized in IGPFs at 10–15% of total intake in New Zealand, 15–20% in Australia and the United States, and 20% in Ireland.

### 7.3.3 THE N CYCLE AND FERTILIZER N EFFICIENCY

Intensively grazed dairy and beef farming has considerably higher earning potential per hectare than extensive grazing, but requires ongoing fertilizer inputs. The N cycle, optimum soil fertility status, and fertilizer inputs for IGPFs have been studied for over 60 years (Walker 1956), and have been accurately defined for many different soils and productivity levels, and made available through farm extension services. Some countries have developed computerized soil fertility management programs that calculate nutrient inputs, including N, required for different levels of productivity on a wide range of soil types, soil fertility status, and farming systems, and considering the nutrient content of brought-in feed. Some also calculate the losses of nutrients from the soil in produce and soil fixation and so on, and, in the case of N, nitrate leaching and gaseous N losses. The modeling behind such systems is often considerably more complex than most models used in cropping. An example is New Zealand's Overseer® program (Wheeler 2009).

The N cycle in IGPFs has changed greatly in New Zealand from the clover-based system of the 1970s (Quin 1982) to the fertilizer N based system of today. An irrigated dairy IGPFm or a non-irrigated farm in >1200 mm annual rainfall, running 3 Holstein-cross cows ha$^{-1}$ with low (10%) use of feed supplements, may need 250, 60, 80, 40 kg/ha annually of N, P, K, and S, respectively, to produce 1500 kg milksolids (MS) per ha annually. A simplified, annualized, N-balanced N cycle for intensively grazed, high fertilizer-N input system is shown in Figure 7.3. In these systems, nitrate leaching, ammonia volatilization, and $N_2O$ emissions all become increasingly serious (Ledgard et al. 1999; Jarvis 2000; Saggar et al. 2009, 2013; Gourley and Weaver 2012), posing difficult decisions if sustainable water quality and emission standards are to be maintained. The numbers attached to the various flows have been derived from studies referenced in this chapter.

Despite the huge importance of fertilizer input, relatively little research has been conducted into improving its efficiency. Application of precision farming technology enabled by either direct or remote sensing-driven variable rate fertilizer application (VRFA), unlike VR irrigation, has been very slow to be adopted on IGPFs compared to cropping farms (where adoption rates are about 30% in the United States, for example). It is now commercially available on spreading aircraft and trucks in New Zealand (Morton et al. 2016). VRFA is predominantly used on steeper areas under cattle and sheep grazing, principally to reduce fertilizer application in areas of both lower production potential

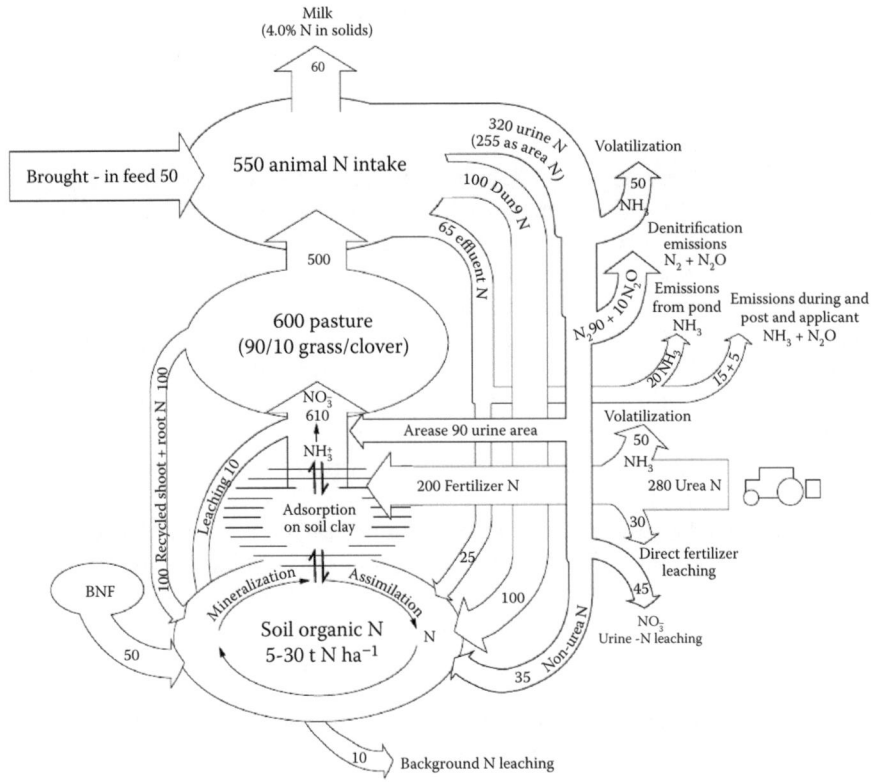

**FIGURE 7.3** An N-balanced, simplified example of annual N cows (kg Nha$^{-1}$) in an intensively grazed irrigated pasture in New Zealand; pasture production 15 t DM ha$^{-1}$, pasture utilization 85%, 1500 kg milk solids ha$^{-1}$. In this example, total annual losses as nitrate leaching, ammonia volatilization, and nitrous oxide emissions are N 85, 135, and 15 kg N ha$^{-1}$, respectively. The relative losses of these emissions (and N$_2$) are very dependent on soil type, moisture, and pH. Some of these systems may be in a negative soil organic N balance (see text).

and high-fertility stock camping areas, and to help avoid direct ingress of fertilizer N and P into waterways (Morton et al. 2016).

Even more importantly perhaps, intensive grazing has fallen behind cropping in the optimization of the chemical and physical form of nutrients used, particularly regarding efficiency and reducing losses of N and P to the environment. Few cropping farmers would consider the use of standard granular urea fertilizer as a surface application because of its known susceptibility to volatilization; it is, however, still the most commonly used fertilizer N on intensively grazed farms in many countries. Reasons for this include relatively low nutrient, transport, and application costs per unit N. However, volatilization losses of 5–30% globally on pasture are similar to those under cropping (Saggar et al. 2009; Cameron et al. 2013).

Fertilizer urea coated with polymers, sulfur, and other materials are widely used in cropping, horticulture, and the turf industry (Nash et al. 2015), but the cost premium has greatly restricted their use on IGPFs. However, coating granular urea with the urease inhibitor N-(n-butyl) thiophosphoric triamide (nbpt) has been demonstrated to produce gains in N use efficiency (NUE) of approximately 25% (Watson et al. 1990a, 1990b; Blennerhassett et al. 2006; Zaman et al. 2013a, 2013b). With cost premiums of only 10% per kg N, the use of nbpt-coated granular urea (SustaiN®) has moved steadily toward domination of the urea market in New Zealand since its introduction in 2002. Much greater improvements again are possible (Quin et al. 2015); this is discussed in Section 7.4.

Nevertheless, N recommendations provided by industry and private extension services to IGPF operators rarely differentiate between N form for use on pasture. Conversion factors of kg urea N to kg pasture dry matter normally range from 8–20 depending on the season, roughly equating to NUEs of 15–40%. However, NUEs below 10% are not uncommon in cooler, wetter situations.

### 7.3.4   Implications for Soil Organic N

As with urine patches, the N unrecovered in nutrient budgets (i.e., not measured in plant uptake, mineral soil N levels, leaching, or ammonia and $N_2O$ emissions) has traditionally been assigned to "soil immobilization" or "incorporation into soil organic matter" (implicitly assumed to be long-term if not permanent) by advisory services (Wheeler 2009). While assimilation of fertilizer N by soil microorganisms occurs to some degree, this N pool is rapidly recycling; there is no scientific evidence that soil organic N stocks on IGPFs are actually increasing long term (Scott et al. 2015). This topic is now the subject of considerable research, as outlined by several speakers at the 2016 joint New Zealand and Australian Soil Science Societies conference. There are increasing signs that soil organic N levels may be slowly declining, especially under irrigation on free-draining soils. This is likely to be driven by a combination of the large N losses occurring from urine patches and unevenly distributed inputs of organic matter in dung. This situation is not being helped by the use of N fertilizers that are far less efficient than they should be. Leached nitrate takes with it companion cations, particularly $K^+$ and magnesium cations ($Mg^{2+}$); these losses ultimately need to be replaced.

### 7.3.5   A Closer Look at N Losses from the Urine Patch

Within the urine-deposition area, N application rates range from 300–1200 kg N ha$^{-1}$ depending on the season and quality of feed, with an average of 600–800 (Saggar et al. 2009; Moir et al. 2011; Saggar et al. 2013). Typically, 60–85% of it is present as urea (Bristow et al. 1992), with similar amounts of K. These rates are too high to be efficiently recovered by the pasture for growth, meaning large losses of N can occur as (1) leaching of urea itself below the root zone by rainfall and/or irrigation (0–30% of the total N deposited), (2) ammonia volatilization after hydrolysis of urea to ammonium-N by bacterial and extracellular urease enzyme (5–30% of the N), (3) nitrate leaching after oxidation by nitrifying microorganisms (20–70% of the N), and (4) nitrous oxide (<1–3%) and nitrogen gas emissions (5–40%) formed by denitrifying microorganisms in anaerobic soils, or anaerobic microsites present in generally aerobic soils, such as inside soil aggregates (Sexstone et al. 1985; Saggar et al. 2009). Although urine patches have significantly higher soil pasture growth, microbial activity and elevated soil mineral N levels for only 2–6 months before returning to background levels of growth and activity (Moir et al. 2011), they make up 80% or more of nitrate leaching and nitrous oxide emissions per year from IGPFs (de Klein et al. 2001; Cameron et al. 2013). Following increasing global concerns regarding $N_2O$ emissions from agriculture, and the role of ammonia emissions in the formation of particulate ($PM_{2.5}$) smog (Hristov 2011), the importance of the bovine urine patch in these emissions from grazed pastures has been demonstrated (Zaman et al. 2009; Saggar et al. 2013). Urine patches are also a major source of nitrous oxide emissions in IGPFs (Saggar et al. 2009). The contribution of urine-N losses is illustrated in Figure 7.3. The relative amounts of nitrate leaching on the one hand, and denitrification to $N_2O$ and $N_2$ on the other, are known to be largely a function of (1) soil particle aggregation and drainage, and (2) soil moisture levels, especially the degree of saturation of pore spaces (Bateman and Baggs 2005). The extent of ammonia emission is controlled largely by soil pH, soil moisture, and wind friction (Rotz 2004).

Measurements of N budgets in urine patches have often found 20–25% of the N unaccounted for (Ball et al. 1979; Selbie et al. 2015). This has typically been ascribed to "soil immobilization" or "incorporation into soil organic matter" (Wheeler 2009), stating or at least implying a continuing net sequestration of organic N. However, there is no scientific proof that intensive grazing leads to continuous increases in soil N beyond the normal time-frame of 10–30 years following conversion from any less intense land use; in fact, the opposite may be true of both organic C and N in IGPF systems in temperate climates (Scott et al. 2015).

Some studies ascribe the missing N to emissions of nitrogen gas, but quantifying this has been difficult due to the high background $N_2$ level. Measuring $N_2O$ loss in the absence and presence of a high-pressure acetylene atmosphere, with the increase in $N_2O$ measured being ascribed to what would have been converted to $N_2$ in the absence of $C_2H_2$, has long been known to be subject to technical limitations to its accuracy (Felber et al. 2012). The recent introduction of totally new methods for measuring transient and longer-term gaseous N losses of ammonia and $N_2O$ have greatly increased understanding of the drivers of these processes such as temperature and wind in feedlots (Mount et al. 2002), but considerable N still remained missing from many mass balances.

Recently, previous indications of a separate soil N pathway, known as co-denitrification, have been confirmed as a likely explanation the missing N, using comparisons of real versus calculated N isotope ratios (Selbie et al. 2015). Thought to be driven by fungi, this is the reaction between nitric oxide (NO) and the $NH_2$ group of amino acids and protein to form what is considered to be mainly $N_2$ (Selbie et al. 2015). This is a positive environmentally in the sense that most of the previously missing N is not in an environmentally adverse form. However, every molecule of $N_2$ emitted still requires the energy in 16 molecules of ATP in the nitrogenase enzyme (Ward 2012), or an equivalent or greater amount in the industrial Haber–Bosch process, to convert another molecule of $N_2$ to fixed N. It therefore represents a significant economic cost to the farmer, either directly in increased requirement for fertilizer N, or indirectly via the additional costs or lower returns involved in growing more leguminous crops.

Although dung contributes 25–40% of the N in excreta under intensive grazing (Rotz 2004; Luo and Kelliher 2010), it is responsible for a considerably smaller proportion of nitrate leaching, ammonia volatilization, and $N_2O$ emissions (Luo et al. 2008); the N in dung, being present in bacterial and other organic forms, acts as a slow-release form of N for the growth of soil microorganisms and pasture.

However, these returns in organic N may be insufficient to offset large losses of N from urine patches, removals in milk and meat produce, and reduced legume N fixation; as stated earlier there is evidence that soil organic C and N levels are slowly declining on IGPFs in temperate systems, especially where inputs of organic N such as through clover N fixation are low (Scott et al. 2015).

### 7.3.6 N Losses from Dairy Parlors and Effluent Ponds

The manure from holding yards and milking parlors (once called milking sheds, now dairies in New Zealand and Australia) is collected in liquid form, partly because of the relatively high urine output from grazed animals compared to housed animals receiving dry feed, and partly because of the large amount of water used to wash down the facility after each milking operation, typically twice a day. This effluent is usually stored in ponds prior to land application. In many countries, farms are required to have sufficient pond volume to store up to 8 months of effluent output, to enable land application to be restricted to warmer and drier periods of the year. In a typical IGPF, approximately 15% of total excreta is deposited in the dairy parlor (Figure 7.3).

Effluent ponds support large concentrations of bacteria which deaminate protein, releasing ammonium-N in the process. Given the high biological oxygen demand (BOD) of effluent, anaerobic conditions are generally present in ponds below the surface (Li et al. 2015). Typically, 50% of the total N is volatilized (Li et al. 2015), along with methane. Covers can be used to trap these gases; the ammonia can be scrubbed from it with, for example, sulfuric acid, and the ammonium sulfate formed used as a fertilizer (Melse et al. 2009; Jiang et al. 2010). The methane can then be burned to create heat or electricity. However, the high capital costs and low economic returns of these technologies have limited their uptake. Other methods for retaining the N in the ammonium form, such as by addition of alum or urease inhibitors reduce these losses, but ongoing chemical costs, monitoring requirements, and waste disposal limitations have again limited the commercial uptake of these mitigation options.

Where suitably free-draining and well aggregated soils are available, dairy effluent can be irrigated directly onto land, provided rotational grazing allows effluent-treated pastures to be withheld from grazing for typically a one-month period to ensure non-survival of pathogens on the pasture leaves and the soil surface (Wang et al. 2004). Typically, only the 10–20% of the farm closest to the pond is used for effluent application to reduce infrastructure, but this can lead to undesirably high soil K levels accumulating over time. The higher proportion of the farm treated, the better for the environment, and the bigger the saving in fertilizer inputs. This system is not suitable for areas with cold, wet winters; these require ponds to store the effluent during the entire cool/wet period, to avoid leaching and runoff of N (and P) and deterioration in soil drainage and aeration.

### 7.3.7 N Losses from Effluent, Slurry, and Manure Application

Grazed beef animals deposit 90% or more of their excreta directly onto the pasture; the animals spend only a small (5–10%) of their time in handling yards. Manure is regularly removed from these by machinery, typically in damp solid form containing 30–50% moisture, and stored prior to land application, as with feedlots (Rotz 2004). As in feedlot manures, high ammonium concentrations can be reached during storage due to the anaerobic conditions, leading to gaseous ammonia volatilization losses.

   Much larger losses (typically 50% of the total remaining N content) typically occur during and after application of dairy parlor effluent, slurries, and manure to the soil, mainly as ammonia (Li et al. 2015). $N_2O$ emissions can range from 0.01–4.9% of effluent N depending on soil conditions (Luo et al. 2008). Subsurface placement technology can considerably reduce gaseous losses, but once again at significant additional cost (de Klein et al. 1996).

## 7.4 EXISTING AND DEVELOPING N LOSS MITIGATIONS FOR IGPFS

### 7.4.1 Options for Reducing Urinary N Output via Animal Intake

The greatly increased level of pasture production in urine patches in the 2–4 months after deposition, and the N losses from these patches, has lead researchers to investigate how the inefficiencies in, and losses of N from, this micro-system could be mitigated via the feed type, its contents, and amendments to it, to reduce N losses to the environment and if possible to increase production.

#### 7.4.1.1 Altering the Type of Feed

The N content of diet of grazing animals that is utilized by bacteria for growth in the rumen is deaminated into ammonium-N for reassimilation into protein for animal growth and milk production; waste from this production process is excreted as organic N (mainly bacteria and protein) in the dung. Excess N in the diet that is not utilized in this process is converted to urea in the liver and voided in the urine (Rotz 2004; Hristov et al. 2013). Typically, 50–80% of the dietary N from cows and cattle grazing improved (fertilized ryegrass/clover) pastures is voided in the urine; most of this (80–90%) as urea (Bristow et al. 1992). As in CAFOs, research has focused on reducing urinary N output by reducing the N content of the diet, and modifying the ratio of rumen degradable protein (RDP) to total protein (Rotz 2004; Misselbrook et al. 2013). While the use of alternative low N content forages such as plantain and chicory has been successfully demonstrated to reduce urine-N output, it has been much harder to implement this strategy at a farm level, for a variety of reasons. Most importantly, alternative forages with lower N content can suffer from one or more of (1) lower production (measured either as metabolizable feed or animal production per ha), especially in cooler months (2) grazing management difficulties such as treading damage and poor post-grazing regrowth, (3) reduced longevity and associated increased reseeding costs, and (4) lower tolerance to water and/or temperature stress (Crush et al. 2007).

   Varieties of ryegrass selected specifically for higher carbohydrate to protein content have to date produced little benefit for increased growth at a farm scale (Edwards et al. 2007; Rasmussen et al. 2009; Proctor et al. 2015). Genetic modification may be necessary to achieve worthwhile increases in sugar to protein ratios, although the use of GMOs in IGPFs is banned in some countries such as New Zealand.

### 7.4.1.2    The Use of Feed Amendments and Dosing to Reduce Urinary N Losses

Treatment of feed, or direct dosing of animals, with urease and nitrification inhibitors (or the insertion of stomach boluses to slow-release them) have been investigated with the intention of their being voided in the urine. All methods been shown to be effective in reducing ammonia volatilization and nitrate leaching losses from urine (Luo et al. 2015); ingested nitrification inhibitors have also reduced $N_2O$ emissions (Welten et al. 2014; Luo et al. 2015). However, more research is required regarding side-effects on the animal and adverse effects of any impurities present. For example, the commercial-grade (97% pure) nitrification inhibitor dicyandiamide (DCD) can contain small amounts of impurities of the chemically related but far more toxic compound melamine.

A variation on this approach is the addition of common salt (sodium chloride) to drinking water, deliberately to increase thirst and therefore higher consumption, and therefore urinary output. First proposed by Quin 1982 (Table 7.3) this technique has the effect of increasing the number of urinations, which have the same average volume but lower N content, thus spreading the urinary N over a larger proportion of the pasture during each grazing effect. This can reduce N leaching by up to 15% (Ledgard et al. 2015), but the increased sodium content of the milk could reduce its value in some markets.

### 7.4.1.3    The Selection of Forages Containing Natural N Process Inhibitors

There is growing research interest in the potential for the selection of breeding or forages that contain natural compounds that inhibit either urea hydrolysis or nitrification. Extracts of the *Neem* tree, a native of India, have long been considered to inhibit nitrification (Santhi et al. 1986), but attempts to commercialize it have had little success, due to unreliable effects. There are many compounds found in other plants that potentially have N inhibition properties (Gardiner et al. 2016) but research into these, and into transferring any successful traits found to the common animal forages, is very much in its infancy.

### 7.4.2    TREATMENT OF PASTURE WITH NITRIFICATION INHIBITORS

Because of the difficulty in locating urine patches in freshly grazed pasture, commercial options for treatment have until recently been limited to spraying inhibitors over the entire paddock (Di and Cameron 2005). However, this is extremely wasteful of chemicals, given that (1) in any one grazing event, typically only 4% of the area receives urine deposition, and (2) the timing of application before or after grazing (urine deposition) is very limited, in turn meaning that several applications must be made per year, particularly in soils with high biological activity (Ledgard et al. 2014). Early research drew conclusions regarding reductions in nitrate leaching achievable with single DCD applications from small lysimeter studies, in which heavy applications of water (typically 10 mm) were made shortly afterward to depict a worst-case scenario for N leaching (Di and Cameron 2002). Early estimations of reductions in nitrate leaching of up to 75% achievable with a single DCD application (Di and Cameron 2004) seem to have disregarded the fact that each area of the farm is regrazed every 20–40 days, by which time the previously applied inhibitor has largely decomposed on many soils (Watkins et al. 2013).

More recently it has been demonstrated that preferably three applications of DCD (depending on the soil and climate), applied during the key autumn period for nitrate accumulation (before autumn rainfall brought soils to their water-holding capacity), reduces nitrate leaching losses by up to 35% (Ledgard et al. 2014). Nitrous oxide emissions were also reduced by DCD (Luo and Kelliher 2010; Di and Cameron 2007; Luo et al. 2013). However, high spreading costs of the DCD suspensions by specialized trucks limited market adoption of DCD application to less than 5% of New Zealand dairy farmers, even in the most highly nitrate-sensitive catchments.

A further limitation with all-paddock treatment is the risk of inhibitor residues in the pasture being passed into the milk of dairy cows (Pal et al. 2015), particularly if the DCD application is not restricted to immediately post-grazing, to allow 20–40 days for its decomposition. DCD is chemically related to melamine, although it is not considered to be remotely as toxic. Nevertheless, because of the current lack of a globally accepted (e.g., CODEX) safe upper limit on DCD in infant formula and other dairy products, the use of DCD in New Zealand was ceased in 2013. This limitation also applies of course to feeding of DCD directly to animals. Alternative nitrification inhibitors such as Nitrapyrin® and sodium and calcium thiosulfate are also being investigated.

### 7.4.3 Mitigating N Losses by Fitting of Urine-Treatment Devices to Cows

Devices fitted externally to cows have been proposed as a means of either injecting N inhibitors and other products into the urine stream or onto the urine patch, and/or dispersing the urine flow so that a larger area of soil receives urine, increasing the opportunity for plant uptake (Quin 2016). The anatomy of cows and their frequent impacts with stalls, milking yard fences, and each other, unlike horses, for example, does not easily lend itself to the attachment of low-cost, practical, and durable devices. Most proposed designs have relied on the action of the tail being raised to urinate to trigger the product release or urine dispersal action (Quin 2000, 2016). This means they are also triggered by defecation. While this represents a relatively small addition to the total treatment costs, and can reduce ammonia volatilization from dung pats, dung collection on the devices means daily cleaning (e.g., prior to milking) can be required. A new design that avoids dung contact is under development. Moisture sensors attached near the animal's urethra have been used to temporarily GPS-map urine patches for research purposes (Betteridge et al. 2013), but are unlikely to have any commercial application. Urine dispersal devices inserted inside the urethra have technical potential, but risks of irritation and infection mean that animal ethics approval is unlikely to be given.

### 7.4.4 Mechanized Location and Treatment of Individual Fresh Urine Patches

#### 7.4.4.1 The Spikey® Urine Detection Technology (UDT)

Urine patches on pasture can normally be detected by eye, chlorophyll fluorescence, near infra-red (NIR), or multispectral radiometry (Pullanagari et al. 2012) from 7 days after deposition. However, this is far too late for effective treatment with urease inhibitors, as all the urine-urea has been hydrolyzed to ammonium-N by this time, and up to half of this can have been converted to nitrate, greatly reducing the benefit of nitrification inhibitors (Bates et al. 2015). In addition, with every day, the wetting front of the urine, and the ammonium-N in it that has not yet been adsorbed onto clay particles or converted to nitrate, moves increasingly out of potential contact from a surface-applied application of a nitrification inhibitor.

The recently developed Spikey® urine detection technology (UDT) enables accurate detection of fresh (same day) urine patches, using electrical signals related to the shallow-soil (0–5 mm) conductivity and surface resistance (Bates et al. 2015). UTC consists of an array of paired spiked metal wheels which act as electrodes. The wheels are spiked to ensure the wheels cut through pasture leaves and surface organic residues to make contact with the surface soil (0–5 mm). Spacing of the wheel pairs, and of the treatment spray nozzles, is designed to facilitate accurate location and treatment of individual fresh urine patches (Bates et al. 2015). Fresh urine is highly conductive, due to the high content of the cations $K^+$ and $Na^+$, and the anions $HCO_3^-$ and $Cl^-$, even though the N itself is largely present as non-conductive urea and amines. These signals are easily discernable from the background, typically by a factor of 5–10 (Figure 7.4; Bates et al. 2015). The signals degrade quickly from day 2 or 3 after deposition, due to vertical and lateral movement of the urine, adsorption of $K^+$ onto clay particles, and dilution of the ions by surrounding soil water.

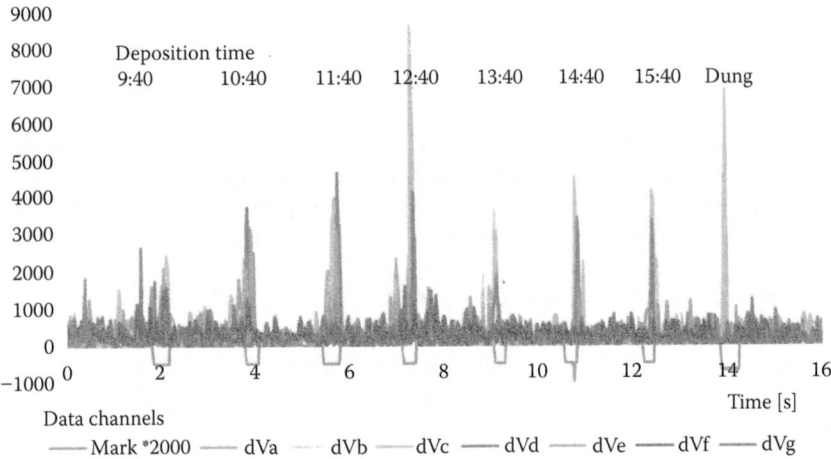

**FIGURE 7.4** (**See color insert.**) Detection of several manually placed urine patches (position indicated by bracket below axis), with dung patch placed at the end. Detection data taken at 5 pm on the same day as deposition. Note the single channel detection of the dung patch, compared to the multichannel detection of urine patches.

The increase in conductivity deeper in the soil following conversion of urine-urea to ammonium-N and nitrate could be tracked by using deeper-driven electrodes, but while this is very relevant to N research, it has little direct relevance to the practical use of the technology on dairy farms to detect and treat the fresh urine patches. In practice this means within a maximum of 2 days of a grazing event, while the urine-N is still present largely as urea, enabling both urease and/or nitrification inhibitors to be used effectively (Saggar et al. 2009; Zaman et al. 2009; Saggar et al. 2013). In high rainfall or on extremely free-draining soils, it may be desirable to inject nitrification inhibitors 1–5 cm below the soil surface to catch the wetting front of the urine. This could be achieved by optimizing droplet size and velocity (shallow penetration) or an array of high-pressure pulses of the treatment solution into the urine patch for deeper penetration.

The equipment is planned to be released commercially in a range of widths from 8 m upwards (Photo 7.6), towable or mounted on a tractor 3-point linkage, with the outer modules foldable for

**PHOTO 7.6** (**See color insert.**) The first Spikey-8 fresh urine patch detection and treatment device in use on an IGPF in New Zealand, September 2017.

passage down farm lanes and gateways. In the 8-m configuration, towed at 12–15 km/h, it takes about 10 minutes per hectare to detect and treat; the response to the signal triggers an essentially instantaneous spray application of treatment from the closest spray nozzle (or nozzles) to the urine patch; it is a "one-pass" operation. On a typical New Zealand dairy farm of 150 ha, grazing 2–4 ha per day, this daily or every second-day operation is expected to be acceptable to most dairy farmers, particularly those having to meet newly introduced nitrate leaching restrictions.

Alternative future UDT approaches may include the sensing of fresh urine patches (and cowpats) using real-time sensing of sudden spikes in ammonia gas concentrations just above ground level.

### 7.4.4.2  Urine Treatment Products (UTPs) for Application with UDT

Current research is focused on selecting the urine treatment products (UTPs) for application with UDT in a variety of soil and climatic conditions. The product ORUN® contains the urease inhibitor NBPT and the growth promotant gibberellic acid variant 3 (GA3; Quin et al. 2016). A major advantage of the urease inhibitor (compared to a nitrification inhibitor) is that, by retaining the urine-N in the form of urea for 5–10 days (Watson et al. 2008), the urea in the urine patch was calculated to be able to spread laterally by 44% in surface diameter (Bishop and Quin 2010). Recent testing, where rings were not used to confine the urine during deposition, gave a 70% increase in the size of the average urine patch over three intensive dairy farms (Table 7.4; Quin et al. 2016), with similar increases in N uptake. This represents a 5–15% increase in pasture production annually, based on measurements of normal urine patch frequency, N content, and sizes by Moir et al. (2011). This is easily sufficient to make the technology cost-effective for use by the farmer, a major factor in its rate of adoption by farmers (Quin et al. 2016). Other UTP options or additions include DCD, which has been demonstrated to reduce $N_2O$ emissions and further reduce nitrate leaching (Zaman et al. 2009). Treatment of individual urine patches, soon after grazing to maximize the period before the herd returns, combine to reduce UTP application rates and minimize the presence of UTP residues.

---

### TABLE 7.4

### Increase in the Pasture Dry Matter Grown in the 1 m² Area Centered on Fresh Dairy Cow Urine Patches, in the Absence and Presence of ORUN® UTP Spray (nbpt Plus Gibberellic Acid) on Three Sites in New Zealand from Mid-April to Early June 2015

|  | Trial 1 | Trial 2 | Trial 3 |  |
|---|---|---|---|---|
| Location | Southland | Canterbury | Waikato | Ave. |
|  | Dry Matter Production (kg DM ha⁻¹) | | | |
| Control (no urine) | 1732 | 2025 | 2388 | 2048 |
| Increase above 1200 kg DM/ha grazing residual (GR) | 532 | 835 | 1188 | 848 |
| Urine only | 2111 | 2717 | 3332 | 2720 |
| Incr. above 1200 GR | 911 | 1517 | 2132 | 1520 |
| Urine plus UTP | 2763 | 3160 | 3686 | 3203 |
| Incr. above 1200 GR | 1563 | 1960 | 2486 | 2003 |
| *Increase in DM above urine with UTP application* | *652* | *443* | *354* | *483* |
| *LSD 5%* | *169* | *301* | *247* | *210* |

*Source:*  Quin, B.F., Bates, G., and P. Bishop 2016. Locating and treating fresh cow urine patches with Spikey®: the platform for practical and cost-effective reduction in environmental N losses. In *Integrated Nutrient and Water Management for Sustainable Farming,* Eds. L.D. Currie and R. Singh. Occasional Report No. 29. Fertilizer and Lime Research Centre, Massey University, Palmerston North, New Zealand. 8 pp. http://flrc .massey.ac.nz/publications.html.

In high rainfall environments, particularly on soils with very high infiltration rates, existing nbpt-based UTP treatments are likely to be less effective, as the urine-urea can be leached vertically before it can spread laterally (Bishop 2016). Application of low-cost readily mineralizable carbon sources such as raw vegetable oil to the urine patch may assist by temporarily greatly stimulating the assimilation of urine N by soil microbes (Bishop 2016). As this becomes slowly released during bacterial decomposition, plant N uptake is better able to keep pace, thereby reducing losses from the soil-plant system.

It is proposed that the daily or every second day operation be combined with other operations, such as spreading of fertilizer urea by the addition of a fertilizer hopper and spinning discs to the Spikey® equipment, and measurement of residual pasture levels and soil moisture, both very important in efficient grazing management.

### 7.4.4.3   The Potential for Autonomous Robotic Tow Vehicles for UDTs

The urine detection technology lends itself to the use of fully autonomous robotic tow vehicles, as a labor-free alternative to tractor towing (Bates and Quin 2013). It is envisaged that small robots would towing a 2-m wide UDT would be sufficient, as the robot can work all day and/or night if necessary, following preprogrammed object-free pathways for individual paddocks. The slow speed (6 m/sec) of the robots also facilitates the on-the-go collection of data for precision farming and soil nitrate governance.

### 7.4.5   Options for Improving Fertilizer N Efficiency

### 7.4.5.1   Alternatives to Fertilizer Urea

Given the considerable potential for direct environmental N losses (leaching and gaseous) from the considerable amounts of fertilizer N applied to many dairy farms (Ledgard et al. 1999; Jarvis 2000), there has been relatively little scientific study of alternative N fertilizers on IGPFs (Chen et al. 2008). Fertilizer N efficiency on grazed pastures has traditionally been given a very low priority by scientific research institutions; the conduct of field trials under grazing conditions is far more complicated, subject to increased variability, and far more expensive to conduct than those under cropping. For this reason, the great majority of fertilizer N trials on pastures are conducted using small-plot trials which deliberately exclude the animal. Grazing, animal treading, and urine-N deposition all have very important effects; omitting them reduces the value of the data and increases the difficulty of interpretation.

On either poorly drained soils which minimize nitrate leaching, enhance lateral movement of nitrate, and promote denitrification losses of $N_2O$ and $N_2$, or in moist conditions which minimize ammonia volatilization, differences in performance between nitrate and ammonium forms of N fertilizer will be minimized, as will particle size effects (Watson and Kilpatrick 1991; Harty et al. 2016). However, in warmer, windy, and free-draining soil environments, differences in effectiveness between types of fertilizers can be far greater (Quin et al. 2015). Cameron et al. (2013) reported that ammonia volatilization losses from the application of fertilizer urea to pasture and crops both ranged from about 5–30% in most countries. It is generally accepted in the fertilizer industry that because ammonium fertilizers such as ammonium sulfate and diammonium phosphate (DAP) have an acid reaction in the soil (unlike urea which has an alkaline effect initially), they minimize ammonia volatilization losses, and consequently are typically 20% more efficient than urea. Although calcium ammonium nitrate (CAN) and ammonium nitrate (AN) are more susceptible to nitrate leaching than straight ammonium fertilizers, some farmers prefer it because of the immediate plant availability of the nitrate component.

A significant weakness in many fertilizer N comparisons is the use of a single rate of application of the products being compared. Without definition of the shape of the response curve, and no information where the chosen N rate lies on the response curve, it is exceedingly difficult to make comparisons of NUEs.

### 7.4.5.2  Improving the Efficiency of Fertilizer Urea

Granular urea (46% N) has traditionally been a major form of fertilizer N used on IGPFs. It has the advantages of being cheap per unit N, especially on an applied cost basis when the lower transport and spreading costs per unit N are factored in. However, its efficiency is disturbingly low. Granular urea coated with sulfur or polymers, as well as slow-release isobutylidene diurea and applying Nitrapyrin® nitrification inhibitor and urea together, have been compared with single or divided applications of granular urea on ryegrass or pasture, difficulties in predicting both the length of the delay in N release and the pattern of this release, and the  unproven cost-effectiveness of any improvement in overall NUE (Halevy 2007; Bishop et al. 2008), have greatly restricted their use on IGPFs. However, the 53% reduction in farm-scaled $N_2O$ losses from corn using polymer-coated urea on a poorly drained soil with subsurface drainage (Nash et al. 2015) is an indication of future potential beneficial environmental benefits for pastures.

In windier, hotter, or drier conditions, the NUE of granular urea can be improved by 25–30%, at a cost premium of less than 10%, simply by adding via spraying 0.025–0.05% of nbpt; it acts primarily by reliably minimizing ammonia volatilization (Watson et al. 1990a,b, 2008; Blennerhassett et al. 2006; Saggar et al. 2013; Zaman et al. 2013a,b). Some researchers have also reported an additional (indirect) benefit of reduced nitrate leaching, but it has little apparent effect on $N_2O$. This product is rapidly becoming the biggest-selling fertilizer N in New Zealand. The efficiency of granular urea was found to be increased by much more—by a factor of 1.7–2.4—by fluidizing urea (crushing the urea into a thick flowable paste of 10–15% water content) and adding the urease inhibitor NBPT (Quin 2008; Quin and Findlay 2009; Zaman and Blennerhassett 2009). However, higher manufacturing costs, on-truck processing costs, and higher application costs resulting from the much narrower spread achievable compared to solid products limited adoption of the technology.

More recently, it has been demonstrated that ONEsystem®, a process using prilled urea (instead of granular) sprayed with a dilute solution with nbpt during spreading, gave the same benefits as fluidized urea plus nbpt, but with little or no increase in processing or spreading costs compared to granular urea (Figure 7.5, from Quin et al. 2015). Scientific assessment of reduction in gaseous emissions and nitrate leaching is currently in progress, but it seems safe to assume from initial fine-urea studies that the doubling in efficiency is a direct consequence of greatly reduced environmental losses (Dewar et al. 2010). Moreover, prilled urea is approximately 5–10% cheaper to produce

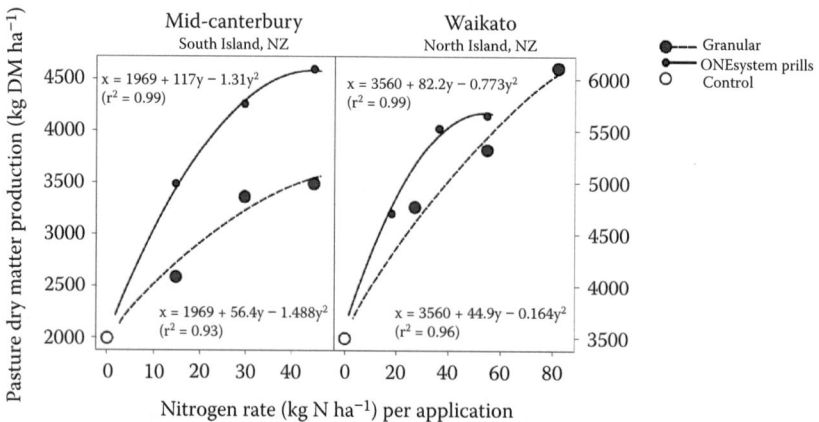

**FIGURE 7.5**  Pasture production (kg DMha⁻¹) from control (nil N; circle), 3 rates of granular urea (large dots) and 3 rates of ONEsystem® wetted, nbpt-treated prilled urea (small dots). (From Quin, B.F., Gillingham, A.G., Baird, D., Spilsbury, S., and M. Gray 2015. A comparison under grazing of pasture production, pasture N content and soil mineral N levels between granular urea and ONE*system*® on two contrasting dairy farms in New Zealand. *J. of New Zealand Grasslands* 77:251–258.)  Four N application were made at each site.

than granular urea, and now comprises over 50% of global urea manufacture. Prills have a typical average particle diameter of 1.5 mm (range typically 0.5–2-8), compared to 4–5 mm for granular. Both have a bulk density of 0.8 tonnes per $m^3$. At any given N application rate, prills supply about 10 times as many particles as granular urea. This is extremely important at the low individual N application rates (typically 30–50 kg N/ha used on pasture); at 30 kgN/ha granular urea is applying only 35 particles per square meter compared to 350–400 for prills. Pasture has a high plant density, typically 400 plants/$m^2$; it is not physically or chemically feasible for each granule of granular urea to provide an even supply of N to 10 individual plants. Other advantages of wetted, nbpt-treated prilled urea include a degree of foliar uptake and reduced elevation of pH in the soil solution surrounding fertilizer particles (Quin et al. 2015).

Spreader design, particularly in terms of fertilizer delivery from the hopper to the spinning discs and the disc design and rotation speeds, needs to be optimized for reliable and accurate delivery and spreading of urea prills at the very low rates per ha needed. These improvements have also enabled spreading vehicle paths of 20–25 m, acceptable to most farmers and contractors. In addition, the distribution pattern of prills has a far more distinct cut-off than granules, reducing the risk of direct input of fertilizer N into drains and streams. UDT and nbpt-treated prilled urea application can be combined into a single piece of machinery for single-pass operation.

### 7.4.6   Options for Improving the Efficiency of Effluent N Recycling

#### 7.4.6.1   Covering Ponds or Using Fully Contained Pods

Ammonia emissions from slurry and effluent storage can be reduced by as much as 90% by covering storage ponds or using fully enclosed tanks, bags, or pods for storage. The emitted ammonia can be removed by passing the gaseous emissions through a variety of acidifying chemicals or ammonia adsorbing agents such as zeolite. However, until recently the cost of implementation and lack of accepted financial benefit to the farmer has restricted adoption of this technology.

#### 7.4.6.2   Reducing Wash Water Use in Dairy Parlors

There are clearer financial benefits to reducing water use in washing and flushing in dairy parlors; up to 800 liters are used per head per day on fully enclosed systems; fully grazed systems use only 80 liters per head per day in washing the holding yard and milking platform. More efficient yard washing systems such as the Dungbuster® reduce water use (and staff time) by 30% (G. Bates, pers. comm.); a more recent option called greenwash reduces this even more by using effluent pond water to do the initial wash. The incorporation of dry-scrape technology could get wash volumes to as low as 10–20 l $hd^{-1}$ $day^{-1}$. Reduced volumes of effluent reduce the size of the pond required and reduces effluent irrigation times. Future technology using robotically driven effluent spreading tanks and spray booms daily become feasible (Bates and Quin 2013).

#### 7.4.6.3   Reducing N Losses with Stand-Off Pads, Fenceless Stock Control, and On-Demand Robotic Milking

Grazed dairy and beef cows typically need to be grazing no more the 50% of the time to maintain 95% of maximum production; they spend the rest of the time resting while chewing cud, walking to and from the milking parlor, waiting for milking, and being milked. Many dairy farmers feed out high-energy supplementary feed rations during milking.

Cows can be taken off the pasture onto hard-based "stand-off" pads for two periods of several hours each with little loss of production. The potential environmental benefit of this is that a roughly proportionate percentage of the animal's excreta is deposited on the surface of the pad, from where it can be collected and flushed into the dairy parlor effluent pond, and then spread evenly over the land, rather than be subject to leaching as nitrate from urine patches. Reductions in nitrate leaching of up to 43% were estimated (Christensen et al. 2011). While this is beneficial in nitrate-sensitive catchments, it is not likely to result in reduced total N losses to the environment, due to ammonia volatilization from the pad surface, in the same way as a feedlot. $N_2O$ losses could be increased or decreased depending on factors such as soil type and rainfall. The use of stand-off pads essentially moves some of the N loss from one part of the N cycle to another, rather than assist in attaining true sustainability. It also involves significant capital expenditure and increased animal management. They are, however, more likely to be economically viable in soil and rainfall combinations susceptible to soil pugging damage during late autumn and through winter, reducing denitrification losses in the process (Cameron et al. 2013).

The concept of fenceless stock control using GPS collar-sensors is not new, but as yet has only been commercialized for dog control. Animals are guided with GPS and network-linked collars which have both incentive and warning (sounds) as well as punishment (electric shock) tools to guide the animals. This technology has the potential to stagger departure of the cows from the pasture and delay their arrival at the parlor until a short time before milking, thus reducing the excreta deposited in the milking yard, meaning less cleaning and less effluent to store and spread.

The adoption of on-demand robotic milking on IGPFs is increasing, however. One robot stall is usually available per 50 cows. When individual cows can choose when and how often to be milked, much less time is wasted waiting to be milked. In addition, less excreta are deposited in the milking yard, meaning less cleaning and less effluent to store and spread.

### 7.4.7 ALTERING OR SELECTING THE WAY THE ANIMAL DIGESTS FEED

Research is also being conducted into improving the feed protein utilization ability of the animal, *via* genetic selection for beneficial digestive traits and manipulation of rumen bacterial populations (Reynolds and Kristensen 2008; Hristov et al. 2011). The commercialization of "Low-N" semen has recently commenced in New Zealand. Good progress is being made in the genetic and biochemical understanding of how reductions of enteric methane gas emissions can be managed (RNS 2015), and reduced with various dietary additives (Gerber et al. 2013).

### 7.4.8 INCREASING BIOLOGICAL N FIXATION (BNF)

A global study of microbial cycling genetic traits in soil microorganisms (Nelson et al. 2016) has provided a fascinating insight into the enormous genetic potential for N fixation and other N processes that exist in soils around the world, the great majority of it unused. As ways of interpreting the mass of genetic data being produced are developed, methods of harnessing this latent ability for BNF will follow.

## 7.5 N IN DAIRY AND BEEF CAFOS

### 7.5.1 ENVIRONMENTAL N LOSS ISSUES

Beef production in the United States is now very predominantly in CAFOs (Table 7.1), typically confining 10,000–30,000 head (and up to 100,000 in the Harris Farm in California). They are stocked at 1 head per 9–30 $m^2$ depending on state regulations, such as 17 $m^2$ in the case of Texas (Todd et al. 2006). Housed dairy operations are generally smaller (150–3000 head), and provide individual, open, or (in the case of smaller operations) tied stalls. Beef feedlots are increasingly located in clusters around large meatpacking (aka meatworks) facilities, which are owned by a small number of huge companies which control ex-feedlot prices and the distribution of meat to the retail sector.

After weaning, most cattle in the United States spend a short adaptation period grazing alfalfa before going to feedlots. Their diet needs to be very carefully controlled during the transition to the acid-producing grain-based feedlot diet. Antibiotics are commonly used to reduce adverse rumen bacterial reactions and reduce the incidence of density-driven disease. While 27 m ha of land are still classified as being used for controlled grazing, mainly in the mid-west, it is mainly used for either (1) growing what are called "stockers," namely beef cattle that are moved from feedlots at about 330 kg live weight to set-stock graze on unfertilized summer grass for 160 days from April to September, during which time they reach the 590 kg "finishing" weight for slaughter; or (2) "winter grazers," which are moved from specialized "pre-conditioning" feedlots at a heavier 385 kg, to graze unfertilized summer-saved pasture for about 130 days over the November to April winter period. It is estimated that over 75% of the 90 million beef cattle in the United States are confined full-time in CAFOs from weaning. Solely CAFO animals reach finishing weights in 350–400 days from birth, compared to 500–650 days for cattle grazed on pasture since weaning. In most of Europe, China, and Russia, on the other hand, most dairy and beef cattle are housed in small barns with individual stalls, and are put out to set-stock or extensively graze pastures for spring and summer, feed permitting; finishing weights typically take 800 days to achieve.

The increasing intensification of animal confinement operations impacts on the environment in several ways. Manure excretion, its collection, storage, and land application invariably become both cost and environmental issues, particularly regarding methane, ammonia, and nitrous oxide emissions, and nitrate leaching. The N utilization of CAPOs ranges from 23–27% (Hristov et al. 2011). While probably no worse than the conversion of dietary N to dairy products and meat in other systems, the problem is that the excess 73–77% is far more susceptible to gaseous N emissions because excreta from the CAFO system is separated from the soil until 60% or more of N emissions have already occurred, placing far greater need for the adoption of N loss mitigations which can retain the N in the manure until it is eventually returned to the soil.

Animal health (not addressed in this chapter) is also impacted; for example, the use of antibiotics is much higher per head in feedlots than under grazing, for a combination of reasons.

A conceptual diagram showing the components, pools, flows, and losses of N in dairy and cattle operations is shown in Figure 7.6, modified from Del Prado et al. (2013). It helps to illustrate the increased susceptibility of CAFOs to environmental N and C losses, by reducing the opportunity for soil N incorporation and processing of excreta N and C.

### 7.5.1.1 Methane ($CH_4$) Emissions

$CH_4$ emission from housing and feedlots was the first gaseous emission after carbon dioxide to receive widespread investigation, largely because of the early demonstration of methane's role as a greenhouse gas (GHG). Although not an N compound, it is covered here briefly because of its importance. Methane is emitted mainly from beef cattle and dairy cows due to enteric fermentation in the rumen, which is belched into the air, and decomposition of manure, both in storage and during and particularly after land application. Emissions have been characterized in considerable detail in a wide variety of different systems (Rotz 2004; Hristov et al. 2011; Gautam et al. 2016). An excellent review of technical options for mitigating non-$CO_2$ emissions from livestock is provided by Gerber et al. (2013). The discovery that feeding cattle as little as 2% kaolinite clay in the diet can considerably reduce enteric methane emission in some conditions may have great potential.

### 7.5.1.2 Ammonia ($NH_3$) Emissions

Ammonia was initially regarded as relatively harmless. However, concern has greatly increased as the role of ammonia in the chemistry behind the formation of lethal particulate emissions ($PM_{2.5}$) by reaction with atmospheric $NO_x$ radicals has been gradually elucidated (Hristov et al. 2011). $NH_3$ volatilization from feedlot operations originates from three sources:

1. Urine patch deposition onto the feedlot floor; losses average close to 50% of dietary N (Cole et al. 1995; Todd et al. 2006; Hristov et al. 2011; Cooprider et al. 2014; Gautam et al. 2016)

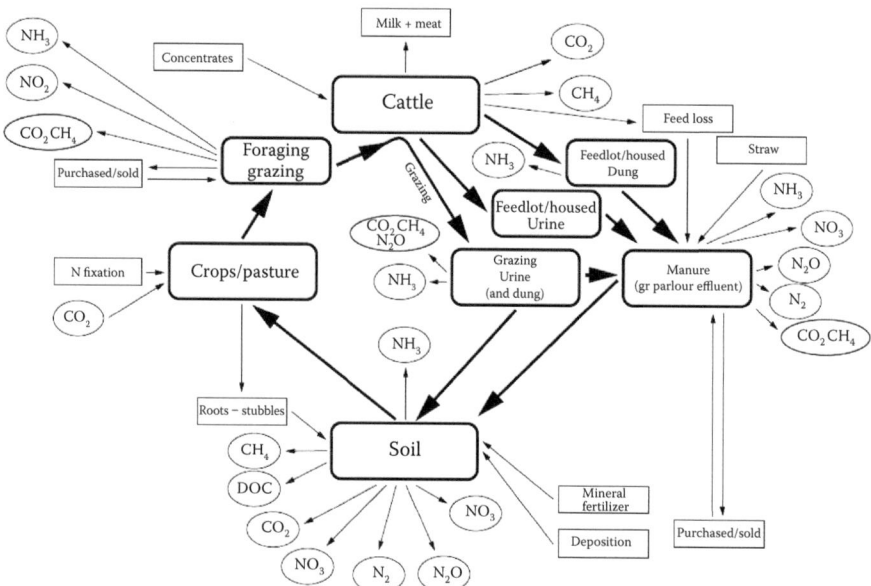

**FIGURE 7.6** Conceptual diagram showing the components, pools, gains and losses of C and N that are effected after changes in dietary protein in intensive beef and dairy systems (Modified from Del Prado, A., Crosson, P., Olesen, J.E., and C.A. Rotz 2013. Whole-farm models to quantify greenhouse gas emissions and their potential use for linking climate change mitigation and adaptation in temperate grassland ruminant-based farming systems. *Animal* 7(2):373–385. DOI 10.1017/S1751731113000748.) In the CAFO and MCGF systems, much greater opportunities for N loss occur before manure/effluent is finally returned to the soil.

2. From manure during storage (Hao et al. 2001; Rotz 2004; Hristov et al. 2004)
3. From manure during land application (Eghball and Power 1994; Rotz 2004; Hristov et al. 2011; Sakirkin et al. 2011)

Many factors significantly alter the ratio between these losses and the total amounts; these include feedlot design, cow house design and air flow (Monteny and Erisman 1998), manure storage conditions and depth (Muck et al. 1984), manure treatment such as turning or addition of water for flushing and/or slurrifying for storage and spreading (Ogink and Kroodsma 1996), land application methodology (Rubaek et al. 1996), and air temperature, wind speed, and humidity during application (Sommer et al. 1991).

Hristov et al. (2011) tabulated the results of 26 feedlot studies of ammonia emissions, including total ammonia emissions as percentages of dietary N intake. An animal N efficiency of 26% is used (Bequette et al. 2003; Huhtanen and Hristov 2009). A summary of ammonia losses from feedlot floors, as a percentage of total N loss from feedlots, is presented in Table 7.5. Nitrous oxide losses from feedlot floors are minimal. Combined losses of ammonia plus nitrous oxide during manure storage and from land after application are combined in Table 7.5.

Percentage losses for ammonia and $N_2O$ combined from housed dairy floors, ponds, and field application are also given in Table 7.5, derived largely from daily ammonia emission rate data from the studies tabulated by Hristov et al. (2009); other data were sourced from Thompson and Meisenger (2004), McGinn et al. (2008), and Rumberg et al. (2008). Note that in both beef and dairy operations, nitrous oxide is a very minor loss from floors, but can reach 5% of total N losses from storage and land application.

In the case of dairy houses, greater consumption of water and volume of urine per animal, and the extra water used for washing equipment, mean that the total excreta has a higher water content

## TABLE 7.5

### Combined Ammonia and $N_2O$ Loss Percentages from Confined Beef and Dairy Operations (Range in Brackets)[a]

| | % of Dietary N | % of Total Excreted N (26% Feed Efficiency)[1] | Most Practical Mitigations |
|---|---|---|---|
| **Beef Cattle Feedlots** | | | |
| *Animal production efficiency*[a] | 26 (20–27) | – | |
| Feedlot floors[b] | 48 (10–72) | 65 (30–80) | Mnthly lignite appl |
| Manure/slurry storage[c] | 8 (5–12) | 11 (7–16) | air-tight covers |
| Manure/slurry composting/applcn[d] | 18 (15–25)[i] | 24 (20–33)[i] | deep placement |
| **Dairy Barns/Houses** | | | |
| *Milk*[e] | 26 (24–27) | | |
| Parlour floors[f] | 40 (20–50) | 54 (26–72) | Freq. flushing to pods |
| Manure/slurry/effluent storage[g] | 12 (8–20) | 16 (12–27) | air-tight covers |
| Slurry/effluent application[h] | 22 (15–35)[i] | 30 (20–47)[i] | Deep placement (slurry) Light irrigns (effluent) |

*Note:* Almost all N lost from floors is as ammonia and smaller quantities of volatile organic N compounds, but that up to 5% of N lost from manure, slurry, and effluent storage and application can be as $N_2O$.

[a] Bequette et al. 2003; Huhtanen and Hristov 2009.
[b] From 26 studies tabulated by Hristov et al. 2011.
[c] From Rotz 2004.
[d] From Rotz 2004; Thompson and Meisenger 2004.
[e] From Bequette et al. 2003; Huhtanen and Hristov 2009.
[f] Est. from ammonia emissions per cow from 27 studies tabulated by Hristov et al. 2011.
[g] Deduced from daily emissions reported by McGinn et al. 2008; Rumberg et al. 2008; Hristov et al. 2009.
[h] Est. from annual per cow emissions reported by Thompson and Meisinger 2004; Rumberg et al. 2006.
[i] Note that some of this N applied to land as manure, slurry, or effluent can be leached as nitrate N.

than in beef feedlots; the waste is therefore virtually always a liquid effluent, which is washed or pumped into holding ponds prior to land application (Rotz 2004; Rumberg et al. 2008). This reduces the percentage of urine-urea volatilized as ammonia, but much of this saved N can be subsequently lost during effluent storage and land application or irrigation unless targeted mitigation strategies are implemented (Thompson and Miesinger 2004; McGinn et al. 2008; Rumberg et al. 2008; Hristov et al. 2011).

In both beef feedlot and housed dairy operations, total N losses as ammonia plus $N_2O$ make up an average of 74% off dietary N (Table 7.5). The systems differ in how the losses are distributed; in feedlots a greater percentage is lost directly from the floor, almost all as ammonia. In dairy housing, there is less floor ammonia lost due to relatively frequent flushing and/or scraping and/or the use of grooved or slatted floors. This appears to have little net benefit, however, as the systems are not designed to efficiently utilize the increased proportion of N that is passed to storage.

### 7.5.1.3 Nitrous Oxide ($N_2O$) Emissions

$N_2O$ emissions occur during manure and effluent/slurry storage (Saggar et al. 2013; Bai et al. 2015), and from the soil during and after land application (Li et al. 2015). Nitrifying soil organisms obtain energy by converting the ammonium-N through to nitrite then nitrate (Cameron et al. 2013; Nelson et al. 2016). Nitrate is rarely present in large concentrations on the surface of feedlots or in manure or dairy cow effluent because of the relatively anaerobic conditions present; $N_2O$ losses are consequently low from these sources (Bai et al. 2015). Losses from manure during storage or during and after land application can be much higher (Bai et al. 2015), but rarely reach above 5% of total losses from these sources. If soils used for land application become anaerobic due to excess application of manures and effluents and/or rainfall, nitrate N present can be converted to $N_2O$ and/or $N_2$ and lost to the atmosphere (Cameron et al. 2013).

### 7.5.2 Mitigation Options for N Losses from Cow Houses and Feedlots

#### 7.5.2.1 Optimizing Dietary Protein Content

The importance of optimizing dietary protein concentration and types to minimize urine N output has been established by many feedlot and dairy house studies (Cole et al. 1995; Tomlinson et al. 1996; Rotz 2004; Todd et al. 2006; Hristov et al. 2011; Cooprider et al. 2014). The animal biochemistry, microbiology, and physiology involved is not a topic of this chapter.

#### 7.5.2.2 Mitigation of Ammonia Volatilization Losses from Floors

A variety of methods for reducing ammonia volatilization directly from urine deposited on the floors of cow houses and feedlots has been investigated. These range from high-technology means such as the installation of "acid waterfalls" to trap the ammonia (Melse et al. 2009); the use of tie stalls and/or slatted or grooved floors and/or washing (with water or formaldehyde solution) to enable more efficient capture and recovery of excreta and/or separation or urine and dung (Braam and Swierstra 1997; Monteny and Erisman 1998; Rotz 2004); the application to feedlot floors of amendments such as zeolite, alum, humate material, urease inhibitors and lignite, a low-grade coal (Varel 1996; Shi et al. 2001; Chen et al. 2015). Much research has been conducted on reducing the total and/or rumen metabolizable protein content of the feed to reduce urinary N output (Cole et al. 1995; Tomlinson et al. 1996; Rotz 2004; Todd et al. 2006). The addition of N inhibitors to the feed has also been studied (Varel 1997). Bovine metabolism is being investigated to determine whether it is possible to identify and select for metabolism traits that increase the conversion of feed protein to milk or beef protein, thereby reducing methane emissions and/or urea excretion in the urine (Reynolds and Kristensen 2008; Hristov et al. 2011; Reynolds et al. 2013). Informative industry-orientated summaries of environmentally orientated feedlot design (Adams et al. 1995; Spellman and Whiting 2007) and manure management (Sweeten 2000) are available.

#### 7.5.2.3 Application of Lignite and Other Products to Feedlot Floors

Perhaps the most practicable option demonstrated recently is the application of low-grade lignite coal to feedlot floors (Chen et al. 2015). Lignite was shown to reduce ammonia volatilization by 40% over the 6-week trial period in outdoor pens stocked with beef cattle. Lignite will adsorb about 2% N as ammonium (the same as peat); it must be replaced once the saturation point is reached, every 4–6 weeks, less frequently than required for nbpt application. It is proposed that the saturated lignite could be sold at a profit (Chen et al. 2015); however, the costs and inconvenience of transportation and spreading such a low N-content product are likely to pose considerable barriers to its use except in the close vicinity of the feedlot operation, in the same way as do the disposal of large amounts of low nutrient content organic wastes from any point source.

#### 7.5.2.4 Mitigation of N Losses during Manure, Slurry, and Effluent Storage

N losses during storage prior to spreading occur mainly as ammonia, and to a much lesser extent nitrous oxide and nitrogen gas (Rotz 2004). The quantities lost and their ratios are very dependent on storage conditions and treatment (Rotz 2004); from 10% to 30% of total N can be lost as ammonia. Nitrous oxide losses increase as the manure becomes increasingly anaerobic. Turning of the manure to aerate it reduces nitrous oxide loss, but can increase ammonia volatilization by exposing more surface. In the case of slurry and effluent, methane and ammonia emissions can be reduced by 80–90% by using covered ponds or pods instead of open ponds (Sommer 1997; Sommer et al. 1993;

Rotz 2004). The trapped gas can be burned and turned into electricity or used directly for heating. Ammonia needs to be stripped from the gas before burning to avoid emission and nitrogen oxide ($NO_x$) formation; contact with sulfuric acid can be used to produce ammonium sulfate fertilizer (Melse et al. 2009). Open manure can be treated with urease inhibitors (Varel et al. 1999). Virtually no feedlots have adopted these technologies because of perceived and/or real lack of cost-effectiveness.

### 7.5.2.5 Mitigation of N Losses during Land Application of Manure, Slurries, and Effluent

The land application of manure and effluents returns nutrients to the soil to promote growth of pasture, forages, or cut and carry crops (Li et al. 2015). However, the process of land application is also susceptible to significant N losses. Ammonia volatilization in the first few days after each application can reach 70% of the applied N (Hristov et al. 2011; Li et al. 2015). Subsequent nitrate leaching occurs from free draining soils, and nitrous oxide emissions occur where soils become anaerobic through excess application or rainfall, and in anaerobic microsites in otherwise aerobic soils (Sexstone et al. 1985). Much of the nitrate leached through or from the soil ends up in waterways or groundwater aquifers (Cameron et al. 2013). The former can cause eutrophication, leading to excess weed and algal growth, rendering the water less suitable for recreational pursuits. Nitrate leaching into aquifers can affect the suitability of the water for human consumption, as it can be reduced in the body to nitrite and nitric oxide (NO), precursors to the formation of various carcinogens (Habermeyer et al. 2014; SKLM 2014). Leached nitrate can also be denitrified to nitrous oxide (and nitrogen gas) in the vadose (non-saturated) zone above aquifers, especially if sources of metabolizable organic C (MOC) have been leached with nitrate; this is much more likely to occur where manure or effluents are being applied, due to their high MOC content (Stenger et al. 2008).

Methods for mitigating these losses include the avoidance of excess application for the soil type and crop N uptake capability (Rotz 2004); sub-surface application (de Klein et al. 1996); and the application of urease and nitrification inhibitors (Varel et al. 1999). The biggest limitation is almost invariably the lack of sufficient suitable land for application within an economically viable transport distance. In essence, the assessed economic advantages of ever-larger CAFOs have not factored in to a remotely sufficient extent the adverse economic, practical, environmental, and societal effects of all aspects of dealing with the excreta from these operations. Consequently, many face an uncertain future; on both full cost-accounting and environmental bases they have no inherent advantages over IGPFs, apart from slightly lower methane production per kilogram of meat produced (Del Prado et al. 2015). Their only real advantage is the physical capability, in the absence of environmental considerations, to produce ever-increasing quantities of meat for human consumption.

### 7.5.3 MODELING OF REDUCTIONS IN N LOSSES

Modeling different scenarios based on optimizing diet and including some of the proven, practical mitigation strategies technologies discussed in the following indicate that reductions in total N losses of over 50% are achievable (Hristov et al. 2011; Cooprider et al. 2014; Costa et al. 2014; Del Prado 2016). However, except for modifying the N content of the diet, and the quite rare use of slatted or grooved floors in dairy parlors, few of these options have come into use commercially, for a variety of reasons including cost, impracticality, and a lack of perceived advantage to the owner. Even diet modification is problematic in many cases; in the United States, for example, the increased use of crops such as corn for the production of ethanol fuel has meant an increasing supply of relatively high N content by-product becoming available to farmers as a competitively priced feed (Hristov et al. 2011). Also, the potential advantages of strategies to reduce ammonia emissions from feedlot floors and other surfaces are only realized if coupled to efficient treatment and land application of the manure or effluent.

Despite the existence of proven mitigation strategies, in the current political climate, and the structure of the beef and dairy industry structuring in the United States, it is difficult to see how more regulatory pressure is going to be brought to bear on housed dairy and beef feedlot operations and meatpacking operations to bring about meaningful steps toward true sustainability. Consequently, the same applies to countries that export grain-fed beef to the United States.

## 7.6 DISCUSSION AND CONCLUSIONS: THE FUTURE FOR COVERED HOUSING, FEEDLOTS, AND INTENSIVELY GRAZED FARMING

It is clear from this assessment that intensive beef and dairy systems operated in temperate climates—MCGFs, IGPFs, and CAFOs—have been pushed well past soil-plant-atmosphere limits to true sustainability in many areas, through increasing demand for food occurring simultaneously with intensifying global trading and competitiveness. In this situation, there are declining opportunities for farmers and producers to pass on the costs of achieving and maintaining true environmental sustainability, except by targeting niche higher income, environmentally aware markets. True sustainability is defined here as a system where there are no emissions or soil accumulation of any chemical, biochemical, or biological agents at levels that pose a real threat either to human health or the environment; a tall order.

### 7.6.1 CONCENTRATED ANIMAL FEEDING OPERATIONS (CAFOs): THE UNITED STATES SITUATION

Considerable progress has been made in the scientific understanding of the processes responsible for N losses to the environment from dairy houses and beef feedlots. Mitigation strategies have been developed that are capable of considerably reducing the main N losses—ammonia volatilization losses from urine and dung deposition, and manure, slurry, and effluent storage and application to land, by—50% or more. The most promising in terms of effectiveness and ease of implementation are (1) the use of only highly protein-efficient feeds, (2) the use of low-grade lignite coal spread on feedlot floors to absorb ammonia and its monthly collection and (possibly free) distribution to neighboring farms as a low-concentration (2% N) fertilizer, (3) the storage of manure as covered slurries followed by its subsurface injection application to soils, and (4) the use of grooved floors in dairy operations, designed to physically separate urine from dung prior to separate storage and subsurface land application, and/or frequent scraping of housing surfaces. These strategies must be installed on a whole-system basis; if not, there are risks that savings in N losses in one part of the operation will simply be lost by increases in another.

The essential problem is that, unless clear financial benefit of important N mitigation strategies can be demonstrated to the CAFO operator, in the absence of sufficient societal and political will to enforce adoption through regulation, the effects of housed dairy operations and feedlots on the climate and environment will continue to worsen. In the United States, Canada, and China, the three largest CAFO-operating countries, there are currently few regulations restricting the installation and design of these operations, apart from minimum animal space requirements to reduce the risk of heat stress, and in some cases setting maximum allowable ammonia emissions. In the United States and Canada, there is ever-increasing pressure on CAFOs to reduce production costs wherever possible, in order remain viable at the meat prices offered by the small group of large meatpacking companies that control an increasing majority of meat supply to retailers. The search for cost-efficiency has in turn meant that more and more of the total cattle production is coming from ever-declining numbers of increasingly large CAFOs. The smaller and family-operated CAFOs are rapidly being lost from the industry.

In the United States, the individual states have long empowered the federal Environmental Protection Agency (USEPA) to control adverse environmental effects, particularly emissions to the atmosphere, on their behalf. The EPA is legally obliged to collect on-site, scientifically robust, ammonia or $N_2O$ emission or nitrate leaching data to bring a prosecution where considered warranted. But there also must be the political will to finance what can be expensive, drawn-out legal disputes. Some large CAFOs are increasingly avoiding this risk of prosecution by resorting to the installation of very large effluent/slurry ponds which, because of high evaporation rates in dry environments, may not reach capacity for 25 years or more, thereby avoiding the need for land application for a generation. Total gaseous emissions from these huge ponds per unit of N are of course very large, but stay within current USEPA regulatory ammonia emission concentration limits per m$^2$ as a consequence of their huge surface area. They present a hugely expensive, highly environmentally risky sludge deposal problem for future generations.

Additional difficulty comes from the impossibility of achieving industry-wide agreement for maximum desirable stock capacities for beef feedlots and dairy cows. Limits such as 10,000 head for beef and 2000 head for dairy cows would certainly reduce the risk of disease outbreaks or catastrophic storage failures. Other conditions should include the requirement that they be built only in locations where they can be fully surrounded by sufficient areas of suitable soils, with adequate availability of irrigation water, to sustainably utilize the entire manure output to grow crops such as corn. Feedlots will need to have an ammonia-absorbing material such as lignite coal added regularly; the N saturated material should be required to be supplied to neighboring farms as a free, low N-content fertilizer. Dairy operations should have grooved floors to help separate dung and urine. All manure should be fully covered during storage and be sub-surface injected in suitable weather and soil conditions. This is likely to require at least 0.5 ha per head of cattle under confinement. In combination, these mitigations are likely to add up to 50% to the cost of meat and dairy products.

Given current societal meat price expectations, and the current political push to reducing regulations, it seems clear that nothing short of an extremely serious health outbreak or environmental catastrophe is likely to bring about the acceptance by society that changes to CAFO regulations must be made, and that higher retail prices of meat, and therefore lower consumption, are necessary. By contrast in China, where scientific evidence, and the growing government acceptance of it, has resulted in the adoption in 2016 of policy to reduce meat consumption per capita by 50% in that country. Whether this reduction is achieved by restrictions on the rapidly growing number of CAFOs or by other strategies remains to be seen, but the intended result seems clear.

### 7.6.2 MIXED CROPPED AND GRAZED DAIRY AND BEEF FARMS (MCGFs): THE NETHERLANDS WARNING

The problems facing MCGFs are by no means trivial, but do seem more capable of being overcome, for a variety of reasons. Put simply, regulations covering best management practices seem to be slowly reducing environmental losses from the amount of excreta being produced and applied to land in any given country (with the probable exception of the Netherlands) to environmentally acceptable levels. The absence of any apparent focus on improving the NUE of fertilizer N is a curious exception. The EU has a classic, well-documented and understood example in the case of the Netherlands of the perils of excessive inputs combined with low fertilizer NUE. There is growing political and societal acceptance in the EU that true sustainability must be reached and maintained, even if this increases cost of dairy products and beef meat to the consumer. This is automatically a slow process in any democracy, to allow time for both societal acceptance of increased meat and dairy product prices, and for farmers to change their farming operations and management.

### 7.6.3 INTENSIVELY GRAZED PASTURE FARMS (IGPFs): NEW ZEALAND TAKES STOCK

IGPFs are in a more similar situation to MCGFs than to CAFOs. There is less pressure on the equilibrium between farmer earnings and protection of the environment than in CAFOs. However, there are two fundamental issues that must be dealt with. The average NUE of fertilizer N needs to be greatly improved, as does the recovery of urine N by pasture. The latter can be achieved by reducing N intake in feed, improved distribution of urine, and by detecting and treating the urine patches to improve N recovery.

These technologies are expected to be cost-effective for the farmer eventually. However, to gain sufficient adoption to bring about the necessary economies of scale, more widespread regulated limits on nitrate leaching and gaseous emissions are likely to be required. As this regulatory approach gains momentum, IGPFs will increasingly adopt the most cost-effective technologies in fertilizer N efficiency and in detecting and treating urine patches to improve N recycling, to avoid penalties and/ or enforced destocking resulting from excessive nitrate leaching and gaseous emissions. Nbpt application to fresh urine patches is likely to be a part of the solution for reducing ammonium volatilization. The introduction of much safer nitrification inhibitors such as sodium or calcium thiosulfate and/or carefully regulated reintroduction of DCD (for urine patch treatment only, not broadcast) has perhaps the greatest potential for reducing nitrate leaching. Denitrification losses of nitrous oxide can be reduced by avoiding treading damage in wet conditions and by urine patch treatment with sources of readily mineralizable organic C to help immobilization of N and subsequent slow-release from microbes, and/or to promote denitrification fully through to $N_2$. Longer term, lower protein, higher energy content, persistent grasses, and other forages better matched to the rumen metabolism of ruminant animals will be introduced. More erect clovers that are more competitive with ryegrasses for light, and thereby fix more atmospheric N, will be found.

Dairy products and meat produced under grazing are progressively being seen as premium products to both housed cow milk and non-dairy "milk" products. The premiums are slowly reducing pressure on IGPFs to meet bedrock commodity prices. Intensive grazing is generally situated in more densely populated areas than are CAFOs; this has the benefit of increased consumer awareness of the good and bad aspects of IGPFs; this increases pressure for environmental (and animal welfare) improvements on IGPFs, and increases consumer understanding of the challenges facing farmers. Over time, this will slowly lead society to accept that IGPF production cannot simply keep increasing and in fact probably needs by 20 or 30% on a global basis. Regardless, the need for increasing prices of dairy products and meat will slowly become accepted.

For this to be successful, however, there will be a continuing need for improved dialog between farmers, consumers, researchers, and environmental governance organizations. The steadily increasing percentages of the human population living in cities increases the challenge to ensuring that this dialog takes place, and overcomes simplistic, misleading propaganda by self-interest groups.

*A final comment.* All three systems are exceeding to varying degrees the ability of available soils to accept and process the excreta and waste from animal food production while keeping emissions within truly sustainable limits. Proven N loss mitigation practices must be adopted, and the search to find new ones that reduce the excess loadings we place on soil processes must be continued. Regardless of what steps are taken to temporarily remove CAPO outputs from the soil-plant system, the soil needs to be the final reservoir; the alternative is an environment that will become increasingly incapable of supporting humankind. There is an urgent need for a far more coordinated and multidisciplinary approach between countries to find agreement on the limits to dairy and meat production in the various systems.

## REFERENCES

Adams, R.S., Comerford, J.W., Ford, S.A. et al. 1995. *Dairy Reference Manual,* 3rd ed. NRAES-63. Natural Resources, Agricultural and Engineering Service, Cooperative Extension.

Bai, M., Flesch, T.K., McGinn, S.M., and D. Chen. 2015. A snapshot of greenhouse gas emissions from a cattle feedlot. *J. Environ. Qual.* Short Comm. DOI 10.2134/jeq2015.06.0278.

Ball, R., Keeney, D.R., Theobald, P.W., and P. Nes. 1979. Nitrogen balance in urine-affected areas of a New Zealand pasture. *Agronomy J.* 71:309–314.

Bateman, E.J., and E.M. Baggs. 2005. Contributions of nitrification and denitrification to $N_2O$ emissions from soils at different water-filled pore space. *Biol. Fertil. Soils* 41:379–388. DOI 10.107/s00374-005-0858-3.

Bates G., and B.F. Quin. 2013. Robotic urine-patch treatment and effluent application—Technology to support intensification of New Zealand dairy farming while protecting the environment. *Proceedings of the New Zealand Grassland Association* 75:125–130.

Bates, G., Quin, B.F., and P. Bishop. 2015. Low-cost detection and treatment of fresh cow urine patches. In *Moving Farm Systems to Improved Attenuation,* Eds. L.D. Currie and L.L Burkitt. Occasional Report No. 28. Fertilizer and Lime Research Centre, Massey University, Palmerston North, New Zealand. 10 pp. http://flrc.massey.ac.nz.publications.html.

Bequette, B.J., Hanigan, M.D., Lapierre, H., and J.P.F. D'Mello. 2003. Mammary uptake and metabolism of amino acides by lactating ruminants. In Amino Acids in Farm Animal Nutrition, 2nd ed., pp. 347–365. CABI Publishing, Wallingford, UK.

Betteridge, K., Costall, D.A., Fi, D., Luo, D., and S. Ganeshi. 2013. Why we need to know what and where cows are urinating—A urine sensor to improve nitrogen models. Proc. N.Z. Grasslands Ass. 75:119–124.

Bishop, P.A., Liu, H.Y., Hedley, M.J., and P. Loganathan. 2008. New Zealand made controlled release coated urea increases winter growth rates of Italian ryegrass with lower N leaching than uncoated urea. *Proc. N.Z. Grassland Ass.* 70:85–89.

Bishop, P.A., and B.F. Quin. 2010. Modelling of the effect of combined DCD and urease inhibitors on the postdeposition size of urine patches: Implications for decreased N losses and increased pasture production using the 'Taurine' tail-attached dispenser. In *Farming's Future: Minimising Footprints and Maximising Margins,* Eds. L.D. Currie and C.L. Christensen. Occasional Report No. 23. Fertilizer and Lime Research Centre, Massey University, Palmerston North, New Zealand pp. 268–273.

Bishop, P.A. 2016. Survey of potential nitrification inhibitors to replace DCD for targeted application to urine patches. In *Soil, a Balancing Act Downunder.* Joint Conf. N.Z. Soc. Soil Science and Soil Science Australia, Queenstown, New Zealand, in press.

Blennerhassett, J.D., Quin, B.F., Zaman, M., and C. Ramakrisnan. 2006. The potential for increasing nitrogen responses using Agrotain® treated urea. *Proc. N.Z. Grassland Ass.* 68:297–301.

Braam, C.R., Detelaars, J.J.M.H., and M.C.J. Smits. 1997. Effects of floor design and floor cleaning on ammonia emission from cubicle houses for dairy cows. *Neth. J. Agric. Sci.* 45:49–64.

Bristow, A.W., Whitehead, D.C., and J.E. Cockburn. 1992. Nitrogenous constituents in the urine of cattle, sheep, and goats. *J. Sci. Food Agric.* 59:387–394.

Bussink, D.W., and O. Oenema. 1998. Ammonia volatilization from dairy farming systems in temperate areas: A review. *Nutr. Cycl. Agroecosystems* 51(1):19–33.

Cameron, K.C., Di, H.J., and J.L. Moir. 2013. Nitrogen losses from the soil/plant system: A review. *Ann. Appl. Biol.* 162:145–173.

Chen, D., Suter, H., Islam, A., Edis, R., Freney, J.R., and C.N. Walker. 2008. Prospects of improving efficiency of fertilizer nitrogen in Australian agriculture: A review of enhanced efficiency fertilizers. *Australian J. of Soil Res.* 46:289–301.

Chen, D., Sun, J., Bai, M. et al. 2015. A new cost-effective method to mitigate ammonia loss from intensive cattle feedlots: Application of lignite. *Sci. Rep.* 5:16689. DOI 10.1038/srep16689.

Christensen, C.J., Hanly, J.A., Hedley, M.J., and D.J. Horne. 2011. Nitrate leaching and pasture production from two years of duration-controlled grazing. In *Adding to the Knowledge Base for the Nutrient Manager,* Eds. L.D. Currie and C.L. Christensen. Occasional Report No. 24, Fertilizer and Lime Research Centre.

CIWF.org 2012. Statistics: Dairy cows. Compassion in world farming. 8 pp. Updated July 1, 2012.

Clark, D.A., and S.L. Harris. 1996. White cover or nitrogen fertilizer for dairying? *Agronomy Society of New Zealand* Special Publication No. 11/Grassland Research and Practice Series no. 6, 107–114.

Cole, N.A., Clark, R.N., Todd, R.W., Richardson, C.R., Gueye, A., Greene, L.W., and K. McBride. 1995. Influence of dietary crude protein concentration and source on potential ammonia emissions from beef cattle manure. *J. Anim. Sci.* 83:722–731.

Cooprider, K.L., Mitloehner, F.M., Famula, T.R. et al. 2014. Feedlot efficiency implications on greenhouse gas emissions and sustainability. *J. Anim. Sci.* 89(8):2643–2656.

Costa, C. Jr., Li, C., Cerri, C.E., and C.C. Cerri. 2014. Measuring and modelling nitrous oxide and methane emissions from beef cattle feedlot manure management: First assessments under Brazilian condition. *J. Environ. Sci. Health B.* 49(9):696–711. DOI 10.1080/03601234.2014.922856.

Crush, J.R., Easton, H.S., Waller, J.E., Hume, D.E., and M.J. Faville. 2007. Genotypic variation in patterns of root distribution, nitrate interception and response to moisture stress of a perennial ryegrass *(Lolium perenne L.)* mapping population. *Grass For. Sci.* 62(3):265–273.

Del Prado, A., Crosson, P., Olesen, J.E., and C.A. Rotz. 2013. Whole-farm models to quantify greenhouse gas emissions and their potential use for linking climate change mitigation and adaptation in temperate grassland ruminant-based farming systems. *Animal* 7(2):373–385. DOI 10.1017/S1751731113000748.

Dewar, K., Zaman, M., Rowarth, J.S., Blennerhassett, J.D., and M.H. Turnbull. 2010. The impact of urease inhibitor on the bioavailability of nitrogen in urea and in comparison with other N sources in ryegrass *(Loilium perenne L.). Crop and Pasture Sci.* 61: 214–221.

Di, H.J., and K.C. Cameron. 2002. The use of a nitrification inhibitor, cyanadiamide (DCD), to decrease nitrate leaching and nitrous oxide emissions in a simulated grazed and irrigated grassland. *Soil Use and Management* 18: 395–403.

Di, H.J., and K.C. Cameron. 2004. Treating grazed pasture soil with a nitrification inhibitor, eco-n™, to decrease nitrate leaching in a deep sandy soil under spray irrigation—A lysimeter study. *NZ J. Agric. Res.* 47:353–361.

Di, H.J., and K.C. Cameron. 2005. Reducing environmental impacts of agriculture by using a fine particle suspension nitrification inhibitor to decrease nitrate leaching from grazed pastures. AEE 109: 202–212.

Di, H.J., and R.C. Cameron. 2007. Nitrous oxide emissions from two dairy pasture soils as affected by different rates of a fine particle suspension nitrification inhibitor, dicyandiamide. *Biol. Fert. Soils* 42:472–480.

Drewry, J.J., Cameron, K.C., and G.D. Buchan. 2008. Pasture yield and soil physical property responses to soil compaction from treading and grazing—a review. *Aust. J. Soil Res.* 46(3):237–256.

Eckard, R.J., and D.R. Franks. 1998. Strategic nitrogen fertiliser use on perennial ryegrass and white clover pasture in north-western Tasmania. *Aust. J. Exp. Agric.* 38:155–160.

Eckard, R.J., Chapman, D.F., White, R.E. et al. 2001. Nitrogen balances in high rainfall, temperate dairy pastures of south-eastern Australia. In *Proceedings of the XIX International Grasslands Congress,* Sao Pedro, Brazil, Feb. 2001, pp. 169–170. Brazilian Society of Animal Husbandry, Piracicaba, Brazil.

Eckard, R.J., Chen, D., White, R.E., and D.F. Chapman. 2003. Gaseous nitrogen loss from temperate perennial grass and clover dairy pastures in south-eastern Australia. *Aust. J. Agric. Res.* 54:561–570.

Edwards, G.R., Parsons, A.J., Rasmussen, S., and R.H. Bryant. 2007. High sugar ryegrasses for livestock systems in New Zealand. *Proc. N.Z. Grassland Ass.* 69:161–171.

Eghball, B., and J.F. Power. 1994. Beef cattle feedlot manure management. *J. Soil Water Cons.* 49:189–193.

Felber, R., Conen, F., Flechard, C.R., and A. Neftel. 2012. Theoretical and practical limitations of the acetylene inhibition technique to determine total denitrification losses. *Biogeosciences* 9:4125–4138. doi 10.5194/bg-4125-2012.

Field, R.R.O., and P.R. Ball. 1978. Tactical use of fertiliser nitrogen. *Proc. Agron. Soc. N.Z.* 8:129–133.

Field, T.R.O., and P.R. Ball. 1982. Nitrogen balance in an intensively utilised dairy farm system. *Proc. N.Z. Grassland Ass.* 43:64–69.

Fowler, D., Coyle, M., Skiba, U. et al. 2013. The global nitrogen cycle in the twenty-first century. *Phil. Trans. R. Soc.* B 368:20130164. DOI 10.1098/rstb.2013.0164.

Frame, J., and P. Newbould. 1984. Herbage production from grass/white clover swards. *Occasional Symposium of the British Grassland Society* 16:15–35.

Fraters, B., van Leeuwen, T., Boumans, L., and J. Reijs. 2015. Use of long-term monitoring data to derive a relationship between nitrogen surplus and nitrate leaching for grassland and arable land on well-drained sandy soils in The Netherlands. *Acta. Agr. Scand.* B-S P 65 (suppl. 2), pp 144–154.

Gautam, D.P., Rahman, S., Borhan, M.D.S., and C. Engel. 2016. The effect of feeding high fat diet to beef cattle on manure composition and gaseous emission from a feedlot pen surface. *J. Anim. Sci. Tech.* 58:22. DOI 10.1186/s40781-016-0104-6.

Gerber, P.J., Hristov, A.N., Henderson, B. et al. 2013. Technical options for the mitigation of direct methane and nitrous oxide emissions from livestock: A review. *Animal* 7:220–234.

Gillingham, A.G., Morton, J.D., and M.H. Gray. 2007. Pasture responses to phosphorus and nitrogen fertilisers on East Coast hill country: (1) Total production from easy slopes. *N.Z. J. Agric. Res.* 50:307–320.

Gilmanov, T.G. 1995. The state of rangeland resources in the newly independent states of the former USSR. In *Rangelands in a Sustainable Biosphere.* Proceedings of the Fifth International Rangeland Congress, Ed. N.E. West. Denver CO: Society for Range Management, Vol. 2, pp. 10–13.

Godfray, H.C.J., Beddington, J.R., Crute, I.R. et al. 2010. The challenge of feeding nine billion people. *Science* 327:812–818.

Gourley, C.J.P., and D.M. Weaver. 2012. Nutrient surpluses in Australian grazing systems: management practices, policy approaches and difficult choices to improve water quality. *Crop and Past. Sci.* 63(9):80–85.

Habermeyer, M., Roth, M.A., Gith, S. et al. 2014. Nitrate and nitrite in the diet: How to assess their benefit and risk for human health. *Mol. Nutr. Food Res.* 59(1):106–128. DOI 10.1002/mnfr.201400286.

Halevy, J. 1987. Efficiency of isobutylidene diurea, sulphur-coated urea and urea plus nitrapyrin, compared with divided dressings of urea, for dry matter production and nitrogen uptake of ryegrass. *Experimental Agric.* 23(2):167–179. DOI 10.1017/s0014479700016963.

Hao, X., Chang, C., Larney, F.J., and G.R. Travis. 2001. Greenhouse gas emissions during cattle feedlot manure composting. *J. Environ. Qual.* 30:376–386.

Harper, L.A., Catchpoole, V.R., and I. Vallis. 1983. Ammonia loss from fertilizer applied to tropical pastures. In *Gaseous Loss of Nitrogen from Plant-Soil Systems,* Eds. J.R. Freney and J.R. Simpson. pp. 195–214. Martinus Nijhoff/Dr. W. Junk Publishers, The Hague, Netherlands.

Harris, S.L., Penno, J.W., and A.M. Bryant. 1994. Effects of high rates of nitrogen fertilizer on dairy pastures and production. *Proc. N.Z. Grasslands Assn.* 56:27–31.

Harty, M.A., Forrestal, P.J., Carolan, R. et al. 2016. Temperate grassland yields and nitrogen uptake are influenced by fertilizer nitrogen source. *Agron. J.* 109: 9. doi 10.2134/agronj2016.06.0362.

Hilhorst, G., Oenema, J., and H. Van Keulen. 2001. Nitrogen management on experimental farm 'De Marke': Farming system, objectives and results. *Netherlands J. Agric. Sci.* 49:135–151.

Hoglund, J.H., Crush, J.R., Brock, J.L., Ball, R., and R.A. Carran. 1979. Nitrogen fixation in pasture. XII. General discussion. *N.Z. J. Exp. Agric.* 7:45–51.

Houlbrooke, D.J., Horne, D.J., Hedley, M.J., Hanly, J.A., and V.O Snow. 2003. The impact of intensive dairy farming on the leaching losses of nitrogen and phosphorus from a mole and pipe drained soil. *Proc. N.Z. Grassland Assn.* 65:179–184.

Hristov, A.N., Etter, R.P., Ropp, J.K., and K.L. Grandeen. 2004. Effect of dietary crude protein level and degradability on ruminal fermentation and nitrogen fertilization in lactating dairy cows. *J. Anim. Sci.* 82(11):3219–3229.

Hristov, A.N., Zaman, S., Vander Pol, M., Ndegwa, P., Campbell, L., and S. Silva. 2009. Nitrogen losses from dairy manure estimated through N mass balance or using markers. *J. Environ. Qual.* 38:2438–2448.

Hristov. A.N. 2011. Contribution of ammonia emitted from livestock to atmospheric $PM_{2.5}$ in the United States. *J. Dairy Sci.* 94(6):3130–3136.

Hristov, A.N., Hanigan, M.D., Cole, N.A. et al. 2011. Review: Ammonia emissions from dairy farms and beef feedlots. *Can. J. Anim. Sci.* 91:1–35.

Huhtanen, P., and A.N. Hristov. 2009. A meta-analysis of the effects of protein concentration and degradability on milk protein yield and milk N efficiency in dairy cows. *J. Dairy Sci.* 92:3222–3232.

Hutchinson, G.L., Mosier, A.R. and C.E. Andre. 1982. Ammonia and amine emissions from a large cattle feedlot. *J. Environ. Qual.* 11:288–293.

Jarvis, S.C. 2000. Progress in studies of nitrate leaching from grassland soils. *Soil Use and Management* 16:152–156.

Jiang, A., Zhang, T., Zhao, Q., Frear, C., and S. Chen. 2010. Integrated ammonia recovery technology in conjunction with dairy anaerobic digestion. In *Climate Friendly Farming,* CSANR Research Report 2010-001, Chapter 8, 19 pp.

Kebreab, E., Johnson, K.A., Archibeque, S.L., Pape, D., and T. Wirth. 2008. Model for estimating enteric methane emissions from United States dairy and feedlot cattle. *J. Anim. Sci.* 86: 2738–2748. doi 10.2527/jas.2008-0960.

de Klein, C.A.M., van Logtestijn, R.S.P., van de Meer, H.G., and J.H. Guerink. 1996. Nitrogen losses due to denitrification from cattle slurry injected into grassland soil with and without a nitrification inhibitor. *Plant Soil* 183:161–170.

de Klein, C.A.M., Sherlock, R.R., Cameron, K.C., and T.J. van der Weerden. 2001. Nitrous oxide emissions from agricultural soils in New Zealand—A review of current knowledge and directions for future research. *J. Royal Soc. N.Z.* 31: 543–574.

Lampe, C., Dittert, K., Wachendorf, M., Sattlemacher, B., and F. Taube. 2004. 15N traced nitrous oxide fluxes from fertilised grassland soils. *Grassland Sci. Eur.* 9:334–336.

Ledgard, S.F., and K.W. Steele. 1992. Biological nitrogen fixation in mixed legume/grass pastures. *Plant Soil* 141:137–153.

Ledgard, S.F., Penno, J.W., and M.S. Sprosen. 1999. Nitrogen inputs and losses from clover/grass pastures grazed by dairy cows, as affected by nitrogen fertiliser application. *J. Agric. Sci.* 132: 215–225.

Ledgard, S.F., Luo, J, Sprosen, M.S., Wyatt, J.B., Balvert, S.V., and S.B. Lindsey. 2014. Effects of the nitrification inhibitor dicyandiamide (DCD) on pasture production, nitrous oxide emissions and nitrate leaching in Waikato, New Zealand. *N.Z. J. Agric. Res.* 57 (4). http://dx.doi.org/10.1080/002288233.2014.928642.

Ledgard, S.E., Welten, B., and K. Betteridge. 2015. Salt as a mitigation option for decreasing nitrogen leaching losses from grazed pastures. *J. Sci. Food Agric.* 95(15):3033–3040. doi 10.1002/jsfa.7179.

Li, J., Luo, J., Shi, Y. et al. 2015. Nitrogen gaseous emissions from farm effluent application to pastures and mitigation measures to reduce the emissions: A review. *NZ J. Agric. Res.* doi 10.1080/00288233.2015.1028651.

Luo, J., Ledgard, S.F., and S.B. Lindsay. 2007. Nitrous oxide emission from application of urea on New Zealand pasture. *N.Z. J. Agric. Res.* 50:1–11.

Luo, J., Saggar, S., Bhandral, R., Ledgard, S., Lindsey, S., and W. Sun. 2008. Effects of irrigating dairy-grazed grassland with farm dairy effluent on nitrous oxide emissions. *Plant Soil* 309(1):119–130.

Luo, J., and F. Kelliher. 2010. Partitioning of animal excreta N into urine and dung and developing the $N_2O$ inventory. New Zealand Ministry of Agriculture and Fisheries Policy paper 0910-11528, 09-03, 21 pp.

Luo, J., Ledgard. S.F., and S.B. Lindsey. 2013. Nitrous oxide and greenhouse gas emissions from a grazed pasture as affected by use of a nitrification inhibitor and a restricted grazing regime. *Sci. Total Environ.* 465:107–114. http://dx.doi.org/10.1016/j.scitotenv.2012.12.075.

Luo, J., Ledgard, S., Wise., B., Welten, B. et al. 2015. Effect of dicyandiamide (DCD) delivery method, application rate, and season on pasture urine patch nitrous oxide emissions. *Biol. Fert. Soils* 51:453–464.

McGinn, S.M., Coates, T., Flesch, T.K., and B.P. Crenna. 2008. Ammonia emissions from dairy cow manure stored in a lagoon over summer. *Can. J. Soil Sci.* 88: 611–615.

Meneer, J.C., Ledgard, S.F., McLay, C.D.A., and W.B. Silvester. 2005. The effects of treading by dairy cows during wet soil conditions on white clover productivity, growth and morphology in a white clover-perennial ryegrass pasture. *Grass For. Sci.* 60:46–58.

Melse, R.W., Ogink, N.W.N., and W.H. Rulkens. 2009. Air treatment techniques for abatement of emissions from intensive livestock production. *Open Agric. J.* 3:6–12.

Misselbrook, T., Del Prado, A., and D. Chadwick. 2013. Opportunities for reducing environmental emissions from forage-based dairy farms. *Agric. Food Sci.* 22:93–107.

Moir, J.L., Cameron, K.C., Di, H.J., and U. Fertsak. 2011. The spatial coverage of dairy cattle urine patches in an intensively grazed pasture system. *J. Agr. Sci.* 149:473–485.

Monteny, G.J., and J.W. Erisman. 1998. Ammonia emission from dairy cow buildings: A review of measurement techniques, influencing factors and possibilities for reduction. *Neth. J. Agric. Sci.* 46:225–247.

Morton, J.D., Stafford, A.D., Gillingham, A.G., Old, A., and O. Knowles. 2016. The development of variable rate application of fertilizer from a fixed wing topdressing aircraft. In *Hill Country—New Zealand Grassland Research and Practice Series* 16: 163–168.

Mount, G.H., Rumburg, B., Havig, J. et al. 2002. Measurement of atmospheric ammonia at a dairy using differential optical adsorption spectroscopy in the mid-ultraviolet. *Atmos. Env.* 36(11):1799–1810.

Muck, R.E., Guest, R.W., and B.K. Richards. 1984. Effects of manure storage design on nitrogen conservation. *Agric. Wastes* 10:205–220.

Nash, P., Motavalli, P., Nelson, K., and R. Kremer. 2015. Ammonia and nitrous oxide gas loss with subsurface drainage and polymer coated urea fertilizer in a poorly drained soil. *J. Soil Wat. Cons.* 70(4):267–275. doi 10.2489/swc.70.4.267.

Nelson, M.B., Martiny, A.C., and J.B.H. Martiny. 2016. Global biogeography of microbial nitrogen cycling-traits in soil. *PNAS* 113(29):8033–8040. doi 10.1073/pnas.1601070113.

O'Connor, M.B. 1982. Nitrogen fertilizers for production of out-of-season grass. In *Nitrogen Fertilisers in New Zealand Agriculture,* Ed. P.B. Lynch. Ray Richards Publisher, Auckland, New Zealand. pp. 65–74.

Ogink, N.W.M., and W. Kroodsma. 1996. Reduction of ammonia emission from a cow cubicle house by flushing with water or a formalin solution. *J. Agric. Eng. Res.*63:197–204.

Pal, P., McMillan, A.M.S., and S. Saggar. 2015. Routes of dicyandiamide uptake in pasture plants: A preliminary study. In *Moving Farm Systems to Improved Nutrient Attenuation,* Eds. L.D. Currie and L.L. Burkitt. Occasional Report No. 28, Massey University, Palmerston North, New Zealand, 9 pp.

Pilgrim, E.S., Macleod, C.J.A., Blackwell, M.S.A. et al. 2010. Interactions among agricultural production and other ecosystem services delivered from European temperate grassland systems. *Adv. Agron.* 109:117–154.

Proctor, L.E., Craig, H.J.B., Mclean, N.J. et al. 2015. The effect of grazing high-sugar ryegrass on lamb performance. *Proc. N.Z. Soc. Anim. Prod.* 75:235–238.

Pullanagari, R.R., Yule, I.J., Tuohy, M.P, Hedley, M.J., Dynes, R.A., and W.M. King. 2012. Proximal sensing of the seasonal variability of pasture nutritive value using multispectral radiometry. *Grass and Forage Sci.* 68:110–119.

Quin, B.F. 1977. The fate of sheep urine-nitrogen on surface irrigated pasture in New Zealand. *N.Z. Soil News Supplement* 25(4).

Quin, B.F. 1982. The influence of grazing animals on nitrogen balances. In *Nitrogen Balances in New Zealand Ecosystems: Proceedings of a Workshop May 1980,* Eds. P.W. Gandar and D.S. Bertaud, Palmerston North, New Zealand. Publ. Dept. of Scientific and Industrial Research, NZ, pp. 95–102.

Quin, B.F. 2000. The 'Taurine' cow-mounted, tail-activated nitrogen inhibitor dispensing device. *New Zealand Patent Office* No. NZ 506883.

Quin, B.F. 2008. One-step, on-truck fluidisation and spreading of fertiliser optimises pasture response while minimising nutrient losses to the environment. In *Carbon and Nutrient Management in Agriculture,* Eds. L.D. Currie and L.J. Yates. Occasional Report No. 21. Fertilizer and Lime Research Centre, Massey University, Palmerston North, New Zealand, pp. 243–247.

Quin, B.F., and G. Findlay, 2009. On-truck fluidisation and spreading: Removing obstacles to achieving greater efficiency and environmental protection with fertilizer N. In *Nutrient Management in a Rapidly Changing World*, Eds. L.D. Currie and C.L. Lindsay, Occasional Report No. 22. Fertilizer and Lime Research Centre, Massey University, Palmerston North, New Zealand, pp. 250–256.

Quin, B.F., Gillingham, A.G., Baird, D., Spilsbury, S., and M. Gray, 2015. A comparison under grazing of pasture production, pasture N content and soil mineral N levels between granular urea and ONE*system*® on two contrasting dairy farms in New Zealand. *J. of New Zealand Grasslands* 77:251–258.

Quin, B.F. 2016. Reduction of environmental harm from urine patches. *New Zealand Patent Office* No. NZ 705824.

Quin, B.F., Bates, G., and P. Bishop. 2016. Locating and treating fresh cow urine patches with Spikey®: The platform for practical and cost-effective reduction in environmental N losses. In *Integrated Nutrient and Water Management for Sustainable Farming*, Eds. L.D. Currie and R. Singh. Occasional Report No. 29. Fertilizer and Lime Research Centre, Massey University, Palmerston North, New Zealand, 8 pp. http://flrc.massey.ac.nz/publications.html.

Rasmussen, S., Parsons, A.J., Xue, H., and J.A. Newman. 2009. High sugar grasses—harnessing the benefits of new cultivars through growth management. *Proc. N.Z. Grassland Ass.* 71: 167–175.

Reid, D. 1983. The combined use of fertilizer nitrogen and white clover as nitrogen sources for herbage growth. *J. Agric. Sci. Cambr.* 100:612–623.

Reynolds, K., and N.B. Kristensen. 2008. Nitrogen recycling through the gut and the nitrogen economy of ruminants: An asynchronous symbiosis 2008. *J. Anim. Sci.* 86(14):E293–E305.

Reynolds, C.K., Crompton, L.A., Mills, J.A.N., Barratt, C.E.S., and D.G. Barber. 2013. Improving the efficiency of feed nitrogen utilization. In *Feed Efficiency in Cattle*. International Dairy Nutrition Symposium, Wageningen, Netherlands.

RNS. 2015. Ruminant Nutrition Symposium: Use of genomics and transcriptomics to identify strategies to lower ruminal methanogenesis. *J. Anim. Sci.* 93(4):1431–1449.

Rose, C., Parker, A., Jefferson, B., and E. Cartmell. 2015. The characterization of feces and urine; a review of the literature to inform advanced treatment technology. *Crit. Rev. Environ. Sci. Technol.* 45(17):1827–1879. doi 10.1080/10643389.2014.1000761.

Rotz, C.A. 2004. Management to reduce nitrogen losses in animal production. *J. Anim. Sci.* 82 (E):E119–E137.

Rotz, C.A., Oenema, J., and H. van Keulen. 2006. Whole farm management study to reduce nutrient losses from dairy farms: a simulation study. *Appl. Eng. Agric.* 22:773–774.

Rubaek, K., Henriksen, K., Petersen, J., Rasmussen, B., and R.G. Sommer. 1996. Effects of application techniques and anaerobic digestion on gaseous nitrogen loss from animal slurry applied to ryegrass *(Lolium Perenne)*. *J. Agric. Sci.* 126: 481–492.

Rumberg, B., Mount, G.H., Filipy, J. et al. 2008. Measurement and modelling of atmospheric flux of ammonia from dairy milking cow housing 2008. *Atmos. Environ.* 42:3364–3379.

Saggar, S., Luo, J., Giltrap, D.L., and M. Maddena. 2009. Nitrous oxide emission processes, measurements, modelling and mitigation. In *Nitrous Oxide Emissions Research Progress*, Eds. A.J. Sheldon and E.P. Barnhart, pp. 1–66, Nova Science Publishers Inc., New York.

Saggar, S., Singh, J., Giltrap, D.L. et al. 2013. Quantification of reductions in ammonia emissions from fertiliser urea and animal urine in grazed pastures with urease inhibitors for agriculture inventory: New Zealand as a case study. *Sci. Tot. Env.* http://dx.doi.org/10.1016/j.scitotenv.2012.07.088.

Sakirkin, S.L.P., Cole, A.A., Todd, R.W., and B.W. Auvermann. 2011. Ammonia emissions from cattle-feeding operations. Part 1: Issues and emissions. *Air Quality Education in Animal Agriculture* http://www.extension.org/pages/15538/air-quality-in-animal-agriculture.

Santhi, S.R., Palaniappan, S.P., and D. Purushothaman. 1986. Influence of neem leaf on nitrification in a lowland rice soil. *Plant Soil* 93(1):133–145.

Scott. J.T., Lambie, S.M., Stevenson, B.A., Schipper, L.A., Parfitt, R.L., and A.C. McGill. 2015. Carbon and nitrogen leaching under high and low fertility pasture with increasing nitrogen inputs. *Agric. Ecosystems Env.* 202:139–147. doi 10.1016/j:agee.2014.12.021.

Selbie, D.R., Lanigan, G.J., Laughlin, R.J. et al. Confirmation of co-denitrification in grazed grassland. Scientific Reports 5: 17361. doi 10.1038/srep17361.

Sexstone, A.J., Revsbech, N.P., Parkin, T.B., and J. M. Tiedje. 1985. Direct measurement of oxygen profiles and distribution rates in soil aggregates. *Soil Sci. Soc. Am. J.* 49: 645–651.

Shi, Y., Parker, D.B., Cole, N.A., Auvermann, B.W., and J.E. Mehlhorn. 12001. Surface amendments to minimize ammonia emissions from beef cattle feedlots. *Trans ASAE* 44:677–682.

SKLM (DFG Senate Commission on Food Safety, Germany). 2014. Nitrate and nitrite in the diet: How to assess benefit and risk for human health. SKLM Commission Secretariat, Hannover, Germany, 118 pp.

Sommer, S.G., Olesen, J.E., and B.T. Christiensen. 1991. Effects of temperature, wind speed and air humidity on ammonia volatilization from surface applied cattle slurry. *J. Agric. Sci.*117:91–100.

Sommer, S.G., Christensen, B.T., Nielsen, N.E., and J.K. Schjorring. 1993. Ammonia volatilization during storage of animal and pig slurry: Effect of surface cover. *J. Agric. Sci.* 121: 63–71.

Sommer, S.G. 1997. Ammonia volatilization from farm tanks containing anaerobically digested animal slurry. *Atmos. Environ.* 31:863–868.

Spellman, F.R., and N.E. Whiting. 2007. *Environmental Management of Concentrated Feeding Operations (CAFOs)*, CRC Press/Taylor & Francis Group, Boca Raton, FL.

Stenger, R., Barkle, G., Burgess, C., Wall, A., and J. Clague. 2008. Low nitrate contamination of shallow groundwater in spite of intensive dairying: The effect of reducing conditions in the vadose zone-aquifer continuum. *Journal of Hydrology (NZ)* 47 (1): 1–24.

Sweeten, J. 2000. Manure management for cattle feedlots. *Great Plains Cattle Handbook*. Cooperative Extension Service, Great Plains States.

Thomas, R.J. 1992. The role of legumes in the nitrogen cycle of productive and sustainable pastures. *Grass For. Sci.* 47:133–142.

Thompson, R.B., and J.J. Meisinger. 2004. Gasous nitrogen losses and ammonia volatilization measurement following land application of cattle slurry in the Mid-Atlantic region of the USA. *Plant Soil* 266:231–246.

Todd, R.W., Cole, N.A., and R.N. Clark. 2006. Reducing crude protein in beef cattle diet reduces ammonia emissions from artificial feedyard surfaces. *J. Environ. Qual.* 35:404–411.

Tomlinson, A.P., Powers, W.J., van Horn, H.H., Nordstedt, R.A., and C.J. Wilcox. 1996. Dietary protein effects on nitrogen excretion and manure characteristics of lactating cows. *Trans. ASAE* 39:1441–1448.

Trott, H., Wachendorf, M., Ingwesen, B., and F. Taube. 2004. Performance and environmental effects of forage production on sandy soils. I. Impact of defoliation system and nitrogen input on performance and N balance of grassland. *Grass For. Sci.* 59:41–55.

USDA 2017. World cattle inventory by country. 4 pp. Updated by R. Cook. Downloaded Jan 6, 2018. http://beef2live.com/story-world-cattle-inventory-usda-130-106898.

Vallis, I., Harper, L.A., Catchpoole, V.R., and K.L. Weier. 1982. Volatilization of ammonia from urine patches in a subtropical pasture. *Aust. J. Agric. Res.* 33:97–107.

Van Grinsven, H.J.M., Rabl, A., and T.M. de Kok. 2010. Estimation of incidence and social cost of colon cancer due to nitrate in drinking water in the EU: A tentative cost-benefit assessment. *Environ. Health* 9:58.

Van Grinsven, H.J.M, Tiktak, A., and C.W. Rougoor. 2016. Evaluation of the Dutch implementation of the nitrates directive, the water framework directive and the national emission ceilings directive. *NJAS-Wageningen J. Life Sciences* 78:69–84. doi 10.1016/j.njas.2016.03.010.

Varel, V.H. 1996. Effect of urease inhibitors on reducing ammonia emissions in cattle waste. In *Proceedings of the International Conference on Air Pollution from Agricultural Operations*, Ed. A. Heber. Midwest Plan Service, Ames, IA, pp. 459–465.

Varel, V.H. 1997. Use of urease inhibitors to control nitrogen loss from livestock waste. *Bioresour. Technol.* 62:11–17.

Varel, V.H., Nienaber, J.A., and H.C. Freetly. 1999. Conservation of nitrogen in cattle feedlot waste with urease inhibitors. *J. Anim. Sci.* 77:1162–1168.

Vitousek, P.M., Menge, D.N.L., Reed, S.C., and C.C. Cleveland. 2013. Biological nitrogen fixation: Rates, patterns and ecological controls in terrestrial ecosystems. *Phil. Trans R. Soc. B* 368: 20130119. doi 10.1098/rstb.3013.0119.

Wachendorf, M., Buchter, M., Trott, H., and F. Taube. 2004. Performance and environmental effects of forage production on sandy soils. II. Impact of defoliation system and nitrogen input on nitrate leaching leaching losses. *Grass and Forage Sci.* 59:56–68.

Wachendorf, M., and P. Golinski. 2006. Towards sustainable intensive dairy farming in Europe. *Pastos* 36(2):159–174.

Walker, T.W. 1956. The nitrogen cycle in grassland soils. *J. Sci. Fd. Agric.*7(1):66–72.

Wang, H., Magesan, G.N., and N.S. Bolan. 2004. An overview of the environmental effects of land application of farm effluents. *N.Z. J. Agric. Res.* 47(4):389–403.

Ward, B. 2012. The global nitrogen cycle. In *Fundamentals of Geobiology*, Eds. A.N Knoll., D.E. Canfield, and K.O. Konhauser. Blackwell Publishing Ltd, pp. 36–48.

Watkins, N.L., Schipper, L.A., Sparling, G.P., Thorrold, B., and M. Balks. 2013. Multiple small monthly doses of dicyandiamide (DCD) did not reduce denitrification in Waikato dairy pasture. *N.Z. J. Agric. Res.* 56(1):37–48. doi 10.1080/00288233.2012.736389.

Watson, C.J, Akhonzada, N.A., Hamilton, J.T.G., and D.I. Matthews. 2008. Rate and mode of application of the urease inhibitor N-n-butyl thiophsophoric triamide on ammonia volatilization from surface-applied urea. *Soil Use Man.* 24:246–253.

Watson, C.J., and D.K. Kilpatrick. 1991. The effect of urea pellet size and rate of application on ammonia volatilization and soil nitrogen dynamics. *Fert. Res.* 28(2);163–172.

Watson, C.J. and H. Miller. 1996. Short-term effects of urea amended with the urease inhibitor N-(n-butyl) thiophosphoric triamide on perennial ryegrass. *Plant and Soil* 184:33–45.

Watson, C.J., Stevens, R.J., Garrett, M.K., and C. H. McMurray 1990a. Efficiency and future potential for urea in temperate grassland. *Fert. Res.* 26:341–357.

Watson, C.J., Stevens, R.J., and R.J. Langhlin. 1990b. Effectiveness of the urease inhibitor NBPT (N-(n-butyl) thiophosphoric triamide) for improving the efficiency of urea for ryegrass production. *Fert. Res.* 24:11–15.

Watson, C.J, Akhonzada, N.A., Hamilton, J.T.G., and D.I. Matthews. 2008. Rate and mode of application of the urease inhibitor N-n-butyl thiophsophoric triamide on ammonia volatilization from surface-applied urea. *Soil Use Man.* 24:246–253.

Welten, B.G., Ledgard, S.F., and J. Luo. 2014. Administration of dicyandiamide to dairy cows via drinking water reduces nitrogen losses from grazed pastures. *J. Agric. Sci.* 152:S150–S158. doi 10.1017/S0021859613000634.

Wheeler, D.M. 2009. OVERSEER scenarios—'what if' analysis. In *Nutrient Management in a Rapidly Changing World,* Eds. L.D. Currie and C.L. Lindsay. Occasional Report No. 22. Fertilizer and Lime Research Centre, Massey University, Palmerston North, New Zealand, pp. 335–340.

Wilman, D., and P.T. Wright. 1983. Some effects of applied nitrogen on the growth and chemical composition of temperate grasses. *Herbage Abstrs.* 53:387–393.

Wright, I.A. 2005. Future prospects for meat and milk from grass-based systems. http://www.fao.org/ag/AGp/agpc/doc/Newpub/perspectives/wright.html.

Zaman, M., and J.D. Blennerhassett. 2009. Can fine particle application of fertilisers improve N use efficiency in grazed pastures? In *Nutrient Management in a Rapidly Changing World*, Eds. L.D. Currie and C.L. Lindsay, Occasional Report No. 22, Fertilizer and Lime Research Centre, Massey University, Palmerston North, New Zealand, pp 257–264.

Zaman, M., Saggar, S., and A.D. Stafford. 2013a. Mitigation of ammonia losses from urea to a pastoral system: The effect of timing and amount of irrigation. *Proceedings of the New Zealand Grassland Association* 75:209–214.

Zaman, M., Zaman, S., Adhinarayanan, C., Nguyen, M.L., Nawaz, S., and K.M Dawar. 2013b. Effects of urease and nitrification inhibitors on the efficient use of urea for pastoral systems. *Soil Sci. Pl. Nutr.* doi 10.1080/00380768.2013.812490.

# 8 Efficient Nitrogen Management in the Tropics and Subtropics

*Rajendra Prasad and Peter R. Hobbs*

## CONTENTS

## 8.1    INTRODUCTION

Discovery of synthetic ammonia coupled with hybrid corn (Zea mays) led to a phenomenal increase in corn production in the United States and production of most cereals throughout the world (Prasad and Shivay 2015). Erisman et al. (2008) observed that without this discovery about half of the world population would have remained hungry. Similarly, Synder (2010) observed that 40% of the population on Earth owes its existence to increased food production made possible by fertilizer nitrogen (N). In the mid-twentieth century (1954), world cereal production (wheat, rice, maize) was only 446.8 Tg (million metric tons) (FAO 1955) but increased 4.5 times in the next 50 years to 2068 Tg during the triennium 2005–2007 and is further estimated to increase to 3009 Tg in 2050 (Alexandratos and Bruinsma 2012) (Table 8.1). This is necessary to meet the increasing food demands of the increasing human population, which was 6.6 billion during the triennium 2005–2007 and is estimated to reach 9.7 billion by 2050 (UN 2015). To achieve this, per hectare cereal yield needs to increase by about 130% from 3.32 Mg ha$^{-1}$ in 2005–2007 to 4.30 Mg ha$^{-1}$ in 2050. Since arable land increase will be minimal, fertilizer N has to play a key role and is likely to increase in use to 225–250 Tg by 2050 (Tilman et al. 2011) from the 2016 estimated consumption of 116 Tg (FAO 2012). During the triennium 2005–2007, developing countries accounted for 79.4% of the population and this is estimated to increase to 84.2% in 2050.

Conant et al. (2013) observed that from 1961 to 2007, worldwide fertilizer N input has increased faster (+134%) than crop yields (+120%) and among the cereals, wheat (*Triticum aestivum*) yields have increased the most (+143%), followed by maize (+140%) and rice (*Oryza sativa*) (+103%). Based on thousands of on-farm trials in India, Tandon (1989a) reported that increase in yield due to fertilizer N was 59% in wheat, 64% in *kharif* (rainy season) maize, 68% in *kharif* sorghum (*Sorghum bicolor*), 58% in *kharif* pearl millet (*Pennisetum glaucum*), and 27–28% in rice. Heffer (2013) reported that globally in 2010–2011, fertilizer N consumption by cereals was 55.2% (wheat 18.1%, rice 16.8%, maize 15.4%, and other coarse cereals 4.8%) of the total N consumption.

Of the estimated fertilizer N demand of 116 Tg for the year 2016, South, Southeast Asia, and China are estimated to consume nearly two-thirds of the total global demand; as a contrast, the estimate for

---

**TABLE 8.1**

**Population and Food Production in the World for 2005–2007 and Estimates for 2050**

| Item | 2005–2007 | Estimated 2050 |
|---|---|---|
| Population (millions) | 6569 | 9110 |
| Population in Developed countries | 1351 | 1439 |
| Population in Developing countries | 5218 | 7671 |
| Global Arable land (million hectares) | 1592 | 1661 |
| Cereal Production (Tg) | 2068 | 3009 |
| Cereal yield (Mg/ha) | 3.32 | 4.30 |
| Cereals as food (kg/capita/year) | 158 | 160 |
| Cereals all uses (kg/capita/year) | 314 | 330 |

*Source:*  Alexandratos, N. and J. Bruinsma, 2012. World Agriculture Towards 2030/2050. The 2012 Revision. ESA working Paper No. 12–03. International Atomic Energy Commission, Vienna. pp. 3–22. Agriculture Development and Economic Division, FAO, Rome (www.fao.org/economic/esa).

---

Africa is only 4.5% of the global demand (FAO 2012). Ladha et al. (2016) constructed a top-down global N budget for maize, rice, and wheat for a 50-year period (1961–2010). Cereals harvested a total of 1551 Tg of N, of which 48% was supplied through fertilizer-N and 4% came from net soil depletion. An estimated 48% (737 Tg) of crop N, equal to 29, 38, and 25 kg N ha$^{-1}$ yr$^{-1}$ for maize, rice, and wheat, respectively, is contributed by sources other than fertilizer- or soil-N. Non-symbiotic N$_2$-fixation appears to be the major source of this N, which is 370 Tg or 24% of total N in the crop, corresponding to 13, 22, and 13 kg N ha$^{-1}$ yr$^{-1}$ for maize, rice, and wheat, respectively. Manure (217 Tg or 14%) and atmospheric deposition (96 Tg or 6%) are the other sources of N. Crop residues and seed contribute marginally.

In cereals in general, the recovery of applied fertilizer N is below 50%. Of special interest in tropical and sub-tropical regions is the fact that the dominant crop is rice, the most important cropping system is rice-wheat (Prasad 2005a), and the N use efficiency values are fairly low (Cassman et al. 2002; Prasad 2013a). The downstream effects of this low utilization of fertilizer N have led to environmental issues such as eutrophication of water bodies through nutrient leaching and increased production of nitrous oxide (N$_2$O), an important greenhouse gas (GHG) (Haworth et al. 2002). N$_2$O has a global warming potential around 300 times greater than that of carbon dioxide (CO$_2$) over a 100-year period (Forster et al. 2007), as well has having the potential to damage the ozone layer (Cicerone 1987). Furthermore, soils in the tropics and sub-tropics are generally lower in soil organic matter and organic N (Joergensen 2010) as compared to the soils in the temperate regions and to obtain good yields adequate N fertilization is a must. In addition to this tropics and sub-tropics have, in general, small holder farmers with little funds at their disposal resulting in inadequate N application and N deficiencies are rampant in most crops (Figure 8.1). Some of the crops, such as, sugarcane may need 170–400 kg N ha$^{-1}$ (Bahrain and Shomeli 2008; DSD 2013; McCray et al. 2016), while others, such as finger millet (*Eleucine coracana*) or sweet potato (*Ipomoea batatas*) may need

**FIGURE 8.1**  Nitrogen deficiency for various crops grown in the tropics and sub-tropics. **Maize:** Nitrogen deficiency symptoms move up the corn plant starting with the older leaves. Typical yellow "V" in the center of the leaf blade. (Personal photo by Ivan Monasterio, CIMMYT, Mexico). **Wheat:** Nitrogen deficiency symptoms first appear on the oldest leaves as shown, with new leaves initially remaining relatively green. The older leaves become paler, with chlorosis (marked yellowing) beginning at the tip and gradually merging into light green further down the leaf. (Photo from CIMMYT flickr photo archives). **Rice:** N deficient crops are stunted and discolored. Older leaves or whole plants are yellowish green and old leaves and sometimes all leaves become light green and chlorotic at the tip. Entire field may appear yellowish. Reduced tillering and grain number. (Photo from IRRI Flickr photo archives). **Sorghum:** Deficiency first seen on older leaves, irregular necrotic patterns intermingled with red pigmentation. Streaked patterns on the interveinal tissue symptoms at tips and margins move towards the base. (Photo from Vikaspedia, India). **Pearl millet:** Little new growth, yellow leaves, this being more pronounced in older leaves and leaf drop. Plants stunted, spindly pale yellow or deep yellow color near the tips and margins progresses toward the base, heads small, seed numbers reduced. (Photo from Vikaspedia, India). **Sweet potato:** Typically, the oldest leaves turn completely yellow before necrosis spreads. Numerous yellow leaves on a young crop are indicative of N-deficiency. (Photo by Jane O'Sullivan, University of Queensland, Australia.)

as little as 30–50 kg N ha⁻¹ (Goron et al. 2015; Voswai et al. 2015). Even yam (*Dioscorrea spp.*) needs heavy (more than 200 kg Nha⁻¹) N fertilization (Diby et al. 2011). Therefore, there is an urgent need for efficient management of N in the tropics and sub-tropics, where N losses are the most.

## 8.2  NITROGEN USE EFFICIENCY (NUE)

### 8.2.1  DEFINING NITROGEN USE EFFICIENCY

NUE means different things to different people. Some of the most commonly used terms related to NUE are given below (Prasad 2013a).

$$\text{Agronomic efficiency (AEn)} = (Yf - Yc)/N = \Delta Y/\Delta N \text{ (kg grain/kg N applied)}$$

$$\text{Apparent Recovery efficiency (AREn)} = 100(Uf - Uc)/N$$
$$= \Delta U/\Delta N \text{ (\% of applied N)}$$

$$\text{Physiological efficiency (PEn)} = (Yf - Yc)/Uf - Uc$$
$$= \Delta Y/\Delta U$$
$$= AE/RE \text{ (Kg grain/Kg N absorbed by the crop)}$$

$$\text{Partial factor productivity of fertilizer (PFPn)} = Y/N \text{ (Kg grain/Kg N applied)}$$

In the above expressions, Yf and Yc are grain yield in kg/ha in fertilized and control (check) plots, while Uf and Uc refer to the N uptake (kg/ha) in fertilized and control plot, respectively, and N refers to the N applied (kg/ha). PFPn does not consider the control plot yield vis-a-vis the inherent nitrogen supplying capacity of the soils but still is a useful term because it is easy to calculate its value from the experiments on tillage including conservation agriculture experiments, sowing dates, seed rates, weed control, water management, and so on, where a blanket fertilizer application is made. PFPn can also be used for comparing NUE in different countries or in different parts of a country. When tracers such as ¹⁵N are used, the recovery is known as true recovery efficiency (TREn).

### 8.2.2  APPARENT (AREn) AND TRUE RECOVERY (TREn) USE EFFICIENCY OF NITROGEN

One of the major concerns in N management is its low and declining AREn. Many reports provide data on AREn in different crops, but only a few are cited here to focus the attention on this important issue. Average AREn of nitrogen from research trials throughout the world was reported to be 63% in maize (corn), 54% in wheat, and 44% in rice, while the values for on-farm trials were 37% in maize and only 31% in rice (Doberman 2005) (Table 8.2). However, when N application rates were higher as in China (181–219 kg N ha⁻¹), AREn even in research trials was 35.7% for wheat, 30.5% for maize, and only 24.8% for rice. The average AREn value for cereals including the Chinese data works out to 40%.

Nevertheless, for wheat in India, Jat et al. (2014a) reported an AREn of 36.3–79.4%, based on experiments conducted on 22 sites during 1970–1998, showing that if N is properly managed, higher AREn values can be achieved. Furthermore, Yan et al. (2014) pointed out that while due to continuous high N input for the last 30 years (1980–2010) AREn of fertilizer N in China has declined to below 30% in recent years, the residual effect has been continuously increasing with the result that 40–68% of applied fertilizer N is taken up by the crops sooner or later.

TREn values, as determined by ¹⁵N studies, are generally lower than AREn values (Ladha et al. 2005). TREn values in rice, wheat, maize, and sugarcane from some studies are given in Table 8.3 and generally range from 19–39%.

**TABLE 8.2**

**Average Apparent Recovery Efficiency (AREn) of Applied Fertilizer-N in Cereals**

| Crop Region/Country (Number of Observations) | Average N Rate (kg ha⁻¹) | AREn (%) |
|---|---|---|
| Maize, research trials world[a] (36) | 102 | 63 |
| Maize, on-farm United States[b] (55) | 103 | 37 |
| Maize, research trials, China[c] (70) | 219 | 30.5 |
| Rice, research trials world[a] (307) | 113 | 44 |
| Rice, on-farm Asia[d] (179) | 117 | 31 |
| Rice, research trials, China[c] (51) | 187 | 24.8 |
| Wheat, research trials, China[c] (30) | 181 | 35.7 |
| Wheat, research trials, world[a] (507) | 117 | 54 |
| Cereals, research trials, world (1235) | 102–219 | 40 |

[a] Ladha et al., 2005
[b] Cassman et al., 2002
[c] Jin, 2012
[d] Dobermann and Cossman, 2002

**TABLE 8.3**

**TREn in Different Crops in the Tropics and Subtropics**

| Crop | Country | kgN ha⁻¹ | Split/Single | TREn% | Reference |
|---|---|---|---|---|---|
| Rice | India | 180 | Split | 28.9 | Bandyopadhyay & Sarkar (2005) |
| | China | 240–300 | Split | 26–30 | Zhang et al. (2012) |
| | China | 135–270 | Single | 13.6 | Xiao-hui et al. (2006) |
| | | | Split | 28.1 | |
| Wheat | China | 240 | Single | 18.0 | Chen et al. (2016) |
| | | | Split | 31.7 | |
| Maize | China | 300 | Split | 25–31 | Yong et al. (2015) |
| Sugarcane | Australia | 160 | Surface | 18.9 | Prasartek et al. (2002) |
| | | | Sub-soil | 28.8 | |
| | Brazil | 40–120 | Single total 3 ratoons | 21–31 | Otto et al. (2015) |

## 8.2.3 AGRONOMIC EFFICIENCY OF NITROGEN (AEN) (KG GRAIN KG⁻¹N)

Ladha et al. (2005) reported that based on 411 trials all over the world for N applications varying from 112–123 kg N ha⁻¹, AEn (kg grain kg⁻¹ N) was 24 for maize, 22 for rice, and 18 for wheat. Jin et al. (2012) from China reported that for rates of N applications varying from 181–219 kg N ha⁻¹ AEn was only 10.8–12.2 in rice, wheat, and maize. Averaged over a number of experiments at the Indian Agricultural Research Institute, New Delhi, Prasad (2009) reported that AEn in rice decreased from 22 to 13 kg grain kg⁻¹ N and that in wheat from 28.8 to 15.9 kg grain kg⁻¹ N as the level of N was increased from 40–60 to 121–180 kg N ha⁻¹. The data from research centers under the Project Directorate of Cropping System Research, Modipuram showed an average AEn value of 9.5 kg grain kg⁻¹ N for cereals (Sharma and Batra 2011). Besides the low AEn in India more disturbing is its decline over time. Biswas and Sharma (2006) reported that during 1970–2005 the AEn in irrigated areas declined from 13.4 to 3.4 kg grain kg⁻¹ N. This is supported by the fact that the farmers in the rice-wheat cropping system belt of India are forced to apply more and more fertilizer to obtain the same yields as in the preceding years; essentially a drop in nitrogen total factor productivity.

### 8.2.4 PARTIAL FACTOR PRODUCTIVITY OF NITROGEN (PFPN)

Scott Murrel (2011) reported that in the United States, PFPn in maize has increased from 28 kg kg grain kg⁻¹ in 1980 to 42 kg grain kg⁻¹ in 2010. As a contrast, at New Delhi, PFPn values were the same for rice and wheat and decreased from 84 kg grain kg⁻¹ N at 40–60 kg Nha⁻¹ to 32 kg grain kg⁻¹ N at 120–180 kg N ha⁻¹ (Prasad 2009). Amanullah and Almas (2009) from Pakistan also reported that PFPn in maize decreased from 36 to 22 kg grain kg⁻¹ as the level of N was increased from 60 to 180 kg N ha⁻¹.

### 8.2.5 PHYSIOLOGICAL EFFICIENCY OF NITROGEN (KG GRAIN KG⁻¹ N ABSORBED)

Ladha et al. (2005) reported global average PEn values of 37 kg grain kg⁻¹ N for maize, 29 kg grain kg⁻¹ N for wheat, and 53 kg grain kg⁻¹ N for rice. The PEn value was the highest for a tropical cereal rice crop and this explains why the PEn value was 47 kg grain kg⁻¹ N for Asia as compared to 28 kg grain kg⁻¹ N in the Americas as reported by Ladha et al. (2005). At New Delhi PEn values were 44.4 and 40.4 kg grain kg⁻¹ N for rice and 47.8 and 24.0 kg grain kg⁻¹ N for wheat at 40–60 kg N ha⁻¹ and 120–180 kg N ha⁻¹, respectively (Prasad 2009). Of course, these values can change over time due to changes in soil quality and other factors. It is interesting to note that even though physiological efficiency of nitrogen is higher in tropical and sub-tropical regions, the recovery efficiency and agronomic efficiency of fertilizer N is lower in the tropics and sub-tropics due to higher N losses. A brief discussion on mechanisms and amounts of losses of applied fertilizer N follows.

## 8.3 LOSSES OF FERTILIZER NITROGEN APPLIED TO FARM-FIELDS

All conventional nitrogen fertilizers are soluble in water and therefore are subject to various loss mechanisms, including: surface run-off, ammonia volatilization, leaching, denitrification, and dissimilatory nitrate reduction to ammonium (DNRA) (Prasad 1998, 2009). Estimates of N loss through different processes are discussed below. Most of the data discussed are from tropical and sub-tropical regions, however, some data from the United States and other countries are included to make a specific point and for comparison.

### 8.3.1 SURFACE RUNOFF

Chen and Hong (2011) from Wuchuan sub-watershed in southeast China reported a runoff loss of 68 kg N ha⁻¹yr⁻¹ from N application of 690 kg N ha⁻¹yr⁻¹; 75% of which was during the wet season (March–September). Yan et al. (2005) reported a runoff loss of 6 kg N ha⁻¹ from Baiyangdian and 14.8 kg N ha⁻¹ from Taihu lake basin in China. Shan et al. (2015) from a 3-year study (2010–2012) in a cabbage field in Taihu lake basin of China, reported a runoff loss of 10.4–22.7 kg N ha⁻¹ (3.5–7.6% of applied fertilizer N) of which 49.3–71.8% was as nitrate-N. In this study surface runoff N loss was 58–61% lesser with sulfur coated urea, a slow-release N fertilizer. Mandal et al. (2012) reported an annual runoff loss of 10.2 kg N ha⁻¹yr⁻¹ from semi-arid tropical regions in India. Urea being a non-polar fertilizer is not retained by the clay particles and is prone to surface runoff losses and is reported to persist in floodwater in rice growing areas (Daigh et al. 2014). Thus, the source and rate of N application and amount of surface runoff water determined by the intensity and frequency of rainfall are the major factors determining the surface runoff of applied fertilizer N.

Nitrogen, mostly nitrate but may also include urea, lost as runoff in floodwaters and sediments is the main cause of nitrogen pollution of inland lakes in China (Le et al. 2010), Bay of Bengal (Chowdhury, 1989), and Gulf of Mexico (Turner et al. 2008) and other water bodies and has already reached alarming levels from the viewpoint of marine life (Vitousek et al. 2009). Van Drecht and Bouman (2003) using IMAGE (Integrated Model to Assess the Global Environment) reported that in Western Europe and Japan 70–75 kg N ha$^{-1}$ was exported from the point of application by surface run-off; the suggested value for South Asia, East Asia, Canada, and the United States was 30–40 kg Nha$^{-1}$. It may be desirable to fix a monetary value per unit of N exported by a country to seawaters and the environment and fine that country for this damage.

### 8.3.2 AMMONIA VOLATILIZATION

Ammonia volatilization takes place from ammonium-containing or ammonium-producing fertilizers, such as ammonium sulfate, ammonium nitrate, and urea, when these are surface applied on a moist soil. Urea has emerged as the leading N fertilizer in the world, due to several advantages, including the highest analysis (46% N) solid N fertilizer, free flowing characteristics, non-polar nature permitting foliar application alone or in association with pesticides including herbicides. In India about 80% of fertilizer N used is urea. Once urea is applied on a moist soil surface, it is hydrolyzed by the enzyme urease, which is produced by several heterotrophs in the soil (Mobley and Hausinger 1989; Zimmer 2000), to ammonium carbamate and then to ammonium carbonate, which breaks down to produce ammonia and carbon dioxide. Consequently, soil pH in the vicinity of the applied urea rises to 8.5 or higher (Figure 8.2) and ammonia is lost by volatilization. Pathak et al. (2006) from simulation studies found that ammonia volatilization losses from the rice-wheat cropping system in the IGP could range from 16–62 kg N ha$^{-1}$ with an average value of 30 kg N ha$^{-1}$. Li et al. (2015) from China reported an ammonia volatilization loss of 57 kg N ha$^{-1}$ with an application of 312 kg N ha$^{-1}$. Prasertsak et al. (2002) from Queensland, Australia, reported an ammonia volatilization loss of 37.3% from a sugarcane field when $^{15}$N urea was surface applied. From Wisconsin, Oberley and Bundy (1987) reported a loss of 12–16% of surface applied urea N in corn. As a contrast, ammonia volatilization losses are reported much less with ammonium sulfate. For example,

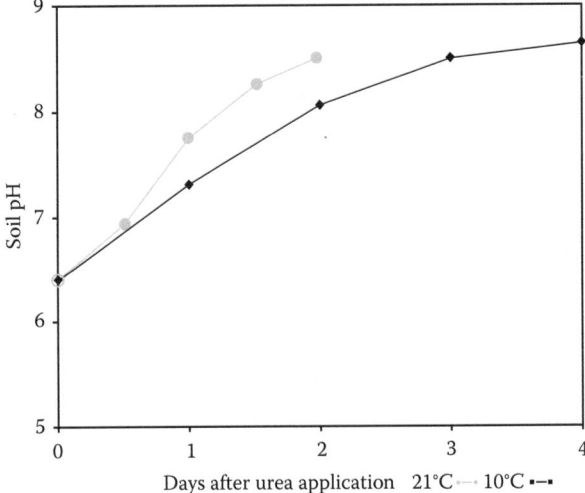

**FIGURE 8.2** Changes in soil pH after application of urea. (LG Bundy, Department of Soils, University of Wisconsin, Madison, with permission from Professor Alfred Hartemink, Chair, Department of Soils.)

using $^{15}$N ammonium sulfate, Wetsellar and Ganry (1982) reported a loss of only 4.3 kg N ha$^{-1}$ from a rice field in Thailand. In China, ammonium bicarbonate is used on a fairly large scale and Gui-Xen et al. (1986) reported that in flooded rice ammonia volatilization loss was 18.2% of applied N with ammonium carbonate as compared to 8.8% with urea.

### 8.3.3 LEACHING

All N fertilizers are subject to leaching; nitrate fertilizers leach faster than ammonium containing or producing N fertilizers (urea). This is one reason why ammonium nitrate consumption in the United States has declined from 15% (of the total N consumption) in 1960 to only 2% in 2011, while urea consumption has increased from 2% in 1960 to 22% in 2011 (USD-ERS 2016).

After ammonium containing fertilizers or urea are applied to a moist soil, they get readily oxidized to nitrate, which is a two-step process involving *Nitrosomonas* and *Nitrococcus* bacteria for conversion of ammonium to nitrite and then involving *Nitrobacter* for the conversion of nitrites to nitrates (Prasad and Power 1997). Nitrification is affected by soil pH, moisture, aeration, and availability of ammonium, however, ammonium concentrations above 800 µg g$^{-1}$ soil could inhibit nitrification (Prasad and Power 1997). Once the nitrates are formed they dissolve in soil water and are subject to leaching with downward movement of water in soil, which is affected by the intensity and frequency of rainfall or irrigation.

Leaching of N in the form of $NO_3$ beyond the root zone has received considerable attention in developed countries because of its linkage to nitrate pollution of drinking water (Prasad and Power 1995; Zhang et al. 1996), which is mostly drawn from groundwater and is responsible for methaemoglobinemia or blue baby syndrome (Knobeloch et al. 2000). Burrow et al. (2010) reported that, in general, nitrate content in U.S. groundwater ranged from 1–8.5 mg $NO_3$-N L$^{-1}$ with a median value of about 2.5 mg $NO_3$-N L$^{-1}$, however, higher values are reported from some zones, such as 30 mg $NO_3$-N L$^{-1}$ from Nebraska (Spalding et al. 2001). The safe limit of nitrate in drinking water as per Environmental Protection Authority (EPA) is 10 mg $NO_3$-N/L.

About tropical and sub-tropical regions, from India, Kundu et al. (2008) reported a value of 0.01–4.56 mg $NO_3$-N L$^{-1}$ in groundwater of Hooghly district in West Bengal. Bharadwaj et al. (2012) reported that 28% of shallow well waters in Ludhiana district contained 11–25 mg $NO_3$-N L$^{-1}$, which is above the EPA safe limit. Vinod et al. (2015) reported that in Karnataka, the average nitrate-N leached from irrigated areas varied from 51.2–74.9 kg N ha$^{-1}$. Using crop simulation modeling, Pathak et al. (2006) calculated that N losses due to leaching from the rice–wheat cropping systems of the Indo-Gangetic plains could account for 10% of applied N. The situation, however, is pretty serious in China, where very heavy nitrogen rates are applied. Zhang et al. (1996) reported that in northern China, half of the 69 samples of drinking water had nitrate levels well above 10 mg $NO_3$-N L$^{-1}$, the USEPA safe limit. Further, in vegetable growing areas, where fertilizer N levels applied ranged from 500 to 1900 kg N ha$^{-1}$, the nitrate level in drinking water was 70 mg $NO_3$-N L$^{-1}$ or more. Recently, Zhou et al. (2016) have reported that based on 7000 samples from 141 sites in semi-humid regions of China during 1994–2015, N content (kg N ha$^{-1}$) in groundwater from 0–4 m depth was 453 ± 38 kg N ha$^{-1}$ in wheat, 749 ± 75 kg N ha$^{-1}$ in maize, 1191 ± 89 kg N ha$^{-1}$ in vegetable fields, 1269 ± 114 kg N ha$^{-1}$ in solar plastic-roof green houses and 2155 ± 330 kg N ha$^{-1}$ in orchards. Kahrl et al. (2010) reported that nitrate levels in groundwater in some parts of north China may be as high as 30 times the safe limit of USEPA. Not only fertilizer nitrogen but large applications of manure can also lead to heavy nitrate leaching. For example, Dahen et al. (2014) from Israel reported nitrate content of 400 mg L$^{-1}$ at 10 m depth in organic farms as compared to 50 mg L$^{-1}$ in conventional farms. Pathak et al. (2004) also pointed out that sporadic reports of high $NO_3$-N contamination in groundwater from different parts of India may be related more to sewage and dumping of organic wastes rather than agricultural use of fertilizer nitrogen.

In addition to heavy application of fertilizer N, the amount of irrigation water applied is also partly responsible for nitrate leaching. For example, in a 5-year (1991–1996) study in Nebraska,

Spalding et al. (2001) found that nitrate content in shallow groundwater (at 16 m depth) remained at 30 mg $NO_3$-N $L^{-1}$ in the conventional surge irrigated corn fields, while it was only about 13 mg $NO_3$-N $L^{-1}$ (until fall 1996) in the center pivot irrigated corn fields.

### 8.3.4 NITROGEN LOSS AS NITROUS OXIDE ($N_2O$)

Agricultural soils are believed to contribute as much as 84% of global $N_2O$ emissions (Smith et al. 2007), primarily through microbial soil processes such as denitrification (Yan et al. 2003; Stehfast and Bouwman 2006) and dissimilatory nitrate reduction to ammonium (DNRA) (Rutting et al. 2011). Denitrification produces $N_2O$ when abiotic conditions prevent its reduction to $N_2$ and DNRA releases $N_2O$ as a by-product of the reduction process (Gile et al. 2012). However, DNRA has a lesser role in the production of $N_2O$. For example, Silver et al. (2005) reported that only 3% of the nitrogen mineralized in a tropical forest system resulted in $N_2O$ production by DNRA. Both bacteria and fungi are capable of $N_2O$ production by denitrification as well as DNRA (Takaya 2002; Philippot 2005). The bacteria involved in denitrification could be heterotrophs, such as *Parasococcum denitrificans* and various *Pseudomonas* (Carlson and Ingraham 1983) as well as by autrotrophs, such as *Thiobacillus denitrificans* (Baalsrud and Baalsrud 1954). Many abiotic factors are known to control both denitrification and DNRA (Rheinbaben 1990; Gile et al. 2012) including: water saturation and oxygen concentration (Bateman and Baggs 2005), nitrate concentration (Smith and Tiedje 1979), amount of carbon present (Burford and Bremner 1975), pH (Van den Heuvel et al. 2011), and temperature (Wolfe and Bruman 2002). It may be mentioned that nitrous oxide ($N_2O$) has about 300 times the global warming power compared to carbon dioxide ($CO_2$) and an atmospheric life span of a little over 100 years. It is estimated that 38% of nitrous oxide in the atmosphere is of anthropogenic origin out of which agriculture may account for about 67–84% and out of the total contribution of agriculture, fertilizer N may contribute up to 25% (6.36% of total anthropogenic contribution) (IPCC 2007; Smith et al. 2007; Smith 2010). In 1990, 12.7 Tg $N_2O$-N $y^{-1}$ was produced and the estimates for 2100 are at 25.7 Tg $N_2O$-N $y^{-1}$ (Kroetze 1994). In addition to its global warming effect, nitrous oxide is also detrimental to the ozone layer (Portman et al. 2012). The amounts of nitrous oxide produced will of course vary from region to region depending upon crops and cropping systems, rates of fertilizer N applied, and the environment.

From India, Aulakh et al. (2001) reported that 22–33% of N applied to rice was lost by denitrification, while Pathak et al. (2004) reported that average N loss by denitrification was 10–15 kg N $ha^{-1}$ in rice and 5–10 kg N $ha^{-1}$ in rice-wheat cropping system of north India. From Brazil, Soares et al. (2015) reported that in sugarcane ratoon, $N_2O$-N loss was 1.93 kg $ha^{-1}$, 0.7% of applied urea-N and the use of NIs dicyandiamide (DCD) and 3, 4-dimethylpyrazole phosphate (DMPP) reduced N loss as $N_2O$-N by 90%. From China, Lin et al. (2013) also reported that application of DCD and DMPP along with urea reduced $N_2O$-N loss and increased crop yield and N uptake in wheat-maize cropping system.

### 8.3.5 AMMONIA VOLATILIZATION LOSS FROM PLANT FOLIAGE

In addition to nitrogen losses by different processes in soils, a fairly large amount of N is also lost from the plant foliage. Generally, ammonia concentration in the atmosphere is much lower than in the sub-stomatal cavities of leaves and emission of ammonia from the leaves takes place; mostly post-anthesis. The reverse process is also possible when ammonia concentration in the atmosphere is higher than in the plant leaves (Denmead et al. 1976) and it may partly compensate for the ammonia loss from the plant foliage (Sharpe and Harper 1997). Many biochemical processes are responsible for ammonia emission from the plant foliage. These include: (1) photo-respiratory nitrogen cycling in which ammonia is released during the decarboxylation of glycine in mitochondria (Keys et al. 1978); (2) nitrate reduction (Lea and Ireland 1999); (3) lignin biosynthesis (Nakashima et al. 1997); and (4) protein degradation and amino acid deamination in the cytosol (Olea et al. 2004). Reported post-anthesis losses of N range from 5.9–80 kg N ha$^{-1}$ (Daigger et al. 1976; Harper et al. 1987; Papkosta and Gagianas 1991) in wheat, from 45–81 kg N ha$^{-1}$ in irrigated corn (Francis et al. 1993), and 7.3–59.8 kg N ha$^{-1}$ in rice (Hayashi et al. 2008; He et al. 2014). Ammonia volatilization losses from plant foliage may reach up to 5% of the total N content in a shoot (Schjoerring et al. 2000). Nitrogen losses from plant foliage are higher when large amounts of fertilizer N are top dressed.

## 8.4  FERTILIZER NITROGEN MANAGEMENT

As already explained, NUE is really low in rice, which is the main crop of South Asia, Southeast Asia, and China, the region that uses about two-thirds of the global nitrogen consumption. Rice is the most important cereal in Asia and provides 35–80% of calorie intake (IRRI 1997). Nitrogen management is a serious problem in rice paddies because of the various mechanisms of N loss just discussed. The 4Rs of nutrient stewardship "right source, right rate, right method, and right time of application" as advocated by IPNI (Fixen 2007) as a global framework of FBMP (fertilizer best management practices) apply to nitrogen management as well.

### 8.4.1  RATE OF NITROGEN APPLICATION

For most plant nutrients, soil testing has long been considered as the sound basis for deciding the rate of application. Many soil tests for determining potentially available N have been proposed from time to time. These include estimation of N mineralized on incubating a moist soil sample for 2–16 weeks (Stanford and Smith 1972), alkaline permanganate oxidizable N (Subbiah and Asija 1956), calcium hydroxide hydrolysable N (Prasad 1965), 6 N hydrolysable N (Gallagher and Bartholomew 1964), and the estimation of amino sugars released on hydrolysis with 0.2 M Na OH (Kahn et al. 2001). In view of the use of large applications of fertilizer N to crops, soil tests for nitrate N in the soil profile were also suggested, such as PPNT (pre-plant nitrate test) (Fox and Piekielek 1983) and PSNT (pre-side-dress nitrate test) (Fox et al. 1992). However, getting information on the exact amount of N to be applied to a crop based on soil tests remains a problem (Prasad 2013b).

Latest efforts to recommend fertilizer N application to crops involve the use of chlorophyll or a SPAD meter or leaf color chart (LCC) that will measure the intensity of chlorophyll in plant leaves and is referred to as "Need Based Nitrogen Management" (NBNM). Gopal et al. (2010) found that N applications in rice should be made at a critical LCC value of 4.0 for high-yielding varieties and hybrids, while the critical LCC value for Basmati types was 3.0. NBNM results not only in saving of fertilizer N, but also in higher grain yield, and higher AEn and PFPn (Bijay-Singh et al. 2002; Sen et al. 2011; Shukla et al. 2003; Varinderpal-Singh et al. 2010; Peng et al. 2010). Some data on corn from on-farm demonstrations in 6 provinces of China (Peng et al. 2010) and in rice from Ludhiana district in India are given in Table 8.4. Bijay-Singh et al. (2002) reported that NBNM resulted in 12.5–25% saving in nitrogen application in rice.

In general, nitrogen rates are recommended based on response curves (Heady and Shrader 1953). There are two optima on a response curve of a crop to nitrogen application. These are: yield optimum (Yield$_{opt}$) and economic optimum (Econ$_{opt}$). Yield$_{opt}$ refers to the nitrogen rate at which maximum yield is attained, while Econ$_{opt}$ refers to the rate at which economic returns are maximized and is the point on the response curve where the cost of the yield increase equals the cost of the unit increment in nitrogen. Most fertilizer recommendations are based on Econ$_{opt}$ rate. However, even Econ$_{opt}$ rates of nitrogen in cereals are fairly high and lead to large nitrogen losses to the environment and it has been suggested that there is a need for an ecological optimum (Ecol$_{opt}$) rate of nitrogen application to cereals (Ju et al. 2009; Chen et al. 2011; Gao et al. 2012; Grassini and Cassman 2012; Xia and Yan 2011; Prasad et al. 2016c). In a study in 12 counties of Henan province in China, 91 on-farm trials were conducted on response of maize to nitrogen (Wang et al. 2014). Based on the results from these trials, Yield$_{opt}$, Econ$_{opt}$, and Ecol$_{opt}$ rates were found to be 289, 237, and 171 kg N ha$^{-1}$, respectively. Thus, the Ecol$_{opt}$ rate was 66 kg N/ha lower than the economic optimum rate and

**TABLE 8.4**

**Grain Yield, N Rate, Agronomic Efficiency of N (AEn) and Partial Factor Productivity of N (PFPn) as Influenced by SPAD Reading/LCC Recommendation (NBNM) in Corn in China and in Rice in India**

| Crop | Treatment | Grain Yield (Mg ha$^{-1}$) | N Rate (kg ha$^{-1}$) | AEn (kg grain kg$^{-1}$ N) | PFPn (kg grain kg$^{-1}$ N) |
|------|-----------|---------------------------|----------------------|---------------------------|-----------------------------|
| Corn | Farmers practice | 7.08b | 195a | 7.1b | 36.3b |
|      | NBNM | 7.46a | 133b | 13.4a | 56.2a |
| Rice | Farmers Practice | 6.87 ± 1.13 | 120 | 11.3 | 57.2 |
|      | NBNM | 7.05 ± 0.96 | 100 | 19.2 | 70.5 |

*Source:* Peng et al. (2010) for corn (average of 107 on-farm trials in 6 provinces in China) and Bijay-Singh et al. (2002) for rice (average of 25 trials in Ludhiana district, India); values followed by the same letter do not differ significantly.

this reduced nitrous oxide emission, ammonia volatilization, and nitrate leaching losses by 47%, 65%, and 38%, respectively. The ecological optimum rates of nitrogen reduced maize yield by 0.3 t ha$^{-1}$ only and maintained a high level of return on investment for fertilizer nitrogen. Results from many field trials suggest that N applications to grain crops in China can be reduced by 20–30% (Kaherl et al. 2010).

From a global crop-wise analysis for the three major cereals (maize, wheat, rice), Mueller et al. (2014) reported that maize received 16.6 Tg of N (12 Tg fertilizer N, 4 Tg manure N, and 0.6 Tg atmospheric N deposit) of which 8.6 Tg (51%) was lost to the environment. Similarly, wheat received 20.1 Tg N (14.7 Tg fertilizer N, 4.6 Tg manure N, and 0.8 Tg atmospheric N deposit) of which 10.4 Tg N (52%) was lost. Again, rice received 18.8 Tg N (12.8 Tg fertilizer N, 5.4 Tg manure N, and 0.6 Tg atmospheric N deposit) of which 12 Tg N (64%) was lost; nitrogen losses were most from the rice fields. They concluded that if fertilizer nitrogen is properly managed, the year 2000 level of cereal production can be achieved with 50% less fertilizer nitrogen and about 60% less nitrogen losses.

### 8.4.2 Method of Nitrogen Application

Most of the famers in tropical and subtropical regions especially in Asia are smallholder farmers and fertilizer N at seeding and in standing crop is generally broadcast by hand. This is the most inefficient way of applying fertilizer N. Method of fertilizer N application is important for its higher efficiency and is briefly discussed.

#### 8.4.2.1 Deep Placement

Due to their nature of rapid reaction with calcium, iron, aluminum, and other cations, deep placement of phosphate fertilizers near the likely root system has received considerable attention (Prasad et al. 2016d). Deep placement of urea and ammonium containing N fertilizers is equally important from the viewpoint of reducing ammonia volatilization losses (Prasad and Shivay 2016a). Using $^{15}$N urea, Prasertsak et al. (2002) showed that in a sugarcane ratoon crop, drilling nitrogen in soil reduced ammonia volatilization by 5.5% as compared to surface application, where it was 37.3% (Table 8.5). This also resulted in reducing total N losses from 59.1% in surface applied urea to 45.6% when urea was drilled in soil. Drilling of urea also increased TREn to 28.8% as compared to 18.9% obtained with surface application. Similar results were obtained by Wetsellar and Ganry (1982) for rice. Devasenapathy and Palaniappan (1995) from India reported that band placement of urea solution resulted in higher rice yield and recovery efficiency than the conventional surface split application of urea.

**TABLE 8.5**

**Effect of Method of Nitrogen Application on Ammonia Volatilization Loss, Total N Loss and TREn Using $^{15}$N**

| Crop | Method of Application | Ammonia Volatilization (%) | Total N Loss (%) | TREn (%) |
|---|---|---|---|---|
| Sugarcane | Surface application | 37.3 | 59.1 | 18.9 |
| | Drilled in soil | 35.5 | 45.6 | 28.8 |
| Rice | Surface application | 4.3 | 59.0 | 19.1 |
| | Deep placed as mudballs | 0.7 | 45.3 | 30.2 |

*Sources:* Prasetak et al. (2002) for sugarcane, and Wetsellar and Ganry (1982) for rice.

Deep placement of N was done by smallholder farmers in some South Asian countries by making mudballs of a mixture of N fertilizer with soil or by making a hole in the mudballs and placing the fertilizer N in it and sealing the hole. The mudballs were dried in the shade and then placed at a desired spacing in the rice fields, generally after transplanting. Ventura and Yoshida (1978) from the International Rice Research Institute, Los Banos, Phillipines using $^{15}$N ammonium sulfate, reported that a 60 kg N ha$^{-1}$ basal application as mudballs produced 11% higher grain yield of rice, 12.6% higher N uptake, and 10% higher TREn as compared to basal incorporation of fertilizer N. This was possibly due to the reduction of ammonia volatilization losses as indicated by data from Wetsellar and Ganry (1982) (Table 8.5).

Taking a lead from mudballs, the International Fertilizer Development Center (IFDC) at Muscle Shoals, Alabama, developed urea supergranules (USG) of 1.8 g each or heavier for deep placement in rice. Superiority of USG over prilled urea (PU) in rice has been reported from several countries, such as India (Prasad and Prasad 1980; Prasad et al. 1984), China (Rong-ye and Zhao-Liang 1982), Thailand (Snitwongse et al. 1988), Bangladesh (Kabir et al. 2009), and Burkino Faso (Alimata et al. 2016; Bandaogo et al. 2015). In India, based on trials at several research stations, Tandon (1989b) concluded that 26–59% of fertilizer N can be saved by using USG as compared to urea for different targeted yields (Table 8.6). Savant and Stangel (1990) reviewing the data from many experiments on rice observed that USG can help save on average 33% of fertilizer N and can increase grain yield of rice on average by 15–20% as compared to urea. USG has been widely accepted by rice farmers in Bangladesh, where it is considered a key factor in agricultural productivity (Sikdar and Jian 2014).

Deep placement of a basal dose of fertilizer N can be easily done by combining with the application of P and K fertilizers at seeding. About top-dressing or side-dressing of N, a light irrigation

**TABLE 8.6**

**Nitrogen Required as USG (kg N/ha) to Produce Same Yield of Rice as Prilled Urea Taken as 100 at Different Targeted Yields at Different Research Centers in India**

| Targeted Yield (Mg ha$^{-1}$) | Bhubaneswar (Odisha) | Coimbatore (Tamil Nadu) | Cuttack (Odisha) | Hyderabad (Andhra Pradesh) | Pantnagar (Uttarakhand) |
|---|---|---|---|---|---|
| 4 | 60 | 39 | 41 | 52 | 55 |
| 5 | – | 46 | 74 | 49 | 64 |
| 6 | – | – | – | 54 | 63 |

*Source:* Tandon. 1989. In *Fertility and Fertilizer Use. Vol. III.* Eds. V. Kumar, G.C. Shrotriya, and S.V. Kaore, pp. 10–22, New Delhi: Indian Farmers' Cooperative (IFFCO).

soon after application of fertilizer can considerably reduce ammonia volatilization (Oberley and Bundy 1987). Katyal et al. (1987) from Ludhiana, India, recommended that in wheat grown on light sandy soils, top dressing of nitrogen should be done before irrigation.

### 8.4.2.2 Foliar Application

Plants can also absorb nitrogen as urea, when applied to foliage or even roots (Gooding and Davies 1992). However, in tropical and sub-tropical soils, urea gets hydrolyzed rapidly (Reddy and Prasad 1975) and is not available for absorption. Kuykendall and Wallace (1953) reported that foliar application of urea on citrus leaves increased their N content from 2% to 6% in 24 hours; however, most of the N in the leaf at that time was not present as urea, indicating its rapid hydrolysis in the leaf. Smith et al. (1991) using $^{15}$N labelled urea found that when applied to foliage at full emergence of the inflorescence stage in the main stem of wheat, 69% of applied N was in the above-ground portion of the plant. Thus, absorption of foliar applied urea is fairly rapid and N so applied avoids its loss through various N loss mechanisms associated with soil applied fertilizer N. Efficiency of foliar applied urea is reported to be higher than its soil application (De 1971; Balasubramanian et al. 2004). Bhuyan et al. (2012) from Bangladesh reported that AEn of foliar applied urea in rice was 93.8 kg grain kg$^{-1}$ N as compared to 43.7 kg grain kg$^{-1}$ N for soil applied urea. Saleem et al. (2013) from Pakistan reported that soil application of 60 kg N/ha plus two foliar sprays of 2% urea solution at tillering and booting in wheat was as good as soil application of 120 kg N/ha as urea. Foliar application of urea is the best way to apply nitrogen to a crop, when its deficiency is detected at later stages of crop growth, since soil applied N takes more time for its absorption and translocation to the targeted plant part. Foliar application of urea is the only way to supply N to flooded rice fields. The only drawback is that using large volume knap-sack sprayers available with smallholder farmers only about 10 kg N/ha can be applied in a single spray; 3% urea is the maximum concentration possible in such foliar sprays (Prasad and Shivay 2016). Foliar application of N can be combined with the foliar application of Zn, Fe, and other micronutrients to reduce application cost. Again for iron, only foliar application is recommended. Urea is compatible with most insecticides and herbicides and can be applied mixed with them.

### 8.4.3 Time of Nitrogen Application

Applying all N at sowing of a crop including direct seeded rice is recommended only for dryland farming areas, where the crop has to grow only on stored soil moisture. In irrigated agriculture in India, nitrogen is generally applied in two or more split applications, the fractions of total N dose at different applications differ with crops and soil (Tandon 1989a). In rice in India, three split applications of N are recommended: 33–50% at transplanting, 25–33% at active tillering, and 25–33% at panicle initiation (Kaushal et al. 2010). In China, in general, 70% of the N dose is applied to rice at transplanting and 30% one week after transplanting (Shihua and Wenqiang 2000); this obviously leads to high N losses and low NUE. As a contrast, in a joint China-Japan study (Katsura et al. 2008), application of 140 kg N ha$^{-1}$ in four split doses (22% one week after transplanting, 28% at panicle initiation, 28% before heading, and 22% at heading) recorded an ARn of 47% at Yunnan in China and 37% at Kyoto in Japan.

In wheat, in India, the general recommendation is application of N in two split applications, 50% at sowing and the rest at crown root initiation (21–25 days after sowing), when the first irrigation is applied. However, in a study at New Delhi (Abdin et al. 1996), application of 40 kg N ha$^{-1}$ at anthesis increased wheat grain yield of some genotypes of wheat by 38.5%. Application of some N at first node formation was also recommended by Hobbs et al. (1998). In view of these findings, Singh et al. (2016) from India and Attaur Rahman et al. (2011) from Bangladesh have recommended application of N to wheat in three split doses (1/3 at sowing + 1/3 at crown root initiation + 1/3 at anthesis).

In maize, application of N is recommended in three split doses: 1/3 dose each at sowing, 8–10 leaf stage and tasseling in Iran (Nemati and Sharifi 2012) or 50% 2 weeks after sowing (WAS),

25% at 4 WAS, and 50% at 8 WAS in Nigeria (Kwaga 2014). In summer maize, in Hebei province in northern China (Zhao and He 2012) split application of nitrogen (half at planting and half at the 10-leaf stage) of 180 kg N ha$^{-1}$ gave the highest grain yield, which was significantly superior to a single application of 240 kg N ha$^{-1}$ at planting (Table 8.7). Data in Table 8.7 also show that ARn and AEn significantly improved with split application as compared to a single application both at 120 and 180 kg Nha$^{-1}$. The advantage of two or three split applications over a single application was also reported for maize in semi-arid regions of China (Wang et al. 2016).

Probably the best way to decide about timings of nitrogen applications in a standing crop is need based nitrogen management (NBNM) using a leaf color chart or SPAD meter (Yadvinder-Singh et al. 2004; Singh et al. 2008; Varinderpal-Singh 2010). Some data on rice from India are shown in Table 8.8. Six equal applications of 20 kg N ha$^{-1}$ at LCC ≤5 gave significantly higher grain yield and AEn than application of the same amount of N at the recommended three split applications

## TABLE 8.7
## Effect of Level and Timing of Application of Nitrogen to Summer Maize at Xinji (Hebei Province, China) on Grain Yield, AREn, and AEn

| Rate of N (kg N ha$^{-1}$) | Timing of Application (Planting + 10-Leaf Stage) | Grain Yield (Mg ha$^{-1}$) | AREn (%) | AEn (kg grain kg N$^{-1}$) |
|---|---|---|---|---|
| 0 | 0 + 0 | 6.16c | – | – |
| 120 | 0 + 100% | 6.30b | 17.5e | 1.2d |
| 120 | 25% + 75% | 6.91b | 41.9a | 6.2ab |
| 120 | 50% + 50% | 6.77b | 36.3b | 5.1b |
| 180 | 0 + 100% | 6.68bc | 19.0e | 2.9c |
| 180 | 25% + 75% | 7.23ab | 26.2c | 5.9b |
| 180 | 50% + 50% | 7.43a | 28.3c | 7.7a |
| 240 | 0 + 100 | 6.9b | 21.8d | 3.1c |

*Source:* Zhao and He. 2012. *Better Crops* 96(2):8–10.

*Note:* Means followed by the same letter do no differ significantly.

## TABLE 8.8
## Effect of Real Time Nitrogen Management Using Leaf Color Chart in Rice on Grain Yield and AEn

| Treatment | kg N ha$^{-1}$ | Applications | Grain (Mg ha$^{-1}$) | AEn (kg grain kg N$^{-1}$) |
|---|---|---|---|---|
| Check (no N) | 0 | 0 | 4.1 | – |
| General recommendation[a] | 120 | 3 | 7.1 | 20.5 |
| LCC≤ (20 kg N ha$^{-1}$ each application) | 80 | 4 | 6.8 | 33.8 |
| LCC≤ (30 kg N ha$^{-1}$ each application | 90 | 3 | 7.0 | 31.2 |
| LCC≤ (20 kg N ha$^{-1}$ each application) | 100 | 5 | 7.4 | 32.6 |
| LCC≤ (30 kg N ha$^{-1}$ each application) | 120 | 4 | 7.6 | 28.8 |
| LCC≤ (20 kg N ha$^{-1}$ each application) | 120 | 6 | 7.9 | 31.7 |
| LCC≤ (30 kg N ha$^{-1}$ each application) | 150 | 5 | 8.2 | 27.1 |
| LSD (0.05) | – | – | 0.34 | 2.4 |

*Source:* Bhat et al. 2015. *Indian Journal of Agronomy* 60(1):70–75.

[a] General recommendation is 120 kg N ha$^{-1}$, 40 kg N ha$^{-1}$ each at transplanting, early tillering, and panicle initiation.

(at transplanting, early tillering, and panicle initiation); the AEn increased from 20.5 kg grain kg$^{-1}$ N with the recommended three applications to 31.7 kg grain kg$^{-1}$ N with six applications as per LCC values. For rice–wheat, cotton–wheat, pearl millet–wheat, and cluster bean–wheat rotations in Haryana, India, Covetry et al. (2011) recommended application of N to wheat in three split applications determined by LCC values and this resulted in a savings of 21–25 kg N ha$^{-1}$ as compared to general recommendations. Thus, application of fertilizer N to most crops in two or more split doses is a proven technology for increasing NUE and needs to be encouraged. On the other hand, application of all or a major share of N applied to a crop at sowing/transplanting should be discouraged.

### 8.4.4 Sources of Nitrogen

In the tropical and subtropical countries of Asia, only solid fertilizers are used and urea is the number one N fertilizer; in India it supplies about 80% of N applied to crops. In China, ammonium bicarbonate is also used in large amounts, although its contribution is decreasing in favor of urea. Thus, for all practical purposes, urea is the principal N fertilizer in the tropics and subtropics. However, due to its solubility and associated leaching, ammonia volatilization, and other losses, there is an increasing demand for urea based enhanced efficiency fertilizers (EEF). The American Association of Plant Food Control Officials has described EEFs as "fertilizer products with characteristics that minimize the potential of nutrient losses to the environment, as compared to 'reference soluble' products." This description says nothing about agronomic effectiveness, but it can be generally assumed that in most cases EEFs have higher agronomic efficiency. The earlier term used for many of these materials was "slow-release" fertilizers, and nitrification or urease inhibitor stabilized materials.

Research on EEFs has been done throughout the world for the last 60 years or so, yet these have not made a big dent in the fertilizer market, because of their high cost. Presently EEFs are evaluated only based on grain or biological yield or nutrient uptake. However, the EEFs also reduce the pollution of the atmosphere and groundwater and this needs to be monetarily accounted for. In the United States, during the last 10 years, experimentation has started on nitrous oxide reducing capability of EEFs. Halvorson et al. (2014) from Colorado reported that polymer coated urea (PCU) reduced nitrous oxide emission by 42% compared to urea under no-till and strip-till conditions.

The environmental impacts of agricultural practices are the costs that are typically not measured (Tilman et al. 2002) and it is high time that environmental pollution savings from EEFs are given a proper evaluation and some monetary values are allotted for the reduction of environmental N (Prasad et al. 2016c). Once monetary estimates of environmental protection provided become available, the EEF producers or users as per decisions of different national governments can file their claim for refund to concerned authorities.

Commercially available EEN fertilizers generally fall into one of the following three categories: (1) Physical coating or barrier around soluble N fertilizer, such as sulfur coated urea (SCU), polymer coated urea (PCU), and neem coated urea (NCU), generally known as "coated fertilizers"; (2) synthetic organic compounds containing N having low inherent solubility, for example, urea-aldehyde condensation products [urea-formaldehyde (UF) and isobutylidene diurea (IBDU)], triazines, and so on, and (3) N fertilizers stabilized with nitrification and urease inhibitors. EEFs have received considerable attention from time to time and have been reviewed (Prasad et al. 1971; Trenkel 1997; Shaviv 2001; Lindquist et al. 2013; Timilsema et al. 2015). EEFs are more expensive than urea (Shaviv and Mikkelsen 1993), and on N basis may cost two to four times of urea or even more.

#### 8.4.4.1 Coated N Fertilizers

*Sulfur coated urea (SCU)*: SCU was the first coated EEF developed by the Tennessee Valley Authority (TVA) researchers around 1961 for controlled release of nitrogen and is a popular turf fertilizer in the United States (Prasad et al. 1971; Shaviv 2001). It is currently manufactured by many companies in the United States, Canada, Japan, and China using different techniques and sulfur content may vary from 4 to 30% (Trenkel 1997). SCU has also been widely tested for rice, where nitrogen use efficiency is fairly low (Prasad 2013a). In an international study conducted by the International Rice Research Institute, Philippines in Asia under the International Soil Fertility and Fertilizer Evaluation for Rice Program (INSFER), 22–25% less N was required as SCU to produce the same rice yield as urea (Flinn et al. 1984). From Iran, Malkouti et al. (2008) reported that in field experiments at 14 locations, SCU recorded significantly higher grain and protein yield in wheat than urea. In the rice-wheat cropping system at New Delhi, SCU recorded 15.6% increase in grain yield over urea (Prasad et al. 1981). From China, Xian-Ju et al. (2016) reported that use of a mixture of 20% urea and 80% SCU or PCU (polymer coated urea) reduced N application in rice-wheat cropping system by 24.3%. They further observed that the benefits of SCU or PCU were more in wheat than in rice.

SCU has been mostly evaluated as an EEF for nitrogen and little attention was paid into its contribution to S nutrition of crops, which is now recognized as the fourth important essential plant nutrient (Dick et al. 2008; Tandon 2011) and responses of crops to sulfur fertilization are reported from several parts of the world including the United States (Mitchell and Mullins 1990), Canada (Malhi et al. 2012), Europe (Messick 2003; Gallejones et al. 2012), China (Messick et al. 2003), India (Biswas et al. 2004), and Pakistan (Rashid et al. 1992). A recent field study for two years conducted on wheat at New Delhi showed that urea coated with 4–5% S significantly increased wheat grain yield to the tune of 9.6–11.2%, nitrogen uptake by 19.1–23.9% and sulfur uptake by 21.8–29.3% over uncoated urea (Shivay et al. 2016). However, high sulfur requirement for manufacturing SCU is a big barrier for countries like India, which has no S deposits and must presently import all the S it needs for the manufacture of phosphate fertilizers.

*Polymer coated urea* (PCU): Coating materials in PCU could be a polymer, polyurethane, polyethylene, or an alkyl resin and the amount of coating material could be 3–4% in general, but may go up to 15% in some PCUs. PCU has an advantage over other coated fertilizers, such as SCU, NCU, and so on in that its performance is not affected by soil properties, such as texture, pH, salinity, redox potential, microbial activity, and etc. PCU has a slower dissolution rate than urea (Salman, 1988), which reduces leaching losses of N (Blaise and Prasad 1997). Blaise and Prasad (1995) also reported reduced ammonia volatilization loss with PCU as compared to urea. From India, Kondo et al. (2005) reported a 12.9% increase in rice yield with PCU over four split applications of urea; ARn was 32% with PCU and 20% with urea. Similarly, Patil et al. (2010) reported that in rice fertilizer dose of nitrogen could be reduced to 50% with 50–70% mixtures of PCU and urea. In a field experiment at New Delhi, although PCU did not give a significant increase in rice yield, it recorded higher agronomic efficiency and apparent recovery of applied N (Blaise and Prasad 1996). From a greenhouse study in Japan, Fashola et al. (2002) reported an ARn of 45% for PCU as compared to 32% with urea. Wang et al. (2015) from a study in southeast China reported that an increase in rice yield with a single basal application was obtained with 12% coated PCU but not with 6 or 8% coated PCU over conventional three split applications of nitrogen, however, all PCUs (with 6% or 8% or 12% coating) reduced N losses and improved NUE. Carreras et al. (2003) reported that PCU (43%N) was better than IBDU and DCD for flooded rice. Advantages of PCU over urea in rice has also been reported from Japan (Ando et al. 2000; Saigusa 2005). Rose (2016) from Australia reported that PCU delayed growth and accumulation of key nutrients in aerobic rice but did not affect grain yield and mineral concentration. Golden et al. (2009) from the United States, however, failed to obtain any advantage with PCU over uncoated urea.

*Neem Coated Urea*: Neem coated urea (NCU), developed at the Indian Agricultural Research Institute, New Delhi is primarily based on the nitrification inhibiting properties of neem (*Azaduracta indica* Juss) (Reddy and Prasad 1975; Thomas and Prasad 1983). Initially neem cake left after the extraction of oil from neem seeds was used for coating due to the energy crisis that hit the world during the early 1970s, when this research was initiated, and the product was referred to as neem cake coated urea (NCCU). Reddy and Prasad (1975) reported that after 3 weeks of incubation, percent inhibition of nitrification by N-Serve (Nitrapyrin), ST (sulfathiazole), and NCCU was 90,

49, and 5, respectively; further, neem cake coating of urea inhibited nitrification only for 2 weeks. Joseph and Prasad (1993) compared the inhibition of nitrification by DCD (dicyandiamide) and NCCU under wheat field conditions. Percent inhibition of nitrification was 48–53% at 10% DCD (10% of N as DCD) and 69–75 at 20% DCD at 30 days after application of 120 kg N/ha. The values of nitrification for 10% NCCU (neem cake at 10% of urea) and 20% NCCU were 7–16 and 33–42%. Thus, neem cake is a weak nitrification inhibitor as compared to standard nitrification inhibitors, such as N-Serve, DCD, and ST. Nevertheless, a number of researchers (Chhonkar and Misra, 1978; Sahrawat and Parmar 1975; Tiwari 1989; Subbiah and Kothandaraman 1980; Gnanavelarajah and Kumaragamage 1998) have confirmed nitrification inhibition of fertilizer N by neem seed extract/cake/oil coating of urea; of course, the results vary with soil type and experimental conditions. Santhi et al. (1986) reported that acetone extracts of neem cake and fresh and dried neem leaves controlled the populations of *Nitrosomonas* and *Nitrobacter* sp in soil and Suresh et al. (1995) reported that at all stages of rice growth, neem cake coated urea (NCCU) maintained significantly lower population of *Nitrosomonas* sp and higher ammonium-N concentration in a rice field having sandy clay soil of pH 8.1. However, since 15–20% neem cake was needed to make NCCU, the Indian fertilizer manufacturers did not accept this technology. Since nitrification inhibiting chemicals in neem are lipid associated, a "neem oil-urea-water emulsion" was developed for coating urea (Prasad et al. 1994; Suri et al. 2004). Only 0.5 kg neem oil per ton of urea is needed in this technique. Neem coated urea thus made was found superior to urea in field trials all over India (Prakasa Rao et al. 2016; Thind et al. 2010). Increase in rice yield with NCU over urea may vary from 6.1 to 11.9% (Figure 8.3). Neem coated urea is also reported to reduce methane emission in rice. Naik et al. (2015) reported that during the entire rice growing season methane emission was 41.2 kg ha$^{-1}$ with urea and 37.9 kg ha$^{-1}$ with NCU and the difference was statistically significant. This further shows the ecofriendly quality of NCU. The technique of coating urea with neem oil emulsion developed at the Indian Agricultural Research Institute (Prasad et al. 1994; Suri et al. 2004) or its modifications are currently used in India by most fertilizer manufacturers and the product is now referred to as neem coated urea (NCU) and is commercially available in India. The cost of coating urea with neem oil emulsion is only US$ 2–3 t$^{-1}$ urea (Prasad 2013a). All urea produced in India now has to be neem coated (Kumar 2015). The government of India is the first national government for subsidizing an EEF, a first step for increasing NUE and for reducing environmental pollution due to fertilizer nitrogen.

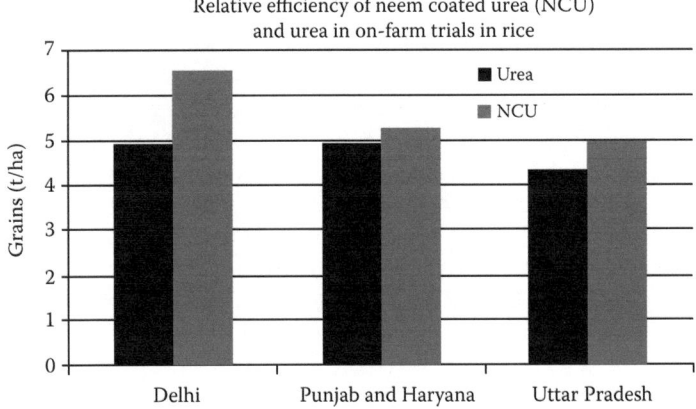

**FIGURE 8.3** Relative efficiency of neem coated urea (NCU) and urea in on-farm trials at 120 kg N/ha (t/ha is Mgha$^{-1}$). (From Prasad, R. 2013a. *Proceeding Indian National Science Academy* 79(4):997–1010.)

In addition to N regulation, neem is globally known for its fungicidal, bactericidal, nematicidal, and insecticidal properties (Bhattacharya and Goswami 1988; Gowda and Setty 1978; Rossner and Zebitz 1987; Moline and Locke 1993; Achimu and Schlosser 1992; Amadioha 2000; Saxena and Khan 1985) and NCU will impart some of these benefits to crops to which it is applied. Although scientific data have not been collected, the farmers using NCU have reported lesser incidence of termites in wheat and nematodes in eggplant (Prasad et al. 2007).

### 8.4.4.2   Inherently Low Solubility N Fertilizers

These fertilizers include oxamide, guanylurea, glycoluril, and urea-aldehyde condensates (Prasad et al. 1971; Alexander and Helme 1990; Trenkel 1997; Shaviv 2001). Of these urea-aldehyde condensates, such as, ureaform (urea-formaldehyde), isobutylidene diurea (IBDU) or urea-butyraldehyde, and crotolydine diurea (CDU) or urea-crotonaldehyde have received considerable attention and have been tested for rice and other crops. Rajale and Prasad (1970) studied the release of N from many EEFs both under field capacity moisture and submergence. After 20 days of incubation mineral-N (ammonium + nitrate) was 67, 43, 31, and 27 mg/kg soil with urea, oxamide, IBDU, and SCU, respectively; of the EEFs oxamide released N the fastest and SCU the slowest. They opined that mineralization of urea-N in SCU is slow due to formation of a sulfide coating on urea under reduced conditions. From a field study, Rajale and Prasad (1974) at New Delhi reported that IBDU gave higher rice yield than urea stabilized with nitrapyrin or AM-fertilizer. Hamamoto (1966) reported that in the field experiments conducted by the National Institute of Agricultural Science, Japan, IBDU gave 20% higher yield than two split applications of ammonium sulfate. However, Rubio and Hauck (1986) reported that oxamide gave higher rice yield than IBDU.

### 8.4.4.3   Nitrification Inhibitors/Urease Inhibitors Stabilized Urea

*Nitrification inhibitor (NI) stabilized urea:* Nitrification inhibitors (NIs) are chemicals that inhibit or retard oxidation of ammonium to nitrate N. Use of NIs increases the efficiency of N-fertilizers because nitrates formed by the oxidation of ammonium are easily lost by leaching and denitrification. Many chemicals are known to have nitrification inhibiting properties. These include N-Serve or nitrapyrin (2-chloro-6-(trichloromethyl) pyridine), DCD (dicyndiamide), AM (2-amino-4-chloro-6-methyl pyrimidine), CMP (1-carbamoyle-3-methylpyrazole), terrazole (etridiazole), CP (2-cyanimino-4-hydroxy-6-methyl pyrimidine), AT/ATc (4 aminotriazole), ST (sulphatiazole or sodium thiosulphate), ATS (ammonium thiosulphate), ZPTA (thiosulphoryl triamide) (Prasad and Power 1995; Trenkel 1997; Prasad 2005b), 3,4-dimethypyrazol phosphate (DMPP) (Zerulla et al. 2001), neem extractives (containing epinimbin, deacetylnimbin, azadirachtin) (Prasad et al. 1993), *Mentha speciata* oil (Patra et al. 2002) and pyrites (Blaise and Prasad 1993; Blaise et al. 1997). Of this long list of NIs only nitrapyrin, DCD, and DMPP have been widely tested.

Nitrapyrin is a volatile compound and is difficult to use with solid fertilizers. Its use is therefore restricted with the use of anhydrous ammonia in the United States, which is injected into the soil. Many researchers in the United States have reported increased corn yield and apparent recovery of N, when it was applied as anhydrous ammonia along with nitrapyrin (Randall et al. 2003; Randall and Vetsch 2005; Parkin and Harfield 2010). Parkin and Hatfield (2010) also reported reduced $N_2O$ emissions due to nitrapyrin application. Liu et al. (2013) from China reported increased crop yields in wheat-maize rotation and reduced $N_2O$ emissions due to application of DCD and DMPP; cumulative $N_2O$ emission (kg N ha$^{-1}$) was 4.49 ± 0.21 from urea, 2.93 ± 0.06 from urea + DCD and 2.78 ± 0.016 from urea + DMPP. Again, from China, Liu et al. (2016) reported an increase of 6–7% in rice yield and a reduction of 25–38% in $N_2O$ emission, when urea was applied with DCD. Sharma and Prasad (1995) from India reported that application of DCD increased maize yield and left a significant residual effect that increased the yield of succeeding wheat in maize-wheat rotation. However, in the United States, Wilson et al. (1990) found no increase in rice yield due to DCD application. Duncan et al. (2016) in Australia also failed to obtain an increase in wheat yield due to nitrapyrin,

DCD, or DMPP application. Use of NIs has been recommended for reducing $N_2O$ losses by many researchers including Cole et al. (1997) and Synder et al. (2009).

Blaise and Prasad (1993) reported nitrification-inhibitory property of iron pyrites. However, significant nitrification inhibition comparable to the commercial nitrification inhibitors was observed only at high rates of application (Blaise and Prasad 1996; Blaise et al. 1997). Besides the nitrification inhibition, pyrites retarded $NH_3$ volatilization from fertilizer urea (Blaise and Prasad 1995; Blaise et al. 1997). Shivay et al. (2005) also reported reduction in $NH_3$ loss by mixing urea and pyrite.

*Urease inhibitor (UI) stabilized urea*: Many urease inhibitors have been reported which include phenyl phosphorodiamide (PPD/PPDA), N-(n-butyl) thiophosphoric triamide (NBPT or NBTPT), thiophosphoryl triamide (TPT), cyclohexyl phosphoric triamide (CHTPT), cyclohexyl phosphoric triamide (CNPT), phosphoric triamide (PT), hydrquinone (HQ), amino thiosulphate (ATS), and hexaamidocyclo triphosphazine (HACTP) (Trenkel 1997; Kiss and Simihan 2002). Urease inhibitors reduce the rate of hydrolysis of urea to ammonium and can reduce loss of N due to ammonia volatilization which occurs when urea is surface applied (Sarkar et al. 1991; Fox and Piekielek 1993; Abadlos et al. 2014). Some data on this from Frame et al. (2012) from Blacksburg, VA, on the effect of NBPT on ammonia loss are in Figure 8.4. Ding et al. (2011) from China reported that application of urea with UI NBPT, NIs DCD and NBPT+DCD reduced $N_2O$-N loss by 37.7%, 39.0%, and 46.8%, respectively, as compared to urea. The PPDA and NBTP have been tested for rice and the results did not show their advantage in increasing grain yield (Buresh et al. 1988; Raju et al. 1989). However, Spindula et al. (2013) from Brazil reported that application of NBPT with urea increased wheat yield by 14–15% at 90 or 120 kg N ha⁻¹ and when averaged over 30, 60, 90, 120, and 150 kg N ha⁻¹ the increase in AEn was 15.6%.

*General Remarks on EEFs:* A meta-analysis of results from many trials on the effects of NIs, UIs, and slow-release N fertilizers including DCD, Neem, NBPT, PPD, IBDU, SCU, PCU, and so on, by Lindquist et al. (2013) in the rice systems showed an increase of 5–7% in grain yield and 8% in N uptake. Similarly, another meta-analysis of results from many trials by Abaldos et al. (2014) in several crops including maize, wheat, rice, barley, cotton, rapeseed, and some vegetable and forage crops showed that DCD, DMPP, and NBPT increased crop yields by 7.5% and NUE by 12.9%.

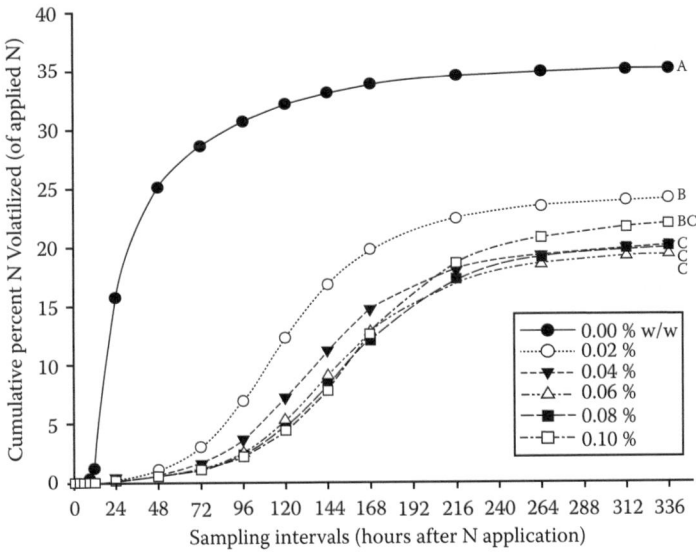

FIGURE 8.4   Cumulative N loss (percent of applied) as NH3 from granular urea treated with varying rates of the urease inhibitor N-(n-butyl) thiophosphoric triamide (NBPT). Different letters to the right of the volatility curves indicate significantly different cumulative N loss between treatments at α = 0.05. (*Source*: Frame et al. 2012. *Agronomy Journal* 104(5):1201–1207. With permission.)

Thus, in general, these materials hold promise. Of course, the performance of different EEFs will depend upon soil, crop, and climatic conditions. Another major factor determining their large-scale adoption is the benefit:cost ratio involved in their use.

## 8.5 AGRONOMIC MANAGEMENT

Crop yield directly or indirectly is the numerator in all the NUE terms and the crop and soil management practices that increase crop yield also increase NUE. Some data on the potential yield, on-station, and on-farm yields in several regions of India, where research stations are located are summarized in Table 8.9. It shows a gap of 37–52% between potential and on-station yields and a 35–70% gap between potential and on-farm yields. The gap between on-station and on-farm yields varied from 6–44%. In general, the gaps are wider in rice than in wheat. The available farm technology can at least reduce the on-station–on-farm gap and this can increase rice and wheat production by 15–20%. What is good to note is that in rice, the on-station–on-farm yield gap is zero in the Ludhiana region of Punjab and that in wheat it is even slightly negative in the Pantnagar region of Uttarakhand. This shows that farmers have already applied the available technology in these regions. Thus, with good extension efforts it can be replicated in other parts of the country. This also applies to all other countries. Some important components of improved agronomic farm technology are briefly discussed.

### 8.5.1 CROP VARIETIES

The Green Revolution in India was initiated with the introduction of high yielding, fertilizer responsive dwarf Mexican varieties of wheat (Swaminathan 2006), which not only gave higher yields but also higher AEn (Sharma and Singh 1966). As shown by the data in Table 8.10, dwarf Mexican wheat varieties responded well up to 90 kg N ha$^{-1}$ and there was no yield increase when the level of

## TABLE 8.9
## Potential On-Research Station and On-Farm Yields (Mg ha$^{-1}$) of Rice and Wheat in Different Zones of Indo-Gangetic Plains (IGP)

| Research Center | Potential Yield (A)[a] | On-Station Yield (B) | On-Farm Yield (C) | Yield Gap 100 (A–B)/A (%) | Yield Gap 100 (A–C)/A (%) | Yield Gap 100 (B–C)/B (%) |
|---|---|---|---|---|---|---|
| *Rice* | | | | | | |
| Ludhiana | 10.7 | 5.6 | 5.6 | 47.6 | 47.6 | 0 |
| Karnal | 10.4 | 6.8 | 3.8 | 34.6 | 63.5 | 44.1 |
| Kanpur | 9.5 | 4.5 | 2.8 | 52.1 | 70.5 | 37.8 |
| Pantnagar | 9.0 | 5.5 | 4.2 | 38.9 | 53.2 | 23.6 |
| Varanasi | 9.2 | 4.1 | 3.2 | 55.4 | 65.2 | 21.9 |
| Faizabad | 9.1 | 4.2 | 2.8 | 53.8 | 69.2 | 33.3 |
| *Wheat* | | | | | | |
| Ludhiana | 7.9 | 4.7 | 4.3 | 40.5 | 45.6 | 6.3 |
| Karnal | 7.3 | 4.6 | 3.6 | 37.0 | 50.7 | 21.7 |
| Kanpur | 7.0 | 4.6 | 2.8 | 34.3 | 60.0 | 39.1 |
| Pantnagar | 6.5 | 3.9 | 4.2 | 40.0 | 35.4 | −0.07 |
| Varanasi | 7.0 | 3.8 | 3.2 | 45.7 | 54.3 | 15.8 |
| Faizabad | 6.7 | 3.4 | 2.8 | 49.2 | 58.2 | 17.6 |

*Source:* Prasad, R. 2009. *J Trop Agric* 47(1–2):1–17.

[a] Aggarwal et al. (2000); all other data from Ladha et al., 2003.

**TABLE 8.10**

**Grain Yield (t ha⁻¹) Response of Wheat to Nitrogen and AEn (kg grain Kg⁻¹N) (in parentheses) in India**

| Varieties | 0 kg N ha$^{-1}$ | 45 kg N ha$^{-1}$ | 90 kg N ha$^{-1}$ | 135 kg N ha$^{-1}$ |
|---|---|---|---|---|
| Dwarf Mexican (Sonora 63, Sonora 64, Lerma Rojo) | 2.09 | 3.56 (32.7) | 4.54 (27.2) | 4.06 (14.6) |
| Tall Indian (NP876, NP887, C 306) | 1.75 | 2.69 (21.0) | 2.67 (10.2) | 2.12 (2.8) |

*Source:* Sharma and Singh. 1966. *Fertiliser News (India)* 11(9): 18–22.

N was increased to 135 kg N ha⁻¹; the economic optimum dose worked out as 120 kg N ha⁻¹. At each level of N fertilization, tall Indian wheat varieties gave lower yield and AEn; in tall Indian wheat varieties there was no response to increasing N level beyond 45 kg N ha⁻¹. Similar data exist to show the yield advantage and increased efficiency of fertilizer nitrogen by dwarf rice varieties over tall Indian rice varieties (Lakhdive and Prasad 1970) and of hybrids over dwarf high yielding varieties of rice (Kumar and Prasad 2004; Kumar et al. 2007). The introduction of rice hybrids promises 1 metric ton additional yield over that obtained with current high yielding varieties and an increase in AEn and REn (Chang et al. 1988; Siddiq 2006). Similarly, maize hybrids give higher yield and AEn (Barthakur et al. 1975). Thus, choice of the right variety of a crop is important for higher yields and AEn. Large benefits are expected from genetically modified (GM) crop plants (Gupta 2008) and Bt cotton is an excellent example in India (Singh and Kaushik 2007).

In view of low NUE by plants, the genetics of NUE in crop plants has received considerable interest (Han et al. 2015). While manipulations of genes, such as NR, NiR, GS, and GOGAT, have been hypothesized to affect NUE, greenhouse and field experiments of plants with modifications of these enzymes have not produced consistent NUE phenotypes (McAllister et al. 2012). Meanwhile, it has been observed that crop plants over-express AlaAT enhanced NUE (Good et al. 2007; Shrawat et al. 2008). Analysis of AlaAT over-expression conducted in rice, utilizing a rice btg26 homologue OsAnt1, showed that rice plants overexpressing AlaAT and grown in N-limiting conditions showed bushier plants with increased tiller number, biomass, and yield, as well as increases in total N and key metabolites (Gln, Glu, and Asn) (Shrawat et al. 2008). It is hoped that in the future more efficient N using plant types are developed, which may reduce the rates of N applied to crop plants and increase NUE.

### 8.5.2 Soil Management

Both chemical amendments such as lime and gypsum and physical management involving tillage are important for increasing crop yields and in doing so, they improve NUE. Also important is adequate availability of other essential plant nutrients in soil.

#### 8.5.2.1 Liming Acid Soils

One-third of the tropics or 1.7 billion hectares is acid enough for soluble Al to be toxic for most crops (Sanchez and Logan 1992). Many parts of sub-tropical soils are also acid and crop yields are adversely affected. In India, nearly 51 million ha of soils in eastern and northeastern states and northwestern regions have a pH 5.5 or less (Mahapatra and Pattanayak 2008; Maji et al. 2008). Large areas of acid soils occur in Hainan, Guangdong, Yuannan, Fujian, Guangxi, Jiangxi, Hunan, Sichuan, Guizhon, Zheijiang, and Anhui provinces in China (He et al. 2004), in Kalimanthan, Sumatra, and Sulavesi regions of Indonesia (Leiwakabessy 1989), in northern Queensland in Australia (Edwards and Bell 1989), and in large tracts in western, central, and southern Africa (Spaargaren 1989). Since acid soils are formed due to leaching of bases, especially calcium, due to heavy rains, lime minerals such as limestone and gypsum deposits are not present in these regions. This makes liming acid soils in these regions a very costly affair and it is not practiced. Fortunately, lowland rice is the main crop in most South, Southeast, and East Asian countries and flooding itself changes the pH to near neutrality (Ponnamperuma 1972) and the farmers do not feel the negative impact of soil acidity. However, in upland areas and for growing crops in rice fallows and for aerobic rice (Prasad 2011), the need for liming has been proved. Data from long-term fertilizer experiments from Ranchi in eastern India (Nambiar 1994) showed that over a period of 13 years, maize yield was higher under NPK + lime compared to no-lime treatments. Continuous application of FYM also maintained reasonably good yields (71.7% of that obtained with NPK + lime). Similarly, without lime, wheat yield declined from 3.4 t ha$^{-1}$ in 1970–1971 to 0.84 t ha$^{-1}$ in 1983–1984, while in limed plots the wheat yield remained steady at 4.0 t ha$^{-1}$. Thus, liming acid soils can increase crop yield and NUE. Advantages of liming include optimizing the soil environment for plant growth, better microbial activity including nitrification, increased availability of plant nutrients, such as P, and preventing Al and Fe toxicities (Prasad and Power 1997).

### 8.5.2.2 Gypsum and Other Soil Amendments in Sodic Soils

As a contrast to acid soils in the wetter regions of the tropics and sub-tropics, large areas in the semi-arid and arid regions in the sub-tropics have salt affected soils including 23.8 m ha in India, 10.5 m ha in Pakistan, 3.0 m ha in Bangladesh, and 36.7 m ha in China (Massoud 1977). There are three kinds of salt affected soils: (1) Saline soils [ECe > 4.0 dSm$^{-1}$, saturation extract (SE) pH < 8.5, sodium absorption ratio (SAR) <13–15], (2) saline sodic soils [Ece > 4.0 dSm$^{-1}$, SE pH < 8.5, SAR > 13–15] and (3) sodic soils [ECe < 4.0 dSm$^{-1}$, SE pH >8.5, SAR >13–15] (USDA 1954). Growth of crop plants is poor and crop yields are drastically reduced on these soils and can be largely increased by reducing the degree of sodicity (Eynard et al. 2005). The reclamation proce-dures include scraping, flushing, leaching, drainage, and application of soil amendments (Prasad and Power 1997). The soil amendments include gypsum, finely ground sulfur, and pyrites (Verma and Abrol 1990); gypsum being the most favored amendment in India (Abrol et al. 1988). Data from Karnal, India (Singh and Abrol 1988) illustrated that grain yield and NUE in wheat almost doubled following gypsum application. It may be added that gypsum is also receiving considerable attention in the United States as a source of nutrients like calcium and sulfur and as a soil conditioner even in acid soils (Chen and Dick 2011).

### 8.5.2.3 Tillage

Rice is the principal crop of the tropics and sub-tropics in Asia and a major part is lowland rice grown as a transplanted crop on puddled soil. Despite destroying soil structure (Sharma and DeDatta 1985), puddling reduces percolation of water and leaching of fertilizers, especially N, and helps in weed control (De Dutta 1981). The net result is increased rice yield and NUE. Several new tillage implements such as laser aided land levelers, mechanical rice transplanters, and drum seeders have now become available and their use in rice cultivation is on the increase (Bouman 2014). There is also an attempt to move rice to a direct seeded, non-puddled system to help save water and labor, but this change needs to also be evaluated in terms of rice yield and NUE (Prasad 2011). The weeds that are partially controlled by ponding water in puddled soils also need to be assessed and ways found to control them. Use of herbicide tolerant rice would help greatly with this issue. A breakthrough in direct seeding of rice would enable conservation agriculture (CA) described below to be used in both wheat and rice and can lead to improved soil properties, water savings, and increased yields as long as nitrogen issues are also resolved.

One of the most significant developments in the crop production technology has been the intro-duction of CA. The three key principles of CA are: (1) Permanent residue cover, (2) minimal soil disturbance (zero or no tillage), and (3) crop rotations (Hobbs et al. 2008). Comparison of conven-tional tillage and CA for some issues is shown in Table 8.11. Zero-till machines have become par-ticularly relevant in the rice-wheat cropping system, where wheat sowing is generally delayed if the

## TABLE 8.11
## Comparison of Conventional Tillage and Conservation Agriculture for Some Issues

| Issues | Conventional Tillage | Conservation Agriculture |
|---|---|---|
| Soil disturbance | High | Minimum |
| Soil surface | Bare surface | Permanently covered |
| Erosion | High wind and soil erosion | Low wind and soil erosion |
| Water infiltration | Low | High |
| Weeds | Kills established weeds, but also stimulates more weed seeds to germinate | Weeds are a problem in the early stages of adoption but decrease with time |
| Diesel use and costs | High | Low |
| Production costs | High | Low |
| Timeliness Operations | Can be delayed | Timeliness of operations, tends to be optimal |
| Yield | Can be lower where planting is delayed | Same or higher if planting timelier |
| Soil fertility | Generally becomes poorer | Improves with time, especially SOM |
| Soil moisture | Rapid loss | Conserved better |
| Fertilizer N | Less immobilization | More immobilization |
| Fertilizer P | Less accumulation | More accumulation |

*Source:* Hobbs, P.R., K.D. Sayre, and R.K. Gupta, 2008. The role of conservation agriculture in sustainable agriculture. *Phil Trans Royal Soc* B 363: 543–555; Prasad, R. and J.F. Power. 1991. Crop residue management. *Adv Soil Sci* 15: 205–251; Prasad, R., B. Gangiah, and K.C. Aipe. 1999. Effect of crop residue management in a rice-wheat cropping system on growth and yield of crops and on soil fertility. *Exp Agric* 35:427–435.

conventional method of pre-sowing irrigation and land preparation is adopted (Mehla et al. 2000). It is now possible to sow wheat soon after rice harvest without primary cultivation, which permits timely sowing and ensures increased grain yields, since the crop escapes the heat stress during grain filling (Singh et al. 1998; Hobbs and Gupta 2003; Jat et al. 2014b; Gopal et al. 2010). In 338 front line demonstrations in India conducted during 1998–2001, an increase in wheat yield of 8.1 to 35.5% was obtained with zero-till machines (Singh and Kharub 2000) (Table 8.12). CA builds up soil fertility, especially soil organic matter (SOM) (Singh et al. 1998; Prasad et al. 1999; Sharma and Prasad 2002, 2008; Lal 2005), phosphorus, and other nutrients (Prasad and Power 1991), which is important for sustainable food production in a changing climate (Lal 2013a, b). However, Verhurst et al. (2014) observed that NUE efficiency in CA may be lowered due to immobilization of some

## TABLE 8.12
## Wheat Yields (Mg ha$^{-1}$) in Zero Tillage Over Conventional Tillage in 338 Frontline Demonstrations on Farmers' Fields in India (averaged over 3 years of data from 1998–1999 to 2000–2001)

| Zone | Zero Tillage | Conventional Tillage | Yield Gain (%) |
|---|---|---|---|
| Northwest plains (states of Punjab, Haryana, western UP, Delhi, and Rajasthan) | 4.95 | 4.58 | 8.1 |
| Northeast plains zone (states of Bihar, West Bengal, Orissa, Assam, and eastern UP) | 3.69 | 2.72 | 35.5 |
| Central zone (states of MP Gujarat, and parts of Rajasthan) | 5.05 | 4.52 | 11.8 |

*Source:* Singh, A., and A. S. Kharub. 2000. Performance of zero tillage in wheat, evidences from participatory research. *Fertil. Market News*, India 32(11): 1–5.

fertilizer N and increased ammonia volatilization and Yadvinder-Singh et al. (2005) reported that additional 20–40 kg N ha$^{-1}$ is needed in the initial few years in the rice-wheat cropping system in India. Some major benefits of CA are given in Table 8.11.

Another new tillage technology for growing crops is permanently raised beds (PRB) or furrow irrigated raised beds (FIRB) developed by CIMMYT, Mexico, which covers about 90% of the wheat area in the Yaqui Valley, Sonora, Mexico (Sayre and Moreno Ramos 1997; Sayre et al. 2005). This technology is reported to increase water use efficiency (Kaur et al. 2003; Naresh et al. 2012) as well as NUE (Limon-Ortega et al. 2000; Majeed et al. 2015). However, this technology has met with mixed success in many field experiments in India (Kaur et al. 2003; Kukal et al. 2005; Naresh et al. 2012) and China (Wang et al. 2004). The main reasons for their large-scale non-adoption by smallholder farmers in Asia are the lack and cost of suitable farm equipment for making of raised beds and then seeding on them. This technology is being demonstrated to Indian farmers with the use of trained service providers who provide the tractor power and implements needed. This technology is likely to expand in India and elsewhere in the future.

### 8.5.2.4 Crop Rotations

Crop rotations are one of the pillars of CA as well as sustainable agriculture. Crop rotations have been known and practiced in the tropics and subtropics for ages and the reference to them can be found in Indian agriculture back to the Chalcolithic period (Raychaudhari and Mira 1993) and in Chinese agriculture during the Han dynasty in the second century (Hsu 1980). In temperate regions crop rotations were practiced in Greece and Rome (White 1970) and in colonial United States (Karlen 1994). The key to successful crop rotations was inclusion of a soil restorer crop of legumes, such as faba beans (*Vicia faba* L), clovers (*Medicago* sp), lupins (*Lupinus album,* L), and vetch (*Vicia sativa,* L). The benefits of including a legume in a crop rotation have been known for a long time, although bacteria responsible for N fixation were identified by Beijerinck only in 1988 (Waksman 1952). Now it is well known that growing of legumes in crop rotations not only adds N, but also increases organic C (Sharma et al. 2000), improves soil physical conditions (Mandal et al. 1999), increases microbial populations (Sharma et al. 2000), and helps in soil moisture conservation (Singh et al. 2009). Bullock (1992) observed that the replacement method overestimated the N contribution of legumes grown in rotation with non-legumes and underestimated the overall rotation effect. For example, fertilizer recommendations for corn following alfalfa in the Unites States credit alfalfa with N contribution of 100–125 kg N ha$^{-1}$ (Fox and Piekielek 1988) based on replacement methodology, while the actual N contribution, as measured by $^{15}$N techniques, showed only 24 kg N ha$^{-1}$ (Harris and Hestermann 1990). $^{15}$N studies in India have also shown that inclusion of summer mungbean in a wheat-maize rotation considerably reduced N losses (Arora et al. 1980). Inclusion of a summer mungbean (*Phaseolus radiatus*) in the rice-wheat rotation in India has shown significant advantages from a soil fertility buildup point of view in addition to the increased grain yield; the nitrogen saving in this rotation was 60–90 kg N ha$^{-1}$ (Sharma and Prasad 2002, 2008; Sharma et al. 2000). Carefully selected cropping systems can result in increased fertilizer use efficiency as determined by crop equivalents or net returns (Gill and Ahlawat 2006).

### 8.5.2.5 Adequate Supply of Plant Nutrients Other than N

It is well known that the efficiency of any plant nutrient depends upon the adequate availability of other essential plant nutrients (Liebig's law of the minimum) and N is no exception The concept of balanced NPK fertilization was first mooted by Bray as early as 1945 (Bray 1945) and has been well written and talked about (Tiwari 2001; Tandon 2012). As the deficiencies of secondary nutrient S (Lin and Liu 2004; Tandon 2011; Choudhary et al. 2013; Prasad and Shivay 2016b) and micronutrients Zn (Cakmak et al. 1996; Takkar et al. 1997; Prasad et al. 2016a), B (Singh and Goswami 2014), and others (Alloway 2008) emerged in large areas, the concept of balanced NPK fertilization has been replaced by Site Specific Nutrient Management (SSNM) (Tiwari et al. 2006; Khurrana et al. 2008). However, even SSNM was not adequate to prevent crop yield decline in intensive cropping

**TABLE 8.13**
**Effect of N, P, K, and Farmyard Manure (FYM) on Grain Yield,**
**N Uptake, AEn, and AREn in Rice at Pantnagar, India**

| Treatment | Grain Yield (Mg ha$^{-1}$) | N Uptake (kg ha$^{-1}$) | AEn (kg grain kg$^{-1}$ N) | AREn (%) |
|---|---|---|---|---|
| $T_1$_Check (no fertilizer) | 3.03 | 44.8 | – | – |
| $T_2$_120 kg N ha$^{-1}$ | 3.49 | 64.9 | 3.8 | 16.7 |
| $T_3$_$T_2$+40kg $P_2O_5$ ha$^{-1}$ | 4.75 | 93.1 | 14.3 | 40.2 |
| $T_4$_$T_3$+40kg $K_2O$ ha$^{-1}$ | 5.22 | 104.3 | 18.2 | 59.5 |
| $T_5$_$T_4$+5tha$^{-1}$ FYM | 6.20 | 125.1 | 26.4 | 66.9 |
| LSD (P = 0.05) | 0.35 | 7.8 | | |

*Source:* Paul, T., P.S. Bisht, P.C. Pandey et al. 2013. Rice productivity and soil fertility as influenced by nutrient management in rice (*Oryza sativa*)—wheat (*Triticum aestivum*) cropping system. *Indian J Agron* 5894: 495–499.

systems such as rice-wheat (Ladha et al. 2003) and other crops (Ray et al. 2012; Mondall 2011). Therefore, attention focused on adding soil organic matter and building soil organic C (Lal 2013b) and led to the development of the concept of Integrated Nutrient Management (INM) involving chemical fertilizers, organic manures, crop residues, and biofertilizers (Chivenge et al. 2009; Prasad et al. 2014). An example of a stepwise increase in grain yield, N uptake, AEn, and AREn in rice using N alone, NP, NPK, and NPK along with farmyard manure is shown in Table 8.13 (Paul et al. 2013).

### 8.5.3 Crop Management

A volume of literature is available on various aspects of crop management. However, only three aspects, namely, plant population, weed control, and water management are briefly discussed here to point out the importance of crop management in affecting NUE.

### 8.5.3.1 Plant Population

Optimal plant population is essential for high yields; both sub-optimal and above-optimal plant populations affect crop yield adversely. This has been well brought out by SRI (System of Rice Intensification) method of rice cultivation developed by the French Jesuit Father Henri de Laulenie' in Madagascar in 1983 and promoted by Dr. Norman Uphoff of Cornell University (Dass et al. 2015). The key components of the SRI method are the use of 10- to 12-day-old rice seedlings and a wider spacing of 25–30 cm in transplanting (108,000–160,000 plants ha$^{-1}$) as compared to conventional 15 cm × 15 cm (396,000 plants ha$^{-1}$) or 20 cm × 10 cm (500,000 plants ha$^{-1}$) spacing recommended for high yielding rice varieties and hybrids. The SRI method of rice cultivation leads to better root development (Barison and Uphoff 2011), profuse tillering (Menete et al. 2008), higher rate of photosynthesis (Uphoff 2002), and higher grain yield and NUE (Barison and Uphoff 2011; Weijabhandara et al. 2011; Dass and Chandra 2012). The SRI method of rice cultivation is now widely practiced in the world. In general, higher plant populations also demand higher N levels. A positive interaction between plant population and N in maize (corn) is well known and has recently been reviewed (Rao et al. 2014). As an example, Ameda and Dhakar (2000) reported that 120 kg N ha$^{-1}$ in winter maize was adequate for plant populations of 65,000 and 75,000 plants ha$^{-1}$ but not for plant populations of 85,000 and 100,000 plants ha$^{-1}$, which needed 150 kg N ha$^{-1}$ at Banwara, Rajasthan, India. Similarly, in direct seeded rice increasing seed rate from 100 to 125 kg ha$^{-1}$ with adequate N increased rice yield by 240 kg ha$^{-1}$ (Tripathy and Mohapatra 2007).

### 8.5.3.2   Weed Control

Weeds compete with crop plants for plant nutrients including nitrogen, water, and light. Some weeds such as *Echinochloa crusgalli* have a high nitrophily index (Moreau et al. 2013) and therefore grow faster than cereals and take away a large part of applied N. Some estimates of yield losses due to weeds are: 43–56% in rice (Johnson et al. 2007; Rashid et al. 2012), 33–56% in maize (Yang and Lu 1994), and 25–30% in wheat (Banga et al. 2003). Estimates of nitrogen loss due to weeds in some crops in India are: 65 kg N ha⁻¹ in rice (Rana et al. 2000), 72 kgN ha⁻¹ in wheat (Jat et al. 2007), 52 kg N ha⁻¹ in maize (Singh et al. 2007), 32 kg N ha⁻¹ in sorghum (Mishra et al. 2012), and 48 kg N ha⁻¹ in soybean (Chander et al. 2013). As an example, a few data from experiments in India show that yield with weed control was 1.5 to 2 times of that in check plots (Table 8.14). This obviously increased PFPn from 24.1 to 38.1 in rice, from 24.2 to 36.7 in wheat, and 17.4 to 36.4% in maize. Interestingly, polymer coated urea ESN was reported to reduce weed growth and N removal by weeds in canola (Blackshaw et al. 2004).

### 8.5.3.3   Water Management

For good crop husbandry adequate water and sunshine and warm temperatures are essential and these were available in the Tigris-Euphrates River valley in the Fertile Crescent (now Iraq and adjoining areas), Indus valley in India, Yangtze-Yellow River Valley in South China, and Upper Nile Valley in Egypt in Africa. Thus, it was the tropical and sub-tropical regions of the world that contributed most to the evolution of agriculture about 13,000 to 10,000 BP (before present) (Braidwood 1960; Bender 1975; Prasad et al. 2016b). Even today, agriculture is responsible for 70% of the world's total consumption of water (WBCSD 2005). Rice consumes 50% of all agricultural water used for irrigation in Asia (Barker et al. 1999) and efforts are therefore under way to develop completely aerobic rice production systems (Prasad 2011) or partially aerobic rice production systems, such as those using alternate wetting and dry (AWD) conditions (Bouman and Tuong 2001). Aerobic rice systems are gaining importance in China (Huaqi et al. 2002; Bouman et al. 2002) and Brazil (Pinheiro et al. 2006). The AWD rice production system is gaining importance in Bangladesh (Tuong et al. 2009) and Senegal (Mendoza 2010). However, nitrogen management is going to be a serious problem in these aerobic rice cropping systems. For example, Zhou et al. (2012) reported that N loss in rice through denitrification was 38.9% of applied N in the AWD (alternate wet and dry conditions) conditions as compared to 9.9% under continuous flooding.

### TABLE 8.14
### Effects of Chemical Weed Control on Grain Yield and PFPn from Some Experiments at Research Centers in India

| Crop (Center) | Treatment | Grain (Mg ha⁻¹) | PFPn (kg grain kg N⁻¹) | Reference |
|---|---|---|---|---|
| Rice (Varanasi) | Check (no herbicide) | 2.89 | 24.1 | Sahu et al. |
| | 750 g ha⁻¹ Petrilachlor + 85 g ha⁻¹ fb Azimsulfuron | 4.57 | 38.1 | (2015) |
| Wheat (Varanasi) | Check (no herbicide) | 2.90 | 24.2 | Meena and |
| | 25 g ha⁻¹ Sulfosulfuron + 4 g ha⁻¹ Metsulfuron | 4.4 | 36.7 | Singh (2013) |
| Maize (Udaipur) | Check (no herbicide) | 2.09 | 17.4 | Choudhary |
| | 500 g ha⁻¹ Atrazine + 1500 g ha⁻¹ Alachlor | 4.37 | 36.4 | et al. (2013) |

*Note:*   Nitrogen applied in each experiment was 120 kg N ha⁻¹.

**TABLE 8.15**
**Effects of Irrigation Scheduling on Yield and PFPn in Winter Maize and Sugarcane in India**

| Crop (Center) | Treatment (IW/CPE)[a] | Number of Irrigations | Water Depth (mm) | Yield[b] (Mg ha$^{-1}$) | PFPn (kg kg$^{-1}$N) | Reference |
|---|---|---|---|---|---|---|
| Maize (Udaipur) | 0.6 | 6 | 450 | 4.01 | 23.1 | Singh et al. |
| | 0.9 | 8 | 600 | 4.07 | 37.4 | (2015) |
| | 1.2 | 9 | 675 | 5.09 | 40.7 | |
| Sugarcane (Pusa) | 0.5 | 2 | 160 | 75.4 | 503 | Kumar et al. |
| | 0.75 | 3 | 240 | 89.7 | 598 | (2013) |
| | 1.0 | 4 | 320 | 98.0 | 653 | |

[a] IW/CPE—Irrigation water/cumulative pan evaporation ratio.

[b] Millable canes in the case of sugarcane and grain in the case of maize. Maize received 125 kg N ha$^{-1}$, while sugarcane received 150 kg N ha$^{-1}$.

Many studies have been made to optimize irrigation and nitrogen. For example, Singh et al. (1980) from India recommended irrigation at 50% of available moisture in a 60-cm soil profile and an application of 120 kg N ha$^{-1}$ for wheat to obtain a yield of 5.09 t ha$^{-1}$ giving a PFPn value of 42.4. Similarly, Alkaisi and Yin (2003) from China reported that a combination of irrigation at 0.8 of the estimated evapotranspiration (ET), 140 to 250 kg N ha$^{-1}$ and 57,000 to 60,000 plants ha$^{-1}$ was found to be the best for maize production. Some data from irrigation experiments on winter maize and sugarcane are in Table 8.15, which show that at the same level of N, yield as well as PFPn increased with an increase in irrigation. Adequate soil moisture optimizes yield and NUE.

## 8.6   SUMMARY

Fertilizer nitrogen has been responsible for augmenting food production and alleviating hunger in the world and the demand for it by the year 2050 is estimated at 225–250 million metric tons (Tilman et al. 2011), as against estimated consumption of 116 Mt in 2016. However, the losses of applied fertilizer N are fairly severe and the global figures for the apparent recovery of N are only 63% in maize, 54% in wheat, and 44% in rice; the true recovery of applied nitrogen as determined by $^{15}$N studies could be as low as 21% in lowland rice.

Nitrogen management is thus a serious problem especially in the tropics and sub-tropics, where heavy rains can come in short spells of time and lead to large runoffs and soil erosion; in well N fertilized fields about 5–8% of applied N could be lost from the point of placement to inland waters and estuaries. Loss of fertilizer N from agricultural fields is thus not only a monetary loss, but contributes to environmental pollution. Nitrogen lost as surface runoff is responsible for the eutrophication of inland waters (lakes), estuaries, and the sea leading to the mortality of fish and other marine life. Nitrogen leached from agricultural fields leads to nitrate pollution of ground-waters leading to ailments such as Blue Baby Syndrome. Nitrous oxide produced due to deni-trification leads to depletion of the ozone layer and global warming. Finally, ammonia released after ammonia volatilization contributes to acid rain. It is therefore important that improved N management practices, such as placement at seeding, split applications especially using LCC or SPAD meters, and foliar application need to be encouraged for increasing NUE in the tropics and sub-tropics.

Many enhanced efficiency fertilizers (EEFs) with lesser N losses and higher NUE have been developed in recent years. These include sulfur coated urea (SCU), polymer coated urea (PCU), neem coated urea (NCU), urea-formaldehyde or Ureaform, isobutylidene diurea (IBDU), and conventional N fertilizers stabilized with nitrification inhibitors Nitrapyrin, DCD, and so on or urease inhibitors PPDA, NBTP, and so on. However, EEFs are expensive and have not reached the farmers even in developed temperate countries, let alone poor farmers in the developing tropical and sub-tropical countries. The only exception is the neem-coated urea (NCU) in India, which is subsidized by the government of India.

Also important for increasing NUE is the use of improved soil and crop management practices, such as liming acid soils, adding gypsum to sodic soils, adequate plant population, efficient water management, effective weed control and adoption of good cropping systems. Some efforts are underway to develop more efficient N efficient genotypes in crop plants and it is hoped that in the future more efficient nitrogen is possible using plant types leading to lesser application of fertilizer N. Some new management systems like CA including aerobic rice cultivation show promise for sustainable intensification in the future but more studies are needed to determine how they affect NUE. If NUE is reduced in these new systems, research will be needed to identify ways to correct this issue.

## LIST OF ABBREVIATIONS AND ACRONYMS

| | |
|---|---|
| **AEn** | Agronomic efficiency of N |
| **AREn** | Apparent recovery of N |
| **AWD** | Alternate wet and dry conditions |
| **CA** | Conservation agriculture |
| **EEFs** | Enhanced efficiency fertilizers |
| **ET** | Evapotranspiration |
| **FAO** | Food and Agriculture Organization of UN, Rome, Italy |
| **FIRB** | Furrow irrigated raised beds |
| **FYM** | farmyard manure |
| **IARI** | Indian Agricultural Research Institute, New Delhi, India |
| **IFDC** | International Fertilizer Development Center, Muscle Shoals, AL |
| **IRRI** | International Rice Research Institute, Los Banos, Philippines |
| **LCC** | Leaf color chart |
| **NI** | Nitrification inhibitor |
| **NCU** | Neem coated urea |
| **NUE** | Nitrogen use efficiency |
| **PCU** | Polymer coated urea |
| **PEn** | Physiological efficiency of N |
| **PFPn** | Partial factor productivity of N |
| **SCU** | Sulfur coated urea |
| **SPAD** | Soil-plant analyses development chlorophyll meter |
| **SRI** | System of rice intensification |
| **TREn** | True recovery of N using $^{15}$N |
| **UI** | Urease inhibitors |
| **UN** | United Nations |

## REFERENCES

Abadlos, D., J. Simon, A. San-Cobena, G. Guardia and A. Vallejo. 2014. Meta-analysis of the effect of urease and nitrification inhibitors on crop productivity and nitrogen use efficiency. *Agric Ecosyst Environ* 189:136–144.

Abdin, M.Z., K.C. Bansal and Y.P. Abrol. 1996. Effect of split application on growth and yield of wheat (*Triticum aestivum* L) genotypes with different N-assimilation potential. *J Agron Crop Sci* 176(2):83–90.

Abrol, I.P., J.S.P. Yadav and F.I. Masood. 1988. Salt affected soils and their management. *FAO Soil Bull* 39. 131 p.

Achimu, P. and E. Schlosser. 1992. Effect of neem extract *(Azadirachta indica* A. Juss) against downy mildew *(Plasmopora viticola)* of grapevine. In: *Practice Oriented Results on Use and Production of Neem Ingredients* (Kleeberg H, Ed.) pp. 99–107. Giessen, Germany: Druck and Graphic.

Aggarwal, P.K., K.K. Talukdar and R.K. Mall. 2000. Potential yields of rice-wheat cropping system in the Indo-Gangetic Plains of India. Rice-wheat cropping system in the Indo-Gangetic Plains of India. *Rice-Wheat Consortium Series* 10. 16 p. Rice-Wheat Consortium for the Indo-Gangetic Plains. New Delhi.

Alexander, A. and H. Helm. 1990. Ureaform as a slow-release fertilizer. *J Plant Nutr Soil Sci* 153(4):249–255.

Alexandratos, N. and J. Bruinsma, 2012. World Agriculture Towards 2030/2050. The 2012 Revision. ESA working Paper No. 12–03. International Atomic Energy Commission, Vienna. pp. 3–22. Agriculture Development and Economic Division, FAO, Rome (www.fao.org/economic/esa).

Alimata, B., A. Opoku and F. Bidjokazo et al. 2016. Urea super granules and phosphorus application increases rice yield and agronomic use efficiency in Burkino Faso. *Int J Agric Res* 8(4):35–43.

Alkaisi, M.M. and X. Yin. 2003. Effect of nitrogen rate and plant population on corn yield and water use efficiency. *Agron J* 1475–1482.

Alloway, J. Ed. 2008. *Micronutrient Deficiencies in Global Crop Production.* Springer, Belgium.

Amadioha, A.C. 2000. Controlling rice blast in vitro and in vivo with extracts of *Azadirachta indica. Crop Prot* 19:287–290.

Amanullah and L.K. Almas. 2009. Partial factor productivity, agronomic efficiency, and economic analysis of maize in wheat-maize cropping system in Pakistan. Paper presented at Southern Agricultural Economics Association, Annual Meeting, Atlanta, GA, January 31–February 3, 2009 (via Internet).

Ameda, G.S. and L.L. Dhakar. 2000. Response of winter maize (Zea mays L) to nitrogen levels in relation to varying population density and row spacing. *Int J Trop Agric* 18(4):395–398.

Ando, H., K. Kakuda, M. Nakayama, and K. Yokoto. 2000. Yield of no-tillage direct seeded lowland rice as influenced by different sources and application methods of fertilizer nitrogen. *Soil Sci. Plant Nutr.* 46, 105–115.

Attaur Rahman, M., M.A.Z. Sarker, M.F. Amin et al. 2011. Yield response and nitrogen use efficiency of wheat under different doses and split application of nitrogen fertilizer. *Bangladesh J Agric Res* 36(2):231–240.

Aulakh, M.S., T.S. Khera, J.W. Doran, and K.F. Bronson. 2001. Denitrification $N_2O$ and $CO_2$ fluxes in rice-wheat cropping system as affected by crop residues, fertilizer N and legume green manure. *Biol Fertil Soils* 34(6):375–389.

Baalsrud, K. and K.S. Baalsrud. 1954. Studies on *Thiobacillus denitrificans. Arch Microbiol* 20:34–62.

Bahrain, M.T. and M. Shomeli. 2008. Sugarcane response to irrigation and nitrogen in sub-tropical soils. *Iran Agric Res* 27(2):17–25.

Balasubranian, V., B. Alves., M.S. Aulakh et al. 2004. Crop, environment and management factors affecting N use efficiency. In Agriculture and Nitrogen Cycle: Assessing the Impacts of Fertilizer Use on Food Production and the Environment (AR Mosier, JK Seyers, and JR Freney, Eds.), *SCOPE*, Paris 65:19–33.

Bandaogo, A., F. Bidjokazo, S. Youl et al. 2015. Effect of deep placement with urea supergranules on nitrogen use efficiency of irrigated rice in Souru valley (Burkina Faso). *Nutr Cycl Agroecosyst* 102(1):79–89.

Bandyopadhyay, K.K. and M.C. Sarkar. 2005. Nitrogen use efficiency, 15 N balance and nitrogen losses in flooded rice in an inceptisol. *Commu Soil Sci Plant Anal* 36:1661–1679.

Banga, R.S., A. Yadav and R.K. Malik. 2003. Bioefficiency of flufecacet and sulfosulfuron alone and in combination against weed flora in wheat. *Indian J Weed Sci* 35(3/4):179–182.

Barison, J. and Uphoff, N. 2011. Rice yield and its relation to root growth and nutrient-use efficiency under SRI and conventional cultivation: An evaluation in Madagascar. *Paddy Water Environ* 9:65–78.

Barker, R., D. Dave, T.P. Tuong et al. 1999. The outlook for resources in the year 2020: Challenges for research on water management in rice production. In *Assessment and Orientation Towards 21st Century.* Proceedings of 19th Session, International Rice Commission, Cairo, Egypt, September 7–9, 1998, FAO, Rome, pp. 96–109.

Barthakur, B.C., S. Nath and P.K. Purkayastha. 1975. Effect of dates of sowing dates, rates of nitrogen and planting dates on grain quality of hybrid Ganga 101. *Indian J Agron* 20(3):253–259.

Bateman, E.J. and E.M. Baggs. 2005. Contributions of nitrification and denitrification to $N_2O$ emissions from soils at different water-filled pore space. *Biol. Fertil. Soils* 41:379–388.

Bender, B. 1975. *Farming in Pre-History-From Hunter-Gatherer to Food Production.* London: Billing & Sons, 268 p.

Bharadwaj, A., S. Garg, S.K. Sondhi and D.S. Taneja. 2012. Nitrate contamination of shallow aquifer ground-water in the central districts of Punjab, India. *J Environ Sci Eng* 54(1):90–97.

Bhat, T.A., R. Kotru, K.N. Singh et al. 2015. Real time nitrogen management using leaf color chart in rice (*Oryza sativa*) genotypes. *Indian J Agron* 60(1):70–75.

Bhattacharya, D. and B.K. Goswami. 1988. Effect of different dosages of neem and groundnut oil cakes on plant growth characters and population of root-knot nematode, *Meloidogyne incognita* in tomato. *Indian J Nemat* 18:125–127.

Bhuyan, M.H.M., M. Ferdousiand and M.T. Iqbal. 2012. Foliar spray of nitrogen fertilizer on raised bed increases yield of transplanted aman rice over conventional method. *ISRN Agronomy 2012(2012), Article ID 184953.*

Bijay-Singh, Yadvinder-Singh, J.K. Ladha et al. 2002. Chlorophyll meter and leaf color chart based nitrogen management for rice and wheat in northwest India. *Agron J* 94(4): 821–829.

Biswas, B.C., M.C. Sarkar, S.P.S. Tanwar et al. 2004. Sulfur deficiency in soils and crops and response to fertilizer sulfur in India. *Fertil News (India)* 49(10):13–33.

Biswas, P.P. and P.D. Sharma. 2006. A new approach for establishing fertilizer response ratio-the Indian Scenario. *Indian J Fert* 4(41):59–62.

Blackshaw, R.E., X. Hao, R.N. Brandt et al. 2004. Canola response to ESN and urea in a four year no-till cropping system. *Agron J* 103:92–99.

Blaise, D. and R. Prasad. 1993. Evaluating pyrites as a nitrification inhibitor. *Fertil News (India)* 38(9):43–45.

Blaise, D. and R. Prasad. 1995. Effect of blending urea with pyrite or coating with polymer on ammonia volatilization loss from surface applied prilled urea. *Biology and Fertility of Soils* 20:83–85.

Blaise, D. and R. Prasad. 1996. Relative efficiency of modified urea fertilizers in wetland rice (Oryza sativa). *Indian J Agron* 41(3):373–378.

Blaise, D. and R. Prasad. 1997. Dynamics of N-release from differently coated fertilizers under flooded soil conditions. *Proc Indian Natn Sci Acad* Section B63:73–78.

Blaise, D., A. Amberger, and S. von Tucher. 1997. Influence of iron pyrites and dicyandiamide on nitrification and ammonia volatilization from urea applied to loess brown earths (Luvisols). *Biol Fertility Soils* 24: 179–182.

Bouman, B. 2014. Modernizing Asian rice production. *Rice Today* (January-March 2014) 13(1):32–33.

Bouman, B.A.M. and T.P. Tuong. 2001. Field water management to save water and increase its productivity in irrigate rice. *Agric. Water Manage* 49:11–30.

Bouman, B.A.M., Y. Xiaoguang, Y. Huaqi et al. 2002. Aerobic rice (Han Dao): A new way of growing rice in water-short areas. In "Proceedings of the 12th International Soil Conservation Conference, Beijing, 26–31 May 2002," pp. 175–181. Tsinghua University Press, Beijing.

Braidwood, R.J. 1960. The agricultural revolution. *Scientific American* 203(30):130–152.

Bray, R.H. 1945. Soil-plant relationships. II. Balanced fertilizer use through soil tests for K and P. *Soil Sci* 60:463–473.

Bullock, D.G. 1992. Crop rotation. *Crit Rev Plant Sci* 11:309–326.

Bundy, L.G. 2001. Managing urea containing fertilizers-overcoming nitrogen volatilization losses. Department of Soils, University of Wisconsin, Madison, WI.

Buresh, R., S.K. DeDatta, J. Padalia, and M. Samson. 1988. Field evaluation of two urease inhibitors with transplanted lowland rice. *Agron J* 80(5):763–768.

Burford, J.R. and J.M. Bremner. 1975. Relationships between denitrification capacities of soils and total, water-soluble and readily decomposable soil organic-matter. *Soil Biol. Biochem.* 7:389–394.

Burrow, K.R., B.T. Nolan, M.G. Rupert, and N.M. Dubrovsky. 2010. Nitrate in the groundwater in the United States—1991–2003. *Environ Sci Tech* 44:4988–4997.

Cakmak, I., A. Yilmaz, M. Kalyari et al. 1996. Zinc deficiency as a critical problem in wheat production in Central Anatolia. *Plant Soil* 180:165–172.

Carlson, C.A. and J.L. Ingraham. 1983. Comparison of denitrification by *Pseudomonas stutzei. Pseudomonas aeruginosa* and *Paracoccus denitrificans. Appl Environ Microbiol* 45:1247–1253.

Cassman, K.G., A. Dobermann, and D.T. Walters. 2002. Agroecosystems, nitrogen-use efficiency, and nitrogen management. *Ambio* 31:132–140.

Chander, N., S. Kumar, Ramesh and S.S. Rana. 2013. Nutrient removal by weeds and crops as affected by herbicide combinations in soybean-wheat cropping syste. *Indian J Weed Sci* 45:137–139.

Chang, Q., N. Hu, N. Jing, and F. Huang. 1988. Physiological and biochemical characters of hybrid rice in Dongting Lake region, China. In: *Hybrid Rice*, International Rice Research Institute, Los Banos, Manila, pp. 279–286.

Chen, L. and W.A. Dick. 2011. Gypsum as an agricultural amendment—General use and guidelines. *Ohio State University Bull 945*. 36 p.

Chen, N. and H. Hong. 2011. Nitrogen export by surface runoff from a small agricultural watershed in southeast China: Seasonal pattern and primary mechanism. *Biogeochemistry* 106(3):311–321.

Chen, X.P., Z.L. Cui, P.M. Vitousek et al. 2011. Integrated soil-crop system management for food security. *Proc Natl Acad of Sci, USA* 108:6399–6404.

Chen, Z., H. Wang., X. Liu, D. Lu and J. Zhou. 2016. The fate of 15 N labelled fertilizer in a wheat-soil system as influenced by fertilizer practices in a loamy soil. *Sci Rep* 6 article No. 34754 (doi: 10.1038/srep 34574(2016)).

Chhonkar, P.K. and K.C. Misra. 1978. Possible utilization of neem cake for inhibiting nitrification in soil. *J Indian Soc Soil Sci* 26:90–92.

Chivenge, P., B. Vanlauwe, R. Gentile et al. 2009. Organic and mineral input in management to enhance crop productivity in central Kenya. *Agron J.* 101(5):1266–1275.

Choudhary, P., V. Nepalia and D. Singh. 2013. Effect of weed control and sulfur on productivity of protein maize (*Zea mays*), dynamics of associated weeds and relative nutrient uptake. *Indian J Agron* 58(4):534–538.

Chowdhury, N. 1989. *Consultancy Report on Environmental Pollution*. Department of Forests, Dhaka, Bangladesh.

Cicerone, R.J. 1987. Changes in stratospheric ozone. *Science* 237:35–42.

Cole, C.V., J. Duxbury, A.R. Bishnoi et al. 1997. Global estimates of potential mitigation of greenhouse gas emission by agriculture. *Nutr Cycl Agroecosyst* 49:21–228.

Conant, R.T., A.B. Berdanier, and P.R. Grace. 2013. Patterns and trends in nitrogen use and nitrogen recovery efficiency in world agriculture. *Global Biogeochem Cycles* 27:558–566.

Coventry, D.R., A. Yadav, R.S. Poswal et al. 2011. Irrigation and nitrogen scheduling as a requirement for optimizing wheat yield and quality in Haryana, India. *Field Crops Res* 123(2):80–88.

Dahan, D., A. Babal, N. Lazarovitch, E.E. Russek and D. Kurtzman. 2014. Nitrate leaching from intensive organic farms to groundwater. *Hydrol Earth Syst Sci* 18:331–341.

Daigger, L.A., D.H. Sander, and G.A. Peterson. 1976. Nitrogen content of winter wheat during growth and maturation. *Agron J.* 68:815–818.

Daigh, A.L., M.C. Savin, K. Brye, R. Norman and D. Miller. 2014. Urea persistence in blood water and soil used for flooded rice production. *Soil Use Manage* 30(4):463–470.

Dass, A. and Chandra, S. 2012. Effect of different components of SRI on yield, quality, nutrient accumulation and economics of rice in *tarai* belt of northern India. *Indian J Agron* 49(4):504–523.

Dass, A., Kaur, R., Choudhary K et al. 2015. System of rice (Oryza sativa) intensification for higher productivity and resource-use efficiency review. *Indian J Agron* 60(1):1–19.

De Dutta, S.K. 1981. *Principles and Practices of Rice Production*. John Wiley, New York, pp. 618.

De, R. 1971. A review of fertilization of crops in India. *Fertil News (India)* 16(12):77–81.

Denmead, O.T., J.R. Freney, and J.R. Simpson. 1976. A closed ammonia cycle within a plant canopy. *Soil Biol Biochem* 8:161–164.

Devasenapathy, P. and S. Palaniappan. 1995. Band placement of urea application increases nitrogen use efficiency in transplanted lowland rice. *Int Rice Res Notes* 20.

Diby, L.N., B.T. Tie, O Girardin et al. 2011. Growth and nutrient use efficiencies of yams (Dioscorea spp.) grown in two contrasting soils of West Africa. *International Journal of Agronomy* 2011 (2011), Article ID 175958, 8 pages (http://dx.doi.org/10.1155/2011/175958).

Dick, W.A., D. Kost and L. Chen. 2008. In: *Sulphur: A Missing Link between Soils, Crops and Nutrition* (J. Jezed, Ed.), pp. 59–82. Madison, WI: American Society of Agronomy.

Ding, W.X., H.Y. Yu and Z.C. Cai. 2011. Impact of urease and nitrification inhibitors on nitrous oxide emissions from a Fluo-aquic soil in the Northern China Plain. *Biol Fertility Soils* 47(1):91–99.

Dobermann, A.R. 2005. Nitrogen use efficiency—State of the art, *Agronomy & Horticulture*—Faculty Publications. University of Nebraska, Lincoln, USA, Paper 316. (http://digitalcommons.unl.edu /agronomyfacpub/316)

Dobermann, A., and K.G. Cassman. 2002. Plant nutrient management for enhanced productivity in intensive grain production systems of the United States and Asia. *Plant Soil* 247:153–175.

DSD. 2013. Status paper on sugarcane development. Directorate of Sugarcane Development, Lucknow, India. 16 p.

Duncan, E., C. O'Sullivan., M. Roper and M. Peoples. 2016. The effect of nitrification inhibitors on wheat crop performance on coarse grained soils in Mediterranean environments. Proc. International Nitrogen Initiative Conference "Solutions to Improve Nitrogen Use Efficiency for the World." December 4–8, 2016, Melbourne, Australia.

Edwards, D.G. and L.C. Bell. 1989. Acid soil fertility in Australian tropical soils. In *Management of Acid Soils of the Humid Tropics of Asia*. ET Craswell, E Pushprajah, Eds. pp. 20–31. ACIAR, Australia Mongraph No 13, 118 p.

Erisman, J.W., M.A. Sutton, J. Gallaeway, Z. Kilmont, and W. Winiwarter. 2008. How a century of ammonia synthesis changed the world. *Nature Geo-Science* 1:636–639.

Eynard, A., R. Lal, and K. Wiebe. 2005. Crop response in salt affected soils. *J Sustain Agric* 27(1):6–50.

FAO. 1955. *The State of Food and Agriculture 1955*. Rome: FAO.

FAO. 2012. *Current World Fertilizer Trends and Outlook to 2016*. Rome: FAO.

Fashola, S.S., K. Hayashi and T. Wakatushi. 2002. Effect of water management and polyolefin coated urea on growth and nitrogen uptake of indica rice. *J Plant Nutr Soil Sci* 25:2173–2190.

Fixen, P.E. 2007. Can we define a global framework within which fertilizer best management practices can be adapted to local conditions? Proc VIV AAPRESID Cong. pp. 181–187. Rosario, Argentina, August 8–11, 2006.

Flinn, J.C., C.P. Marmaril, L.E. Valseo and K. Kaiser. 1984. Efficiency of modified urea fertilizers for tropical agriculture. *Fert Res* 13:209–221.

Forster P., V. Ramaswamy, P. Artaxo et al. 2007. Changes in Atmospheric Constituents and in Radiative Forcing, in Climate Change 2007: The Physical Science Basis. Contribution of Working Group I to the Fourth Assessment Report of the Intergovernmental Panel on Climate Change (Solomon, S., Qin, D., Manning, M. et al., Eds.), Cambridge University Press, 212 p.

Fox, R.H. and W.P. Piekielek. 1983. Response of corn to nitrogen fertilizer and the prediction of soil nitrogen availability with chemical tests in Pennsylvania. *Penn Agric Expt Sta Bull* 843.

Fox, R.H. and W.P. Piekielek. 1988. Fertilizer nitrogen equivalence of alfalfa, birds foot trefoil and red clover for succeeding crops. *J Prod Agric* 1:313–317.

Fox, R.H. and W.P. Piekielek. 1993. Management and urease inhibitor effect on nitrogen use efficiency in no-till corn. Access Digital Lib, American Society of Agronomy, Madison, WI (doi 10.2134 /jpa1993.0195).

Fox, R.H., J.J. Meisinger, J.T. Sims, and W.P. Piekielek. 1992. Predicting N fertilizer needs for corn humid regions: Advances in the Mid-Atlantic states. In *Predicting N Fertilizer Needs for Corn in Humid Regions* (B.R. Bock and K.R. Kelly, Eds.), National Fertilizer and Environmental Research Center, TVA, Muscle Shoals, AL. Bull Y-226:43–46.

Frame, W.H., M.M. Alley, G.B. Whitehurst, B.M. Whitehurst, and R. Campbell. 2012. In vitro evaluation of coatings to ontrol ammonia volatilization from surface-applied urea. *Agron J* 104(5):1201–1207.

Francis, D.D., J.S. Schepers, and M.F. Vigil. 1993. Post-anthesis nitrogen loss from corn. *Agron J* 85:659–663.

Gallagher, P.W. and W.V. Bartholomew. 1964. Comparison of nitrate production and other procedures in determining nitrogen availability in southeastern coastal plain soils. *Agron J* 85:179–184.

Gallejones, P., A. Castelloin, A. delPedro et al. 2012. Nitrogen and sulfur fertilization effect on leaching losses, nutrient balance and plant quality in wheat-rapeseed rotation under a humid Mediterranean climate. *Nutr Cycl Agroecosys* 93:337–355.

Gao, Q., C. Li, G. Feng et al. 2012. Understanding yield response to nitrogen to achieve high yield and high nitrogen use efficiency and rainfed corn. *Agron J* 104:165–168.

Gile, M., N. Morley., E.M. Baggs, and T.G. Daniell. 2012. Soil nitrate reducing process–drivers, mechanisms for special variation and significance of nitrous oxide production. *Front Microbiol* 3:407–466.

Gill, M.S. and I.P.S. Ahlawat. 2006. Crop diversification—Its role towards sustainability and profitability. *Indian J. Fert.* 2(9):125–150.

Gnanavelrajah, N. and D. Kumaragamage. 1998. Effect of neem (*Azadirachta indica* A. Juss.) materials on nitrification of applied urea in three selected soils of Sri Lanka. *Trop Agric Res* 10:61–73

Golden, B.R., N.A. Slaton, R.J. Norman et al. 2009. Evaluation of polymer-coated urea for direct-seeded, delayed-flood rice production. *Soil Sci. Soc. Am. J.* 73:375–383.

Good, A.G., S.J., Johnson, M. De Pauw et al. 2007. Engineering nitrogen use efficiency with alanine aminotransferase. *Can. J. Bot.* 85:252–262.

Gooding, M.J. and W.P. Davies. 1992. Foliar urea application of cereals: A review. *Fert Res* 32:209–222.

Gopal, R., R.K. Jat, R.K. Malik et al. 2010. Direct dry seeded rice production technology and weed management in rice based systems. *Technical Bulletin. International Maize and Wheat Improvement Center*, New Delhi, India, 28 pp.

Goron, T., V.K. Bhosekar, C.R. Shears et al. 2015. Whole plant acclimation response to finger millet to low nitrogen stress. *Front Plant Sci* 6:652–690.

Gowda, D.N. and K.G.H Setty. 1978. Comparative efficacy of neem cake and methomyl on the growth of tomato and root-knot development. *Curr Res* 7:118–120.

Grassini, P. and K.G. Cassman. 2012. High yield maize with large net energy yield and small global warming intensity. *Proc Natl Acad Sci, USA* 109:1074–1079.

Gui Xin, C., Z. Zhao Linag, A.C.F. Trevitt, J.R. Freney and J.R. Simpson. 1986. Nitrogen loss from ammonium bicarbonate and urea applied to flooded rice. *Fert Res* 10(3):203–215.

Gupta, S.K. 2008. Genetic modification: A significant biotechnological control. *Indian Farming* 58(3): 7–17.

Halvorson, A.D., C.S. Snyder, A.D. Blaylock and S.J. Del Grosso. 2014. Enhanced-efficiency nitrogen fertilizers: Potential role in nitrous oxide emission mitigation. *Agron J.* 106(2):715–722.

Hamamoto, M. 1966. Isobutylidene Diurea as a slow acting nitrogen fertiliser and the studies in this field in Japan. Lecture 40, The Fertilizer Society, U.K. 77 p.

Han, M., M. Okamoto, P.H. Beatty et al. 2015. The genetics of nitrogen use efficiency in crop plants. *Ann Rev Gen* 40:269–289.

Harper, L.A., R.R. Sharpe, G.W. Langdale and J.E. Giddens. 1987. Nitrogen cycling in wheat crop: Soil, plant and aerial transport. *Agron J* 79:965–973.

Harris, G.H. and O.B. Hestermann. 1990. Quantifying nitrogen contribution from alfalfa to soil and two succeeding crops using nitrogen-15. *Agron J* 82:129–134.

Hayashi, K., S. Nishimura and K. Yagi. 2008. Ammonia volatilization from a paddy field following application of urea: Rice plants are both ab absorber and an emitter for atmospheric ammonia. *Sci Total Environ* 390(2):485–494.

He, Y., S. Yang, J. Xu, Y. Wang and S. Peng. 2014. Ammonia volatilization losses from paddy fields under controlled irrigation with different drainage treatments. *Sci World J* 2014(2014) Article ID 417605 (http://dx.doi.org/10.1155/2014/417605).

He, Z., M. Zhang, and M.J. Wilson. 2004. Distribution and classification of red soils of China. In *Red Soils of China*. Eds. M.J. Wilson et al. pp. 29–33. Dordrech, Netherlands: Kluwer Academic.

Heady, E.O. and W.D. Shrader. 1953. The interrelationships of agronomy and economics in research and recommendations to farmers. *Agron J* 45:496–502.

Heffer, P. 2013. Assessment of Fertilizer Use by Crop at the Global Level 2010-2010/11, International Fertilizer Industry Association, Paris.

Hobbs, P.R. and R.K. Gupta. 2003. Resource conservation techniques for wheat in rice-wheat systems. In *Improving the Sustainability of Rice-Wheat Systems: Issues and Input*. Eds. J.K. Ladha, R.K. Gupta, J. Duxbury and R.J. Buresh, *Am Soc Agron Spec Pub* 65:149–171.

Hobbs, P.R., K.D. Sayre and J.I. Oritz-Monasterio. 1998. Increasing wheat yield sustainability through agronomic uses. CIMMYIT NRG paper 98-01, Mexico.

Hobbs, P.R., K.D. Sayre and R.K. Gupta. 2008. The role of conservation agriculture in sustainable agriculture. *Phil Trans Royal Soc* B 363:543–555.

Howarth, R.W., E.W. Boyer, W.J. Pabich and J.M. Galloway. 2002. Nitrogen use in the United States from 1961–2000 and future potential needs. *Ambio* 31:88–96.

Hsu, C. 1980. *Han Agriculture*. Seattle, WA: University of Washington Press.

Huaqi, W., B.A.M. Bouman, D. Zhao et al. 2002. Aerobic rice in northern China: Opportunities and challenges. In *Water-Wise Rice Production*, Eds. B.A.M. Bouman, H. Hengsdijk, B. Hardy et al. pp. 143–154. Los Banos, Philippines: International Rice Research Institute.

IPCC. 2007. *Climate Change-2007. Mitigation of Climate Change*. New York: Cambridge University Press.

IRRI. 1997. *Rice Almanac,* 2nd ed. Los Banos, Philippines: International Rice Research Institute.

Jat, M.L., Bijai-Singh and B. Gerard. 2014a. Nutrient management and use efficiency in wheat system of South Asia. *Adv Agron* 125:171–259.

Jat, M.L., Yadvinder-Singh, H.S. Sidhu et al. 2014b. Managing crop residues for sustainable crop production systems. In *Textbook of Plant Nutrient Management*. Eds. R. Prasad, D. Kumar, D.S. Rana et al. pp. 326–347. New Delhi: Indian Society of Agronomy.

Jat, R.K., S.S. Punia, and R.K. Malik. 2007. Effect of different herbicide treatments on nutrient uptake behavior of weeds and wheat (*Triticum aestivum* L). *Indian J Weed Sci* 39:135–137.

Jin, J. 2012. Changes in the efficiency of fertilizer use in China. *J Sci Food Agric* 92:1006–1009.

Joergensen, R.G. 2010. Organic matter and microorganisms in tropical soils. In *Soil Biology and Agriculture in the Tropics*. Ed. P. Dion, pp. 17–34. Berlin, Heidelberg: Springer Verlag.

Johnson, D.E., M.C.S. Wopereis, D.M. Bodj et al. 2007. Timing of weed management and yield losses due to weeds in irrigated rice in the Sahel. *Field Crops Res* 85(1):31–42.

Joseph, P. and R. Prasad. 1993. The effect of dicyandiamide and neem cake on the nitrification of urea-derived ammonium under field conditions *Biol Fertility Soils* 15(2):149–152.

Ju, X.T., G.X. Xing, X.P. Chen et al. 2009. Reducing environmental risk by improving N management in intensive Chinese agricultural systems. *Proc Natl Acad Sci, USA* 106:3041–3046.

Kabir M.H., M. Sarkar, and A. Chowdhury. 2009. Effect of urea super granules, prilled urea and poultry manure on the yield of transplanted aman rice varieties. *J Bangladesh Agric Univ* 7(2):259–263.

Kaherl, F., L. Yunju, D. Ronald-Host et al. 2010. Toward sustainable use of nitrogen fertilizers in China. University of California, Berkeley, Giannini Foundation of Agricultural Economics. *ARE Update* 14(2):5–7.

Kahn, S.A., R.L. Mulvaney, and R.G. Hoeft. 2001. A simple test for determining sites that are non-responsive to nitrogen fertilization. *Soil Sci Soc Am J* 65:1751–1760.

Karlen, D.L., G.E. Varvel, D.G. Bullock, and R.M. Cruse. 1994. Crop rotations for the 21st Century. *Adv Agron* 53:1–45.

Katsura, K., S. Maeda, and L. Lubis et al. 2008. The highest yield of irrigated rice in Yunnan in China: A cross-location analysis. *Field Crops Res* 107(1):1–11.

Katyal J.C., B. Singh, P.L.G. Vlek, and R. Buresh. 1987. Efficient nitrogen use as affected by urea application and irrigation sequence. *Soil Sci Soc Am J* 51(2):366–370.

Kaur, R., S.S. Mahal and R.K. Mahey. 2003. Growth, yield and water use efficiency of bed planted late sown wheat under different irrigation and seed rate levels. *Haryana J Agron* 19:48–52.

Kaushal, A.K., N.S. Rana, A. Singh et al. 2010. Response of levels and split application of nitrogen in green manured wetland rice (*Oryza sativa* L.). *Asian J Agric Sci* 2(2):42–46.

Keys, A.J., I.F. Bird, M.J. Cornelius et al. 1978. Photorespiratory nitrogen cycle. *Nature* 275:741–743.

Khurrana, H.S., Bijai-Singh, A. Doberman et al. 2008. Site-specific nutrient management in a rice-wheat cropping system. *Better Crops, India* 1:26–28.

Kiss, S. and M. Simihian. 2002. *Improving Efficiency of Urea Fertilizers by Inhibition of Soil Urease Activity.* Dordrecht, Netherlands: Kluwer Academic.

Knobeloch, L., B. Saha, A. Hogan, J. Pestle, and H. Anderson. 2000. Blue babies and nitrate contaminated well water. *Environ Health Perspec* 108(7):675–678.

Kondo, M.C., C.V. Singh, R. Agbisit and M.V.R. Murty. 2005. Yield response to urea and controlled-release urea as affected by water supply in tropical upland rice. *J Plant Nutr* 28:201–219.

Kroetz, C. 1994. Nitrous oxide and global warming. *Sci Total Environ* 143(2):193–209.

Kukal, S.S., Humphreys, E., Yadvinder-Singh et al. 2005. Performance of raised beds in rice-wheat system in northwestern India. In *Evaluation and Performance of Raised Bed Cropping Systems in Asia, Australia and Mexico,* Proceedings of a Workshop, March 1–3, 2005, Griffith, Australia (CH Roth, RA Fischer, CA Meisner Eds.), Australian Center for International Agricultural Research Canberra, Australia. *ACIAR, Proc No* 121:26–40.

Kumar, A. 2015. Government allows fertilizer firms 100% production of neem coated urea. *The Economic Times* (India) January 7, 2015.

Kumar, N. and Prasad, R. 2004. Nitrogen uptake and apparent N recovery by a high yielding variety and a hybrid of rice as influenced by levels and sources of nitrogen. *Fert. News (India)* 49(5): 65–67.

Kumar, N., H. Singh and V.P. Singh. 2013. Productivity and water use efficiency of spring planted sugarcane (*Saccharum* sp hybrid complex) under various planting methods and irrigation regimes. *Indian J Agron* 58(4):592–596.

Kumar, N., R. Prasad and F.U. Zaman. 2007. Relative response of a high yielding variety and a hybrid of rice to levels and sources of nitrogen. *Proc Indian Natn Sci Acad* 73(1):1–6.

Kundu, M.C., B. Mandal and D. Sarkar. 2008. Assessment of the potential hazards of nitrate contamination in surface and groundwater in a heavily fertilized and intensively cultivated didtyrict of India. *Environ Monitor Assess* 146:183–189.

Kuykendall, J.R. and A. Wallace. 1953. Urea nitrogen as foliar spray application to citrus-studies for effects on plant growth, leaf burn, root activity and fruit quality. *California Agric* March 1953:6

Kwaga, Y.M. 2014. Effect of nitrogen application on the response of maize (*Zea mays* L) varieties at Mubi, North Guinea Savanna of Nigeria. *Am J Res Commu* 2(2):71–81.

Ladha, J.K., A.T. Padre, C.K. Reddy, K.G. et al. 2016. Global nitrogen budgets in cereals: A 50-year assessment for maize, rice, and wheat production systems. *Scientific Reports* 6:19355, doi: 10.1038/srep19355.

Ladha, J.K., D. Dave, H. Pathak et al. 2003. How extensive are yield declines in long-term rice-wheat experiments in Asia. *Field Crops Res* 81:159–180.

Ladha, J.K., H. Pathak, T.J. Krupnik, J. Six, and C. von Kassel. 2005. Efficiency of fertilizer nitrogen in cereal production: Retrospect and perspectives. *Adv Agron* 87:85–156.

Lal, R. 2005. Enhancing crop yields in the developing countries through restoration of soil organic carbon pool in agricultural soils. *Land Degrad Dev* 17:197–209.

Lal, R. 2013a. Food security in changing climate. *Ecohydrol Hydrobiol* 13:8–21.

Lal, R. 2013b. Climate-resilient agriculture and soil organic carbon. *Indian J Agron* 58(4):440–450.

Le, C., Y. Zha, Y. Li, D. Sun, H. Lu and B. Yin. 2010. Eutrophication of lake waters in China: Cost, causes and control. *Environ Manage* 45(4):662–668.

Lea, P.J. and R.J. Ireland. 1999. Nitrogen metabolism in higher plants. *In Plant Amino Acids, Biochemistry and Biotechnology,* Ed. B. Singh, pp. 1–47. New York: Marcel Dekker.

Leiwakabessy, F.M. 1989. Management of humid tropical soils in Indonesia. In *Management of Acid Soils of the Humid Tropics of Asia.* Eds. E.T. Craswell, E. Pushprajah, pp. 54–61. ACIAR, Australia Mongraph No 13.

Li, Y., Y. Chen, and C.Y. Wu et al. 2015. Determination of optimum nitrogen application rates in Zheijing Province, China based on rice yield and ecological security. *J Integrative Agric* 14(12):2426–2433.

Limon-Ortega, A., K.D. Sayre, and C.A. Francis. 2000. Wheat nitrogen use efficiency in a bed planting system in northwest Mexico. *Agron J* 92:303–308.

Lin, C., C. Wang, and X. Zhong. 2013. Effect of nitrification inhibitors DCD and DMPP on nitrous oxide emission, crop yield and nitrogen uptake in wheat-maize cropping system. *Biogeoscience* 10:2427–2537.

Lin, Y. and S. Liu. 2004. Effect of sulfur on the yield and quality of spring wheat. *Soil Fert* 1:14–15.

Lindquist, B.A., L. Liu, C. van Kassel, and K.J. van Groeingen. 2013. Enhanced efficiency nitrogen fertilizers for rice systems: Meta-analysis of yield and nitrogen uptake. *Field Crops Res* 154:244–259.

Liu, C., K. Wang and X. Zhang. 2013. Effect of nitrification inhibitors (DCD and DMPP) on nitrous oxide emission, crop yield and nitrogen uptake in a wheat-maize cropping system. *Biogeoscience* 10:2427–2437.

Liu, G., H. Yu., G. Zhang et al. 2016. Contribution of wet irrigation and nitrification inhibitors to reduced $N_2O$ and methane emissions from a rice cropping system. *Environ Pollut Res Int* 23(17):17425–174236.

Mahapatra, I.C. and S.K. Pattanayak. 2008. Acid soils and their management—A case study with paper mill sludge. *NAAS News* 8(4): 6–7, National Academy of Agricultural Sciences, New Delhi, India.

Majeed, A., A. Muhmood, A. Niaz et al. 2015. Bed planting of wheat (*Triticum aestivum* L.) improves nitrogen use efficiency and grain yield compared to flat planting. *The Crop J* 3(2):118–124.

Maji, A.K., G.P. Obi and S. Meshram. 2008. Acid soil map of India. *NBSS&LUP, Nagpur, India. Ann Rept 2008.*

Malhi, S.S. 2012. Increasing organic carbon and nitrogen fertility in sulfur deficient soil with sulfur fertilization. *Biol Fertility Soils* 48:735–739.

Malkouti, H.J., A. Bazbordi, M. Lotfalahi et al. 2008. Comparison of complete and sulfur coated urea fertilizer with pre-plant urea in increasing grain yield and nitrogen use efficiency in wheat. *Journal of Agricultural Sciences and Tech*nology 10:173–183.

Mandal, U.K., K.L. Sharma, J.V.N.S. Prasad et al. 2012. Nutrient losses by runoff and sediment from an agricultural field in semi-arid tropical India. *Indian J Dryland Agric Res Dev* 27(1):1–9.

Mandal, U.K., S.N. Sharma, G. Singh and U.K. Das. 1999. Effect of *Sesbania* green manuring and green gram (*Phsaseolus radiatus*) residue incorporation on physical properties of soil under rice-wheat cropping system. *Indian J Agric Sci* 69(9):615–620.

Massoud, F.I. 1977. Basic principles and prognosis and monitoring of salinity and sodicity. *Proc Int Conf Managing Saline Water for Irrigation.* Texas Tech University, Lubbock, August 16–20, 1976, pp. 432–454.

McAllister, C.H., P.H. Beatty, and A.G. Good. 2012. Engineering nitrogen use efficient crop plants: The current status. *Plant Biotech J* 10 (9):1011–1015.

McCray, J.M., K.T. Morgan and L. Baucum. 2016. Nitrogen fertilizer recommendations for sufarcane production in Florida sand soils. University of Florida, UF-IFAS Extension SS-AGR-401.

Meena, B.L. and R.K. Singh. 2013. Response of wheat (*Triticum aestivum*) to rice (*Oryza sativa*) residue and weed management practices. *Indian J Agron* 58(4):521–524.

Mehla, R.S., J.K. Verma, R.K., Gupta and P.R. Hobbs (eds.). 2000. *Stagnation in the productivity of wheat in the Indo-Gangetic plains: Zero-till-seed-cum-fertilizer drill as an integrated solution.* Rice–Wheat Consortium for the Indo-Gangetic Plains, New Delhi, India, 9 p.

Mendoza, T.L. 2010. AWD goes top Africa. *Ripple* 5(2), 7.

Menete, M.Z.L., H.M. Vane., R.M.L. Brito et al. 2008. Evaluation of system of rice intensification component practices and their synergies on salt affected soils. *Field Crops Res* 109(1–3):34–44.

Messick, D.L. 2003. Sulfur fertilizers—A global perspective. *Proc TSI/FAI.IFA Workshop on Sulfur in Balanced Fertilization.* Eds. M.C. Sarkar, B.C. Biswas, S. Das et al. February 25–26, 2003, The Fertilizer Association of India, New Delhi, India.

Mishra, J.S., S.S. Rao, and A. Dixit. 2012. Evaluation of a new herbicide for control and crop safety in rainy season sorghum. *Indian J Weed Sci* 44:71–72.

Mitchell, C.C. and G.L. Mullins. 1990. Sources, rates and time of sulfur application to wheat. *Sulfur Agric* 14:20–24.

Mobley, H.L. and R.P. Hausinger. 1989. Microbial ureases: Significance, regulation and molecular character-ization. *Microbiol Rev* 53(1):85–108.

Moline, H.E. and J.C. Locke. 1993. Comparing neem seed oil with calcium chloride and fungicides for con-trolling post-harvest apple decay. *Hort Sci* 28:719–720.

Mondall, M.H. 2011. Causes of yield gap and strategies for minimizing the gaps in different crops of Bangladesh. *Bangladesh J Agric Res* 36:469–476.

Moreau, D., G. Millard and N. Munier-Jolan. 2013. A plant nitrophily index based on leaf arera response to soil nitrogen availability *Agron Sustain Dev* 33:809–815.

Mueller, N.D., P.C. West, J.S. Gerber et al. 2014. A trade-off frontier for global nitrogen use and cereal produc-tion. *Environ Res Letters* 9(2014) 054002 (8pp) (doi:10.1088/1748-9326/9/5/054002).

Naik, S., N. Krishnamurthy and C. Ramachandran. 2015. Effect of nutrient sources on grain yield, methane emission and water productivity of rice (*Oryza sativa*) under different methods of cultivation. *Indian J Agron* 60(2):249–254.

Nakashima, J., T. Awano, K. Takebe, M. Fujita, and H. Saiki. 1997. Immunocytochemical localization of phenylal-anine in differentiating tracheary elements derived from zinnia mesophylls. *Plant Cell Physiol* 38:113–123.

Nambiar, K.K.M. 1994. *Soil Fertility and Crop Productivity under Long Term Fertilizer Use*, New Delhi: Indian Council of Agricultural Research.

Naresh, R.K., B. Singh, S.P. Singh et al. 2012. Furrow irrigated raised bed (FIRB) planting technique for diversification of rice-wheat system for western IGP region. *Int. J. Life Sci Biotech Pharm Res* 1(3):134–141.

Nemati, A.R. and R.S. Sharifi. 2012. Effect of rates of nitrogen application timing on yield, agronomic charracteristics and nitrogen use efficiency in corn. *Int J Agric Crop Sci* 4(9):534–539.

Oberley, S.L. and L.G. Bundy. 1987. Ammonia volatilization from nitrogen fertilizer surface applied to corn (*Zea mays*) and grass pasture (*Dactylus glomerata*). *Biol Fertility Soils* 4(4):185–192.

Olea, F., A. Perez-Garcia, F.R. Canton et al. 2004. Up-regulation and localization of asparagine synthetase in tomato leaves in tomato leaves infected by the bacterial pathogen *Pseudomonas syringae*. *Plant Cell Physiol* 45:770–780.

Papkosta, D.K. and A.A. Gagianas. 1991. Nitrogen and dry matter accumulation, remobilization and loss for Mediterranean wheat during grain filling. *Agron J* 83:864–870.

Parkin, T.B. and J.L. Hatfield. 2010. Influence of nitrapyrin on $N_2O$ losses from soil receiving fall applied anhydrous ammonia. *Agric Ecosyst Environ* 136:81–86.

Pathak, H., J. Tismina, E. Humphreys and D. Godwin. 2004. Simulation of rice crop performance and water and nitrogen dynamics, and methane emission in northwest India using CERES Rice model. *Tech Rept CSIRO Land & Water* 23(4):111.

Pathak, H., C. Li, R. Wasserman and J.K. Ladha. 2006. Simulation of nitrogen balance in the rice-wheat sys-tems for the Indo-Gangetic plains. *Soil Sci Soc Am J* 70:1612–1622.

Patil, M.D., B.S. Das, E. Barak et al. 2010. Performance of polymer coated urea in transplanted rice: Effect of mixing ratio and water input in Nitrogen use efficiency. *Paddy & Water Environ* 8(12):189–198.

Patra, D.D., M., Anwar, S. Chand et al. 2002. Nimin and *Mentha spicata* oil as nitrification inhibitors for optimum yield of Japanese mint. *Commu Soil Sci Plant Anal* 33:451–460.

Patrick, W.H. Jr., F.J. Peterson and F.T. Turner. 1968. Nitrification inhibitors for lowland rice. *Soil Sci.* 105: 103–105.

Paul, T., P.S. Bisht, P.C. Pandey et al. 2013. Rice productivity and soil fertility as influenced by nutrient management in rice (*Oryza sativa*)—wheat (*Triticum aestivum*) cropping system. *Indian J Agron* 5894: 495–499.

Peng, S., R.J. Buresh, J. Hung et al. 2010. Improving nitrogen fertilizers in rice by site specific N management—A review. *Agron Sustain Dev* 30:649–656.

Philippot, L., J. Andert, C. M. Jones, D. Bru, and S. Hallin. 2011. Importance of denitrifiers lacking the genes encoding the nitrous oxide reductase for $N_2O$ emissions from soil. *Glob. Change Biol.* 17:1497–1504.

Pinheiro, B. da S., E. da M. de Castro, and C.M. Guimaraes. 2006. Suitability and profitability of aerobic rice production in Brazil. *Field Crops Res* 97:34–42.

Ponnamperuma, F.N. 1972. The chemistry of submerged soils. *Adv Agron* 24:29–97.

Portmann, R.W., J.S. Daniel, and A.R. Ravishankara. 2012. Stratospheric ozone depletion due to nitrous oxide: Influences of other gases. *Phil Trans Royal Soc* B367: 1256–1264.

Prakasa Rao, E.V.S., Y.S. Shivay, D.D. Patra et al. 2016. Indigenous materials for increasing efficiency of fer-tilizer nitrogen. *Souvenir, 4th International Agronomy Congress*, November 22–26, 2016, New Delhi.

Prasad, M. and Prasad, R. 1980. Yield and nitrogen uptake by rice as affected by variety, method of planting and new nitrogen fertilizers. *Fert Res* 1(4):207–213.

Prasad, R. 1965. Determination of potentially available nitrogen in soils—A rapid procedure. *Plant Soil* 23:261–263.

Prasad, R. 1998. Fertilizer urea, food security, health and the environment. *Current Sci* 75(7):677–683.

Prasad, R. 2005a. Rice-wheat cropping systems. *Adv Agron.* 86:255–339.

Prasad, R. 2005b. Research on nitrification inhibitors and slow-release fertilizers in India—A review. *Proc Natl Acad Sci, India Sect B* 75:149–157.

Prasad, R. 2009. Efficient fertilizer use: The key to food security and better environment. *J Trop Agric* 47(1–2):1–17.

Prasad, R. 2011. Aerobic rice systems. *Adv Agron* 111:207–247.

Prasad, R. 2013a. Fertilizer nitrogen, food security, health and the environment. *Proc Indian Natn Sci Acad* 79(4):997–1010.

Prasad, R. 2013b. Soil test based fertilizer recommendations in India—How realistic? How reliable. *Indian J Fert* 9(3):16–19.

Prasad, R. and J.F. Power. 1991. Crop residue management. *Adv Soil Sci* 15:205–251.

Prasad, R. and J.F. Power. 1995. Nitrification inhibitors for agriculture, health and the environment. *Adv Agron* 54:233–287.

Prasad, R. and J.F. Power. 1997. *Soil Fertility Management for Sustainable Agriculture.* Boca Raton, FL: CRC Press.

Prasad, R. and Shivay, Y.S. 2016a. Deep placement and foliar fertilization of nitrogen for increased use efficiency—An overview. *Indian Journal of Agronomy* 61(4):71–75.

Prasad, R. and Y.S. Shivay. 2015. Fertilizer nitrogen for life, agriculture and the environment. *Indian J Fert* 11(8):47–53.

Prasad, R. and Y.S. Shivay. 2016b. Sulfur in soils, plant and human nutrition. *Proc Natl Acad Sci, Sect B. Biol Sci* (doi 10.1007/s40011-016-0769-0).

Prasad, R. G.B. Rajale and B.A. Lakhdive. 1971. Nitrification retarders and slow-release nitrogen fertilizers. *Adv Agron* 23:357–383.

Prasad, R., B. Gangiah, and K.C. Aipe. 1999. Effect of crop residue management in a rice-wheat cropping system on growth and yield of crops and on soil fertility. *Exp Agric* 35:427–435.

Prasad, R., C. Devakumar, and Y.S. Shivay. 1993. Significance in increasing fertilizer nitrogen efficiency. In: *Neem Research and Development,* Eds. N.S. Randhawa and B.S. Parmar, pp. 97–198. New Delhi: Pesticide Association of India.

Prasad, R., D. Kumar, Y.S. Shivay, and R. Raj. 2014. Integrated nutrient management. In *Textbook of Plant Nutrient Management,* Eds. R. Prasad, D. Kumar, D.S. Rana et al. pp. 348–360. New Delhi: Indian Society of Agronomy.

Prasad, R., G. Singh, Y.S. Shivay, and H. Pathak. 2016c. Need for determining ecofriendly optimum fertilizer nitrogen level for better environment and for alleviating hunger and malnutrition. *Indian J Agron* 61(1): 1–8.

Prasad, R., S. Singh, and R. De. 1984. Effect of method of application on the relative efficiency of prilled urea and urea supergranules for rice, *Indian J Agric Sci* 103(3):539–542.

Prasad, R., S.N. Sharma, M. Prasad, and R.N.S. Reddy. 1981. Efficient utilization of nitrogen in rice-wheat rotation. Indian Society of Agronomy National Seminar, Hisar, Haryana, India.

Prasad, R., Singh, S., Saxena, V.S., and C. Devakumar. 1994. Coating prilled urea with neem (*Azadirachta indica* Juss) oil for efficient nitrogen use. *Naturwissenschaften* 86:538–539.

Prasad, R., Y.S. Shivay, and D. Kumar. 2016a. Interactions of zinc with other nutrients in soils and plants. A review. *Indian J Fert* 12(5):16–26.

Prasad, R., Y.S. Shivay, and Y.L. Nene. 2016b. Asia's contribution to the evolution of agriculture: Creativity, history and mythology. *Asian Agri-History* 20(4):233–250.

Prasad, R., Y.S. Shivay, D. Kumar et al. 2007. Neem for sustainable agriculture and the environment—A review. *Proc Nanl Acad Sci, India Section B* 77:313–330.

Prasad, R., Y.S. Shivay, K. Majumdar, and S. Prasad. 2016d. Phosphorus management. In *Soil Phosphorus.* Eds. R. Lal and A.B. Stewart. pp. 81–113, Boca Raton, FL: CRC Press.

Prasertek, P., J.R. Freney, O. Denmead and J. Reghenzami. 2002. Effect of fertilizer placement on nitrogen loss from sugarcane in tropical Queensland. *Nutr Cycl Agroecosyst* 62(3):229–239.

Rajale, G.B. and R. Prasad. 1974. Relative efficiency of urea, nitrification inhibitor treated urea and slow-release nitrogen fertilizers for rice. *J Agric Sci Cambridge* 83(2):303–308.

Raju, R., K. Reddy, and M. Hussain. 1989. Urease inhibitors and coated urea fertilizers. *Indian J Agric Sci* 59(12):817–819.

Rana, S.S., N.N. Angiras, and G.D. Sharma. 2000. Effect of herbicides and interculture on nutrient uptake by puddled rice and associated weeds. *Indian J Weed Sci* 32:72–73.

Randall, G.W. and J.A. Vetsch. 2005. Corn production on a subsurface-drained mollisol as affected by fall versus spring application of nitrogen and nitrapyrin. *Agron. J.* 97:472i478.

Randall, G.W., J.A. Vetsch, and J.R. Huffman. 2005. Corn production on a sub-surface drained mollisol as affected by time of nitrogen application and nitrapyrin. *Agron J* 95:1213–1219.

Rao, P.V., G. Subbiah, and R.R. Veeraraghaviah. 2014. Agronomic response of maize to plant population and nitrogen availability—A review. *Int J Plant Anim Environ Sci* 4(1):107–116 (www.ijpes.com).

Rashid, M., Bajwa, I.M., Hussain, R. et al. 1992. Rice response to sulfur in Pakistan. *Sulfur Agric* 16:3–5.

Rashid, M.H., M.M. Alam, and J.K. Ladha. 2012. Comparative efficiency of peritachlor and hand weeding in managing weeds and improving the productivity and net income of wet seeded rice in Bangladesh. *Field Crops Res* 128:17–26.

Ray, D.K., N. Ramankutty, N.D. Mueller et al. 2012. Recent patterns of crop yield growth and stagnation. *Nature Communications* 3. Article No 1293 (2012) (doi: 10. 1038/ncommuns2296).

Raychaudhari, S.P. and R. Mira. 1993. Agriculture in in Ancient India—A Report. New Delhi: Indian Council of Agricultural Research.

Reddy, R.N.S. and R. Prasad. 1975. Studies on the mineralization of urea, coated urea and nitrification inhibitor treated urea in soil. *J Soil Sci* 26(3):217–223.

Rheinbaben, W.V. 1990. Nitrogen losses from agricultural soils through denitrification—A critical review. *J Plant Nutr* 153(3):157–166.

Rong-Ye, C. and Z. Zhao Wang. 1982. Characteristics of the fate and efficiency of nitrogen in supergranules of urea. *Fert Res* 3(1):63–71.

Rossner, J. and C.P.W. Zebitz. 1987. Effect of neem products on nematodes and growth of tomato *(Lycopersicun esculentum)*. In *Natural Pesticides from the Neem Tree (Azadirachta indica A. Juss) and Other Tropical Plants*. Eds. H. Schmutterer and K.P.S. Ascher, pp. 611–621 (Proc 3rd Int. Neem Conf., Nairobi), Eschbom, Germany, GTZ.

Rubio, J.L. and R.D. Hauck. 1986. Uptake and use pattern of nitrogen from urea, oxamide and isobutylidene diurea by rice plants. *Plant Soil* 94(1):109–123.

Rutting, T., P. Boeckx, C. Mueller, and L. Klemedtsson. (2011). Assessment of the importance of dissimilatory nitrate reduction to ammonium for the terrestrial nitrogen cycle. *Biogeosciences* 8:1779–1791.

Sahrawat, K.L. and B.S. Parmar. 1975. Alcohol extract of neem *(Azadirachta indica)* seeds as nitrification inhibitor. *J Indian Soc Soil Sci* 23:131–134.

Sahu, R., M.K. Singh and L. Yadav. 2015. Yield and economics as influenced by nitrogen scheduling, weed management and rice establishing methods in transplanted rice. *Indian J Agron* 60(2):261–266.

Saigusa, M. 2005. New fertilizer management to maximize yield and minimize environmental effects in rice culture. Rice Is Life: Scientific Perspectives for the 21st Century. *Proc World Rice Res Conf,* November 4–7, 2004, Tsukuba, Japan, pp. 372–373.

Saleem, I., S. Javid, R.A. Sial et al. 2013. Substitution of soil application of urea with foliar application to minimize the wheat yield losses. *Soil Environ* 32(2):141–145.

Salman, O.A. 1988. Polymer coating of urea prills to reduce dissolution. *J Agric Food Chem* 36:661–621.

Sanchez, P.A. and T.J. Logan. 1992. Myths and science about the chemistry and fertility of soils in the tropics. In *Myths and Science of Soils of the Tropics*. Eds. R. Lal and P. A. Sanchez, pp. 34–46. Madison, WI: Soil Science Society of America Special Publication No. 29.

Santhi, S.R., S.P. Palaniappan, and D. Purshottaman. 1986. Influence of neem leaf on nitrification in a lowland rice soil. *Plant Soil* 93:133–135.

Sarkar, M.C., N.K. Banerjee, D.S. Rana, and K.S. Uppal. 1991. Field measurement of ammonia volatilization loss of urea applied to wheat. *Fertil News (India)* 36(11):25–28.

Savant, N.K. and P.J. Stangel. 1990. Deep placement of urea super granules in transplanted rice: Principles and practices. *Fert Res* 25(1):1–83.

Saxena, R.C. and Z.R. Khan. 1985. Effect of neem oil on survival of *Nilaparvata lugens (Hornoptera: Delphacidae)* and on grassy stunt virus transmission. *J Econ Ent* 78:647–665.

Sayre, K.D. and O.H. Moreno Ramos. 1997. Application of raised bed planting systems to wheat. *CIMMYT, Mexico Wheat Spl Rept No 31.*

Sayre, K.D., A. Limon-Ortega, and B. Govaerts. 2005. Experiments with permanent raised bed planting system, CIMMYT, Mexico. In *Evaluation and Performance of Raised Bed Cropping Systems in Asia, Australia and Mexico,* Proceedings of a Workshop, March 1–3, 2005, Griffith, Australia (CH Roth, RA Fischer, CA Meisner Eds.), Australian Center for International Agricultural Research Canberra, Australia. *ACIAR, Proc No 121:12–40.*

Schjoerring, J.K., S. Husted, G. Mack et al. 2000. Physiological relation of plant—Atmospheric ammonia exchange. *Plant Soil* 211:95–102.

Scott Murrel, T. 2011. Nitrogen use efficiency: A midwest perspective. Nutrient Efficiency and Management Conference, Rochester, MN, February 2011.

Sen, A., V.K. Srivastava, M.K. Singh et al. 2011. Leaf colour chart vis-a-vis nitrogen management in different rice genotypes. *Am J Plant Sci* 2:223–236.

Shan, L., Y. He., J. Chen et al. 2015. Nitrogen surface runoff losses from a Chinese cabbage field under different nitrogen treatments in the Taihu lake basin, China. *Agric Water Manage* 159:255–263.

Sharma, K.C. and V. Singh. 1966. Response of Mexican wheat to nitrogen. *Fertil News (India)* 11(9):18–22.

Sharma, N.K. and P.K. Batra. 2011. A study on fertilizer response ratios for various crops and cropping sequences. Project Report. New Delhi: Indian Agricultural Statistics Research Institute.

Sharma, P.K., and S.K. De Datta. 1985. Effect of puddling on soil physical properties and processes. In: *Soil Physics and Rice*. Los Banos, Philippines: IRRI, pp. 217–234.

Sharma, S.N. and R. Prasad. 2008. Effect of crop residue management on the production and nitrogen use efficiency in a rice-wheat cropping system. *J. Plant. Nutr. Soil Sci.* 171:295–302.

Sharma, S.N. and R. Prasad. 2002. Effect of crop residue with and without a culture of cellulolytic fungi on yield, NPK uptake and soil fertility under rice-wheat cropping system. *Arch. Acher. Pfl. Boden.* 48:363–370.

Sharma, S.N. and R. Prasad. 1995. Use of nitrification inhibitors (neem and DCD) to increase nitrogen efficiency in maize-wheat cropping system. *Fert Res* 44(3):169–175.

Sharma, S.N., R. Prasad, S. Singh and P. Singh. 2000. On-farm trials of the effect of introducing a summer green manure of mungbean on the productivity of rice-wheat cropping system. *J Agric Sci, Cambridge* 134:169–172.

Sharpe, R.R. and L.A. Harper. 1997. Apparent atmospheric nitrogen loss from hydroponically grown corn. *Agron J* 89:605–609.

Shaviv, A. 2001. Advances in controlled-release fertilizers. *Adv Agron* 71:1–49.

Shaviv, A., and R.L. Mikkelsen, 1993. Controlled-release fertilizers to increase efficiency of nutrient use and minimize environmental degradation: A review. *Fert. Res.* 35:1–12.

Shihua, T., and F. Wenqiang. 2000. Nutrient management in the rice–wheat cropping system in the Yangtze river flood plain. In Soil and Crop Management Practices for Enhanced Productivity of the Rice–Wheat Cropping System in the Sichuan Province of China. Eds. P. R. Hobbs and R. K. Gupta, pp. 24–34. Rice–Wheat Consortium for the Indo-Gangetic Plains, New Delhi, India.

Shivay, Y.S., D. Kumar, and R. Prasad. 2005. Iron pyrites for reducing ammonia volatilization losses from fertilizer urea applied to a sandy clay loam soil. *Curr Sci* 89(5):742–743.

Shivay, Y.S., V. Pooniya, R. Prasad et al. 2016. Sulphur-coated urea as a source of sulphur and an enhanced efficiency of nitrogen fertilizer for spring wheat. *Cereal Res Commu* 44(3):513–523.

Shrawat, A.K., R.T. Carroll, M. DePauw et al. 2008. Genetic engineering of improved nitrogen use efficiency in rice by the tissue-specific expression of alanine aminotransferase. *Plant Biotechnol. J.* 6:722–732.

Shukla, A.K., J.K. Ladha, V.K. Singh et al. 2003. Calibrating leaf color chart for nitrogen management in different genotypes of rice and wheat in a system perspective. *Agron J* 96(6):1606–1621.

Siddiq, E.A. 2006. Genetic improvement of rice in perspective. *Indian Farming* 56(7):10–16.

Sikdar, R. and X. Jian. 2014. Urea super granules (USG) as key conductor in agricultural productivity development in Bangladesh. *Home* 4(6):132–139.

Silver, W.L., A.W. Thompson, A. Reich et al. 2005. Nitrogen cycling in tropical plantation forests: Potential controls on nitrogen retention. *Ecol. Appl.* 15:1604–1614.

Singh, A., and A.S. Kharub. 2000. Performance of zero tillage in wheat, evidences from participatory research. *Fertil. Market News*, India 32(11):1–5.

Singh, A.K., P. Naraon, and J.S. Samra. 2009. Soil and water conservation. In *Handbook of Agriculture*. pp. 304–306, New Delhi: Indian Council of Agricultural Research.

Singh, G., P.N. Singh, and L.S. Bhushan. 1980. Water use and wheat yields in northern India under different nitrogen regimes. *Agric Water Manage* 3(20):107–114.

Singh, J. and S.K. Kaushik. 2007. *Bt* cotton in India—Percent scenario and future prospects. *Indian Farming* 56(11):26–28.

Singh, K.B., S.K. Jalota, and R.K. Gupta. 2015. Soil water balance and response of spring maize (Zea mays) to mulching and differential irrigation in Punjab. *Indian J Agron* 60(2):279–284.

Singh, M.V. and V. Goswami. 2014. Boron management. In *Textbook of Plant Nutrient Management,* Eds. R. Prasad, D. Kumar, D.S. Rana et al. pp. 188–213. New Delhi: Indian Society of Agronomy.

Singh, P., K.C. Aipe, R. Prasad et al. 1998. Relative effect of zero and conventional tillage on growth and yield of wheat (*Triticum aestivum*) and soil fertility under rice (*Oryzae sativa*)—Wheat cropping system. *Indian J. Agron.* 43:204–207.

Singh, R., A.R. Sharma, and U.K. Behera. 2007. Tillage and crop establishment practices for improving productivity of maize (*Zea mays*) under different weed control methods. *Indian J Agric Sci* 41:207–212.

Singh, R.K., P. Kumar, B. Bansal et al. 2016. Effect of split application of nitrogen on performance of wheat
(*Triticum aestivum* L). *Int J Agric Sci* 12(1):32–37.

Singh, S.B. and I.P. Abrol. 1988. Long-term effect of gypsum application and rice-wheat cropping on changes
in soil properties and crop yield. *J Indian Soc Soil Sci* 36:316–319.

Singh, V.K., R. Tiwari, M.S. Gill et al. 2008. Economic viability of site specific nutrient management in rice-
wheat cropping system. *Better Crops-India* 1:16–19.

Smith, C.J., D.B. Nedwell, L.F. Dong, and A.M. Osborn. 2007. Diversity and abundance of nitrate reductase
genes (narG and napA), nitrite reductase genes (nirS and nrfA), and their transcripts in estuarine sediments. *Appl. Environ. Microbiol.* 73:3612–3622.

Smith, M.S. and J.M. Tiedje. 1979. Effect of roots on soil denitrification. *Soil Sci. Soc. Am. J.* 43:951–955.

Smith, C.J., J.R. Freney, R.R. Sherlock, and I.E. Galbally. 1991. The role of urea nitrogen applied in a foliar
spray to wheat at heading. *Fert Res* 28:129–138.

Smith, K. (ed). 2010. *Nitrous Oxide and Climate Change*. London, UK: Earthscan Ltd.

Snitwongse, P., A. Satrusajang, and P.R. Buresh. 1988. Fate of nitrogen fertilizer applied to lowland rice on a
sulfic Tropaquept. *Fert Res* 16(3):227–240.

Soars, T., H. Cantarella, V.P. Vargas et al. 2015. Enhanced efficiency fertilizers in nitrous oxide emissions
from urea applied to sugarcane. *J Environ Qual* 44:423–430.

Spaargaren, O. 1989. Management of acid soils in Africa. In *Management of Acid Soils of the Humid Tropics
of Asia*. Eds. E.T. Craswell and E. Pushprajah, pp. 20–31. ACIAR, Australia Monograph No 13. 118 p.

Spalding, R.F., D.G. Watts, J.S. Schepers et al. 2001. Controlling nitrate leaching in agriculture. *J Environ
Qual* 30:1184–1194.

Spindale, M.C., V.S. Roche, and M.A. deSouza et al. 2013. Rates of urea with or without urease inhibitor for
top-dressing wheat. *Chilean J Agric Res* 73(2):160–166.

Stanford, G. and S.J. Smith. 1972. Nitrogen mineralization potential of soils. *Soil Sci Soc Am Proc*
36:465–472.

Stehfest, E. and L. Bouwman. 2006. N₂O and NO emission from agricultural fields and soils under natural
vegetation: Summarizing available measurement data and modelling of global annual emissions. *Nutr.
Cycl. Agroecosyst.* 74:207–228.

Subbiah, B.V. and G.L. Asija. 1956. A rapid method for assessment of available nitrogen in rice soils. *Curr
Sci* 25:259–260.

Subbiah, S. and G.V. Kothandaraman. 1980. Effect of neem cake blending of urea on ammonium and nitrate-N
in soils under field conditions. *J Indian Soc Soil Sci* 28:136–137.

Sudhakara, K. and R. Prasad. 1986. Relative efficiency of prilled urea, urea super granules (USG) and USG
coated with neem cake or DCD for direct seeded rice. *J. Agric. Sci. Cambridge* 106:185–190.

Suresh, S., V. Velu, and R.R. Kennedy. 1995. Effect of nitrogen sources on *Nitrosomonas* population, inorganic nitrogen forms and yields of rice in typic ustopept. *J Indian Soc Soil Sci* 43:274–275.

Suri, I., R. Prasad, and C. Devakumar. 2004. Neem coating of urea-present status and future trends. *Ferti
News (India)* 49(8): 21–24.

Swaminathan, M.S. 2006. *Sustainable Agriculture—Towards an Evergreen Revolution*. New Delhi: Konark
Pub. Pvt. Ltd.

Synder, C.S. 2010. World fertilizer nitrogen consumption and challenges. Paper presented at Nitrogen Use
Efficiency Conference, Stillwater, OK, August 3, 2010 (via Internet).

Synder, C.S., T.W. Bruulsema, T.L. Jensen, and P.E. Fixen. 2009. Review of greenhouse gas emissions from
crop production systems and fertilizer management effects. *Agric Ecosyst Environ* 133:247–266.

Takaya, N. 2002. Dissimilatory nitrate reduction metabolisms and their control in fungi. *J. Biosci. Bioeng.*
94:506–510.

Takkar, P.N., M.V. Singh, and A.N. Ganeshnurthy. 1997. A critical review of plant nutrient supply needs, efficiency and policy issues for Indian agriculture for the year 2000: Micronutrients and trace elements. In
*Plant Nutrient, Supply, Efficiency and Policy Issues: 2000–2005*. Eds. J.S. Kanwar and J.C. Katyal, pp.
238–264. New Delhi: National Academy of Agricultural Science.

Tandon, H.L.S. 1989a. Nitrogen recommendations for higher efficiency. *Fertil News(India)* 34(12):63–77.

Tandon, H.L.S. 1989b. Urea super granules for increasing nitrogen soil use efficiency. In *Fertility and Fertilizer
Use. Vol. III*. Eds. V. Kumar, G.C. Shrotriya, and S.V. Kaore, pp. 10–22. New Delhi: Indian Farmers'
Cooperative (IFFCO).

Tandon, H.L.S. 2011. *Sulfur in Soils, Crops and Fertilizers*. New Delhi: Fertilizer Development and Consultation
Organization (FDCO).

Tandon, H.L.S. 2012. *Fertilizer Management: Balance, Efficiency and Profitability*. New Delhi: Fertilizer Development and Consultation Organization (FDCO).

Thind, H., R. Pannu, R. Gupta et al. 2010. Relative performance of neem (*Azadirachta indica*) coated urea *vis-a-vis* ordinary urea applied to rice on the basis of soil test or following need based nitrogen management using leaf color chart. *Nut Cycl Agroecosyst* 87(1):1–8.

Thomas, J., and R. Prasad. 1987. Relative efficiency of prilled urea, urea super granules, sulphur coated urea and nitrification N-Serve blended urea for direct seeded rice. *J. Agron. Crop Sci.* 15:302–307.

Thomas, J. and R. Prasad. 1983. Mineralization of urea, coated urea and nitrification inhibitor treated urea in different rice growing soils. *Z Pflanzen Bodenk* 146(3):341–347.

Tilman, D., C. Blazer, J. Hill and B.L. Befort. 2011. Global food demand and sustainable intensification of agriculture. *Proc Natn Acad Sci, USA*. 108(50):260–264.

Timilsema, Y., R. Adhikari, P. Casey et al. 2015. Enhanced efficiency; a review of formulations and nutrient release patterns. *J Sci Food Agric* 95(6):1113–1142.

Tiwari, S.C. 1989. Effect of neem cake blending of urea on ammonium and nitrate content in vertisol. *J Indian Soc Soil Sci* 37:830–831.

Tiwari, K.N. 2001. Balanced fertilizer use in India. *Better Crops Int* 15(2):24–27.

Tiwari, K.N., S.K. Sharma, V.K. Singh et al. 2006. *Site Specific Nutrient Management for Increasing Crop Productivity in India: Results with Rice-Wheat Cropping Systems*, Project Directorate for Cropping Systems Research, and Phosphate and Potash Institute of India Programme. Gurgaon, India.

Toyo Koatsu Industries, Tokyo, Japan. 1965. *Toyo Koatsu Nitrification Inhibitor AM Tech Bull* 1&2.

Trenkel, M. 1997. *Controlled-Release and Stabilized Fertilizers in Agriculture*. Paris, France: Fertilizer Industry Association.

Tripathy, S.K. and S. Mohapatra. 2007. Maximising direct sown rice production by reducing cost of beaushan and weeding operation. *Indian Farming* 57(2):3–6.

Tuong, T.P. 2009. Promoting AWD in Bangladesh. *Ripple* 4(3):1–2.

Turner, R.E., N.N. Rabalais, and D. Justice. 2008. Gulf of Mexico hypoxia—Alternate states and a legacy. *Environ Sci Tech* 42(7):2323–2327.

Uphoff, N. 2002. System of rice intensification for enhancing the productivity of land, labor and water. *J Agric Resources Manage* 1:43–49.

USAD-ERS. 2016. Fertilizer use and market. United States Department of Agriculture-Economic Research Service. Updated November 9, 2016 (via Internet).

USDA. 1954. Diagnosis and improvement of saline and alkali soils. USDA, Washington, DC. Bull 60.

Van den Heuvel, R.N., S.E. Bakker, M.S. Jetten et al. 2011. Decreased $N_2O$ reduction by low soil pH causes high $N_2O$ emissions in a riparian ecosystem. *Geobiology* 9:294–300.

Van Drecht, G. and A.F. Bouwman. 2003. Global modelling of the fate of nitrogen from point and non-point sources in soils, groundwater and surface water. *Global Geochem Cycles* 17:1115.

Varinderpal-Singh, Bijai-Singh, Yadsvinder-Singh et al. 2010. Need based nitrogen management using the chlorophyll meter and leaf color chart in South Asia: A review. *Nutr Cycl Agroecosyst* 88(3):3861.

Ventura, W. and T. Yoshida. 1978. Distribution and uptake by rice plants of [15] N–labelled ammonium sulfate in mudballs in paddy soils. *Soil Sci Plant Nutr* 24(4):473–479.

Verhurst, N., I. Francois, K. Grahmann et al. 2014. Nitrogen use efficiency and optimization of nitrogen fertilizer in conservation agriculture. CIMMYT, Mexico.

Verma, K.S. and I.P. Abrol. 1990. Effect of gypsum and pyrites on yield and chemical composition of rice and wheat grown in a highly sodic soil. *Indian J Agric Sci* 50:935–942.

Vinod, P.N., P.N. Chandramouli and M. Koch. 2015. Estimation of nitrate leaching in groundwater in agriculturally used area in the state, Karnataka, India using existing models and GIS. *Int Conf Water Resource, Coastal & Ocean Eng*, March 12–14, 2015, National Institute of Technology, Karnataka, Survatha, India.

Vitousek, P.M., R. Naylor, T. Crews et al. 2009. Nutrient imbalances in agricultural development. *Science* 324: 1519–1520.

Vosawai, P., R.A. Halim, and A.R. Shakur. 2015. Yield and nutrition quality of five sweet potato varieties in response to nitrogen levels. *Adv Plants Agric Res* 2(5):00067.

Wang, F., Wang, X., and Sayre, K.D. 2004. Comparison of conventional flood irrigation and raised bed planting for winter wheat in China. *Field Crops Res* 87:35–42.

Wang, G.L., L. Ye, X.P. Chen, and Z.L. Cui. 2014. Determining the optimal nitrogen rate for summer maize in China by integrating agronomic, economic and environmental aspects. *Biogeosciences* 11:3031–3041. (doi:10.5194/bg-11-3031-2041)

Wang, S., S. Luo, X. Li et al. 2016. Effect of split application of nitrogen on nitrous oxide emission from plastic mulching maize in the semi-arid loess plateau. *Agric Ecosyst Environ* 220:21–27.

Wang, S., X. Zhao, G. Xing et al. 2015. Improving grain yield and reducing nitrogen loss using polymer coated urea in southeast China. *Agron Sustain Dev* 35(3):1103–1115.

WBCSD. 2005. Facts and Traits-Water. World Business Council for Sustainable Development. (www.wbcsd .org)

Weijabhandara, D.M., G.S. Dasog, P.L. Patil, and M. Hebbar. 2011. Effect of nutrient levels on rice (*Oryza sativa* L) under system of rice intensification and traditional methods of cultivation. *J Indian Soc Soil Sci* 59(1):67–73.

Wetsellar, R. and F. Ganry. 1982. Nitrogen balance in tropical agrosytems: In *Microbioloy of Tropical Soils and Plant Productivity* (Y.R. Dommergue and H.G. Diem, Eds.), pp. 1–38, The Hague, Netherlands: Martinus Nijhoff/Dr. Junk Publishers.

White, K.D. 1970. Fallowing, crop rotations and crop yields in Roman Times. *Agric Hist* 44:281–290.

Wilson, C.E. Jr., R.J. Norman, and B.R. Wells. 1990. Dicyandiamide influence on uptake of preplant-applied fertilizer nitrogen by rice. *Soil Sci. Soc. Am. J.* 54:1157–1161.

Wolf, I. and R. Brumme. 2002. Contribution of nitrification and denitrification sources for seasonal $N_2O$ emissions in an acid German forest soil. *Soil Biol. Biochem.* 34:741–744.

Xia, Y. and X. Yan. 2011. Nitrogen fertilization rate recommendation integrating agronomic, environmental, and economic benefits for wheat season in the Taihu Lake region. *Acta Pedologica Sinica* 48(1):210–218.

Xian-Ju, X., H.-B. Ma, Y.-W. Wing et al. 2016. Effect of slow-release nitrogen fertilizers with different application patterns on crop yield and nitrogen fertilizer use efficiency in rice-wheat rotation system. *J Plant Nutr Fert* 22(2):307–316 (in Chinese with English abstract).

Xiao-hui, F., S. Yong-Sheng, L. De-Xi et al. 2006.Ammonia volatilization losses and [15]N balance from urea applied to rice on a paddy soil. *J Environ Sci* 18(2):299–303.

Yadvinder-Singh, Bijai-Singh, and J. Timsina. 2005. Crop residue management for nutrient cycling and improving soil fertility in rice based cropping systems. *Adv Agron* 85:269–407.

Yadvinder-Singh, J.S. Bains and B. Singh. 2004. Need based fertilizer nitrogen management using leaf color chart in irrigated rice. *4th Int Crop Sci Cong,* Brisbane, Australia, September 26–October 1, 2004. Poster 649_singhy.

Yan, W., S. Zhang, X. Chen and Y. Tang. 2005. Nitrogen export by runoff from agricultural plots in two basins in China. *Nut Cycling Agroecosyst* 71(2):121–129.

Yan, X.Y., H. Akimoto, and T. Ohara. 2003. Estimation of nitrous oxide, nitric oxide and ammonia emissions from croplands in East, Southeast and South Asia. *Global Change Biology* 9:1080–1096.

Yan, X., T. Chaopu, P. Vitousek et al. 2014. Fertilizer nitrogen recovery efficiencies in crop production systems of China with and without consideration of the residual effect of nitrogen. *Environ Res Lett* 9 (2014)95002 (9 pp), Open Access.

Yang, C.M. and H.S. Lu. 1994. Yield loss in no-till corn due to weed interference and environmental impact. *J Agric Res China* 43(4):391–401.

Zerulla, W., T. Barth, J. Dressel, K. et al. 2001. 3, 4-Dimethylpyrazole phosphate (DMPP)-a new nitrification inhibitor for agriculture and horticulture. *Biolo Fertility Soils.* 34:79–84.

Zhang, Q., Z. Yang, H. Zhang and J. Yi et al. (2012). Recovery efficiency and loss of [15]N labelled urea in a rice-soil system in the upper reaches of the yellow river basin. *Agric Ecosyst Environ* 158:118–126.

Zhang, W., Z. Tian, N. Zhang, and Z. Li. 1996. Nitrate pollution of groundwater in northern China. *Agric Ecosyst Environ* 59:223–231.

Zhao, S. and P. He. 2012. Evaluation of in-season nitrogen management for summer maize in northern China Plain. *Better Crops* 96(2):8–10.

Zhou, J., B. Gu, W.H. Schlesinger and X. Ju. 2016. Significant accumulation of nitrate in Chinese semi-humid croplands. *Scientific Reports* 6.25088(2016) (doi 10.1038/serp 25088).

Zhou, S., Y. Sakiyama, S. Riya et al. 2012. Assessing nitrification and denitrification in a paddy soil with different water dynamics and applied liquid cattle waste using [15]N isotopic technique. *Sci Total Environ* 430:93–100.

Zimmer, M. 2000. Molecular mechanisms evaluating the proposed mechanism for the degradation of urea by urease. *J Biomol Struct Dyn* 17(5):787–797.

# 9 Enhancing Soil Organic Carbon by Managing Nitrogen in China

*Shou-Tian Ma, Xin Zhao, Chao Pu, Yang Liu,*
*Rattan Lal, Jian-Fu Xue, and Hai-Lin Zhang*

## CONTENTS

## 9.1 INTRODUCTION

The climate change is mainly caused by the increased emission of greenhouse gases (GHGs), resulting from anthropogenic activity, for example, fossil-fuel burning, and land use change and management (Edenhofer et al. 2014). World Meteorological Organization (2014) reported that the concentration of atmospheric $CO_2$ increased from 280 ppm in 1750 to 400 ppm in 2013. Great challenges lie ahead for scientists, policy makers, and the public in identifying systems to mitigate climate change, because it is critical to reduce carbon (C) emissions and mitigate global warming. Agriculture has been identified as a source of atmospheric C, and its improvement can play an important role in climate change and its mitigation. Nitrogen (N) and C cycles strongly impact soil fertility, environment quality, and climate change. The stock of soil organic carbon (SOC) and total nitrogen (TN) are estimated at ~1550 Pg (Lal 2004) and ~137 Pg (Batjes 2014) to 1-m depth in the terrestrial environment, respectively. Concentrations of GHGs (i.e., $CO_2$, $CH_4$, and $N_2O$) in the atmosphere can be strongly altered by even a small change in the SOC and TN stocks, with a significant impact on global warming (Batjes 2014; Wang et al. 1999) and soil quality through changes in the SOC and TN concentrations. Rational N and C management can improve soil quality, reduce soil and environment pollution, increase SOC sequestration, and enhance crop productivity. Thus, it is crucial that N and C stocks are prudently managed for advancing food security and mitigating climate change. Some important factors that affect SOC sequestration include specific climate, soil type, cropping system, fertilizer and irrigation, residue management, and so on. Soil N and C cycles interact together and also influence each other in the terrestrial biosphere. The C sink capacity of the terrestrial biosphere is determined substantially by scarcity of N (Gruber and Galloway 2008). In contrast, SOC also plays a critical role in the processes underpinning N mineralization, sequestration, and denitrification (Magill and Aver 2000). Therefore, understanding the interaction of N and C cycles in the terrestrial biosphere is paramount to increasing crop production, improving soil quality, and mitigating climate change.

A range of N fertilizers are used in agri-ecosystems for increasing crop production and meeting the food demands associated with the rapidly increasing global population. Soil application of N (e.g., N fertilizer, N-included in organic matter) can affect physical, chemical, and biological properties; impact soil quality; influence crop productivity; and alter the eco-environment. Soil C:N ratio is an important indicator of soil quality, and the relationship of SOC and TN plays an important role in soil C and N cycles. A range of factors (e.g., climate, soil, vegetation, farming practices) affect the soil C:N ratio (Lou et al. 2012; Xue et al. 2015). The latter strongly impacts the microbial activity, and soil C and N mineralization and sequestration. Soil application of N directly impacts C:N ratio resulting in the change in soil C mineralization and sequestration. However, the processes influencing sequestration or mineralization differ due to differences in N management, for example,

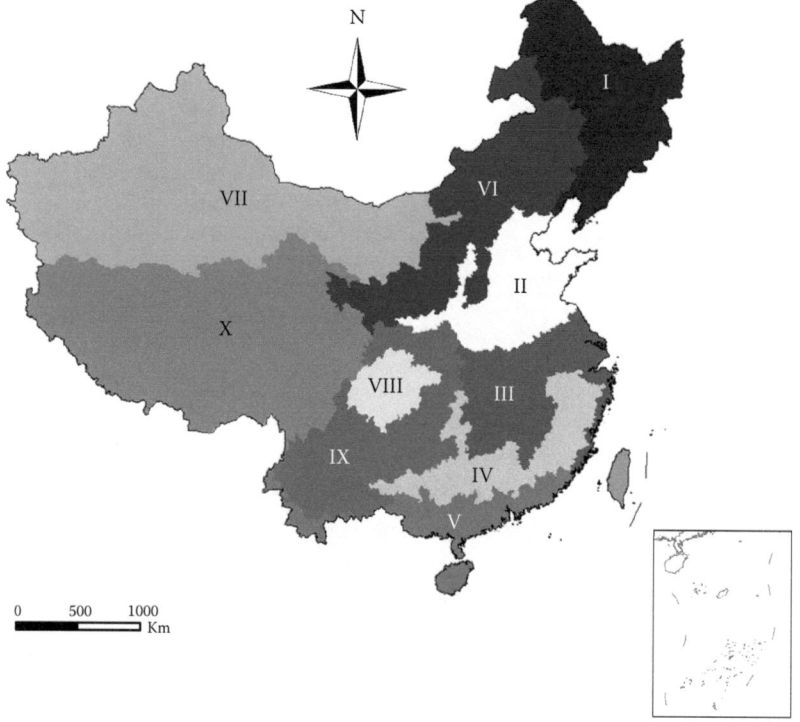

**FIGURE 9.1**   Different agro-ecological regions in China. Ten regions are divided as I. Single cropping region of Northeast China; II. Double cropping region of Huang-Huai-Hai Plain; III. Double or triple cropping region of middle and low reaches of the Yangtze River; IV. Double or triple cropping region of South of Yangtze River; V. South China; VI. Single cropping and pastoral region of Northern China; VII. Dryland farming region of Northwest China; VIII. Double cropping of Sichuan Basin; IX. Double cropping region of Southwest China; X. Single cropping region of Qinghai-Tibet Plateau.

the amount, time, place, and type of N addition. Therefore, understanding the impact of N management on SOC sequestration is important to enhancing soil fertility, increasing crop yields, reducing C emission, and thus mitigating climate change.

Soil C management is also critical to managing soil fertility, agronomic productivity, as well to C sequestration for mitigating the climate change (Lal 2006). Therefore, scientific understanding of soil C stock and its efflux from the soil ecosystem in relation to agro-management practices is essential to managing the global climate change and soil fertility (Lal 2010). Many studies on the effects of N management on SOC turnover and its sequestration have been reported in China. However, the information is often scattered and controversial mainly due to the differences in climate and soil conditions, field operations, cropping systems, and land use types among different agro-ecological regions in China (Figure 9.1). Thus, it is necessary to review and synthesize these studies and understand the relation of SOC to N management. The objective of this article is to review the impacts of N management on SOC sequestration, SOC fractions, and food safety in China.

## 9.2 NITROGEN MANAGEMENT IN CHINA

It is estimated that total chemical fertilizer use increased from 12.7 to 60.0 Tg (Teragram = $10^{12}$ g, million metric ton) during 1980–2014, and that of N fertilizer from ~9.3 to 23.9 Tg for obtaining high crop production in China (Figure 9.2) (NBSC 1981–2015). According to the estimates by the Ministry of Agriculture of the People's Republic of China, the total amount of organic mature input was ~48.8 Pg (Petagram = $10^{15}$ g = 1 giga ton) in China, including 20.4, 20.2, 7, 0.2, and 1 Pg of poultry manure, compost, crop straw, cake manure, and green manure in 2002, respectively (MAPRC 2002). However, the low N use efficiency (NUE) and large N losses have resulted in massive resource consumption and environmental pollution. The NUE for wheat (*Triticum aestivum* L.), paddy rice (*Oryza sativa* L.), and maize (*Zea mays* L.) has increased to 32%, 35%, and 32% between 2005 and 2013, respectively, through implementation of "National Fertilization According to Soil Test Result" and "Enhancement in Soil Organic Matter" programs" (MAPRC 2013). Numerous management practices were gradually popularized to improve fertilizer use efficiency in China

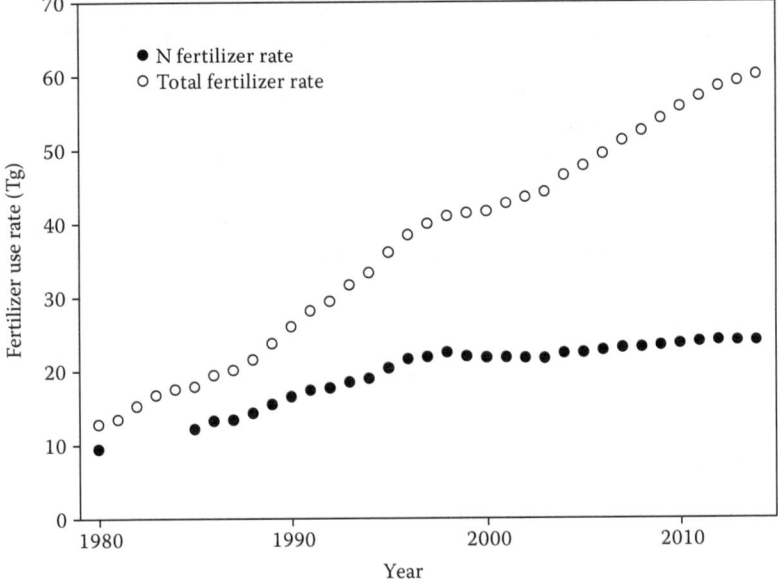

**FIGURE 9.2** The change in the rate of N fertilizer and total fertilizer use from 1980 to 2014. The data sourced from *China Statistical Yearbook* (NBSC 1981–2015).

(e.g., deep tillage and subsoiling, deep fertilization, crop residue retained, fertigation management). Many studies have focused on increasing NUE, improving crop yield, and reducing N loss through the change in N management practices in China (MAPRC 2013). Appropriate rate, time, place, and type of N use can increase crop yield and NUE, improve soil fertility, decrease N loss, and alleviate the environmental pressure in China. These practices are briefly discussed below.

## 9.2.1 THE RATE OF NITROGEN FERTILIZATION

Application of N fertilizer can enhance soil fertility, increase crop yield, and improve crop water use efficiency (Asai et al. 2009; Peng et al. 2006). However, excessive N application may have no significant effect on increasing crop yield. On the contrary, it could decrease the NUE, aggravate N fertilizer loss, and exacerbate the environment pollution (Ju et al. 2003; Lin et al. 2009). However, an appropriate rate of N use could produce high crop yield (Cui et al. 2008; Zhang et al. 2010) and reduce the environmental footprint. The appropriate rate of N application can differ among diverse cropping conditions. In winter wheat-summer maize cropping system, Yang et al. (2013) reported that the rate of N fertilizer at 150–191 and 180 kg ha$^{-1}$ could achieve high yield for winter wheat and summer maize, respectively. Too high or too low N fertilizer could reduce crop yield. For the high-, middle-, and low-yielding fields, the rice yield followed an initial increasing and then decreasing trend with the increase in rate of N application, and the appropriate N application rate was 120–180 kg ha$^{-1}$ for high- and middle-yielding fields, and 180–240 kg ha$^{-1}$ for low-yielding field (Feng et al. 2014). In general, crop yield differs in different regions even for the same N fertilizer rate. For the N fertilizer rate of 120–125 kg ha$^{-1}$ in winter wheat-summer maize cropping system, the yield of summer maize was as much as 7000–7700 kg ha$^{-1}$ in the Guanzhong Region, Northwest China (Zhao et al. 2006) compared with 6000–7500 kg ha$^{-1}$ in the North China Plain or NCP (Liu et al. 2003). Such a differential yield response may be attributed to differences in climate, soil, crop variety, and other site-specific conditions.

The duration between flowering and maturity is a key stage for N uptake and distribution in crops (Lv et al. 2011). The amount of N transfer in different organs of corn and the contribution rate of N to grain N increases with the increase of N fertilizer amount. However, the excessive N supply could decrease the amount of N transferred to the grains (Yi et al. 2006). In the dryland of the Loess Plateau, Zhao and Wang (2010) reported that the amount of N transferred to the grains increased with the increase in rate of N application from 80 to 240 kg ha$^{-1}$. Duan et al. (2012) reported that the amount of N transferred to grains after anthesis, the amount absorbed after anthesis, and the accumulation of N in grains at maturity were the highest at 150 kg ha$^{-1}$ of N fertilizer in the dryland of middle-hilly areas of Shandong. Therefore, the amount of N use had a large influence on N uptake, distribution, and accumulation in plant tissues.

An appropriate rate of N application can increase SOC concentration, while excessive N application can accelerate the rate of decomposition of soil organic matters (SOM). Xie et al. (2015) observed that application of N significantly increased soil TN concentration in the range of 0–180 kg ha$^{-1}$, while TN concentration decreased significantly when the N use amount was more than 180 kg ha$^{-1}$. Li et al. (2014) reported that SOC concentration increased significantly with increase in application of N between 0 and 135 kg ha$^{-1}$, and decreased when the rate of N use exceeded 135 kg ha$^{-1}$.

## 9.2.2 THE TIME OF NITROGEN FERTILIZATION

The basal-topdressing ratio and the time of topdressing of N fertilizer are among the key N management practices affecting crop growth and development under the premise of a certain N amount, which affects directly plant growth physiology and the grain yield. The appropriate time of N application rate improves the NUE, crop yield, and crop quality. Topdressing is a good way to meet the crop N demand at different growth stages to obtain high yield and quality. Numerous studies have suggested that the change of N application from early to later stage improves crop yield

(Cao et al. 2011; Zhang et al. 2010). In general, the N fertilizer should be applied as topdressing in soils of high fertility, in contrast to basal fertilizer plus topdressing at green and jointing periods in soils of low fertility (Du et al. 2009; Wang et al. 2003).

The application of N fertilizer at different ratios for different stages of growth could improve the yield of winter wheat by increasing the physiological and ecological indices during the growing period. Yan et al. (2010) observed that the optimum rate of N application was 225 kg ha$^{-1}$ split into three doses of 55, 100, and 70 kg ha$^{-1}$ at sowing, jointing, and anthesis stage, respectively. However, there are differences in the time and proportion of topdressing of N depending on different site (soil, climate, etc.) conditions. Shi et al. (2014) observed that the topdressing ratio with 1:2–1:3 could ensure high yield and high NUE for the cumulative rate of 225 kg ha$^{-1}$ N fertilizer for the winter wheat-paddy rice cropping system. Lv et al. (2011) reported that the highest yield and NUE for summer maize was observed with the N ratio of 3:5:2 at jointing, tasseling, and anthesis stages for the rate of 240–360 kg ha$^{-1}$ N fertilizer. Wen et al. (2015) concluded that the N ratio of 3:2:5 at sowing, jointing, and spike stages increased the grain number and 1000-grain weight for the rate of 180 kg ha$^{-1}$ of N fertilizer.

### 9.2.3 THE TYPE OF NITROGEN ADDITION

Numerous sources of N, including chemical fertilizers and organic N amendments, are used in agricultural production. Application of organic amendments has been the fundamental system in traditional agriculture of China, and which can provide a range of nutrients and improve soil fertility (Liu et al. 2007). Crop residues are among the most important sources of organic N amendments, which also contain N, phosphorus (P), potassium (K), cellulose, and other nutrients (Ye et al. 2008). Straw mulch or incorporation in soil can supply nutrients for crop growth and improve soil fertility (Ma et al. 2003; Wu et al. 2002).

Several studies have documented that crop yields are enhanced with straw retention, and the beneficial impact increases with the increase in duration of the experiment (Tan et al. 2007; Yu et al. 2004). Straw incorporation can alter soil physical properties, enhance microbial activity, and improve soil nutrients turnover (Qiang et al. 2004). Shen et al. (2011) observed that straw incorporated in the winter wheat and summer maize season not only increased crop yield, but also reduced the incidence of lodging. However, incorporation of residues at an excessive rate did not increase the wheat yield (Lu et al. 2011). Under an optimal rate of fertilization at 150 and 180 kg ha$^{-1}$ of N fertilizer for winter wheat and summer maize season, respectively, 5000 kg ha$^{-1}$ of crop straw incorporated during each season significantly increased NUE in the Guanzhong region, Northwest China (Yang et al. 2013). Similar results were observed for rice (Zhang et al. 2010) and maize (Huo et al. 2015). However, an excessive incorporation of straw could reduce soil microbial activity and curtail the rate of straw decomposition, decrease the release and turnover of nutrients, increase incidence of pests and diseases, reduce germination rate of wheat, and hinder wheat growth (Li et al. 2006).

The stability of crop yield is an important criterion to judge the quality and sustainability of agroecosystems (Dong et al. 2014). For example, Men et al. (2008) did not observe any obvious relationship between the yield stability and the amount of fertilizer applied. However, a significant correlation with the yield stability was observed with the proportional application of N, P, and K fertilizers (Men et al. 2008). A judicious combination of chemical fertilizer and organic amendments is a rational management practice of N resources, which improves soil fertility, while maintaining crop yield, high quality, and efficiency (Pang 2008). Appropriate N management plays an important role in the integrated crop production and eco-environment (Hou et al. 2011).

Some studies have indicated that higher proportion of organic fertilizer is better, while others have indicated that there could be an optimal ratio of organic amendments to chemical fertilizers (Yadav et al. 2000; Zheng et al. 2001). Xu et al. (2008) reported that the combined application of organic amendments and inorganic fertilizers could improve SOM and enhance NUE. Xu and colleagues (2008) observed the increase in SOM content over a 5-year period by 6.5% for chemical fertilizer and by 18.5% for chemical fertilizer plus organic amendments. Nonetheless, there may be no differences in crop yield despite the enhanced soil fertility with higher rates of organic fertilizer input (Figure 9.2) (Hou et al. 2011).

## 9.3 NITROGEN MANAGEMENT ON CARBON SEQUESTRATION IN CHINA

N management can strongly impact soil processes and the attendant changes in soil C and N dynamics (e.g., SOC sequestration, GHGs emission, N losses) (Figure 9.3) along with increase in crop production. Indeed, soil C and N are two types of substrates for soil microbial use, thus, the processes of soil C and N are strongly coupled. Soil C cycle interacts together with the N cycle, and these cycles are strongly accelerated by human activities. Soil C:N ratio is a pertinent indicator of the coupling of soil C and N cycles, which is also moderated by the external changes in C and N management. In general, soil C:N ratio is maintained at a stable range with a mean value of 10:1 (DCCEE 2012). Input of C or N can alter soil C:N ratio and affect microbial activities leading to the change in the process of C and N decomposition and mineralization. Thus, management of soil N leads to strategies that affect the decomposition of SOM and finally change the C output (Davidson and Janssens 2006; Soussana and Lemaire 2014). Indeed, N management in China since the 1980s had a significant effect on soil C sequestration in agroecosystems (Lu et al. 2009; Wang et al. 2015; Zhao et al. 2015). The SOC stock of farmland increased by 24 Tg yr$^{-1}$ since 1980 in China. The increased SOC was significantly related to the increased amount of N fertilizer and total chemical fertilizer use (Figure 9.4). The SOC dynamics, SOC fractions, and C sequestration are strongly affected by N sources (i.e., chemical fertilizer, manure, or crop residue) and methods of management including different N input levels, raw materials, N placement soil depth, combined with other farming operations (Chai et al. 2014; Huang et al. 2013; Wang et al. 2015; Yan and Gong 2010; Zhong et al. 2015).

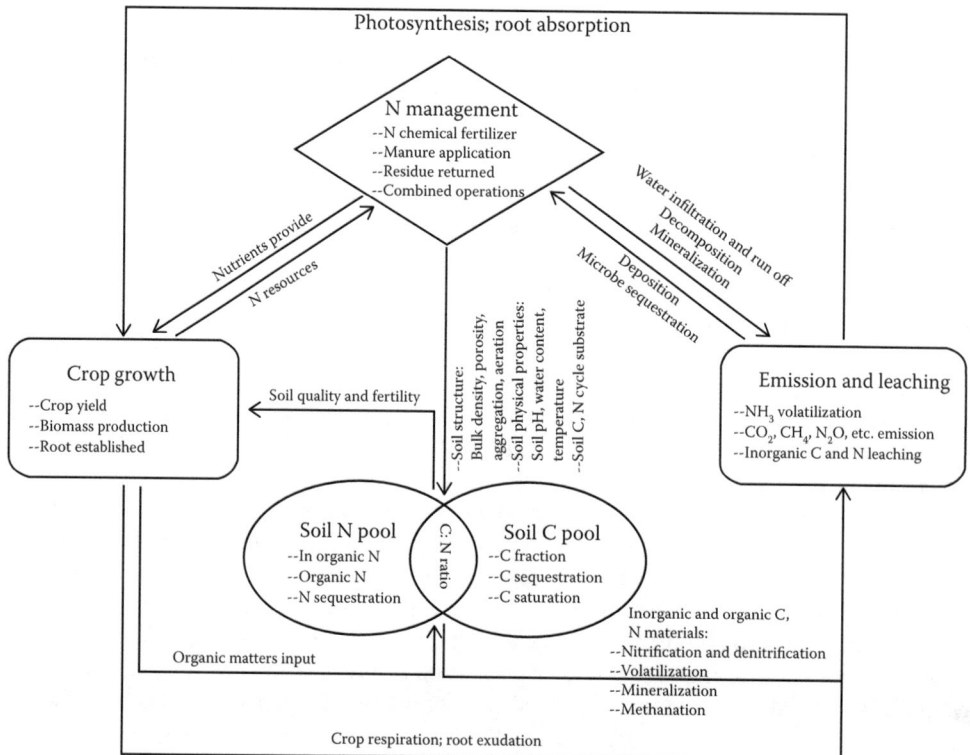

**FIGURE 9.3** Schematic of the effects of N management on coupled cycling of C and N in soil.

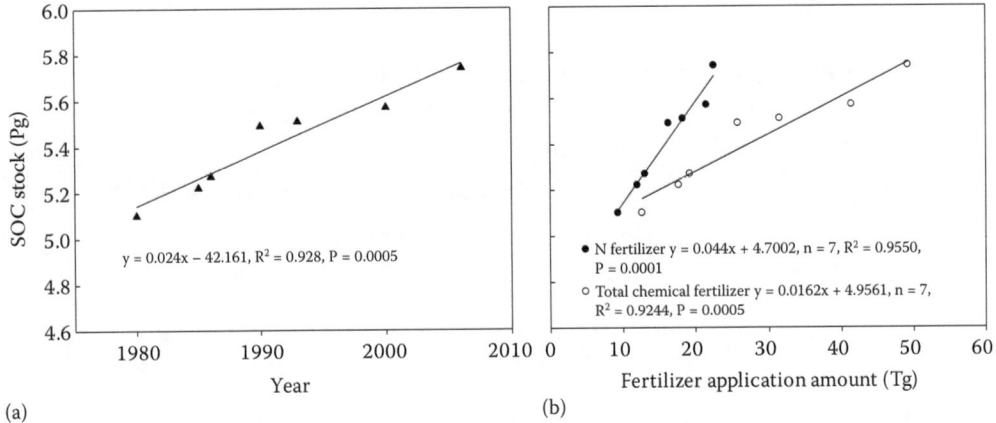

**FIGURE 9.4** Changes in soil organic carbon (SOC) stock (a) since 1980 in farmland soils of China and (b) its relationship with annual fertilizer application amount. Solid points mean N fertilizer and hollow points mean total chemical fertilizer in Figure 9.4b. The data were year of 1980, 1985, 1986, 1990, 1993, 2000, and 2006. SOC stocks of different years were estimated based on studies by Huang and Sun (2006); Liu et al. (2004); Pan et al. (2010); Pan and Zhao (2005); Sun et al. (2010); Wu and Cai (2007); and Xie et al. (2007) and annual fertilizer application amount of corresponding year was sourced from *China Statistical Yearbook* (NBSC 1981–2015).

Additionally, the effect of different N management systems on soil C pool vary among soil types, land uses, cropping systems, climatic regions, and different combinations of farming operations (Ding et al. 2012; Lu et al. 2009; Shi et al. 2016; Zhao et al. 2015; Zhou et al. 2013).

### 9.3.1   Effects of N Management on Carbon Sequestration

Effects of N application on soil C sequestration can differ significantly among regions, land use (paddy rice or upland soil), cropping systems (crop rotation or not), farm operations (organic materials input, no-till, and so on), or even other chemical fertilizers (P or K) and input rates in China and elsewhere (Lu et al. 2009; Qiu et al. 2016; Tian et al. 2015; Wang et al. 2015; Xu et al. 2011). Based on the results of previous meta-analysis conducted at the national scale in China, an increase of $1.11 \pm 0.21$ g kg$^{-1}$ and $2.09 \pm 0.46$ Mg ha$^{-1}$ in SOC concentration and stock have been observed with input of chemical fertilizer compared to that under no fertilizer control (Zhao et al. 2015). Inputs of organic and/or inorganic N materials can significantly increase TN and microbial activity, and thereafter enhance SOC concentration and sequestration capacity. Based on a 33-year-old experiment, Yang et al. (2015) reported a significant positive linear relationship between SOC and TN concentration. However, the increasing trends of TN and SOC fluctuated under different fertilizers and residue management treatments. A similar trend was also observed especially for the top soil under different N fertilizer application rates for wheat production (Zhong et al. 2015) in semi-arid cropland (Zhou et al. 2013) of China. However, the effects of N fertilizer application on SOC depend on a range of factors such as experimental duration, cropping intensity, N fertilizer application rate, and the soil profile characteristics (Lu et al. 2009; Zhao et al. 2015; Zhong et al. 2015).

Based on the data of the national meta-analysis of 37 published peer-viewed papers in China, Zhao et al. (2015) observed a significant increase in both SOC concentration and stock only for 0–20 cm soil depth under N fertilizer application compared to that under unfertilized control ($P < 0.05$, Table 9.1). In addition, differences in depth distribution of SOC concentration were also observed with the increase in soil depth, with a trend of decrease in SOC concentration and stock with increase in depth. However, a smaller mean difference in SOC was observed at 20–40 cm depth because of the plow pan and the attendant changes in soil aeration and water transmission (Zhao et al. 2015).

**TABLE 9.1**

**Results of Meta-Analyses of Changes in SOC Concentration and Stock by Adoption of Nitrogen Fertilizer Compared to No Fertilizer Application in China**

|  |  | SOC Concentration (g kg⁻¹) | | | SOC Stock (Mg ha⁻¹) | | |
|---|---|---|---|---|---|---|---|
|  |  | n | Mean | 95% *CI* | n | Mean | 95% *CI* |
| Soil depth (cm) | 0–20 | 71 | 1.13 | ±0.22 | 75 | 2.12 | ±0.48 |
|  | 20–40 | 23 | 0.18 | ±0.56 | 23 | 0.25 | ±1.02 |
|  | 40–60 | 14 | 0.28 | ±0.41 | 14 | 0.53 | ±0.87 |
|  | 60–80 | 9 | 0.02 | ±0.64 | 9 | −0.01 | ±1.61 |
|  | 80–100 | 9 | −0.04 | ±0.36 | 9 | −0.11 | ±0.91 |
| Experimental | <16 | 20 | 0.53 | ±0.53 | 20 | 1.20 | ±1.12 |
| duration (yr) | 16–25 | 57 | 0.71 | ±0.29 | 61 | 1.65 | ±0.58 |
|  | >25 | 36 | 0.73 | ±0.49 | 36 | 1.09 | ±0.77 |
| Crop intensity | 1 | 43 | 0.61 | ±0.26 | 43 | 1.05 | ±0.43 |
|  | 2 | 52 | 0.88 | ±0.38 | 56 | 1.83 | ±0.71 |
|  | 3 | 18 | 0.25 | ±0.57 | 18 | 0.31 | ±0.92 |

*Source:* Zhao, X., R. Zhang, J. F. Xue, C. Pu, X. Q. Zhang, S. L. Liu, F. Chen, R. Lal, and H. L. Zhang. 2015. Management-induced changes to soil organic carbon in China: A meta-analysis. *Advances in Agronomy* 134: 1–50.

*Note:* n: the number of comparisons in the meta-analysis, Mean: the mean difference between chemical fertilizer adoption and no fertilizer adoption, 95% *CI*: the 95% confident interval, mean ± 95% *CI* don't overlap 0 means $P < 0.05$, Crop intensity: number of crops per year.

Further, no significant effects of fertilization were observed for deeper soil layers. Similarly, soil TN also decreases with increase in soil depth under different N managements (Qiu et al. 2016; Yang et al. 2015; Zhong et al. 2015; Zhou et al. 2013), and a significant relationship is observed between TN and SOC for different soil layers (Zhong et al. 2015). Such a trend indicates that soil N management can affect soil C concentration and sequestration both directly and indirectly. The effect of increase in the rate of N fertilizer on SOC sequestration is limited to the top 0–20 cm depth. Thus, N management has an estimated potential of soil C sequestration at 6.0 Tg C yr⁻¹ under the current situation compared with 12.1 Tg C yr⁻¹ under the recommended N rate vis-à-vis the unfertilized control based on 84 paired data from 28 sites (42 publications) from the official statistical data from China (Lu et al. 2009).

The rate of C sequestration under N fertilizer application is significantly affected by the application rate when compared to no application of N fertilizer in different regions in China (Lu et al. 2009). Such a response is primarily due to a general increase in soil N when N fertilizer application rate is increased, and which affects soil C dynamics. The data in Figure 9.5 show that, on an average, the C sequestration rate increased by 0.64 kg C ha⁻¹ yr⁻¹ with every 1 kg ha⁻¹ increase in N fertilizer rate compared to no N fertilizer application (Figure 9.5a). However, the rate of increase in SOC sequestration can vary from 0.53 to 1.74 kg C ha⁻¹ yr⁻¹ among different regions in China (Lu et al. 2009). Lu et al. (2009) observed that the capacity of SOC sequestration can be doubled if N fertilizer is used appropriately in each region of China. In general, enhancing N fertilizer application rate increases soil microbial activity, promotes crop growth, increases input of organic matter, and leads to a more efficient decomposition process that can increase SOC sequestration. An increase in input of N fertilizer increases with an increase in cropping intensity along with the increase in rainfall and temperature from north to south of China. However, changes in SOC concentration and stock also vary among cropping intensity (Table 9.1). In general, the rate of SOC sequestration is higher

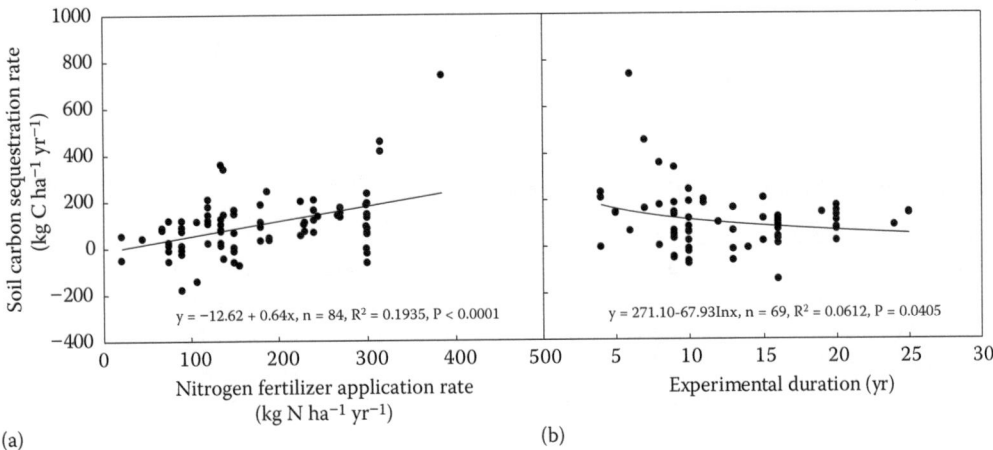

**FIGURE 9.5**    The relationship between soil carbon sequestration rate and (a) nitrogen fertilizer application rate and (b) experimental duration. The data were resourced from Lu et al. (2009).

in the northeastern (single cropping per year) than that under southeastern (double or triple cropping per year) regions of China (Lu et al. 2009). Similar trends have also been reported through meta-analysis for different crops including the paddy rice in China (Tian et al. 2015; Zhao et al. 2015). Differences in soil C response to N application are primarily attributed to the interactive effects between soil microbial processes and the prevailing climate (i.e., rainfall, temperature), input materials (e.g., crop residues or chemical fertilizers) for different soils, and other farm operations (e.g., tillage and irrigation) used in different regions (Eberwein et al. 2015; Lal 2004; Zhang et al. 2014).

Experimental duration is another factor affecting soil C sequestration under N management. As much as 88% of variations in SOC sequestration can be explained by the changes in soil TN. Analyses of the data from a 26-year N fertilizer experiment indicated that soil N storage decreased under treatments without manure but increased with the manure application (Zhou et al. 2013). However, changes in TN, SOC, and C:N ratio fluctuated widely under long-term wheat-corn experiments of chemical and organic inputs in a fluvo-aquio soil (Yang et al. 2015). However, Lu et al. (2009) observed that C sequestration under N fertilizer application decreased with increase in the experimental duration compared to that under no N fertilizer application (Figure 9.5b). A similar trend was also reported in a meta-analysis by Zhao et al. (2015). A phenomena of "C saturation" on SOC concentration and stock was observed with application of N fertilizer (Zhao et al. 2015). Based on the framework proposed by Luo and Weng (2011), C saturation implies a dynamic disequilibrium of the terrestrial C cycles caused by temporal changes in the C source or sink at yearly and decadal scales within one disturbance recovery episode, but without any long-term impact on SOC sequestration unless the disturbance regimes are altered. Additionally, a greater efficiency in soil C sequestration can occur in soils that are not close to C saturation (Stewart et al. 2007). Therefore, understanding the long-term effects of N fertilizer on soil C dynamics is critical to assessing the efficiency of N management in regulating SOC pool and promoting C sequestration. In addition, optimized N management practices contribute to form a new C saturation, thereby enhancing SOC sequestration (Luo and Weng 2011; Stewart et al. 2007). The increase in capacity of C sequestration depends on the disturbance regimes of C saturation upon conversion to N fertilizer combined with different application rates, different chemical fertilizers, manure and residues inputs, and different tillage practices used in different regions of China (Qiu et al. 2016; Tian et al. 2015; Wang et al. 2015; Xu et al. 2011; Zhao et al. 2015; Zhong et al. 2015). Yang et al. (2015) observed higher SOC concentration for N fertilizer application with than without residue return under a 33-year experiment in northern China. Liu et al. (2014) observed that soil C saturation occurred after 12 years of continuous return of crop residues into soil. Interaction of N fertilizer with residue retained, especially when combined with other chemical fertilizers input (K and P) can increase duration of soil C sequestration (~12 years) and accentuate the total potential (Tian et al. 2015). Thus, an integrated management strategy is needed to maintain or improve the long-term effects of N management on SOC sequestration.

### 9.3.2 INTERACTION WITH OTHER FARM OPERATIONS

Several recommended management practices (RMPs), that is, no-till (NT); residue retention; manuring in conjunction with N fertilizer application also impact soil C and N dynamics. These RMPs have positive effects under site-specific conditions on SOC and TN in China (Dikgwatlhe et al. 2014; Qiu et al. 2016; Wang et al. 2011; Xue et al. 2015) and globally (Dalal et al. 2011; Lal 2004; Lal 2010; West and Post 2002). Several studies have also demonstrated positive effects of NT on soil quality and structure, SOC sequestration, reducing GHG emissions and increasing crop production (Lu et al. 2009; Zhang et al. 2014; Zhao et al. 2015, 2016). In general, adoption of NT increases TN in surface layer for most soils in China (Wang et al. 2011; Xue et al. 2015), which in turn should also enhance SOC sequestration. However, the capacity of NT to enhance SOC is smaller than that achievable with increased rate of N fertilizer application in China (Lu et al. 2009; Zhao et al. 2015). For example, a 15-year experiment indicated that reducing tillage intensity can enhance transformation of SOC from the labile into stable pools, and increase SOC accumulation with input of chemicals (Wang et al. 2011). On the other hand, using a system approach to transform NT into conservation agriculture (CA) (Lal 2015) integrated nutrient management can break the vicious cycle by alleviating soil-related constraints under NT, increase nutrient input, enhance crop yield, which over time would result in more biomass input into soil, more humification efficiency, and a higher SOC pool than that under a plow-based system.

Crop residues retention is also a key component of a CA system (Lal 2015; Zhang et al. 2014). Retention of crop residues provides more C and N as substrate for soil microbial processes. As was concluded by the data from a global meta-analysis, straw retention is an effective means to increase SOC accretion, improve plant nutrient reserves, increase crop growth, and add more biomass -C into the system (Liu et al. 2014). Thus, combination of residue retention with judicious management of N can increase SOC concentration, stock, and the overall C sequestration potential (Lu et al. 2009; Xu et al. 2011; Zhao et al. 2015). In addition, increasing the rate of input of crop residues can also enhance the rate of SOC sequestration (Liu et al. 2014; Lu et al. 2009). Furthermore, soil TN has a strong positive linear relationship with SOC concentration. However, a higher slope is observed with the N fertilizer application with than without residue retention (Yang et al. 2015). This trend indicates that soil C and N resources can be improved by judicious input of N fertilizer in conjunction with crop residues (compared adding N fertilizer alone), and leading to a higher decomposition rate of SOM that would enhance the long-term effects on soil C sequestration.

Similarly, TN and SOC are increased with input of manure in soils of North China (Yang et al. 2015; Zhou et al. 2013). Additionally, an improved soil C sequestration capacity and potential have been observed when manure is applied in combination with N fertilizer than that with the input of manure itself (Maillard and Angers 2014; Wang et al. 2015; Xu et al. 2011). Zhao et al. (2015) reported that manure application increased more SOC concentration and stock than that with input of chemical fertilizers. Thus, manuring along with chemical fertilizer application would be a recommended strategy of SOC sequestration in China. In addition, SOC stock increases linearly with cumulative increase in manure-C input (Maillard and Angers 2014). Synthesis of data from site-specific experiments shows that SOC concentration and stock respond differently to different levels of manure input when compared to no fertilizer application (Figure 9.6a, b). The data from long-term site-experiments also indicate that combined application of manure and N fertilizer increases SOC concentration compared to that for other fertilizer treatments (Yan and Gong 2010; Yang et al. 2015). Based on the analysis of a 22-year experiment, Qiu et al. (2016) concluded that combination of manure, straw, and chemical fertilizers is a good strategy to promote C sequestration in northeast China. Furthermore,

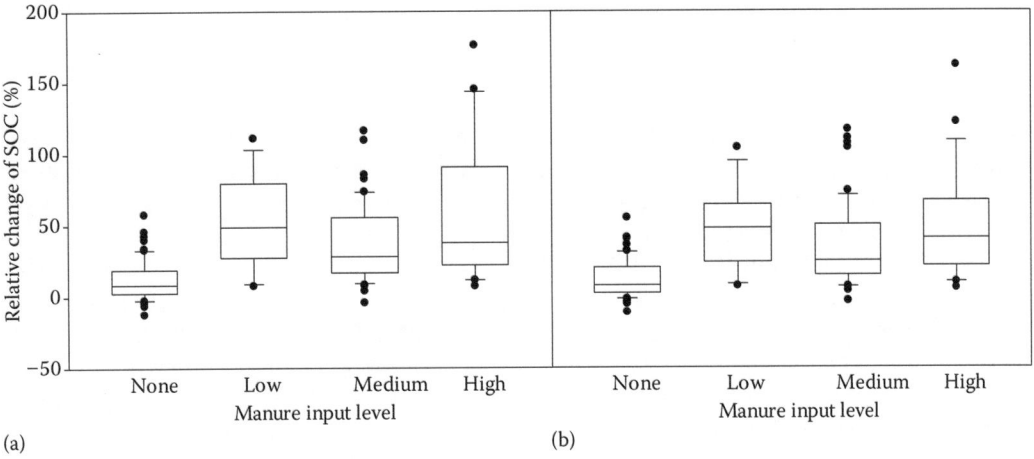

**FIGURE 9.6** Relative change of soil organic carbon (SOC) (a) concentration and (b) stock under different manure input level. The data were sourced from Zhao et al. (2015).

application of manure provides a longer C sequestration duration and higher potential when its input is also combined with chemical fertilizers in China's paddy soils (Tian et al. 2015). Besides, manure combined with chemical fertilizer can improve soil C sequestration along with soil C storage capacity (Ding et al. 2012). Thus, the interactive effect of manure and inorganic N could enhance soil quality to maintain a long-term sequestration and storage of C in soils. In general, a systematic management of N combined with other field operations or even adjusting the rate of inorganic N input can advance the level of soil C saturation, higher potential of C sequestration, and greater rate of SOC sequestration. On the other hand, a comprehensive strategy of farming practices would enhance crop production while increasing organic matter inputs and enhancing the SOC sequestration.

### 9.3.3 Effect of Nitrogen Management on Soil Organic Carbon Fractions

Management of soil fertility, regardless of by adding chemical fertilizer or organic amendments, can increase SOC concentration in soils of arable uplands compared to no fertilizer control (Tian et al. 2014). Traditionally, SOC pool can be divided into three pools: labile, slow, and passive according to the rate of turnover. The labile SOC fractions with fast turnover rate play an important role in the decomposition and mineralization processes of SOC, which could affect the soil C balance and nutrient supply. The labile SOC fractions are good indicators of the response of SOC to change in land use and management (Dou et al. 2008; Plaza-Bonilla et al. 2014; Tan et al. 2007). The passive SOC pool has a slow turnover rate due to the chemical and physical stability associated with physical protection. Farm management practices (e.g., N fertilizer management) can change soil physical, chemical, and biological processes, affect the distribution of different SOC fractions, and alter the SOC turnover and sequestration rate.

The labile SOC fractions strongly impact processes of dissolution, adsorption, desorption, translocation of soil chemical substances that moderate the process of biogeochemistry, soil-formation, the growth and metabolism of micro-organisms, and the diffusion of soil contaminants (Cao et al. 2014). Based on differences in extraction methods, labile SOC fractions are divided into numerous forms: the dissolved organic carbon (DOC), permanganate oxidizable organic carbon (POxC), particulate organic carbon (POC), the water-soluble organic carbon, microbial biomass organic carbon (MBC), and light organic carbon (LOC). These labile SOC fractions respond rapidly to changes in C supply and farm operations. Because N plays an important role in crop growth and development, it strongly affects crop yield and quality. N management practices (e.g., amount, time, place, and type of applied N fertilizer) can change the crop-soil micro-environment, and affect labile SOC fractions distribution and SOC turnover.

With the increase in the amount of N application, labile SOC fractions generally follow an upward trend in soil of the plough layer. Based on a 32-year field experiment with winter wheat cropping, Li et al. (2014) observed that LOC content at 0–5, 5–10, 10–15, and 15–20 cm depths increased gradually with the increase in the amount of N application (0–180 kg N ha$^{-1}$), and declined for 20–30 and 30–40 cm depth after an upward trend in dryland of the Loess Plateau. Under a pot experiment, both MBC and DOC contents tended to a decrease (135–225 kg N ha$^{-1}$) after increasing trend with the increase in N application from 0–135 kg N ha$^{-1}$ (Yang et al. 2016). In contrast, Wang et al. (2016) reported that N fertilization (160 and 320 kg N ha$^{-1}$) had a small influence on MBC content compared to unfertilized at 0–10 cm depth in dryland of the Loess Plateau. No significant difference in DOC and POxC concentrations were observed among different N levels (0–495 kg N ha$^{-1}$ yr$^{-1}$) at 0–20 cm depth in winter wheat-summer maize cropping system in the Northwest China. However, mineral-associated organic carbon (MOC) concentration under N fertilization was significantly higher compared to that of soil under the unfertilized control (Mulati et al. 2012). Overall, effects of the amount of N fertilization on labile SOC fractions differ among different studies. Such a differential response may be attributed to differences in climate, soil, and cropping systems. There also exists a little information on the underlying mechanisms which moderate the impact of the rate of N application on changes in labile SOC fractions as reported in some current studies in China. Further, the depth of soil sampling has always been shallow in most of the studies, which can affect the results. Differences in the rate of N fertilizer can also affect soil nitrate distribution at 0–300 mm depth and the crop N untake (Dai et al. 2016), and change the quantity and composition of crop root exudates and the abundance of rhizosphere bacteria (Zhu et al. 2016). All these factors could affect SOC and N distribution in subsoil layers. Thus, it is necessary to assess the changes in labile SOC fractions by sampling below 300 mm depth to cover the entire root zone.

In general, N fertilizer application increases the liable SOC fractions in upper soil layer compared to no N fertilizer control. However, some differences have been observed with regard to the effect of the rate of N fertilizer plus other chemical fertilizers or organic amendments under diverse experimental conditions. Numerous studies have documented that balanced use of fertilizer (including N, $P_2O_5$, and $K_2O$) plus manure improve the labile SOC fractions (Fan et al. 2013; Zhang et al. 2009, 2014). Based on a 30-year study of a double paddy cropping system, balanced fertilization (including N, $P_2O_5$, and $K_2O$) increased both POC and POxC stocks compared with that for the soil under the unbalanced fertilization treatments (i.e., N and $P_2O_5$, N and $K_2O$) in 0–45 cm soil layer (Sun et al. 2013). In addition, both POC and POxC stocks were also greater for soil receiving chemical fertilization plus pig manure or residues incorporated compared to that in soil receiving only the input of chemical fertilizers (Sun et al. 2013).

The data from a 32-year study showed that fertilizer management may alter the SOC concentration in different soil particle fractions [i.e., sand (>53 μm), coarse silt (5–53 μm), fine silt (2–5 μm), coarse clay (0.2–2 μm), fine clay (<0.2 μm)] in Cumulic Haplustoll of Loess Plateau in China, especially that in the sand (Fan et al. 2013). Fan et al. (2013) reported that SOC concentrations in the sand fraction under inorganic N (90 kg N ha$^{-1}$) and P fertilizer (75 kg $P_2O_5$ ha$^{-1}$) plus residue incorporated (3.75 Mg ha$^{-1}$) or manure (75 Mg ha$^{-1}$) were higher than those under the unfertilized and only N fertilizer (90 kg N ha$^{-1}$) treatments in dryland. Zhang et al. (2014) divided SOC fractions by the physical-chemical fractionation method for soil from an 18-year field experiment in yellow paddy soil. Zhang and colleagues (2014) reported that application of pig manure plus chemical fertilizer (including N, $P_2O_5$, and $K_2O$) significantly improved free coarse POC, physically protected SOC, chemically protected SOC, and biochemically protected clay SOC contents compared to those in soil receiving only chemical fertilizers. In addition, with the increase in mean annual C input, free coarse POC concentration showed a significant linear increase, while soil chemically protected and biochemically protected SOC concentrations and soil total C contents showed a significant increase in the "saturation curve" (Zhang et al. 2014). Most studies have focused on the amount and type of

N fertilizer on the change in SOC fractions. The place and time of N fertilizer application could also affect the distribution of SOC; however, little information exists on the management-induced change in the SOC fractions.

Biochar application is a possible strategy to mitigate global warming through its effect on storage of recalcitrant C in soil (Liu et al. 2016). The application of biochar-based ammonium nitrate fertilizers in agriculture, made by combining biochar with ammonium nitrate, is a new focus of attention of the scientific communities on the improvement of soil fertility and the environment. Zhao et al. (2016) concluded that MOC concentration was increased by 22.1% ($P < 0.05$) and 17.5% under simple mixed type of biochar-based N and biochar at 0–20 cm soil depth, respectively. Further, treatments of ammonium nitrate (AN), solid-liquid adsorption type biochar-based nitrogen (SLBN), and chemical reaction type biocharbased nitrogen (CBN) decreased MOC concentration by 9.1%, 10.9%, and 1.5%, respectively, compared to that in soil under no N fertilizer treatment in a winter wheat-summer maize system for the Northwestern China. Such a trend may be attributed to larger porosity, larger surface area, and stronger adsorption of biochar to conserve soil, water, and nutrients, and promote microbial activity, while stronger capacity of AN adsorption for SLBN and CBN than CBN occupying the porosity of biochar and hindering microbial activity (Zhao et al. 2016). However, there is little relevant information on the effect of biochar-based N fertilizer on SOC sequestration in China. It is necessary to analyze the influence of biochar-based N fertilizer on SOC fractions to understand the mechanisms of SOC turnover and sequestration under different climates, soil types, cropping systems, and agricultural management practices in China.

The data from several studies show that N fertilizer management can change the distribution of SOC concentration, enzyme activity, and soil nutrient reserves in different aggregate fractions for different soil types and cropping systems (Li et al. 2016; Wang et al. 2011; Xiang et al. 2009; Xie et al. 2015; Zhang et al. 2015). Formation and distribution soil aggregates play an essential role in the process of SOC turnover and sequestration (Six et al. 1998, 2000). However, little information exists on the effect of N fertilizer management on SOC and aggregate size fractions. Thus, it is important to understand the SOC turnover and sequestration in relation to changes in different fractions. The reseaerch data on the effect of combined N management with other farming practices on changes in SOC fractions (e.g., irrigation, tillage, cultivation pattern) are scanty.

SOC exists primarily in stable rather than labile forms in terrestrial ecosystems. Therefore, non-labile SOC fractions (i.e., slow and passive SOC) can be used as indicators of the SOC stability, and can be an important determinant of the capacity of SOC sequestration. Most of the on-going studies are focused on the effects of N fertilizer application on labile SOC fractions, and not on the assessment of the non-labile SOC fractions (e.g., MOC) and heavy organic C. Chen et al. (2011) analyzed the effects of fertilization on the recalcitrant organic C content in a 22-year field experiment in red paddy soil in Southern China. Results showed that application of N fertilizer plus other chemical fertilizers or pig manure increase the recalcitrant organic C content compared to that of soil receiving no fertilization or only N application in a double paddy cropping system. However, little attention has been paid to the effect of fertilization on non-labile SOC fractions under other climate and soil conditions.

### 9.3.4 GREENHOUSE GASES EMISSIONS AND N MANAGEMENT

Site-adaptation strategies of N management can enhance the capacity and potential of soil to sequester more C. However, the efficiency of SOC sequestration can be discounted or even negated when the N-induced emission ($CH_4$, $N_2O$) increases in agricultural ecosystems. Global warming potential (GWP) or C footprint (CF) can be used to comprehensively assess the C balance of crop production systems. Scientific evidences show that GHGs emissions and GWP are highly dynamic, and change along with increase in input of N resources (N fertilizers, crop residues, manures, etc.) (Gu et al. 2015; Liu et al. 2014; Pittelkow et al. 2013; Qiao et al. 2014; Zhao et al. 2016). Schlesinger (2010) argued that $CO_2$ emission from production and distribution of N fertilizers and N fertilizer-induced

N$_2$O emission would likely eliminate all the climate benefits of the postulated enhanced C sequestration in China. Additionally, national assessment indicated that CF of N fertilizer and direct N$_2$O emission comprise a high portion of the total CF of crop production in China (Cheng et al. 2011; Lin et al. 2015).

In general, N$_2$O emission is increased with increase of N fertilizer application rate in rice, wheat, and maize production in China (Cui et al. 2014). The data from a global meta-analysis indicated a higher N$_2$O emission factor (ratio of N$_2$O emission to N fertilizer application rate) when soils received organic amendments in combination with synthetic fertilizers (Charles et al. 2017). A long-term experiment, conducted in upland soil in Northeast China, demonstrated that adding N fertilizers, with or without manure input, increased the N$_2$O emission, but reduced the total GWP compared to no fertilizer application due to the changes in SOC pool (Qiao et al. 2014). When taking CH$_4$ emission into consideration in paddy-rice soil, application of chemical fertilizers decreased CH$_4$ as well as the total GWP. However, inputs of manure significantly increased CH$_4$ emission and enhanced the GWP (Wang et al. 2013). Similarly, input of crop residues could also enhance the C balance in upland soils, but the straw-induced CH$_4$ emission in paddy soils offsets the benefits of increase in soil C (Liu et al. 2014). Thus, understanding the C balance and the true C sequestration in agricultural soils is crucial to reducing the GHGs emissions, especially the direct N$_2$O emission from N fertilizer.

Improving N management and enhancing NUE can offset emissions of GHGs in China. N$_2$O emission is strongly related to the amount of N input. Thus, identifying the optimal rate of N fertilizer application is a researchable priority. A large-scale survey of China's cereal production showed that compared to the present level of N management in farm, increasing NUE can strongly decrease N$_2$O emission (Cui et al. 2014). With the use of regionally optimal N rate (114–224 kg N ha$^{-1}$), estimated GHG emissions would be reduced by 11% in China's rice production (Wu et al. 2015). Combined with different field operations could also reduce emissions from N fertilizer application. For example, at a lower rate of N fertilizer application, differences in N$_2$O emission between soil under NT and plow tillage can be narrowed (Zhao et al. 2016). N$_2$O emission under NT could be mitigated compared to plow tillage by changing the N fertilizer placement to 5 cm soil depth (van Kessel et al. 2013). The data from a global meta-analysis showed that N$_2$O emission could be reduced by 39–48% through use of nitrification inhibitors in combination with fertilizers (Qiao et al. 2015). Thus, a proper site-adaptation strategy of how to manage N can have strong positive effects on C balance of crop production systems in China.

## 9.4 CHALLENGES AND OPPORTUNITIES OF NITROGEN MANAGEMENT AND SOIL CARBON SEQUESTRATION

An optimal use of N fertilizer is essential to improving crop yield and ensuring food security in China. Excessive and perpetual application of N fertilizer strongly impacts soil physical, chemical, and biological properties leading to change in SOC storage, soil degradation, and a series of other adverse environmental consequences. Researchable priorities to mitigate climate change by optimizing N management farming practices and SOC sequestration in China are discussed in the following sections.

### 9.4.1 NITROGEN MANAGEMENT AND SOIL ORGANIC CARBON SEQUESTRATION

Application of N has different effects on SOC concentration and sequestration because of differences in physiographic conditions in China, depending on climate, soil type, and farming practices. Thus, it is necessary to assess the principal factors and the underpinning mechanisms of SOC sequestration among different conditions. The rate of SOC sequestration differs among N management practices (e.g., the amount, time, place, and type of N application). Meta-analysis is a useful tool to analyze the effects of factor of N management on SOC change. In addition to determining the rate of SOC sequestration under different N management practices, it is also important to increase the potential of SOC sequestration and optimize N management for climate change mitigation. There are strong influences of N management on SOC concentration, SOC turnover, and sequestration. Other farming management practices also affect soil properties which change rate and magnitude of SOC sequestration. These factors have some interactive influences with N management on SOC sequestration. However, little research information is available on these issues and are a researchable priority for China.

## 9.4.2 Nitrogen Management and Soil Orange Carbon Pools

Labile SOC fractions are strongly sensitive to the farm management practices (e.g., N management). Hence, several studies have been conducted to assess the effects of N management (e.g., the amount, time, place, and type of N application) on concentration of the labile SOC fractions. However, research information about the influences of N management on non-labile SOC fractions (i.e., slow and passive SOC fractions) is scanty. Thus, additional research is needed to analyze the change in non-labile SOC fractions and to understand the turnover of different SOC fractions. With the rapid development of science and technology, the formulations of N fertilizers are more diverse and could result in differences in SOC fractions. There are also some discrepancies in the effects of rate of N management on different SOC fractions under diverse experiment conditions in China (e.g., climatic, soil types, cropping system, and farm operations). Therefore, it is critical to understand the principal factors and the underlying mechanisms of sequestering $CO_2$ in soil and mitigating climate change.

## 9.4.3 Soil Organic Carbon Sequestration and Other Environmental Issues

Appropriate and optimal N management can enhance SOC sequestration, increase crop productivity, and improve soil quality. However, a large amount of N is applied to agroecosystems of China, resulting in severe environmental consequences (e.g., water eutrophication, smog, acid rain, stratospheric ozone depletion, biodiversity loss) (Galloway et al. 2008). Adoption of integrated soil-crop system management practices can produce more grains with lower environmental footprint (i.e., reactive N losses and GHGs emissions) without any increase in N fertilizer input into the main agro-ecological areas of China (Chen et al. 2014). The process of soil N decomposition and mineralization is closely linked with the SOC concentration, and it affects the losses of reactive N associated with numerous environmental issues. Furthermore, more attention should be given to the comprehensive assessment of N management on crop productivity, SOC sequestration, and environmental pollution.

## 9.4.4 Effect of Nitrogen Management on Soil Quality and Food Security

Presently, many studies are focused on assessing the influence of N fertilization on crop yield in China, including the rate, time, and type of N fertilization. However, most of the results depend on the specific site experiment with field plot design, and few studies have analyzed the effect of N fertilization on crop yield at a regional or national scale. Large yield gaps among regions were observed due to the difference in climate, soil texture, and farm management practice. Narrowing the yield gaps at different levels for major food staples would be a potential solution to ensure food safety under the same N fertilization condition in China. In addition, appropriate N management can affect soil physical, chemical, and biological properties (Li et al. 2011; Wang et al. 2016), improve soil quality, and thus enhance crop yield. Moreover, SOC and its fractions are important indicators to soil quality, which can affect crop yield in degraded farmland (Lal 2004; Lal 2010). Thus, it is crucial to study the relationship between SOC pool, soil quality, and crop yield based on N management and to advancing China's food security.

## 9.5 CONCLUSIONS

The SOC sequestration responds strongly to N management because of the coupled cycling of C and N in soil. Numerous studies have focused on the effect of rate of N application, chemical N fertilizer plus other chemical fertilizers or organic amendments on labile SOC fractions in China. But, there is a little research information on the effects of the placement, time, and type of N addition on the changes in SOC turnover and sequestration, and SOC fractions. Yet, the scientific community is faced with a major challenge of mitigating climate change through optimization of N management practices in China. In this context, researchable priorities include: (1) assessing the influence of new formulations of N fertilizers on SOC turnover and sequestration; (2) analyzing the factors influencing and the underpinning mechanisms of SOC change by N management in different agro-ecological regions; (3) strengthening the study of integrated N management with other farming practices on SOC turnover; and (4) studying the relationships of SOC sequestration to soil quality, crop productivity, and environmental impacts under different N managements.

## ABBREVIATIONS AND ACRONYMS

| | |
|---|---|
| **AN** | Ammonium nitrate |
| **C** | Carbon |
| **CA** | Conservation agriculture |
| **CBN** | Chemical reaction type biochar based nitrogen |
| **DOC** | Dissolved organic carbon |
| **GHGs** | Greenhouse gases |
| **GWP** | Global warming potential |
| **K** | Potassium |
| **LOC** | Light organic carbon |
| **MBC** | Microbial biomass organic carbon |
| **MOC** | Mineral-associated organic carbon |
| **N** | Nitrogen |
| **NCP** | North China Plain |
| **NT** | No-till |
| **NUE** | Nitrogen use efficiency |
| **P** | Phosphorus |
| **POC** | Particular organic carbon |
| **POxC** | Permanganate oxidizable organic carbon |
| **RMP** | Recommended management practice |
| **SLBN** | Solid-liquid adsorption type biochar based nitrogen |
| **SOC** | Soil organic carbon |
| **SOM** | Soil organic matter |
| **TN** | Total nitrogen |

## ACKNOWLEDGMENTS

This study was funded by Special Fund for Agro-scientific Research in the Public Interest in China (201503136).

## REFERENCES

Asai, H., B. K. Samson, H. M. Stephan, K. Songyikhangsuthor, K. Homma, Y. Kiyono, Y. Inoue, T. Shiraiwa, and T. Horie. 2009. Biochar amendment techniques for upland rice production in Northern Laos 1. Soil physical properties, leaf SPAD and grain yield. *Field Crops Research* 111: 81–84.

Batjes, N. H. 2014. Total carbon and nitrogen in the soils of the world. *European Journal of Soil Science* 65: 10–21.

Cao, C. C., Y. C. Qi, Y. S. Dong, Q. Peng, X. C. Liu, L. J. Sun, J. Q. Jia, S. F. Guo, and Z. Q. Yan. 2014. Effects of nitrogen deposition on critical fractions of soil organic carbon in terrestrial ecosystems. *Acta Prataculturae Sinica* 23: 323–332 (in Chinese with English abstract).

Cao, Q., M. R. He, X. L. Dai, H. W. Meng, and C. Y. Wang. 2011. Effects of interaction between density and nitrogen on grain yield and nitrogen use efficiency of winter wheat. *Plant Nutrition and Fertilizer Science* 17: 815–822 (in Chinese with English abstract).

Chai, Y. J., X. B. Zeng, S. Z. E., T. Huang, Z. X. Che, S. M. Su, and L. Y. Bai. 2014. Response of soil organic carbon and its aggregate fractions to long-term fertilization in irrigated desert soil of China. *Journal of Integrative Agriculture* 13: 2758–2767.

Charles, A., P. Rochette, J. K. Whalen, D. A. Angers, M. H. Chantigny, and N. Bertrand. 2017. Global nitrous oxide emission factors from agricultural soils after addition of organic amendments: A meta-analysis. *Agriculture, Ecosystems and Environment* 236: 88–98.

Chen, X., Z. Cui, M. Fan, P. Vitousek, M. Zhao, W. Ma, Z. Wang et al. 2014. Producing more grain with lower environmental costs. *Nature* 514: 486–489.

Chen, X., J. Guo, M. Liu, J. Jiang, Q. Huang, T. Lai, H. Li, and F. Hu. 2011. Effects of fertilization on lability and recalcitrancy of organic carbon of red paddy soils. *Acta Pedologica Sinica* 48: 125–131 (in Chinese with English abstract).

Cheng, K., G. Pan, P. Smith, T. Luo, L. Li, J. Zheng, X. Zhang, X. Han, and M. Yan. 2011. Carbon footprint of China's crop production—An estimation using agro-statistics data over 1993–2007. *Agriculture, Ecosystems and Environment* 142: 231–237.

Cui, Z., X. Chen, Y. Miao, F. Zhang, Q. Sun, J. Schroder, H. Zhang et al. 2008. On-farm evaluation of the improved soil $N_{min}$-based nitrogen management for summer maize in North china Plain. *Agronomy Journal* 100: 517–525.

Cui, Z. L., G. L. Wang, S. C. Yue, L. Wu, W. F. Zhang, F. S. Zhang, and X. P. Chen. 2014. Closing the N-use efficiency gap to achieve food and environmental security. *Environmental Science and Technology* 48: 5780–5787.

Dai, J., Z. Wang, M. Li, G. He, Q. Li, H. Cao, S. Wang, Y. Gao, and X. Hui. 2016. Winter wheat grain yield and summer nitrate leaching: Long-term effects of nitrogen and phosphorus rates on the Loess Plateau of China. *Field Crops Research* 196: 180–190.

Dalal, R. C., D. E. Allen, W. J. Wang, S. Reeves, and I. Gibson. 2011. Organic carbon and total nitrogen stocks in a Vertisol following 40 years of no-tillage, crop residue retention and nitrogen fertilisation. *Soil and Tillage Research* 112: 133–139.

Davidson, E. A., and I. A. Janssens. 2006. Temperature sensitivity of soil carbon decomposition and feedbacks to climate change. *Nature* 440: 165–173.

DCCEE. 2012. The carbon farming initiative handbook version 1.0 (Australian Department of Climate Change and Energy Efficiency). http://www.climatechange.gov.au/sites/climatechange/files/files/reducing-carbon/CFI-Handbook-20120403-PDF.pdf.

Dikgwatlhe, S. B., Z. D. Chen, R. Lal, H. L. Zhang, and F. Chen. 2014. Changes in soil organic carbon and nitrogen as affected by tillage and residue management under wheat-maize cropping system in the North China Plain. *Soil and Tillage Research* 144: 110–118.

Ding, X. L., X. Z. Han, Y. Liang, Y. F. Qiao, L. J. Li, and N. Li. 2012. Changes in soil organic carbon pools after 10 years of continuous manuring combined with chemical fertilizer in a Mollisol in China. *Soil and Tillage Research* 122: 36–41.

Dong, C. H., J. S. Gao, X. B. Zeng, Q. Liu, M. G. Xu, and S. L. Wen. 2014. Effects of long-term organic manure and inorganic fertilizer combined application on rice yield and soil organic carbon content in reddish paddy fields. *Journal of Plant Nutrition and Fertilizer* 20: 336–345 (in Chinese with English abstract).

Dou, F., A. L. Wright, and F. M. Hons. 2008. Sensitivity of labile soil organic carbon to tillage in wheat-based cropping systems. *Soil Science Society of America Journal* 72: 1445–1453.

Du, S. Z., C. F. Cao, Y. L. Zhang, Z. Zhao, Y. Q. Qian, Y. H. Liu, and S. H. Zhang. 2009. Effect of basal/topdressing nitrogen ratio on yield and quality of super-high-yielding wheat in Huaibei Area. *Journal of Triticeae Crops* 29: 1027–1033 (in Chinese with English abstract).

Duan, W. X., Z. W. Yu, Y. L. Zhang, D. Wang, and Y. Shi. 2012. Effects of nitrogen fertilizer application rate on nitrogen absorption, translocation and nitrate nitrogen content in soil of dryland wheat. *Scientia Agricultura Sinica* 45: 3040–3048 (in Chinese with English abstract).

Eberwein, J. R., P. Y. Oikawa, L. A. Allsman, and G. D. Jenerette. 2015. Carbon availability regulates soil respiration response to nitrogen and temperature. *Soil Biology and Biochemistry* 88: 158–164.

Edenhofer, O., R. Pichs-Madruga, Y. Sokona, E. Farahani, S. Kadner, K. Seyboth, A. Adler et al. 2014. *Climate Change 2014: Mitigation of Climate Change. Contribution of Working Group III to the Fifth Assessment Report of the Intergovernmental Panel on Climate Change*. Cambridge, United Kingdom and New York: Cambridge University Press.

Fan, T. L., S. Y. Wang, G. Y. Zhou, and N. P. Ding. 2013. Effects of long-term fertilizer application on soil organic carbon change and fraction in Cumulic Haplustoll of Loess Plateau in China. *Scientia Agricultura Sinica* 46: 300–309 (in Chinese with English abstract).

Feng, Y., H. F. Chen, X. M. Hu, H. M. Cai, and F. S. Xu. 2014. Optimal nitrogen application rates on rice grain yield and nitrogen use efficiency in high, middle and low-yield paddy fields. *Journal of Plant Nutrition and Fertilizer* 20: 7–16 (in Chinese with English abstract).

Galloway, J. N., A. R. Townsend, J. W. Erisman, M. Bekunda, Z. Cai, J. R. Freney, L. A. Martinelli, S. P. Seitzinger, and M. A. Sutton. 2008. Transformation of the nitrogen cycle: Recent trends, questions, and potential solutions. *Science* 320: 889–892.

Gruber, N., and J. N. Galloway. 2008. An earth-system perspective of the global nitrogen cycle. *Nature* 451: 293–296.

Gu, B., X. Ju, J. Chang, Y. Ge, and P. M. Vitousek. 2015. Integrated reactive nitrogen budgets and future trends in China. *Proceedings of the National Academy of Sciences of the United States of America* 112: 8792–8797.

Hou, H. J., X. M. Liu, G. R. Liu, Z. Z. Li, Y. R. Liu, Y. L. Huang, J. H. Ji, C. H. Shao, and F. Q. Wang. 2011. Effect of long-term located organic-inorganic fertilizer application on rice yield and soil fertility in Red Soil Area of China. *Scientia Agricultura Sinica* 44: 516–523 (in Chinese with English abstract).

Huang, J. Y., H. Richard, and S. F. Zheng. 2013. Effects of nitrogen fertilization on soil labile carbon fractions of freshwater marsh soil in Northeast China. *International Journal of Environmental Science and Technology* 11: 2009–2014.

Huang, Y., and W. J. Sun. 2006. Changes in topsoil organic carbon of croplands in China over the last two decades. *Chinese Science Bulletin* 51: 1785–1803.

Huo, Z., J. F. Fu, and P. Wang. 2015. Effects of application of N-fertilizer and crop residues as maize on summer maize N recovery rate. *Soils* 37: 202–204 (in Chinese with English abstract).

Ju, X. T., J. R. Pan, X. J. Liu, and F. S. Zhang. 2003. Study on the fate of nitrogen fertilizer in winter wheat/summer maize rotation system in Beijing suburban. *Plant Nutrition and Fertilizer Science* 9: 264–270 (in Chinese with English abstract).

Lal, R. 2004. Soil carbon sequestration impacts on global climate change and food security. *Science* 304: 1623–1627.

Lal, R. 2006. Enhancing crop yields in the developing countries through restoration of the soil organic carbon pool in agricultural lands. *Land Degradation and Development* 17: 197–209.

Lal, R. 2010. Managing soils and ecosystems for mitigating anthropogenic carbon emissions and advancing global food security. *BioScience* 60: 708–721.

Lal, R. 2015. A system approach to conservation agriculture. *Journal of Soil and Water Conservation* 70: 82A–88A.

Li, H., W. J. Ge, X. X. Ma, Q. H. Li, W. D. Ren, X. Y. Yang, and S. L. Zhang. 2011. Effects of long-term fertilization on carbon and nitrogen and enzyme activities of soil microbial biomass under winter wheat and summer maize rotation system. *Plant Nutrition and Fertilizer Science* 17: 1140–1146 (in Chinese with English abstract).

Li, S. K., K. R. Wang, J. K. Feng, R. Z. Xie, and S. J. Gao. 2006. Factors affecting seeding emergence in winter wheat under different tillage patterns with maize stalk mulching returned to the field. *Acta Agronomica Sinica* 32: 463–465 (in Chinese with English abstract).

Li, W. T., Z. P. Li, M. Liu, C. Y. Jiang, M. Wu, and X. F. Chen. 2016. Enzyme activities and soil nutrient status associated with different aggregate fractions of paddy soils fertilized with returning straw for 24 years. *Scientia Agricultura Sinica* 49: 3886–3895 (in Chinese with English abstract).

Li, X. H., F. C. Li, J. S. Liu, M. D. Hao, and Z. H. Wang. 2014. Changes of different carbon fractions caused by long-term N fertilization in dryland soil of the Loess Plateau. *Scientia Agricultura Sinica* 47: 2795–2803 (in Chinese with English abstract).

Lin, J. Y., Y. C. Hu, S. H. Cui, J. F. Kang, and L. L. Xu. 2015. Carbon footprints of food production in China (1979–2009). *Journal of Cleaner Production* 90: 97–103.

Lin, Z. A., B. Q. Zhao, L. Yuan, and B. S. Hwat. 2009. Effects of organic manure and fertilizers long-term located application on soil fertility and crop yield. *Scientia Agricultura Sinica* 42: 2809–2819 (in Chinese with English abstract).

Liu, C., M. Lu, J. Cui, B. Li, and C. Fang. 2014. Effects of straw carbon input on carbon dynamics in agricultural soils: A meta-analysis. *Global Change Biology* 20: 1366–1381.

Liu, J., S. Wang, J. Chen, M. Liu, and D. Zhuang. 2004. Storages of soil organic carbon and nitrogen and land use changes in China: 1990–2000. *Acta Geographica Sinica* 59: 483–496 (in Chinese with English abstract).

Liu, Q., B. Liu, P. Ambus, Y. Zhang, V. Hansen, Z. Lin, D. Shen et al. 2016. Carbon footprint of rice production under biochar amendment—A case study in a Chinese rice cropping system. *Global Change Biology Bioenergy* 8: 148–159.

Liu, R., Z. Wang, and H. Zhu. 2007. Research progress of organic fertilizer in China. *Chinese Agricultural Science Bulletin* 23: 310–313 (in Chinese with English abstract).

Liu, X., X. Ju, F. Zhang, J. Pan, and P. Christie. 2003. Nitrogen dynamics and budgets in a winter wheat–maize cropping system in the North China Plain. *Field Crops Research* 83: 111–124.

Lou, Y., M. Xu, X. Chen, X. He, and K. Zhao. 2012. Stratification of soil organic C, N and C:N ratio as affected by conservation tillage in two maize fields of China. *Catena* 95: 124–130.

Lu, F., X. Wang, B. Han, Z. Ouyang, X. Duan, H. Zheng, and H. Miao. 2009. Soil carbon sequestrations by nitrogen fertilizer application, straw return and no-tillage in China's cropland. *Global Change Biology* 15: 281–305.

Lu, W. T., Z. K. Jia, P. Zhang, W. Wang, X. Q. Hou, B. P. Yang, and Y. P. Li. 2011. Effects of straw returning on soil labile organic carbon and enzyme activity in semi-arid areas of Southern Ningxia, China. *Journal of Agro-Environment Science* 30: 522–528 (in Chinese with English abstract).

Luo, Y. Q., and E. S. Weng. 2011. Dynamic disequilibrium of the terrestrial carbon cycle under global change. *Trends in Ecology and Evolution* 26: 96–104.

Lv, P., J. W. Zhang, W. Liu, J. S. Yang, K. Su, P. Liu, S. T. Dong, and D. H. Li. 2011. Effects of nitrogen application on yield and nitrogen use efficiency of summer maize under super-high yield conditions. *Plant Nutrition and Fertilizer Science* 17: 852–860 (in Chinese with English abstract).

Ma, Y., H. Shi, S. K. Zhang, and R. H. Lv. 2003. Whole maize straw addition: The changes of soil physical and chemical properties and the effect on winter wheat. *Journal of China Agricultural University* 8: 42–46 (in Chinese with English abstract).

Magill, A. H., and J. D. Aver. 2000. Dissolved organic carbon and nitrogen relationship in forest litter as affected by nitrogen deposition. *Soil Biology and Biochemistry* 32: 603–613.

Maillard, E., and D. A. Angers. 2014. Animal manure application and soil organic carbon stocks: A meta-analysis. *Global Change Biology* 20: 666–679.

MAPRC. 2013. Promoting the steady improvement of fertilizers use efficiency, fertilizers use efficiency up to 33% by scientific fertilization in China. http://www.moa.gov.cn/zwllm/zwdt/201310/t20131010_3625203.htm (in Chinese).

Men, M. X., X. W. Li, and H. Xu. 2008. Effects of long-term fertilization on crop yields and stability. *Scientia Agricultura Sinica* 41: 2339–2346 (in Chinese with English abstract).

Mulati, A., T. A. Tong, X. L. Yang, and H. Y. Ma. 2012. Effect of different fertilization on soil organic carbon and its fraction in farmland. *Chinese Journal of Soil Science* 43: 1461–1466 (in Chinese with English abstract).

NBSC. 1981–2015. *China Statistical Yearbook*. Beijing, China: China Statistics Press.

Pan, G., X. Xu, P. Smith, W. Pan, and R. Lal. 2010. An increase in topsoil SOC stock of China's croplands between 1985 and 2006 revealed by soil monitoring. *Agriculture, Ecosystems and Environment* 136: 133–138.

Pan, G. X., and Q. G. Zhao. 2005. Study on evolution of organic carbon stock in agricultural soils of China: Facing the challenge of global change and food security. *Advances in Earth Science* 20: 284–393 (in Chinese with English abstract).

Pang, F. M. 2008. Effects of combined application of organic and inorganic fertilizers on soil ammonia volatilization and nitrate accumulation in winter wheat field. Master thesis. Chinese Academy of Agricultural Sciences, Beijing, China.

Peng, S. B., R. J. Buresh, J. L. Huang, J. C. Yang, Y. B. Zou, X. H. Zhong, G. H. Wang, and F. S. Zhang. 2006. Strategies for overcoming low agronomic nitrogen use efficiency in irrigated rice systems in China. *Field Crops Research* 96: 37–47.

Pittelkow, C. M., M. A. Adviento-Borbe, J. E. Hill, J. Six, C. van Kessel, and B. A. Linquist. 2013. Yield-scaled global warming potential of annual nitrous oxide and methane emissions from continuously flooded rice in response to nitrogen input. *Agriculture, Ecosystems and Environment* 177: 10–20.

Plaza-Bonilla, D., J. Álvaro-Fuentes, and C. Cantero-Martínez. 2014. Identifying soil organic carbon fractions sensitive to agricultural management practices. *Soil and Tillage Research* 139: 19–22.

Qiang, X., H. Yuan, and W. Gao. 2004. Effect of crop-residue incorporation on soil $CO_2$ emission and soil microbial biomass. *Chinese Journal of Applied Ecology* 15: 469–472 (in Chinese with English abstract).

Qiao, C. L., L. L. Liu, S. J. Hu, J. E. Compton, T. L. Greaver, and Q. L. Li. 2015. How inhibiting nitrification affects nitrogen cycle and reduces environmental impacts of anthropogenic nitrogen input. *Global Change Biology* 21: 1249–1257.

Qiao, Y. F., S. J. Miao, X. Z. Han, M. Y. You, X. Zhu, and W. R. Horwath. 2014. The effect of fertilizer practices on N balance and global warming potential of maize-soybean-wheat rotations in Northeastern China. *Field Crops Research* 161: 98–106.

Qiu, S. J., H. J. Gao, P. Zhu, Y. P. Hou, S. C. Zhao, X. M. Rong, Y. P. Zhang, P. He, P. Christie, and W. Zhou. 2016. Changes in soil carbon and nitrogen pools in a Mollisol after long-term fallow or application of chemical fertilizers, straw or manures. *Soil and Tillage Research* 163: 255–265.

Schlesinger, W. H. 2010. On fertilizer-induced soil carbon sequestration in China's croplands. *Global Change Biology* 16: 849–850.

Shen, X. S., J. C. Li, H. J. Qu, F. Z. Wei, Y. Zhang, and W. M. Wu. 2011. Effects of wheat and maize straw returned to the field on lodging resistance of maize in lime concretion black soil region. *Scientia Agricultura Sinica* 44: 2005–2012 (in Chinese with English abstract).

Shi, Y., X. Zhao, X. Gao, S. Zhang, and P. Wu. 2016. The effects of long-term fertiliser applications on soil organic carbon and hydraulic properties of a loess soil in China. *Land Degradation and Development* 27: 60–67.

Shi, Z., D. Gu, C. Zhang, J. Yu, and S. Yang. 2014. Effects of nitrogen rates on vertical distribution of dry matter and nutrient of winter wheat in rice-wheat rotation. *Journal of Triticeae Crops* 34: 114–119 (in Chinese with English abstract).

Six, J., E. T. Elliott, and K. Paustian. 2000. Soil macroaggregate turnover and microaggregate formation: A mechanism for C sequestration under no-tillage agriculture. *Soil Biology and Biochemistry* 32: 2099–2103.

Six, J., E. T. Elliott, K. Paustian, and J. W. Doran. 1998. Aggregation and soil organic matter accumulation in cultivated and native grassland soils. *Soil Science Society of America Journal* 62: 1367–1377.

Soussana, J. F., and G. Lemaire. 2014. Coupling carbon and nitrogen cycles for environmentally sustainable intensification of grasslands and crop-livestock systems. *Agriculture, Ecosystems and Environment* 190: 9–17.

Stewart, C. E., K. Paustian, R. T. Conant, A. F. Plante, and J. Six. 2007. Soil carbon saturation: Concept, evidence and evaluation. *Biogeochemistry* 86: 19–31.

Sun, W., Y. Huang, W. Zhang, and Y. Yu. 2010. Carbon sequestration and its potential in agricultural soils of China. *Global Biogeochemical Cycles* 24: GB3001.

Sun, Y. T., Y. L. Liao, S. X. Zheng, J. Nie, Y. H. Lu, and J. Xie. 2013. Effects of long-term fertilization on soil organic carbon pool and carbon sequestration under double rice cropping. *Chinese Journal of Applied Ecology* 24: 732–740 (in Chinese with English abstract).

Tan, D. S., Jin. J. Y., S. W. Huang, S. T. Li, and P. He. 2007. Effect of long-term application of K fertilizer and wheat straw to soil on crop yield and soil K under different planting systems. *Scientia Agricultura Sinica* 40: 133–139 (in Chinese with English abstract).

Tan, Z., R. Lal, L. Owens, and R. C. Izaurralde. 2007. Distribution of light and heavy fractions of soil organic carbon as related to land use and tillage practice. *Soil and Tillage Research* 92: 53–59.

Tian, K., Y. C. Zhao, X. H. Xu, N. Hai, B. A. Huang, and W. J. Deng. 2015. Effects of long-term fertilization and residue management on soil organic carbon changes in paddy soils of China: A meta-analysis. *Agriculture Ecosystems & Environment* 204: 40–50.

Tian, K., Y. C. Zhao, X. H. Xu, B. Huang, W. X. Sun, X. Z. Shi, and W. J. Deng. 2014. A meta-analysis of field experiment data for characterizing the topsoil organic carbon changes under different fertilization treatments in uplands of China. *Acta Ecologica Sinica* 34: 3735–3743 (in Chinese with English abstract).

van Kessel, C., R. Venterea, J. Six, M. A. Adviento-Borbe, B. Linquist, and K. J. van Groenigen. 2013. Climate, duration, and N placement determine $N_2O$ emissions in reduced tillage systems: A meta-analysis. *Global Change Biology* 19: 33–44.

Wang, B., W. Wang, N. Hu, Z. Gu, B. Chen, Z. Zhang, J. Xu, and L. Zhu. 2016. Short-term effect of different water and nitrogen managements on paddy soil nutrient, enzyme activity and carbon pool under wheat straw-returning fields. *Journal of Nuclear Agricultural Sciences* 30: 957–964 (in Chinese with English abstract).

Wang, H., J. S. Liu, X. L. Hui, J. Dai, and Z. H. Wang. 2016. Responses of soil organic carbon, organic nitrogen and nitrogen supply capacity to long-term nitrogen fertilization practices in dryland soil. *Scientia Agricultura Sinica* 49: 2988–2998 (in Chinese with English abstract).

Wang, J. Y., Z. Z. Chen, Y. C. Ma, L. Y. Sun, Z. Q. Xiong, Q. W. Huang, and Q. R. Sheng. 2013. Methane and nitrous oxide emissions as affected by organic-inorganic mixed fertilizer from a rice paddy in Southeast China. *Journal of Soils and Sediments* 13: 1408–1417.

Wang, W., W. C. Chen, K. R. Wang, X. L. Xie, C. M. Yin, and A. L. Chen. 2011. Effects of long-term fertilization on the distribution of carbon, nitrogen and phosphorus in water-stable aggregates in paddy soil. *Agricultural Sciences in China* 10: 1932–1940 (in Chinese with English abstract).

Wang, Y., R. Amundson, and S. Trumbore. 1999. The impact of land use change on C turnover in soils. *Global Biogeochemical Cycles* 13: 47–57.

Wang, Y., C. Tu, L. Cheng, C. Li, L. F. Gentry, G. D. Hoyt, X. Zhang, and S. Hu. 2011. Long-term impact of farming practices on soil organic carbon and nitrogen pools and microbial biomass and activity. *Soil and Tillage Research* 117: 8–16.

Wang, Y. D., N. Hu, M. G. Xu, Z. F. Li, Y. L. Lou, Y. Chen, C. Y. Wu, and Z. L. Wang. 2015. 23-year manure and fertilizer application increases soil organic carbon sequestration of a rice-barley cropping system. *Biology and Fertility of Soils* 51: 583–591.

Wang, Y. F., D. Jiang, Z. W. Yu, and W. X. Cai. 2003. Effects of nitrogen rates on grain yield and protein content of wheat and its physiological basis. *Scientia Agricultura Sinica* 36: 513–520 (in Chinese with English abstract).

Wen, X. C., X. C. Wang, X. Y. Deng, Q. Zhang, T. Pu, G. D. Liu, and W. Y. Yang. 2015. Effects of nitrogen management on yield and dry matter accumulation and translocation of maize in maize-soybean relay-cropping system. *Acta Agronomica Sinica* 41: 448–457 (in Chinese with English abstract).

West, T. O., and W. M. Post. 2002. Soil organic carbon sequestration rates by tillage and crop rotation. *Soil Science Society of America Journal* 66: 1930–1946.

WMO. 2014. *WMO Statement on the Status of the Global Climate in 2013.* Geneva, Switzerland: World Meteorological Organization.

Wu, L., and Z. Cai. 2007. Estimation of the change of topsoil organic carbon of croplands in China based on long-term experimental data. *Ecology and Environment* 16: 1768–1774 (in Chinese with English abstract).

Wu, L., X. P. Chen, Z. L. Cui, G. L. Wang, and W. F. Zhang. 2015. Improving nitrogen management via a regional management plan for Chinese rice production. *Environmental Research Letters* 10: 095011.

Wu, Z. J., H. J. Zhang, G. S. Xu, Y. H. Zhang, and C. P. Liu. 2002. Effect of returning corn straw into soil on soil fertility. *Chinese Journal of Applied Ecology* 13: 539–542 (in Chinese with English abstract).

Xiang, Y. W., S. X. Zheng, Y. L. Liao, Y. H. Lu, J. Xie, and J. Nie. 2009. Effects of long-term fertilization on distribution and storage of organic carbon and nitrogen in water-stable aggregates of red paddy soil. *Scientia Agricultura Sinica* 42: 2415–2424 (in Chinese with English abstract).

Xie, J. Y., W. J. Yang, C. R. Qiangjiu, W. Xue, J. Li, S. L. Zhang, and X. Y. Yang. 2015. Distribution of soil organic carbon and nitrogen in water-stable aggregates of manurial loess soils under long-term various fertilization regimes. *Journal of Plant Nutrition and Fertilizer* 21: 1413–1422 (in Chinese with English abstract).

Xie, Z., J. Zhu, G. Liu, G. Cadisch, C. Chen, H. Sun, H. Tang, and Q. Zeng. 2007. Soil organic carbon stocks in China and changes from 1980s to 2000s. *Global Change Biology* 13: 1989–2007.

Xu, M. G., D. C. Li, J. M. Li, D. Z. Qin, Y. Kazuyuki, and H. Yasukazu. 2008. Effects of organic manure application combined with chemical fertilizers on nutrients absorption and yield of rice in Hunan of China. *Scientia Agricultura Sinica* 41: 3133–3139 (in Chinese with English abstract).

Xu, S., X. Shi, Y. Zhao, D. Yu, C. Li, S. Wang, M. Tan, and W. Sun. 2011. Carbon sequestration potential of recommended management practices for paddy soils of China, 1980–2050. *Geoderma* 166: 206–213.

Xue, J. F., C. Pu, S. L. Liu, Z. D. Chen, F. Chen, X. P. Xiao, R. Lal, and H. L. Zhang. 2015. Effects of tillage systems on soil organic carbon and total nitrogen in a double paddy cropping system in Southern China. *Soil and Tillage Research* 153: 161–168.

Yadav, R. L., B. S. Dwivedi, K. Prasad, O. K. Tomar, N. J. Shurpali, and P. S. Pandey. 2000. Yield trends, and changes in soil organic-C and available NPK in a long-term rice-wheat system under integrated use of manures and fertilisers. *Field Crops Research* 68: 219–246.

Yan, X. Y., and W. Gong. 2010. The role of chemical and organic fertilizers on yield, yield variability and carbon sequestration—Results of a 19-year experiment. *Plant and Soil* 331: 471–480.

Yang, J., W. Gao, and S. R. Ren. 2015. Long-term effects of combined application of chemical nitrogen with organic materials on crop yields, soil organic carbon and total nitrogen in fluvo-aquic soil. *Soil and Tillage Research* 151: 67–74.

Yang, X. L., Y. L. Lu, Y. A. Tong, W. Lin, and T. Liang. 2013. Effects of long-term N application and straw returning on N budget under wheat-maize rotation system. *Plant Nutrition and Fertilizer Science* 19: 65–73 (in Chinese with English abstract).

Yang, X. Y., X. H. Liu, X. R. Han, P. P. Duan, Y. C. Zhu, and W. Qi. 2016. Responses of soil microbial biomass and soluble organic matter to different application rates of N: A comparison between Liaochun 10 and Liaochun 18. *Scientia Agricultura Sinica* 49: 1315–1324 (in Chinese with English abstract).

Ye, W. P., X. L. Xie, K. R. Wang, and Z. G. Li. 2008. Effects of rice straw manuring in different periods on growth and yield of rice. *Chinese Journal of Rice Science* 22: 65–70 (in Chinese with English abstract).

Yi, Z. X., P. Wang, L. X. Shen, H. F. Zhang, M. Liu, and M. H. Dai. 2006. Effects of different types of nitrogen fertilizer on nitrogen accumulation, translocation and nitrogen fertilizer utilization in summer maize. *Acta Agronomica Sinica* 32: 722–778 (in Chinese with English abstract).

Yu, S. Z., Y. H. Chen, X. B. Zhou, Q. Q. Li, Y. Luo, and Q. Yu. 2004. Effect of straw-mulch during wheat stage on soil water dynamic changes and yield of summer maize. *Journal of soil and water conservation* 18: 175–178 (in Chinese with English abstract).

Zhang, F., Z. Cui, M. Fan, W. Zhang, X. Chen, and R. Jiang. 2010. Integrated soil-crop system management: Reducing environmental risk while increasing crop productivity and improving nutrient use efficiency in China. *Journal of Environmental Quality* 40: 1051–1057.

Zhang, H. L., R. Lal, X. Zhao, J. F. Xue, and F. Chen. 2014. Opportunities and challenges of soil carbon sequestration by conservation agriculture in China. *Advances in Agronomy* 124: 1–36.

Zhang, L., W. J. Zhang, M. G. Xu, Z. J. Cai, C. Peng, B. R. Wang, and H. Liu. 2009. Effects of long-term fertilization on change of labile organic carbon in three typical upland soils of China. *Scientia Agricultura Sinica* 42: 1646–1655 (in Chinese with English abstract).

Zhang, L. M., M. G. Xu, Y. L. Lou, X. L. Wang, S. Qin, T. M. Jiang, and Z. F. Li. 2014. Changes in yellow paddy soil organic carbon fractions under long-term fertilization. *Scientia Agricultura Sinica* 47: 3817–3825 (in Chinese with English abstract).

Zhang, Q., W. Zhou, G. Q. Liang, J. W. Sun, X. B. Wang, and P. He. 2015. Distribution of soil nutrients, extracellular enzyme activities and microbial communities across particle-size fractions in a long-term fertilizer experiment. *Applied Soil Ecology* 94: 59–71.

Zhang, Q. B., L. Yang, W. F. Zhang, H. H. Luo, Y. L. Zhang, and J. Wang. 2014. Effects of agronomic measures on soil organic carbon and microbial carbon content in cotton in arid region. *Scientia Agricultura Sinica* 47: 4463–4474 (in Chinese with English abstract).

Zhang, S. Q., X. H. Zhong, N. R. Huang, and G. A. Lu. 2010. Effect of straw-mulch-incorporation on nitrogen uptake and N fertilizer use efficiency of rice (*Oryza sativa* L.). *Chinese Journal of Eco-Agriculture* 18: 611–616 (in Chinese with English abstract).

Zhang, X., Y. L. Wang, Y. L. Han, and J. F. Tan. 2010. Effects of nitrogen ratio of base fertilizer and topdressing on uptake, utilization of nitrogen and yield in winter wheat. *Acta Agriculturae Boreali-Sinica* 25: 193–197 (in Chinese with English abstract).

Zhao, J., Z. C. Geng, J. Shang, R. Geng, Y. L. Wang, S. Wang, and H. F. Zhao. 2016. Effects of biochar and biochar-based ammonium nitrate fertilizers on soil microbial biomass carbon and nitrogen and enzyme activities. *Acta Ecologica Sinica* 36: 2355–2362 (in Chinese with English abstract).

Zhao, X., S. L. Liu, C. Pu, X. Q. Zhang, J. F. Xue, R. Zhang, Y. Q. Wang, R. Lal, H. L. Zhang, and F. Chen. 2016. Methane and nitrous oxide emissions under no-till farming in China: A meta-analysis. *Global Change Biology* 22: 1372–1384.

Zhao, X., R. Zhang, J. F. Xue, C. Pu, X. Q. Zhang, S. L. Liu, F. Chen, R. Lal, and H. L. Zhang. 2015. Management-induced changes to soil organic carbon in China: A meta-analysis. *Advances in Agronomy* 134: 1–50.

Zhao, X. C., and Z. H. Wang. 2010. Effects of nitrogen fertilizer rates on yield formation and nitrogen utilization of winter wheat. *Agricultural Research in the Arid Areas* 28: 65–70, 91 (in Chinese with English abstract).

Zhao, Y., Y. A. Tong, and H. B. Zhao. 2006. Effects of different N rates on nutrients accumulation, transformation and yield of summer maize. *Plant Nutrition and Fertilizer Science* 12: 622–627 (in Chinese with English abstract).

Zheng, L. J., G. Y. Zeng, and P. F. Wang. 2001. Effects of organic fertilizer plus chemical fertilizer application on rice yield and soil nutrients. *Chinese Agricultural Science Bulletin* 17: 48–50 (in Chinese with English abstract).

Zhong, Y., W. Yan, and Z. Shangguan. 2015. Soil organic carbon, nitrogen, and phosphorus levels and stocks after long-term nitrogen fertilization. *CLEAN - Soil, Air, Water* 43: 1538–1546.

Zhou, Z., Z. Gan, Z. Shangguan, and F. Zhang. 2013. Effects of long-term repeated mineral and organic fertilizer applications on soil organic carbon and total nitrogen in a semi-arid cropland. *European Journal of Agronomy* 45: 20–26.

Zhu, S., J. M. Vivanco, and D. K. Manter. 2016. Nitrogen fertilizer rate affects root exudation, the rhizosphere microbiome and nitrogen-use-efficiency of maize. *Applied Soil Ecology* 107: 324–333.

# 10 Managing Nitrogen in Small Landholder Hill Farms of North Eastern Indian Himalayas

*Anup Das, Jayanta Layek, Gulab Singh Yadav,*
*S. Babu, D. Sarkar, R.S. Meena, and Rattan Lal*

## CONTENTS

## 10.1  INTRODUCTION

The Indian Himalayan Region (IHR), a mega-biological hotspot, covers geographical area of 0.53 × $10^6$ km$^2$ (Palni and Rawal 2010), inhabited by ~51 million people, and represents about 16.2% and 3.86% of total area and population of India, respectively (Singh 2015). The IHR, with width of 250–300 km across, stretches over 2500 km beginning from Arunachal Pradesh in the east to Jammu and Kashmir in the west and spans between 21°57′–37°5′ N latitudes and 72°40′–97°25′ E longitudes (Bhatt et al. 1999). Unsustainable agricultural practices, soil degradation, and cropland scarcity have serious implications to food security and livelihoods of the people in IHR (Partap 1999; Das et al. 2014). Crisis area studies, conducted by the Mountain Farming Systems Programme of International Centre for Integrated Mountain Development (ICMOD), documented evidence of unsustainable mountain agriculture in IHR (Pandey 1992; Jodha and Shrestha 1994). Many of the indicators used to illustrate this are derived from the farmers' responses to the lack of adequate cropland management. For example, marginal rainfed croplands in some areas are abandoned because farmers do not perceive that the benefits of cultivation outweigh the costs (Partap 1999). General trends in mountain farming in the region are listed below:

- Shrinking size of land holdings,
- Expanding farming on marginal lands,
- Increasing risks of soil erosion and declining farmland fertility,
- Growing food deficit due to inadequate production and poor nutrient supplementation,
- Accelerating degradation of support lands,
- Growing trends in rainfed farming and monocropping, and
- Increasing number of small landholders and marginal farmers.

The hill and mountain area of the Himalayas are ecologically fragile and economically less developed facing several problems and imposing severe limitations on resource productivity (Bhatta and Vetaas 2016). Consequently, agriculture is mostly subsistence, and yet is the important source of livelihood of the people of the region (Das et al. 2010). Inadequate nutrient management in agriculture is among the major challenges of sustaining productivity in the region. Among the nutrients, proper management of nitrogen (N) is the key to sustaining productivity and soil quality. It is estimated that without the input of fertilizer N, about half of the world's population would have remained hungry and malnourished (Prasad et al. 2016).

### 10.1.1  North Eastern Himalayan Region

The North Eastern Hill (NEH) region of India, also called north-eastern Indian Himalayas, comprises eight states (Arunachal Pradesh, Manipur, Meghalaya, Mizoram, Nagaland, Sikkim, Tripura, and Hilly tracts of Assam) and covers about 18 million hectare (M ha) area which is ~5.47% of the total geographical area of India (329 M ha). The NEH, together with the whole of the state of Assam, is popularly called the North Eastern Region (NER) of India, which covers 26.3 M ha or ~8% of the total geographical area of India. The NER lies between 22°05′ and 29°30′ N latitudes and 87°55′ and 97°24′ E longitudes and was home to 45.5 million people in 2016. Most livelihoods in the Himalayan region are based on agriculture and land is the most vital component of all socio-economic activities. The NER is marked by the geo-morphological development of a series of ridges and valleys, terraces, scraps, and several planar surfaces at different elevation (15–6000 m above sea level or mASL) (Das et al. 2016b).

The extreme climatic variations in the region range from cold temperate to tropical and moist subtropical. Thus, the NER is comprised of snow-clad mountains in Arunachal Pradesh and Sikkim to tropical and subtropical moist zone in Meghalaya, Tripura, and Assam. Despite the rich endowment of agro-climatic regimes and geo-biodiversity, the NER has lagged in economic development and is projected to face severe food grain deficit, which may be as much as 60% by 2050. Because of the undulating topography and high slope gradient, agriculturally suitable land area is limited in the region (Singh et al. 2014). Further, the agricultural productivity in the NER is also vulnerable to the changing climate. Global warming is predicted to increase the average temperature of the Indian subcontinent by 3–5°C by the end twenty-first century, which can lead to significant changes in rainfall distribution, flood, and drought situations (Ravindranath et al. 2011). The impact is expected to be more severe in the NER due to the small and marginal nature of the farmers and lack of resources for adaptation to the changing climate. The Ministry of Forest and Environment, Government of India, has predicted that the rice (*Oryza sativa* L.) and maize (*Zea mays* L.) productivity in NER (including Assam) may decline by 35 and 40%, respectively, by 2030 (Kumar 2011). Therefore, it is important to adopt sustainable agricultural practices that are favorable to conservation of natural resources and resilient to the climatic aberrations. Nonetheless, there is an ample potential for development with abundance of both biotic and abiotic natural resources in the NER.

## 10.1.2 LAND USES OF NER

The site-specific land use pattern of the NER region is strongly influenced by the elevation, climate, and mountainous terrain (Lenka et al. 2012a,b). The most dominant land use system in this region is forestry followed by agriculture, horticulture, animal husbandry, and non-agricultural uses such as urbanization, road construction, commercial establishments, and so on. Many of the land use systems are hazardous to resources and not conducive to sustain production (Das et al. 2016a). Densely forested area is < 60% of the reported area in most states except Arunachal Pradesh and Mizoram (above 70%), and the net cultivated land is merely 6.8% of the total land area. The area under agricultural use is significantly lower in NER compared to that of India at 46.73%. The agricultural area is the highest in Tripura (22.90%) and the lowest in Arunachal Pradesh (2.04%). The two important agricultural practices of the region are settled farming practiced in the plains, valleys, foot hills, and terraced slopes, and shifting cultivation (slash and burn agriculture locally known as *Jhum*) in hills (Arunachalam and Pandey 2003). The resource-poor and uneducated farmers are cultivating the uncultivable hilly terrains at the cost of resources sustainability. Cultivation of uncultivable terrains with low to moderate conservation measures is not doing any good. As a result, the landscape is slowly but steadily eroding and degrading. Growing crops with low to extremely low average productivity of agricultural commodities with high trade-offs as far as the resources are concerned is the major challenge. What will remain in the long run are rocks and boulders dotting all around the landscape which will be devoid of any vegetation cover.

Agricultural activities in the NER region are practiced up to 3500 m altitude and on slopes of up to 60%. A rice-based cropping system is predominant in the region except for Sikkim, where maize-based systems are popular. Of the total gross cropped area, about 80% is under food grain production, and rice occupies a significant part of the total food grain area and production (89.8% and 93.2%, respectively) (Das et al. 2009a). Around 56% of the cultivated area of the region is under low altitude (valley or lowland), 33% under mid-altitude (flat upland), and the rest under high altitude (upland terrace) terrain. Traditionally, farmers in both upland terrace and valleys practice mono-cropping under rainfed agriculture, where rice is the major crop followed by maize. The cropping intensity of the NER is 134% (Das et al. 2016b). Cropping intensity is the highest in the state of Tripura (184%) and the lowest in Mizoram (106%). The share of total cultivated area under food grain is the highest in Manipur followed by that in Nagaland and Arunachal Pradesh

(Das et al. 2016b). In Assam, Manipur, and Tripura, double-cropping of rice is practiced in plains. In low altitudes of Assam, Tripura, Manipur, and Meghalaya, rice, oilseeds, and vegetables are major crops. Rice, maize, ginger (*Zingiber officinale*), turmeric (*Curcuma longa*), and citrus (*Citrus* spp.) are important crops in hills.

### 10.1.3 SOIL FERTILITY AND LAND DEGRADATION

Acid soils occupy nearly 21.2 M ha (~81% of geographical area) in the region and the majority of these soils (16.2 M ha) is strongly acidic (pH < 5.5) (Sharma et al. 2006a). Singh and Sharma (2002) estimated that about 65% of the total area of NER is affected by extreme forms of soil acidity. Over and above the problem of P deficiency, acid soils are also prone to Al and Fe toxicity, acidity-induced fertility stress, and other nutritional problems causing severe reduction in crop yield (Kumar et al. 2012). Hence, reclamation of soil acidity is needed for reducing toxicity of Al and Fe and improving agronomic productivity. Application of lime (broadcasting or in furrow) along with farmyard manure (FYM) is recommended to increase the availability of essential nutrients and alleviate acidity-induced fertility problems in these areas (Kumar et al. 2012).

Depth of soils on steep slopes in the upper part of the hills varies from extremely shallow to shallow (10–50 cm) and rocks are exposed to the surface at some places. The soils in the valley, which are developed due to the sedimentation of the alluvial materials from the surrounding hills, are deep to very deep (100–150 cm and more) and are of sandy, silt, clay, and gravelly texture (Baishya and Sarkar 2015).

Although the soils in the valleys surrounded by hillocks are fairly rich in soil organic matter (SOM), upland and areas under shifting cultivation have low SOM (Choudhury et al. 2013) and available N contents. The practice of *Jhum* in steep slopes has depleted the SOM content (Sarkar et al. 2015). The amount of clay and soil pH tends to decrease and those of sand, SOM, and cation exchange capacity (CEC) tend to increase with the increase in elevation. Steep slopes accelerate the process of detachment of soil separates and removal of exchangeable cations due to high intensity rainfall and which generate runoff into the valley lands. With a little chance for sedimentation, the topsoil and SOM are lost to the main drainage. Thus, the clay content in soils of narrow valleys is also low. However, clay and SOM are deposited as in the wide valleys of Manipur (Prasad et al. 1981).

Among the various degradation processes, accelerated soil erosion by land misuse and acidification are the most serious problems in the region. Shifting cultivation, especially on very steep slopes, exacerbates the problem of accelerated erosion (Mandal and Sharda 2013). Soil crusting and poor ground cover under shifting cultivation accelerates the process of soil erosion and high soil loss during cultivation after slashing and burning of forest (150 Mg ha$^{-1}$ yr$^{-1}$). But the rate of soil erosion is reduced to about 30 Mg ha$^{-1}$ yr$^{-1}$ during fallowing of the sloping lands (NBSS 1997). On average, ~50 Mg ha$^{-1}$ soil is lost due to removal/burning of biomass from the soil surface and successive cultivation on steep slopes (Baishya and Sarkar 2015). The removal of top soil leads to soil fertility depletion in shallow soils, which cannot be easily restored. Depletion of SOM, available N, P, and exchangeable cations are the common features of *Jhum* soils (Grogan et al. 2012).

The entire soils of Meghalaya, Manipur, Tripura, and Arunachal Pradesh and around 50% of Mizoram, Nagaland, and Tripura are deficient in P (Prasad et al. 1981). The production of crops in the eastern Himalayas region is mainly based on the recycling of nutrients through organic residues from animal feed, feces, and bedding materials and leaf litters from forest either for mulching or to produce organic manure (Patel et al. 2015). In comparison to the present level of NPK consumption in different states of India (all India averages 144.3 kg ha$^{-1}$ yr$^{-1}$), NEH is at the bottom (51.7 kg ha$^{-1}$ yr$^{-1}$) (NEC, Govt. of India 2015).

In Arunachal Pradesh, soil organic carbon (SOC) concentration ranges from 14 g kg$^{-1}$ in Dirang [1475–1775 m above mean sea level (ASL)] to as high as 59 g kg$^{-1}$ in Tawang (2700–2775 m ASL). In Manipur, SOC concentrations range from 1–34 g kg$^{-1}$ in Bishnupur valley to 18.4–31.2 g kg$^{-1}$ in

**TABLE 10.1**

**Distribution of Operational Farm Holdings by Area (×10³ Hectares) in NER**

| States | Marginal | Small | Semi-Medium | Medium | Large | Total | % of Small and Marginal |
|---|---|---|---|---|---|---|---|
| Arunachal Pradesh | 11.9 | 25.9 | 93.9 | 154.9 | 97.3 | 383.9 | 9.8 |
| Assam | 774.8 | 687.2 | 818.0 | 437.4 | 281.8 | 2999.1 | 48.7 |
| Manipur | 40.2 | 62.8 | 55.3 | 13.4 | 0.4 | 172.1 | 59.8 |
| Meghalaya | 61.1 | 96.3 | 87.9 | 37.3 | 4.4 | 287.1 | 54.8 |
| Mizoram | 30.2 | 37.7 | 24.0 | 8.9 | 4.0 | 104.8 | 64.8 |
| Nagaland | 3.9 | 23.3 | 121.8 | 475.8 | 441.2 | 1066.0 | 2.6 |
| Sikkim | 14.8 | 20.4 | 26.9 | 32.2 | 12.3 | 106.7 | 33.0 |
| Tripura | 128.3 | 74.8 | 62.4 | 18.8 | 1.2 | 285.6 | 71.1 |
| NER total | 1065.2 | 1028.4 | 1290.2 | 1178.7 | 842.6 | 5405.1 | 44.3 |
| India total | 11.9 | 25.9 | 93.9 | 154.9 | 97.3 | 383.9 | 85.0 |

*Source:* Agriculture Census, 2010–11, Ministry of Agriculture & Farmers Welfare, Govt. of India.

Tamenglong. In Meghalaya, the SOC concentration is high in most of the soils, and ranges from 5.2–43.0 g kg⁻¹ in East Khasi Hill and 5.2–31.3 g kg⁻¹ in West Khasi Hill districts. The soils of Mizoram have low to high SOC (4.2–13.4 g kg⁻¹) concentration. The SOC concentration is very high in most soils in Sikkim and ranges from 23.5–44.5 g kg⁻¹. On the contrary, the soils of the humid subtropical regions (e.g., Tripura, Assam, and other foothills) are deficient in SOC and K contents (Sharma et al. 2006b; Das et al. 2016a).

### 10.1.4 LAND HOLDING AND FARM SIZE

The proportion of agricultural land (including fallow) to the total geographical area in the NER is 22.2% compared with the national average of 54.5%, and ranges from 4.4% in Arunachal to 37.4% in Assam. Of the total workforce, the share of cultivators (41.6%) and agricultural laborers (13.1%) is very high in comparison with the average share of India (31.65% and 26.55%, respectively). Land ownership is almost universal. The proportion of marginal and small farmers is 44.3% and 85% by area in NER and India, respectively (Table 10.1). Land distribution is mostly egalitarian and follows the principles established by the community (Patel 2013).

## 10.2 NITROGEN IN CROP PRODUCTION

Although N is the most abundant element on Earth, it is widely limiting for growth of most plants due to its unavailability of $N_2$ in air (Graham and Vance 2000). Nitrogen is an essential element for higher plants and a constituent of the *Rubisco* enzyme responsible for photosynthesis (Raines and Lloyd 2007). Healthy plants often contain 3–4% N in their above-ground tissues. Crop photosynthesis is closely associated with chlorophyll and N contents of leaf, both of which depend on the availability of N in soil to plant roots (Lemaire et al. 2007). It is also a constituent of amino acids, the building blocks of proteins. It is also a constituent of energy-transfer compound namely adenosine triphosphate (ATP), which allows cells to conserve and use the energy released in metabolism. Nitrogen is a significant component of DNA, the genetic material of plants. Unfortunately, higher plants cannot transform N directly into proteins (Finzi et al. 2006). Thus, soil microorganisms significantly impact the N cycle in soils and convert organic N into mineral forms, primarily ammonia $\left(NH_4^+\right)$ and nitrate $\left(NO_3^-\right)$, which are taken up by the plants. Plants get N from two main sources:

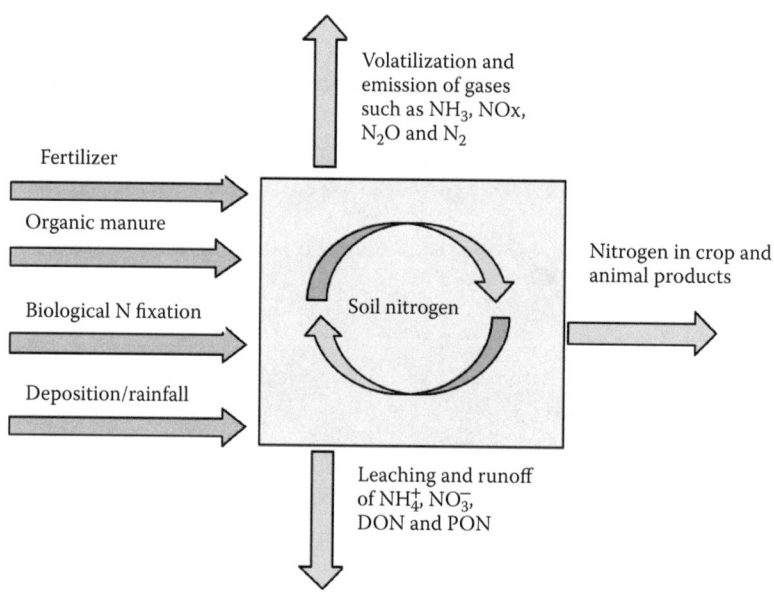

**FIGURE 10.1** Nitrogen budget in crop production and resulting N release to the environment (Zhang et al. 2015). Abbreviations: ammonia ($NH_3$), nitrogen oxides (NOx), nitrous oxide ($N_2O$), dinitrogen gas ($N_2$), ammonium ($NH_4^+$), nitrate ($NO_3^-$), dissolved organic nitrogen (DON), and particulate organic nitrogen (PON).

(1) the soil, through synthetic fertilizers, manures, and/or mineralization of SOM; and (2) the atmosphere through biological N fixation. However, since the 1970s, management of N inputs into agricultural systems has become a contentious issue (Vance 2001). On an average, about 68 Tg (Tg = teragram = $10^{12}$ g = million ton) of N is applied to agricultural land globally per year. Several synthetic N fertilizers are being prepared from ammonia ($NH_3$), and use of N fertilizer is essential to feeding the world population (Erisman et al. 2008). However, excessive and inappropriate application of N fertilizers in croplands can escape into the environment and contaminate water and air. Thus, it is important to understand the impact of different agricultural practices on the dynamics of soil mineral N to improve its use efficiency and reduce the risks of environmental pollution (Williams et al. 2016). An illustration of N balance in crop production is depicted in Figure 10.1.

## 10.3   SOIL EROSION AND NITROGEN LOSS IN HILLY AREAS

Accelerated erosion is among the major causes for soil degradation. It impacts the sustainability of agroecosystems, accelerates the global warming through emission of $N_2O$, and is a threat to the world's food security in the era of global climate change and rapidly increasing population (Lal 1994). Inappropriate land use and subsequent soil erosion is one of the main reasons for loss of fertile soil and plant nutrients (Lenka et al. 2012a). About 5334 Tg (16.4 Mg ha$^{-1}$) of soil is transported annually by erosion in India (Pandey et al. 2009). The high annual precipitation of the NER region (average ~ 2000 mm) also has high inter-annual variability. Intense rainfall is one of the driving forces of soil erosion and nutrient loss, the surface runoff produced by the rainfall carries away soil nutrients in particulate and dissolved states. Thus, erosion not only depletes soil fertility and fertilizer-use efficiency, but also causes the non-point source pollution (Lenka et al. 2012b). Soils of the NER are highly susceptible to accelerated erosion because of its undulating topography, steep slopes, and high precipitation. Deforestation at massive scale and practice of *Jhum* cultivation (slash and burn agriculture or shifting agriculture) on steep slopes accelerated the erosion hazard, which must be controlled (Mandal and Sharda 2013). Recent estimates indicate that about

39% of the area of the IHR has a potential erosion risk of >40 Mg ha$^{-1}$ yr$^{-1}$. The topography and climate of the region are conducive to accelerated soil erosion, which is a serious environmental threat (Lenka et al. 2012a).

About 29.8% of the area in the region falls under the category of very severe erosion, whereas 4.5, 21.2, 16.8, and 13.0% of areas fall under the category of very low, low, moderate, and severe erosion classes, respectively. It is also estimated using the universal soil loss equation (USLE) for Arunachal Pradesh that the total amount of ~669.4 Tg of soil is eroded annually with an average rate of 90.9 Mg ha$^{-1}$ yr$^{-1}$. The very high and intense rainfall (about 2000 mm yr$^{-1}$) and the sparse vegetation cover generate runoff and severe erosion of the downside cultivated lands which jeopardize the sustainability and productivity of agricultural lands (Lenka et al. 2012b). Total amount of nutrient loss is directly proportional to the amount of soil loss (Rajurkar et al. 2004). Soil erosion also leads to the movement of N into the surface water. Nitrogen is absorbed to the surface of clay particles and finer sediments or SOM as ammonium ($NH_4^+$) and as particulate OM in organic-N forms. There is a positive correlation between the corresponding values of soil erosion and nutrient loss. The amount of average loss of annual total N, SOC, and total P in central Himalayan region are reported to be ~267, 33, and 5 kg ha$^{-1}$ (Rai and Sharma 1998). However, Collins and Jenkins (1996) recorded lower amounts of nutrient losses in the middle hills of Nepal Himalayas albeit with similar slopes as that in the Central Himalayan region.

Soil erosion under shifting cultivation has a high inter-annual variability depending on the rainfall pattern. Experiments conducted on steep slopes of >44% have recorded the soil loss of 40.9 Mg ha$^{-1}$ with corresponding nutrient losses of 702.9 kg of SOC, 63.5 kg of P, and 5.9 kg of K ha$^{-1}$ (Ram and Singh 1993). The tolerable soil loss (T) value is 11.2 Mg ha$^{-1}$yr$^{-1}$ in the NER compared with that of 5–12.5 Mg ha$^{-1}$yr$^{-1}$ in North Western Himalayas (Mandal et al. 2006). However, the soil loss from steep hill slopes (60–79%) of NER under first year, second year, and abandoned *Jhum* was 147, 170, and 30 Mg ha$^{-1}$yr$^{-1}$ (Singh and Singh 1908, 1981). The annual loss of top soil, N, P, and K in shifting cultivated areas of the NER is 88346, 10669, 0.372, and 6051 Gg (Sharma 1999). While Singh and Singh (1994) estimated the loss of 6.0 Tg of SOC, 9.7 Mg of available P, and 5.7 Tg of K from the NER.

Loss of N occurs in hilly areas either through leaching or runoff by precipitation and irrigation. Precipitation causes heavy loss of N through leaching and runoff resulting in poor crop productivity of potato in Eastern Himalayas (Sharma 1999). Field experiments conducted on potato for two years on three soils of variable texture and two methods of cultivation, terrace and *bun* (raised bed of about 30 cm height and 1 m width) revealed that N loss through leaching and runoff increased with increase in applied N and was significantly influenced by the nature of soil, amount of rainfall received, and method of cultivation. Application of FYM slightly increased leaching of N in relatively heavier soil while no significant effect was observed on light soils. More than 80% of the variability in N leaching was attributed to the amount of rainfall, and clay and N contents of the soils (Sharma 1996).

## 10.4  NITROGEN IN NORTH EASTERN HIMALAYAS

Nitrogen is often the most yield limiting nutrient in most Indian soils, including those in the NEH region of India. Soil nitrogen plays an imperative role in increasing the food value of crop and its productivity, and it may also influence the health of soil ecology. The soils of the NEH region are rich in total N due to the presence of a high amount of OM. A high annual rainfall and cooler temperature at high altitudes favor greater accumulation and reduced decomposition of OM. Therefore, the content of OM and all forms of N and C:N ratio tend to increase with increase in elevation (Singh and Datta 1987, 1988; Gangopadhayay et al. 1990). Yet, the knowledge about analysis of higher altitude soils for OM may not be adequate to assess the true value of available N. Prasad et al. (1981) reported that the critical limit of SOC for higher elevation (>1000 m) soils of Meghalaya was 1.5%. However, the rate of mineralization is low because of the acidic soils and low temperature, irrespective of the high SOM content. Therefore, supply of N through organic amendments like vermicompost (VC), rural compost, FYM, and fertilizers is necessary to meet the N requirement of crops to realize the agronomic production potential of the region.

### 10.4.1 NITROGEN USE IN AGRICULTURE IN THE NER

There are a variety of anthropogenic sources of N in agriculture such as synthetic fertilizer, animal manure, biological N-fixation (BNF), and crop residues returned to the field after harvest. Agriculture in India, until the middle of the twentieth century, relied mostly on organic manures. After the introduction of urea in 1956 and di-ammonium phosphate (DAP) in 1967, N fertilizer was the most widely used amendment in the country, and its consumption has increased tremendously due to availability of these fertilizers at subsidized rates, spread of high yielding varieties, increased cropping intensity, and availability of irrigation facilities (Prasad et al. 2016). Indian farmers do not have much choice among sources of N, since urea supplies about 80% of all fertilizer N needs in the country. But, agriculture in the NER of India has mostly remained organic by default (Sanwal et al. 2007). The region has a large area under minimum use of fertilizer (12 kg ha$^{-1}$) and pesticides. The consumption of fertilizer in the region is low as the farmers are reluctant to invest in fertilizer due to high risks of losses by runoff and leaching. Yet, the fertilizer use is also low during the winter or pre-rainy season because of the non-availability of the fertilizers, which is a perpetual problem. Lack of financial resources with small and marginal farmers is the main reason for low use of fertilizer. Further, availability of plant biomass like weeds, forest litter and crop residues, and organic manures restrict the use of synthetic fertilizers for growing a wide range of crops (Das et al. 2008). The wide range of biomass from wasteland, bunds, or agro-forestry systems is usually a rich source of N (Das et al. 2006). Since the introduction of N-fertilizer urea to enhance the food grain production and feed the growing population in the NER, its use has increased but at a slower rate than other regions of the country. Further, the use of urea varies widely among states in the NER. For example, the highest rate of fertilizer use in Assam (33.3 kg N ha$^{-1}$) is followed by that in Manipur (23.2 kg N ha$^{-1}$), with the lowest (2–3 kg N ha$^{-1}$) rate in Arunachal Pradesh. The wide variation in the rate of fertilizer use among states of the NER is due to differences in physiography of the region. In recent years, consumption of fertilizer in the NEH region is declining due to popularization of organic farming (OF). Sikkim state of NER has been declared completely organic since 2016. Sharma and Datta (2006) gave an estimate of nutrient potential and actual availability of N (000 Mg) for use through different sources in NER (Table 10.2).

**TABLE 10.2**
**Potential and Actual Availability of N for Use Through Different Sources in NER**

| Source | Potential N Availability (×10³ Mg) | Actual N Available for Use (×10³ Mg) |
|---|---|---|
| Animal excreta (solid/liquid) | 98.51 | 60 |
| Compost | 0.52 | 0.52 |
| Green manure | 0.14 | 0.14 |
| Forest litter | 3.42 | 3.42 |
| Total organic sources | 102.59 | 64.08 |
| Nutrient from organic sources (kg ha$^{-1}$) | 20.91 | 13.07 |

*Source:* Sharma, U.C., M. Datta, and J.S. Samra. 2006. Soils and their management in north east India. ICAR research Complex for NEH region, Umiam, Meghalaya, India, pp. 122–170.

## 10.4.2   Nitrogen Management in Rice and Maize

The N is the most important nutrient affecting productivity of rice and maize in the NER. The optimal response of rice to 60–90 kg N ha$^{-1}$ under lowland and 40–60 kg N ha$^{-1}$ under upland conditions of NER depending upon soil fertility, management practices, and varieties has been reported (Das et al. 2014, 2016a). A linear response of rice yield up to 100 kg N ha$^{-1}$ has also been reported in the NER (Sarkar et al. 2016). However, variable response of rice to N has been reported up to 60–90 kg N ha$^{-1}$ in Sikkim, 80 kg N ha$^{-1}$ + FYM 10 Mg ha$^{-1}$ in Nagaland, and 80–90 kg N ha$^{-1}$ in Manipur and Assam (Kumar and Rao 1992). Further, the N use efficiency (NUE) in the region is low at 30%, mostly due to poor management and inappropriate cultivation practices. High rainfall and flood irrigation also contribute to low NUE. Judicious use of N fertilizers is of the utmost necessity to prevent the loss of the fertilizer, minimize contamination of ground water, and reduce the risks of decline of inherent soil quality. Urea, the most common N fertilizer used in the region, is highly soluble and prone to severe loss. Use of rock-phosphate coated urea (RPU) and periled urea in Meghalaya and gypsum coated urea (GCU) and urea super granule (USG) in Manipur for lowland rice cultivation has been recommend for higher rice productivity (Venkatesh et al. 2006) than other sources (Table 10.3).

Under lowland conditions of Tripura, prilled urea and lac-coated urea (LCU; urea coated with shellac resin) produced similar results. In the plains of Assam, the use of di-ammonium phosphate (DAP) could produce the highest grain yield of rainfed lowland rice (cv. Pankaj) as compared to urea and ammonium sulfate (Deka and Borah 1993). For most parts of the NER, placement of N fertilizer into a direct-seeded rice can be superior to broadcasting of the fertilizer. Grain yield of maize in Sikkim increased up to 150 kg N ha$^{-1}$, but the optimal rate for an economic yield is 60 kg N ha$^{-1}$ (Sharma et al. 2006).

The rate of 80 kg N ha$^{-1}$ is recommended for an optimal yield of maize in Meghalaya, Manipur, Mizoram, and Arunachal Pradesh. Awasthe (2002) reported the highest grain yield of maize in Sikkim for three split applications of N (⅓ at sowing + ⅓ at knee high stage + ⅓ at tasseling) by

## TABLE 10.3
### Rice Productivity as Influenced by Source and Forms of N Fertilizer in NEH Region

| Source and Forms | Method of Application | Meghalaya (980 m ASL) | | Manipur (750 m ASL) | |
|---|---|---|---|---|---|
| | | Yield (Mg ha$^{-1}$) | kg grain (kg$^{-1}$ N) | Yield (Mg ha$^{-1}$) | kg grain (kg$^{-1}$ N) |
| Control (No fertilizer/manure) | | 1.62 | – | 1.84 | – |
| Prilled urea | Basal | 2.97 | 23.27 | 3.89 | 20.84 |
| Prilled urea | 3 splits (25:50:25) | 3.09 | 25.41 | 4.26 | 22.24 |
| Urea super granule | Root zone placement 7 days after sowing | 3.42 | 31.14 | 4.34 | 25.00 |
| Urea super granule | Broad cast at 7 days | 2.98 | 22.71 | 3.86 | 20.16 |
| Gypsum coated urea | Basal | 3.21 | 29.17 | 4.44 | 25.97 |
| Rock phosphate coated urea | Half at 7 days + half at panicle initiation stage | 3.88 | 38.95 | – | – |
| Lsd (p = 0.05) | | 0.21 | | 0.26 | |

*Source:*   Venkatesh, M.S., K. Kumar, M. Raychaudhuri, M. Datta, and U.C. Sharma. 2006. Soil fertility and crop responses to nutrients. In *Soils and Their Management in North East India,* Eds. U.C. Sharma, M. Datta, and J.S. Samra. ICAR Research Complex for NEH Region, Umiam, Meghalaya, India, p. 502.

side band placement. Response of maize to application of N has been reported up to 100 kg ha$^{-1}$ in Nagaland (Sharma and Sharma 1991).

### 10.4.3 INTEGRATED NITROGEN MANAGEMENT

Modern agriculture depends on the intensive use of nutrients from inorganic chemicals and synthetic sources. However, a continuous use of high and imbalanced synthetic N fertilizers in agriculture has led to problems of soil degradation. Excessive use of inorganic N fertilizers increases nutrient imbalance (Vitousek et al. 2009) and contributes to environmental damage from non-point source pollution (Ju et al. 2007). Due to low availability and/or high cost of mineral and organic N inputs, current soil fertility management strategies often entail combined and integrated use of organic and inorganic sources (Yadav and Srivastava 2015). Combined mineral and organic N inputs have the potential for added benefits as a result of positive interactions between them. Integrated N management (INM), by combining organic and inorganic sources, is a better approach for supplying nutrients to the crop (Kumar et al. 2011). The increase in protein content by the combined application of chemical fertilizers (NPK) and vermicompost/FYM (Behera et al. 2007) may be attributed to direct supply of N. Application of chemical N fertilizers with supplementary and complementary use of organic manures and bio-inoculants improves physical, chemical, and biological properties of soil (Gruhn et al. 2000; Sharma et al. 2001). An integrated use of nutrients also enhances the NUE (Cui et al. 2010), supplies micro-nutrients for a longer period (Mishra et al. 2006), proliferates diverse groups of microorganisms in the root zone (Zhong et al. 2010), and maintains soil quality (Sharma et al. 2001; Batabyal et al. 2016). The combined use of organic manure and inorganic N fertilizers in suitable proportion improves agriculture productivity in hilly areas (Yadav and Srivastava 2015). The interaction concept between plant growth factors has recently received new attention with the combined application of organic and synthetic fertilizers. Besides adding nutrients, organic materials also act as substrate for soil organisms. In the context of fertilizer N, which is susceptible to loss if not taken up by a crop, direct interactions are the result of microbial mediated changes in the availability of the fertilizer N due to the addition of available C.

In a multi-locational trial with external inputs of organic matter, the benefit from the INM treatments has been reported at three sites in the NEH region (Table 10.4). The improvement in soil properties and nutrient availability under INM due to the presence of organic material is attributed to greater yield benefits to the crop compared to those under sole application of organic and inorganic fertilizers.

In general, farmers of the region largely use the organic sources of N. Results of field studies have revealed that combined use of different organic sources of N enhances both crop productivity and resources use efficiency. Application of organic manure in soil increases the activity

**TABLE 10.4**
**Added Benefits by INM at Three Sites in NEH Region of India**

| Site | $Y_{INM}$[a] | $Y_{organic}$ | $Y_{80\ urea-N}$ | Added Benefits | References |
|---|---|---|---|---|---|
| | | Mg ha$^{-1}$ | | | |
| Umiam, Meghalaya | 4.87 (0.29) | 4.58 (0.28) | 4.59 (0.26) | 0.28 (0.07) | Das et al. 2014 |
| Lembucherra, Tripura | 4.83 (0.23) | 4.22 (0.22) | 4.31 (0.21) | 0.57 (0.002) | Yadav et al. 2013 |
| Lamphelpat, Manipur | 4.17 (0.25) | 2.82 (0.21) | 4.71 (0.26) | 0.41 (0.01) | Baishya et al. 2014 |

[a] $Y_{INM}$ is the rice grain yield in the "INM" treatment, $Y_{organic}$ that in the "organic" treatments, and $Y_{80\ N\ by\ urea}$ the yield in the "80 urea-N" treatment (values in parentheses are standard error).

of mycorrhizae and other beneficial organisms in the root zone, supplies essential secondary and micronutrients to plants, and improves soil health and crop productivity (Tomar et al. 2001). The NER of India has huge potential of organic manure generation to the tune of 47 Tg including 9 Tg of crop residues (Bujarbaruah 2004). The presence of different leguminous trees with high nutrient content in their leaves (*Acacia auriculiformis, Alnus nepalensis, Cassia siamea, Erythrina indica,* and *Parkia roxburghii*) is in plentiful amount in the region and it can compensate the scarcity of FYM by using them as compost (Saha et al. 2007).

The production of biomass by trees, shrubs, and weeds is high in the region mainly due to high rainfall and moderate temperature. Application of 10 Mg ha$^{-1}$ leaf or weed biomass along with chemical fertilizer can increase rice grain yield by 8–10% over the sole application of inorganic fertilizer (Das et al. 2009b). Further, efficiency of this manure is increased when applied along with animal manure like FYM, VC, and poultry manure (PM) (Das et al. 2010; Babu et al. 2016). The combined application of FYM, VC, and PM is a better option for enhancing rice yield in hilly areas than that for the sole application of these amendments (Babu et al. 2015).

The farmers in different states of NER follow different N management systems. In the valley land of Tripura, farmers leave at least 30–50% of rice residues/standing rice stubbles in the field. Farmers also apply residues of crops/weeds/threshing floor wastages, and so on in rice fields and incorporate them within the soil during ploughing. Besides, they allow grazing of livestocks (e.g., cows, buffalo, goats, and sheep) in the paddy field during the dry season and a good amount of dung is generated in the field which improves soil fertility (Das et al. 2015). The amount of excreta added to rice fields by animals may vary from 1–5 Mg ha$^{-1}$ (average 2 Mg ha$^{-1}$) and depend upon the breed, livestock density, and duration of grazing. The urine of animals also added *in-situ* in the field, is rich in N (0.40%) and K (1.35%) (Yawalkar et al. 2002). Such animal grazing in fallow land is estimated to add about 30–35 kg N, per hectare per year (Das et al. 2015). This additional amount of nutrients available through excreta and urine along with 30–50% residue recycling in rice fields can improve yield of rice. Further, the farmers apply FYM/composts at 5–10 Mg ha$^{-1}$ from outside at least once in three years (average 2.0 Mg ha$^{-1}$ yr$^{-1}$). Higher activity of beneficial bacteria and earthworms are reported from rice fields under *in-situ* residue management compared to fields receiving only inorganic fertilizers (Munda et al. 2006; Das et al. 2008). Farmers believe that local cultivars require less input and produce satisfactory rice grain yield under a low input system.

Numerous important indigenous agricultural practices being followed in the region for centuries are among the best examples of integrated nutrient and resource management (Table 10.5).

### 10.4.4 ORGANIC N MANAGEMENT

#### 10.4.4.1  Source of Nitrogen

Increasing costs of inorganic N fertilizers, declining factor productivity coupled with need for a sustainable supply of nutrients and environmental concerns (Agbenin et al. 1999) have brought into focus the role of organic sources of N. The NER region is blessed with a natural organic source of N. The favorable agro-climatic conditions of the region produce a large amount of plant biomass (e.g., leaves, weeds, forest litter) around the farm. That offers a unique opportunity to farmers of the region to use it as manure or make compost. However, N content varies among the sources, the leaves of *Erythrina indica* contain the highest concentration of N (3.25) as compared with the lowest concentration in rice straw (0.62%). Hence, efficient utilization of these N rich organic resources for soil nutrient replenishment is an opportunity. The suitability of any biomass to be used for crop production is determined by its relative decomposability and mineralization patterns of the nutrient elements (Sarmah and Bordoloi 1994). Therefore, Das et al. (2010) explored the strategies of preparation of nutrient enriched compost to enhance the use efficiency of N in the region. In addition, different types of animal manures are available in the region, because farmers of the region follow the integrated farming system based agriculture.

**TABLE 10.5**

**Indigenous Agricultural Systems of N Management Widely Practiced in NEH of India**

| Indigenous Agricultural Systems | Location | Nitrogen Management |
|---|---|---|
| Rice-based farming system of the *Apatanis*, Arunachal Pradesh | The *Apatani* plateau located at an altitude of about 1525 mASL occupies approximately 27 km² area in the humid tropic climate of the Lower Subansiri district of Arunachal Pradesh, India. | Recycling of agricultural wastes plays a major role in nutrient management in terraces under rice + fish systems. Paddy straw is decomposed and incorporated into the soil during land preparation. The rice husk generated after milling of rice, kitchen waste, ash from fuel woods, weeds, and pig and poultry manure are recycled in the fields every year. The villages are located in a higher elevation than the cultivated land and the domestic sewage is directed to the fields. All these sources add organic matter and nutrient to the water and soil in the terrace. |
| *Zabo*-based terrace wet rice cum fish culture of the Chakhesangs, Nagaland | This is located at an altitude of 1270 m above mean sea level and originated in Kikruma village, Phek district of Nagaland, India having an annual rainfall of 1613 mm. | The runoff from the hilly slopes carries silt and deposit in the water harvesting tanks. These silts are rich in organic matter and nutrients. The cattle enclosures fenced with wood and tree branches or bamboo where the farmers keep their cattle on rotational basis. Such sheds are usually constructed on a little lower side of the water-harvesting pond. The water from the pond is drained through the cattle yard before reaching it to the rice field for irrigation. The water carries dung and urine of cattle to the crop fields which serves as good source of nutrition for the crops. |
| Alder (*Alnus nepalensis*)-based *jhum* system of the Angamis of Nagaland | Khonoma village located about 20 km west of Kohima, the capital town of Nagaland. | Alder is a large deciduous tree that grows well on cooler parts of the NER at high altitudes ranging from 800 to 3000 m. The alder tree has root nodules which fixes atmospheric N in symbiotic association with actinomycetes genus Frankia. The alder trees are grown in crop fields and pollarding is done to reduce shedding and generate fuel wood. The tree sheds its leaves to retain moisture and mulches and adds abundance of humus to the soil. |
| Cattle shed rotation | North Sikkim | Temporary cattle shed are made in field to harvest *in-situ* urine, cow dung and organic litters. The shed is rotated from one place to another until the whole field is covered. |

*Source:* Das, A., G.I. Ramkrushna, B.U. Choudhury, G.C. Munda, D.P. Patel, S.V. Ngachan, P.K. Ghosh et al. 2012. Natural resource conservation through indigenous farming systems: Wisdom alive in North East India. *Indian J. Tradit. Know.* 11(3): 505–513.

Major sources of plant nutrient supply in organic crop production are FYM, VC, GLM, crop residues, poultry manure (PM), pig manure (PGM), forest litter, green manure and green leaf manure, non-edible oil cakes, and so on. Beside this, incorporation of weeds *in-situ*, off farm weeds like *Eupatorium adenophorum, Ambrosia artimisifolia, Mikania micrantha* and so on are also rich sources of nutrient supply in crop production (Tripathi et al. 2007). Due to the favorable climatic conditions, the growth of weeds and shrubs are luxuriant in the NER of India. Most often, these plants are high in nutrient content and can be effectively converted into compost for use in crop production (Das et al. 2006). The *Eupatorium* sp., a weed available in most of the hilly areas of the region, contains high amounts of N (2%) and is easily biodegraded within the soil (Acharya and Kapur 2001). Another fast-growing weed *Lantana camara*, mostly found in the road sides, farm fences, and in wastelands, has no fodder value and can be used as an organic nutrient source (Saha

**TABLE 10.6**
**Organic Natural and Composts as Source of N in the Region**

| Natural Sources | Nitrogen (%) | Compost | Nitrogen (%) |
|---|---|---|---|
| *Erythrina indica* | 3.2 | Rice straw MCE compost | 2.76 |
| *Alnus nepalensis* | 2.2 | Rice straw MCNF compost | 3.40 |
| *Parkia roxburghii* | 2.5 | Eupatorium MCE compost | 2.83 |
| *Acacia auriculiformis* | 3.0 | Eupatorium MCNF compost | 3.48 |
| *Cassia siamea* | 2.5 | Lantana MCE compost | 2.31 |
| *Rice straw* | 0.62 | Lantana MCNF compost | 2.98 |
| *Eupatorium adhenophorum* | 1.85 | Weed mixture MCE compost | 2.63 |
| *Lantana camara* | 1.73 | Weed mixture MCNF compost | 3.18 |
| *Weed mixture* | 1.50 | Farmyard manure | 0.91 |

*Source:* Das, A., J.M.S. Tomar, T. Ramesh, G.C. Munda, P.K. Ghosh, and D.P. Patel. 2010. Productivity and economics of lowland rice as influenced by N-fixing tree leaves under mid-altitude subtropical Meghalaya. *Nutr. Cycl. Agroecosys.* 87: 9–19. 1; Das, A., Munda, G.C., Patel, D.P., Ghosh, P.K., Ngachan S.V., and Baiswar, Pankaj. 2009b. Productivity, nutrient uptake and post-harvest soil fertility in lowland rice as influenced by compost made from locally available plant biomass. Archives of Agronomy and Soil Science. doi: 10.1080/03650340903207907.

et al. 2007). Besides N, a heavy amount of OM, phosphorus (P), potassium (K), and other secondary and micronutrients are also added through green manure. In NER, with a plentiful availability of green biomass (weeds, tree leaves, etc.), the scope of GLM is bright. Further, this eliminates the time taken by green manure crops for growth in the field (Hazarika et al. 2006). The N content of some tree leaves, which can be used as GLM, is given in Table 10.6. The N concentrations in two types of compost [i.e., microbial consortia enriched (MCE) and microbial consortia cum nutrient fortified (MCNF)] prepared from different substrates are also presented in Table 10.6. Although the addition of OM in any form improves soil fertility, its impact depends on the rate of decomposition of OM and the amount of N mineralized (Gupta and Tripathi 1986).

Microbial consortia (*Aspergillus awamorii, Paecilomyces, fusisporus, Azotobacter chrococcum,* and *Bacillus polymyxa*) enriched compost prepared from paddy straw contained 32% higher total N as compared to mineral enriched compost. Application of MCE compost at 10 Mg ha$^{-1}$ reportedly produced 45% higher grain yield of rice over 100% NPK (100, 80–60 N, $P_2O_5$–$K_2O$ kg ha$^{-1}$) due to higher amount of available as well as total N status of the compost that improved soil fertility and crop nutrition (Laxminarayana et al. 2006a). Organic sources of N enhance crop productivity and also improve soil fertility (Singh et al. 1993). The organic manures undergo a mineralization process to make the N available to plants, and nutrient availability for crop production depends largely on N mineralization rates in soil (Azeez and Van Averbeke 2010).

### 10.4.4.2 Role of Biofertilizer

Biofertilizer is a substance containing living microorganisms applied to seeds, plants, or soil to colonize the rhizosphere and promote plant growth and development by increasing the supply of nutrients to the host plant (Malusa and Vassilev 2014; Mahanty et al. 2016). They are biologically active micro-organisms, namely, bacteria, fungi, or algae which provide nutrients to plants (Javorekova et al. 2015; Rao et al. 2015). The majority of biofertilizers are used for N-fixation and solubilization of phosphorous in agriculture. Nitrogen biofertilizers fix atmospheric N into soil in the available forms which can be readily used by plants. Biofertilizers such as *Azolla,* blue green algae (BGA), *Azospirillum, Azotobacter*, and phosphorus solubilizing bacteria (PSB) are also used for N management in agriculture. Application of biofertilizers can save up to 20% of N in rice production (Hazarika et al. 2006).

### 10.4.4.3    Biological N Fixation

Although 78% $N_2$ is present in the atmosphere, it cannot be used directly by the plants. The atmospheric N is converted to ammonia ($NH_3$) in the process of biological N fixation (BNF), which can be easily assimilated by plants (Mahanty et al. 2016). An important enzyme, namely nitrogenase plays a great role in the conversion of N to $NH_3$ (Smith et al. 2013). Nitrogen-fixing organisms can be categorized into two groups, that is, symbiotic and non-symbiotic. Symbiotic organisms include members of *Rhizobiaceae* and make symbiotic relationships with leguminous plants (Ahemad and Khan 2012). While non-symbiotic organisms include free-living and endophytic microorganisms like *Cyanobacteria, Azospirillum, Azotobacter* and so on (Bhattacharyya and Jha 2012). In addition to organic and inorganic sources of N, legumes also play a major role in N economy of NER. They supply N to plants, improve soil and land quality, and reduce soil erosion (Bargali and Bargali 2009). The legume-rhizobium is the most important symbiotic N-fixation system which offers a low input and cost-effective tool in soil fertility restoration. Hence, the role of leguminous N fixing plants for reclamation of degraded land is well known (Bargali et al. 2015). Symbiotic biological N fixation is the major source of N input in traditional agriculture and plays a significant role in improving soil fertility and productivity (Maikhuri et al. 2016). The highest grain yields of lowland rice have been reported in an Ultisol of Mizoram with the application of blue green algae (BGA—a mixture of *Anabaena variabilis, Nostoc muscorum, Aulosira fertilissima,* and *Tolypothrix tenuis*) + 80 kg N ha$^{-1}$ which was higher than 120 kg N ha$^{-1}$. Integrated application of BGA and fertilizer N improved the organic matter and available nutrient especially N status of the soil. BGA inoculation reduced the expenditure on fertilizer cost equivalent to 30–40 kg N ha$^{-1}$ (Laxminarayana 2006a). In another study, application of 50% NPK + *Azolla* dual cropping produced the highest rice yield followed by 100% NPK and Azolla dual cropping alone (Hazarika et al. 2003). Crop response to different biofertilizers in the NER region of India has been 1.8–43.5% yield increase (Table 10.7) over no biofertilizer application (Laxminarayana et al. 2006a).

### 10.4.4.4    Azospirillum

*Azospirillum* belongs to the Spirilaceae family and is heterotrophic in nature. It has the capacity of N fixation of 20–40 kg ha$^{-1}$ (Mahanty et al. 2016). *Azospirillum* can increase the growth (dry matter production, leaf area index, chlorophyll content in leaves, etc.) and yield attributes of various crops. *Azospirillum* can make associative symbiosis with many higher plants, particularly with $C_4$ plants as they grow and fix N on salts of malic and aspartic acid. Thus, it is more beneficial for $C_4$ crops

**TABLE 10.7**

**Crop Response to N Biofertilizer in NER**

| Crop | Biofertilizer | % Increase in Yield Over Control |
|---|---|---|
| Legume | *Rhizobium* innoculation | 8.5–26.2 |
| Rice | *Azospirillum* and PSM | 1.8–43.5 |
| Maize | *Azospirillum* and PSM | 3.5–16.7 |
| Mustard | *Azotobacter* | 5.9–23.9 |
| Potato | *Azotobacter* + PSM | 3.8–12.5 |

*Source:*  Laxminarayana K., M. Datta, Y.P. Sharma, U.C. Sharma, and B. Gopichand. 2006a. Soil microbes and biofertilizers. In *Soils and Their Management in North East India,* Eds. U.C. Sharma, M. Datta, and J.S. Samra. ICAR Research Complex for NEH Region, Umiam, Meghalaya, India, p. 502.

like maize in the hilly region of the NER of India. It is generally used for seed treatment by dipping the roots of rice seedlings in 2% suspension of *Azospirillum* inoculant, which can increase rice productivity by more than 100 kg ha$^{-1}$ (Kennedy et al. 2004).

### 10.4.4.5 Azotobacter

*Azotobacter* is a free-living, aerobic, and heterotrophic microorganism and comes under the family *Azotobacteraceae* (Mahanty et al. 2016). Although the *Azotobacters* are present in a good amount in neutral or alkaline soils, their presence in acidic soils is also reported. The plain areas and valleys of the Eastern Himalayas where rice is grown under stagnated water, due to increase in soil pH by water ponding, *Azotobacter* can play a crucial role in N management. As the OM content in soil in NER is comparatively higher than that of the rest of India, it can positively affect the population of *Azotobacter* and associated N fixation. It can fix approximately 20–25 kg N ha$^{-1}$ in the soil and increase crop yield by 20–30% in N deficient soils. *Azotobacter* is also reported to increase the seed germination and crop growth due to positive response of B vitamins, NAA, GA, and other chemicals (Mazid et al. 2011). Seed treatment and seedling dip methods are most commonly used practices of *Azotobacter* in NER of India. Sometimes *Azotobacter* is also used at 5 kg ha$^{-1}$ for soil application before the last ploughing.

## 10.5   ROLE OF LEGUMES IN N MANAGEMENT IN HILLS

Depletion of soil fertility particularly available N is a major limiting factor for low productivity in a hilly region (Maskey et al. 2001). Although the yield of cereals can be improved with N fertilizer application, most small and marginal farmers of the region cannot afford to purchase them. Diversification of cropping patterns and supply of N through legume BNF have been proposed as ways of sustaining food production and soil fertility restoration. While there is sufficient literature on symbiotic N$_2$ fixation and its benefits for agricultural systems, few studies have been undertaken in the hills of the Himalayan region. Apart from some recent experimental results (Pilbeam et al. 1998), little is specifically known about the amount of N that can be fixed through legumes in the Himalayan region where extremes in topography and climate combine to produce a diverse range of agro-ecosystems. Soil N fixation from 26–224 kg N ha$^{-1}$ by diverse legumes (Table 10.8) has been reported by Maskey et al. (2001) in Nepal hills.

However, the actual value of the fixed N that is returned to the soil N pool as legume residues and soil mineral N would be considerably less than the estimation. This is mainly because harvested grain

**TABLE 10.8**
**Estimated N Fixation (kg ha$^{-1}$) through Legume Crops in Hill**

| Crop | Total N Fixation (kg ha$^{-1}$) | |
| --- | --- | --- |
|  | Range | Mean Value |
| Soybean | 26–197 | 83 |
| Mash bean | 32–66 | 51 |
| Groundnut | 69–224 | 119 |
| Lentil | 28–84 | 57 |
| Soybean | 36–105 | 81 |

*Source:*   Maskey, S.L., S. Bhattarai, M.B. Peoples, and D.F. Herridge. 2001. On-farm measurements of nitrogen fixation by winter and summer legumes in the Hill and Terai regions of Nepal. *Field Crops Res.* 70(3): 209–221.

with a high amount of N is removed from the system and because farmers mostly remove the crop residues from the field for other purposes. However, there is scope for enhancing $N_2$ fixation through the introduction of legumes in fallow or degraded land, relay cropping, or intercropping. Since most soils are acidic in nature in the NER, it is necessary that soil pH matches both the legume and its compatible to *Rhizobia* for successful cultivation of legumes. For that, it is necessary to use acid-tolerant legume cultivars and rhizobia and soil liming to achieve a near neutral soil pH (Graham and Vance 2000).

## 10.6 INTERCROPPING AND N MANAGEMENT

Intercropping of cereals with legumes provides better soil coverage, smothers weeds, and uses different strata of soil for nutrient uptake. A part of the fixed atmospheric N by legume can be used by the associated cereal crops (Layek et al. 2014a). Intercropping of soybean (*Glycine max* L.) and groundnut (*Arachis hypogea* L.) with maize (Photo 10.1) could fix 14.0–55.1 kg N ha$^{-1}$ and 18.1–60.1 kg N ha$^{-1}$, respectively, under different row arrangements and also enhance the system productivity on slopping lands of the NEH region (Choudhary and Choudhury 2016). Suitable intercropping of legumes provides a yield advantage over sole cropping because of the complimentary utilization of natural resources by the component crops and fixation of atmospheric N (Layek et al. 2014b).

## 10.7 HEDGEROW INTERCROPPING FOR N MANAGEMENT

Hedgerow intercropping, though initially developed to restore the fertility of degraded lands in the tropics, is also used presently to provide fodder and reduce soil and water erosion in hills (Lenka et al. 2012a). Hedgerows grown in contours at suitable intervals maintain soil fertility and reduce soil and nutrient losses in sloping lands (Walter et al. 2003; Xu et al. 2012). Leguminous shrub species (e.g., *Indigofera* sp., *Gliricidia* sp., *Crotolaria* sp., *Tephrosia* sp.) are used to form alleys at suitable spacing in sloping lands on the contours. Field crops, spices, and vegetables are grown in between interspaces. Such crop production practices conserve natural resources by reducing soil erosion, enrich soil by adding leaf litters, and so on (Laxminarayana et al. 2006b). Data presented in Table 10.9 indicate the suitable hedgerow tree species and loss of soil nutrients under hill ecosystems.

The *Gliricidia maculcata* grown abundantly in hilly areas of NER contains 2.9% N, 0.5% P, and 2.8% K on a dry wet basis and can be a local source of leaf manure producing 12–15 kg green matter per tree. About 400–500 plants can be grown on the peripheral field areas and bunds yielding 5–6 Mg green manure per hectare. Other tree species like *Crotolaria juncea* and *Tephrosia purpurea* have 2.4–2.7% N, 0.3–0.6% P, and 0.8–2.0% K (Datt et al. 2008).

Six different hedgerow species (*Cajanus cajan, Crotolaria tetragona, Desmodium rensonii, Flemingia macrophylla, Indigofera tinctoria,* and *Tehphrosia candida*) have been tested for soil fertility buildup and soil maintenance at an ICAR experimental farm, Umiam, Meghalaya. The biomass production (leaves and twigs) and nutrient concentration in different leaves are indicated in Tables 10.10 and 10.11. All the species improve soil fertility status especially N (Tomar et al. 2007).

**PHOTO 10.1** **(See color insert.)** Intercropping of groundnut (left) with maize and soybean (right) with rice in Umiam, Meghalaya.

### TABLE 10.9
### Loss of Soil Nutrients (kg ha⁻¹yr⁻¹) under Different Treatments (Mean of Observations during 2002–2004)

| Treatment | Organic C | Available N | Available P |
|---|---|---|---|
| *Indigofera* sp. | 116.25 | 5.84 | 0.38 |
| *Indigofera* sp. + GFS* | 75.62 | 3.22 | 0.24 |
| *Gliricidia* sp. | 96.15 | 4.14 | 0.32 |
| *Gliricidia* sp. + GFS | 80.28 | 3.41 | 0.28 |
| Control | 132.24 | 7.35 | 0.52 |
| Sole GFS | 123.24 | 6.21 | 0.41 |

*GFS: Grass filter strips of *Saccharum* spp. (Lenka et al. 2012a).

### TABLE 10.10
### Nutrient Content (%) in Some Leguminous Tree Leaves

| Tree Species | N% | P% | K% | Zn (mg kg⁻¹) | Cu (mg kg⁻¹) |
|---|---|---|---|---|---|
| *Acacia nilotica* | 1.82 | 0.48 | 1.64 | 40 | 18 |
| *Acacia tortillas* | 1.30 | 0.52 | 1.66 | 40 | 20 |
| *Albizzia labbek* | 1.46 | 0.50 | 1.80 | 38 | 22 |
| *Delbergia sissoo* | 0.76 | 0.38 | 2.10 | 28 | 20 |
| *Leucaena leucocephala* | 3.20 | 0.58 | 1.92 | 68 | 24 |

*Source:* Nitant, H.C., L.S. Bhushan, and I. Pal. 1992. Status of soil properties in Yamuna Ravines under forest cover. *J. Trop. For.* 8(1): 31–37.

### TABLE 10.11
### Production of Total Pruned Biomass and Nutrient Content in Different Hedgerow Species

| Hedge Species | Total Biomass* (Mg ha⁻¹) | Nutrient Concentration (%) in Leaf Biomass | | | Available N (kg ha⁻¹) |
|---|---|---|---|---|---|
| | | N | P | K | |
| C. cajan | 8.94 | 3.29 | 0.67 | 1.43 | 222.5 |
| C. tetragona | 19.55 | 3.47 | 0.48 | 1.63 | 216.6 |
| D. rensonii | 7.06 | 3.63 | 0.48 | 1.56 | 206.7 |
| F. macrophylla | 4.70 | 3.23 | 0.45 | 1.26 | 256.1 |
| I. tinctoria | 12.01 | 3.86 | 0.81 | 1.63 | 210.0 |
| T. candida | 10.86 | 3.57 | 0.32 | 1.67 | 175.0 |

*Source:* Tomar, J.M.S., A. Annaduari, A. Das, and R. Singh. 2007. Agroforestry a viable component of organic farming. In *Advances in Organic Farming Technology in India*, Eds. G.C. Munda, P.K. Ghosh, A. Das, S.V. Ngachan, and K.M. Bujarbaruah. ICAR Research Complex for NEH Region, Umiam, Meghalaya, India.

*Fresh weight basis.

The prunings of *C. tetragona* can add 80, 11, and 38 kg N, P, and K, respectively, per kg ha$^{-1}$ yr$^{-1}$. On average, the prunings from these hedgerow species provide 20–80, 3–4, and 8–38 kg ha$^{-1}$ yr$^{-1}$ of N, P, and K, respectively. Addition of leaf biomass also helps in improving soil fertility and lowering the soil acidity remarkably from its initial level. A significant improvement in soil fertility occurs through addition of pruning of *C. tetragona* and *F. macrophylla* compared to other species (Laxminarayana et al. 2006a).

## 10.8   GREEN LEAF MANURE

Green manuring and GLM in rice with biomass from leguminous and non-leguminous species has had a positive effect on soil fertility and crop productivity (Ashraf et al. 2004; Das et al. 2009; Manjunath and Korikanthimath 2009). As the farmers of NER of India have the apathy to use chemical fertilizer and there is limited availability of livestock excreta-based organic manure, GLM has good potential to supplement N needs of cereal crops. The most common leguminous tree leaves used as N manures in the NER are *Tephrosia* spp, *Gliricidia spp,* and *Alnus nepalensis*. The alder (*Alnus nepalensis*) trees improve soil fertility and have been used by the farmers in Nagaland at varying altitudes (Singh et al. 1993). Use of such locally available plant biomass for building-up of available N content in soil is reported by Behera and Sharma (2010). If the available biomass near farm areas can be effectively used, they can become alternative sources of nutrient supply in place of popular organic manure and inorganic fertilizers. Incorporation of fresh N-fixing tree leaves at 10 Mg ha$^{-1}$ into the rice soil as GLM 20 days before transplanting for crop nutrition can increase rice yield significantly over no GLM application (Das et al. 2010). The N, P, K, and moisture content of some N-fixing tree leaves are given in Table 10.12. The maximum rice productivity of 4.82 Mg ha$^{-1}$ has been reported with recommended NPK (80:60:40 kg ha$^{-1}$) followed by incorporation of *Erythrina* (4.48 Mg ha$^{-1}$) and *Parkia* leaves (4.13 Mg ha$^{-1}$) in the first year. However, in the subsequent years, the gap between yield obtained with NPK and tree leaves incorporation reduces and from the third year onward, all the tree leaves tested except alder surpassed the grain yield level that obtained with recommended NPK (5.13 Mg ha$^{-1}$). Significantly high grain yield of rice in the third year was recorded with incorporation of *Erythrina* leaves (5.67 Mg ha$^{-1}$) that remained on par with *Acacia, Parkia,* and *Casia* leaves. The GLM with N-fixing tree leaves has substantial residual effect in soil and, therefore, improved soil fertility and productivity level (Das et al. 2010).

**TABLE 10.12**

**Effect of Different N-Fixing Tree Leaves on Productivity of Lowland Rice**

| Treatments | Nutrient Composition (%) | | | | Grain Yield (Mg ha$^{-1}$) | | |
|---|---|---|---|---|---|---|---|
| | N | P | K | Moisture | 2003 | 2004 | 2005 |
| *Erythrina* | 3.24 | 0.47 | 1.54 | 73.6 | 4.48 | 4.83 | 5.67 |
| *Alder* | 2.24 | 0.41 | 1.37 | 66.2 | 3.50 | 4.10 | 4.67 |
| *Parkia* | 2.54 | 0.40 | 1.52 | 69.3 | 4.13 | 4.40 | 5.23 |
| *Acacia* | 3.19 | 0.43 | 1.36 | 65.4 | 3.92 | 4.66 | 5.30 |
| *Cassia* | 2.50 | 0.39 | 1.17 | 65.8 | 3.99 | 4.55 | 5.58 |
| Recommended NPK | – | – | – | – | 4.82 | 5.08 | 5.13 |
| Control | – | – | – | – | 2.80 | 3.13 | 3.35 |
| CD (P = 0.05) | – | – | – | – | 0.60 | 0.46 | 0.53 |

*Source:*   Das, A., J.M.S. Tomar, T. Ramesh, G.C. Munda, P.K. Ghosh, and D.P. Patel. 2010. Productivity and economics of lowland rice as influenced by N-fixing tree leaves under mid-altitude subtropical Meghalaya. *Nutr. Cycl. Agroecosys.* 87: 9–19.

## 10.9   NITROGEN DYNAMICS UNDER SLASH AND BURN SYSTEMS OF NORTH EAST INDIA

Slash and burn agriculture or shifting cultivation (locally known as *jhum*) is the major traditional land use practice (about 0.88 million ha area) in the NEH region of India. In jhum fields, as many as 36 crops are cultivated at a time depending on the requirement and size of the farmer's family, and so expected to exert varying pressure on the soil fertility as compared to improved practices of crop production. Therefore, there is a need to understand the nutrients dynamics in *jhum* fields in general and nitrogen in particular for sustainable productivity. Mineralization of organic soil N is influenced by various environmental factors. Hence, proper understanding of soil N dynamics is critical to realize how soil management practice may affect long-term soil and crop productivity, environmental conditions, and soil microorganisms. Crop management practices like tillage, manuring, water, and weed management and so on may change soil health, environmental conditions, and ultimately nitrogen dynamics in various ecosystems. Burning of biomass is a very important ecological tool in slash and burn systems. Hence, there is a need for better understanding of the effect of fire on nutrient availability particularly N-biochemical processes in soils. In general, soils of the *jhum* site are reported to be less acidic than the soils under forest (Arunachalam 2002). Singh and Singh (2014) reported that the initial concentration of $NH_3$-N, $NO_3$-N, and inorganic-N in the un-burned, burned, and control site were the highest in winter months (January and February) and lowest in summer (July). The wider variations in soil inorganic-N concentrations could be due to variation in mineralization rates, uptake by plants and microbes, and N losses. The lower $NH_3$-N, $NO_3$-N, and inorganic-N during the rainy season may be due to greater N-demand by plants and losses of inorganic-N through leaching and denitrification. The higher inorganic N concentrations in the winter season may also be associated with the low demand by plants during the dormant period. Irrespective of the treatments the value of $NH_3$-N was found to be higher than $NO_3$-N might be due to the acidic soil reaction. Similarly, total Kjeldahl N, ammonium-N, and nitrate-N decreased as time elapsed during cultivation in jhum fields. N-mineralization showed significant variations between seasons. Greater amount of N-mineralization was recorded during the rainy season for agroforestry, and winter season for *Jhum*, vegetable, and millet systems. Ammonification rate was positively correlated with soil pH, SOC, $NH_4^+$-N, and $NO_3^-$-N whereas nitrification rate was negatively correlated with $NH_4^+$-N and $NO_3^-$-N concentrations (Bhuyan et al. 2014). Mishra and Ramakrishnan (1983) investigated the effect of slash and burn production systems on soil fertility at Meghalaya province of the NEH region, comparing 15-, 10-, and 5-year jhum cycles and reported that the N level under a 5-year jhum cycle was significantly lower than 10- and 15-year cycles. In general, lower nutrient status under a terrace system after cropping, even when compared to a 5-year jhum cycle, suggested that a terrace system could not be sustained without external nutrients inputs. Mishra and Ramakrishnan (1984) also did the nitrogen budgeting of the jhum system of the region under 15-, 10-, and 5-year fallow and observed that the nitrogen depletion is affected by initial stocks in the soil and vegetation compartment at the time of the slash and burn as well as the rate at which this was lost during the subsequent land use. While nitrogen losses due to the burning was more severe under longer cycles compared to the 5-year cycle, the losses through sediment and water were more under a 15-year cycle compared to 10- and 5-year cycles. Soil N was initially depleted by volatilization losses during slash and burn operations, and subsequently during cultivation processes (Bhadauria and Ramakrishnan 1996). Transfer of nitrogen from soil to the weed biomass increased with shortening of the fallow cycle. The positive role of weeds in conservation of nitrogen in their biomass and subsequent release through organic manure into the agriculture system has been observed. Diversity of soil flora and fauna also greatly influenced the soil nitrogen economy of slash and burn systems. In this regard, Bhadauria and Ramakrishnan (1996) investigated the role of earthworms in the N cycle in a shifting agriculture system under a 5- and a 15-year jhum system fallow and reported that the earthworms participated in the N cycle through diverse processes like mucus production, dead tissue decomposition, and so on. The total soil N made available for uptake by the plant through the activity of earthworms in this agro-ecosystem was higher than the total input of N to the soil through the addition of slashed vegetation, inorganic and organic manure, and recycled crop residue and weeds. Therefore, in highly leached soils of the humid tropics, worm activity is important because of rapid incorporation of litter into the mineral soils and because of local concentrations of nutrients in the surface soil layers. Majumder et al. (2010) studied the litter decomposition and nutrient-release patterns of dominant crop species in a jhum cultivation system in the humid tropics of northeast India and reported that the crop residues had an exponential weight-loss pattern with time and the release of nitrogen followed a similar pattern of weight loss with mineralization constants of 3.46–20.47 (Tawnenga et al. 1996).

### 10.9.1 SOIL NUTRIENTS AND FERTILITY IN *JHUM* FIELDS OF KHONOMA, NAGALAND—A CASE STUDY

Continuous agricultural activities in *jhum* (slash and burn agriculture) fields depletes soil fertility, but soil fertility can be restored by following good fallow management practices. A good fallow management practice involves leaving land fallow for 10–15 years and allowing good biomass growth for fertility buildup. Seeding of leguminous crops such as rice bean (*Vigna umbellata*), wild sunflower, *Mucuna* sp., and so on can build up soil fertility faster than other species. The rate of reclamation from a nutrient deficient condition to nutrient optimum condition is enhanced by growing and maintaining alder (*Alnus nepalensis*) tree. Soil nutrient concentration under 2–5 years old *jhum* fallows have properties ranging between soils of paddy fields and natural forest (Chase and Singh 2014). The soil fertility in *jhum* fallows can be higher than the paddy fields because of N-fixation capacity of alder in all the *jhum* lands. Without the introduction of alder tree, a minimum of 10 years of fallow period is necessary for restoring soil organic C as well as fertility in *jhum* land (Ramakrishnan 1994; Sarkar et al. 2015). The N fixing alder trees improve soil fertility substantially in farmlands because they are fast growing and produce nutrient rich GLM. Besides fixing atmospheric N, the litters of the alder trees supply P, K, Ca, and other nutrients to the soil (Sharma and Singh 1994).

## 10.10  NITROGEN AVAILABILITY AND MANAGEMENT UNDER VARIOUS LAND USE SYSTEMS

Soils under different land use systems have different available N based on the various management practices, residue turnover, fertilizer and manure application, N fixation, and other edapho-climatic factors. Soil available N content of different land use systems including long-term experimental fields of Lembucherra, Tripura (<100 m ASL), Umiam, Meghalaya (950–1000 ASL) and Tadong, Sikkim (>1300 m ASL) was estimated at two depths (0–15 cm) and (15–30 cm). The objective was to recommend best soil management practices for different altitudes for N fertility and soil health.

### 10.10.1  LEMBUCHERRA, TRIPURA (LOW ALTITUDE)

The soil sampling was performed from existing long-term land uses (>10 years) except for jatropha (*Jatropha curcas*) block that was 5 years old for analyzing available N content, bulk density, and SOC concentrations to estimate SOC stock. In the studied soils of different land use systems of Lembucherra, Tripura, the SOC concentrations ranged from 8.6–16.4 g kg$^{-1}$ at 0–15 cm and 7.4–15.2 g kg$^{-1}$ at 15–30 cm layer. The available N was higher in soils under farming system 2-pig based (FS2-PS), alley cropping, and litchi than other land uses (Figure 10.2).

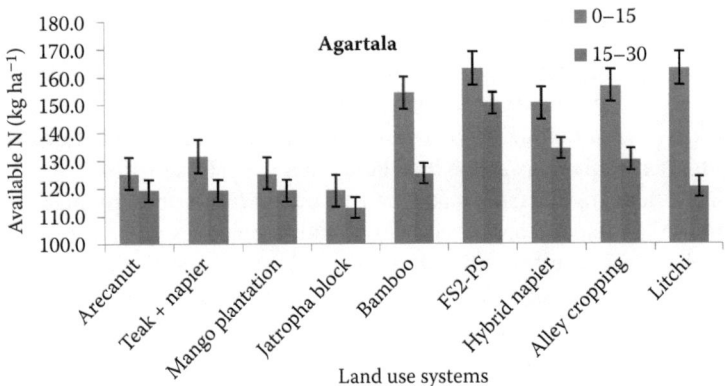

**FIGURE 10.2**   Available N under different land uses in Agartala, Tripura.

### 10.10.2   Umiam, Meghalaya (Mid-Altitude)

At mid-altitude of Meghalaya, land uses comprising conventional tillage (CT), no-till (NT), organic farming (OF), and in-situ residue retention (IRR) in rice, maize, and pine (*Pinus kesiya* Royle ex Gordon) forest (dominant forest), *Jhum* and integrated farming systems (IFS) involving crop-livestock-fodder on terrace risers were evaluated for SOC concentrations and available N status in soils. The SOC concentrations ranged from 16.4–34.7 g/kg at the surface layer (0–15 cm) and 15.5–33.5 g kg$^{-1}$ at the sub-surface layer (15–30 cm) in various land uses of Tripura. The soil available N was higher in rice field under organic sources of nutrient supply (Rice-ORG), rice field under in-situ residue recycling (ISRR), farming system-mid slope (FS-MS), and abandoned jhum fields than other land uses (Figure 10.3). Thus, organic farming, adaptation of the FS approach, fallowing, and ISRR in lowland rice are favorable for building soil fertility, especially carbon and N (Das et al. 2016a).

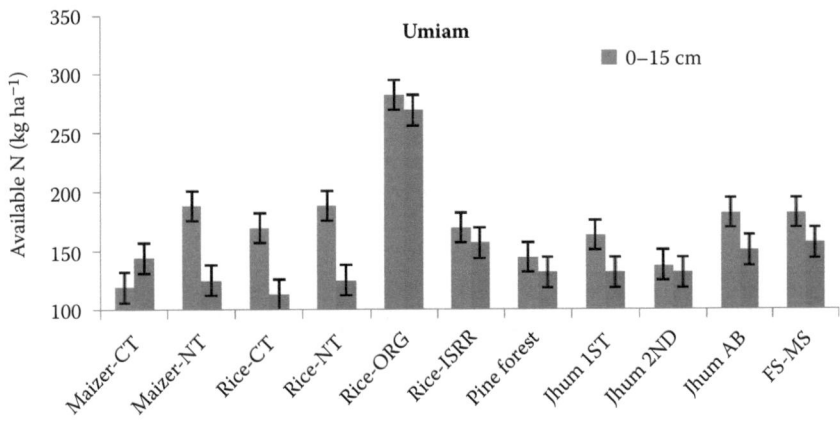

**FIGURE 10.3**   Available N under different land uses in Umiam, Meghalaya.

### 10.10.3 Tadong, Sikkim (High Altitude)

In the studied soils of different land use systems of Tadong, Sikkim, the SOC concentrations varied from 17.6–29.2 g kg⁻¹ at the surface layer (0–15 cm) and 13.5–24.5 g kg⁻¹ at the sub-surface layer (15–30 cm). Alnus + cardamom [*Elettaria cardamomum* (L.) Maton], sala tree (*Shorea robusta*) + cardamom had higher available N status than other land uses (Figure 10.4).

Therefore, it is evident that residue management, IFS, agroforestry systems (AFS), forage crops, multipurpose trees (MPTs), and conservation agriculture practices are some of the potential management approaches for conserving and promoting N buildup and C-sequestration in hill ecosystems of NER of India. Impact of agroforestry (Ramesh et al. 2013), perennial forage crops (Ghosh et al. 2009), crop residue recycling (Das et al. 2008), and conservation agriculture (Ghosh et al. 2010) on improving soil health have been reported by various researchers under eastern Himalayan ecosystems.

Field experiments were conducted in seven consecutive years from 2005 to 2012 under a raised and sunken bed (RSB) land configuration (0.3–0.4 m height, 2 m width, 7 m length) in lowland at Meghalaya (950 m ASL), India. Rice (*Oryza sativa* L.)—vegetable sequences on raised beds and rice (varieties)—fallow (no crop) sequences on sunken beds were assessed under four farming practices in fixed plots. The four farming practices were control (only in-situ recycling of two-thirds crop residues), organic (FYM and rock phosphates on N and P equivalent basis), inorganic (mineral fertilizer), and integrated farming [50/50 organic and inorganic sources (INF)]. After seven cropping cycles, the highest soil available N was observed under rice-French bean (*Phaseolus vulgris* L.) sequence on raised bed. However, soil available N under rice-French bean sequence was significantly higher by 3.7% than that under rice-tomato (*Lycopersicon esculentum* L.). Among farming practices, soil available N was the highest under INF followed by organic, both of which were significantly higher compared to inorganic and control. Soil available N content under INF and organic were 4.77% and 4.36% higher than those under inorganic and 18.2% and 17.8% than control, respectively (Das et al 2016b). In another study on organic production of maize-vegetables cropping systems, organic manures were supplied to crops on N equivalent basis through various amendments such as FYM, VC, and FYM +VC. After five cropping cycles, higher SOC and available N were recorded under maize + soybean (*Glycine max*) – French bean cropping system as compared to the rest of the systems. Among the organic amendments, combined use of FYM and VC as nutrient source registered maximum SOC (23.6 g kg⁻¹) and available P (34.29 kg ha⁻¹) compared to other amendments. However, available N (251.47 kg ha⁻¹) was maximum under FYM as an organic amendment (Patel et al. 2015).

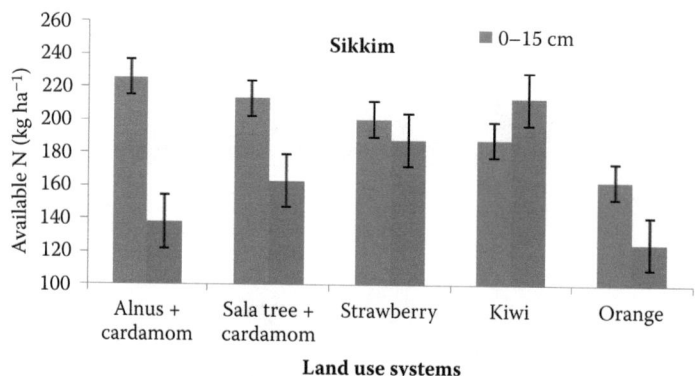

**FIGURE 10.4**   Available N under different land uses in Tadong, Sikkim.

## 10.11 INCREASING NITROGEN USE EFFICIENCY (NUE)

Under field conditions, NUE varies from 25–34% in rice and 40–60% in other crops, with a global average of about 50% (Mosier 2002). However, the efficiency of applied fertilizer nitrogen is hardly 4–14% in NER, hence, the application of coated urea fertilizer such as RPU at 60 kg ha$^{-1}$ is recommended for higher rice yield (Patel 1985). Bharali et al. (2012) screened the nitrogen efficient genotypes of rice and wheat in Assam condition of NER and reported that genotype Kanaklata was found most N efficient followed by Bishnuprasad, Jyotiprasad, Jaymati, and Mulagavaru, respectively. Similarly for wheat, Ankur Omkar was the most nitrogen efficient genotype followed by Sonalika and K-306.

Increasing NUE requires a diverse set of agronomic approaches in addition to molecular and plant breeding approaches of developing an efficient variety. Improvement in NUE thus requires the simultaneous consideration of multiple aspects of crop performance, involving a multidisciplinary approach.

### 10.11.1 Soil Amelioration

Addition of OM to soil maintains good soil physical, chemical, and biological conditions which largely governs NUE. Best management practices such as incorporation of crop residues, GLM, compost, animal excreta based manure and urine, use of cover crops, reduced tillage, and retention of crop residues can significantly improve the level of soil OM and contribute to the sustainability of the cropping systems and higher NUE. Since almost 16.6 million hectare (M ha) of soils having pH <5.5 in NER of India are extremely acidic (Sharma et al. 2006a), liming is an effective way to correct soil chemical constraints. The fixation of atmospheric $N_2$ by free living and symbiotic organisms like rhizobium increases with pH. Due to reduction of subsoil acidity, the roots can penetrate in deeper layers and can extract higher amounts of water and nutrients (Graham and Vance 2000).

### 10.11.2 Improving Physical Conditions of Soil

Tillage, mulching, irrigation, and addition of OM are major management techniques that create suitable physical conditions for crop growth. Tillage systems are reported to change the soil nutrient reserve and root growth which might have direct bearing on the N availability and its use efficiency for crops. Tillage practices such as conventional, conservation, and no-till (NT) can make changes in soil OM, N content, bulk density, water holding capacity, soil temperature, and soil aeration. Greater availability of N has been reported under NT than those under conventional tillage from the eastern Himalayan region of India (Das et al. 2014). Minimum tillage can also increase root weight, length, and density, increasing the nutrient and water use efficiencies (Adkinson 1990). Such improved root parameters contribute to higher yields and uptake efficiencies of N.

### 10.11.3 Fertilizer Management

While applying N fertilizers, selection of the right source of N for application at the right time and in the right place are important for achieving adequate NUE. Fertilizer source, rate, timing, and placement are interdependent and are interlinked with other agronomic management practices involved in crop production.

#### 10.11.3.1 Right Source of Fertilizers

N applied through fertilizers must be in plant available forms or in a form that changes into a plant available form in the soil. The source of N fertilizer should be selected based on soil physical and chemical properties, for example, application of $NO_3$-N to flooded rice soils or broad casting urea on soil with high pH should be avoided. Since nitrates are easily leached and lost by denitrification, retardation of nitrification of ammonium containing or ammonium producing fertilizers by using nitrification inhibitors in submerged rice field helps in increasing NUE. Chemical nitrification inhibitors such as N-serve [2-chloro-6-(trichloromethyl) pyridine], AM [2-amino-4-chloro-6-methylpyrimidine], DCD (Dicyandiamide), 4-amino-1,2,4-6-triazole-HCl (ATC), 3,4-dimethylpyrazole phosphate (DMPP), carbon disulfide ($CS_2$), and ammonium thiosulfate [$(NH_4)_2S_2O_3$] are widely used for improving NUE. Neem (*Azadirachta indica*) cake coated ureas are shown to have nitrification inhibiting properties. Application of PPD (phenyl phosphoro diamidate) and NBPT [N-(nbutyl) thio phosphoric triamide] with urea can reduce the rate of urea hydrolysis and improve its efficiency (Prasad and Power 1995). Nitrification inhibitor DCD has the potential to reduce $N_2O$ emissions and increase NUE of irrigated systems of barley (Delgado and Mosier 1996). Slow and controlled release fertilizers have added advantages in increasing nutrient recovery by plants, lowering $N_2O$ and $NH_3$ emissions, and $NO_3$-N leaching from cropping systems. Further research and development is needed to continue developing new products that can increase the recovery of fertilizers while maintaining and/or increasing yields and preserving the environment. Results of an experiment on direct sown rice in Meghalaya indicated that urea super granule (USG) gave higher yield than prilled urea. In another study, RPU produced the highest yield of rice followed by GCU at Umiam, Meghalaya. RPU and GCU is a superior source of N in clayey soils of Manipur (Patel and Singh 1990). Under lowland conditions of Tripura, prilled urea and LCU gave similar results. DAP resulted in the highest grain yield of rainfed lowland rice (cv. Pankaj) when compared with urea and ammonium sulfate in the plains of Assam (Deka and Borah 1993).

### 10.11.3.2 Right Rate of Fertilizers

Scientific approaches are involved to evaluate soil nutrient supply to crops to work out fertilizer application rates. Soil and plant analysis and crop response studies are useful tools in this regard. Quantity and plant availability of N in all available organic nutrient sources such as manure, composts, bio-solids, crop residues, and irrigation water need to be determined. In estimating N fertilizer application rates, consideration of NUE is important because some loss of nutrients from a soil-plant system is unavoidable. If the removal of N from a cropping system exceeds applied amount for years together, soil fertility depletion may occur (Haileslassie et al. 2005). This aspect also needs to be considered while working out fertilizer application rates. For N unlikely to be retained in the soil, the most economic rate of application is where the last unit of applied N is equal in value to the increase in crop yield, it generates (law of diminishing returns). For N retained in the soil, its residual effects to future crops should be considered. Hence, assessment of probabilities of predicting economically optimum rates of N is crucial (Hong et al. 2007).

### 10.11.3.3 Right Timing of Fertilizers

Fertilizers should be applied to match the crop nutrient demand, which depends on date of planting, crop growth characteristics, and sensitivity to N deficiencies at particular growth stages. For example, top-dressing of N at the tillering stage of wheat has been found to increase both yield and protein content of wheat, especially at low levels of soil N (Singh 2014). Fertilizer application timing is also governed by dynamics of N supply in soil. Mineralization of soil OM makes N available to crops, but if mineralization does not occur during the critical crop growth stages requiring N, deficiencies may limit productivity (Kumar and Goh 2002). Similarly timing of weather factors influencing N loss will govern fertilizer timing (e.g., leaching losses of nitrate tend to be more frequent during monsoon season). Logistics of field operations will also influence the timing of fertilizer application (e.g., multiple applications of N may or may not combine with those of crop protection products and N applications should not delay time-sensitive operations such as planting).

### 10.11.3.4  Placement of Fertilizers

Placement of fertilizer needs to ensure that N is adequately intercepted by plant roots. In conservation tillage systems, subsurface fertilizer application needs to ensure that soil coverage by crop residues is maintained (Morrison et al. 1990). The efficiency of the applied fertilizers can be improved considerably if the rooting habits of various plants during early growth stages are known. If a plant produces a tap root system early, fertilizer can best be placed directly below the seed. On the other hand, if lateral roots are formed early, side dressing of fertilizer would be efficient. In tropical regions, the major soil problems in rainfed systems that affect crop production are low soil fertility, salinity, alkalinity, acidity, Fe toxicity, and P and Zn deficiencies. NUE in these soils can be improved by proper fertilizer placement and adjusting timing to critical crop growth stages (Baligar et al. 2001).

## 10.12  SITE SPECIFIC NUTRIENT MANAGEMENT

Site specific nutrient management is a dynamic field specific nutrient management in particular cropping season to optimize the supply and demand of nutrient based on its cycling through a soil-plant system. Dobermann et al. (2002) reported that field-specific management of N increases yield and NUE in irrigated rice farms in Asia. Site specific technologies in the future might help to develop sound management systems that lead to reduced fertilizer inputs, thereby improving efficiency of costly fertilizer input and minimizing the environmental degradation.

## 10.13  CONCLUSIONS

From the above review of N management, the following conclusions can be drawn:

1. N management is important for sustaining agriculture in eastern Himalayan region and similar other agro-ecosystems.
2. Integrated N management involving judicious use of fertilizer and organic manure and biofertilizer holds promise for enhancing productivity of crops in fragile ecosystems of the eastern Himalayas.
3. Since the farmers are mostly small and marginal, low cost interventions such as the use of locally available resources like recycling of tree leaves, crop and weed biomass, green manuring, use of animal excreta based manure, and biofertilizers are the most viable and suitable options for N management in the eastern Himalayas to sustain soil and crop productivity.
4. Inclusion of legumes in cropping systems and alley cropping is also a low cost and feasible intervention for maintaining N fertility status of hilly soils.
5. Identification of location specific crop genotypes, N management practices along with improved crop production practices are recommended for enhancing NUE in Himalayan ecosystems.

Future studies may focus on the impact of various N management practices on greenhouse gas (GHG) emission especially nitrous oxide and methane and suitable strategies for managing GHG emissions may be developed with special reference to Himalayan ecosystems.

## REFERENCES

Acharya, C.L., and O.C. Kapur. 2001. Using organic wastes as compost and mulch for potato (*Solanum tuberosum*) in low water-retaining hill soils of north-west India. *Indian J. Agric. Sci.* 71 (5): 306–309.

Agbenin, J.O., E.B. Agbaji, I. Suleiman, and A.S. Agbaji. 1999. Assessment of nitrogen mineralization potential and availability from neem seed residue in a savanna soil. *Biol. Fertil. Soils* 29(4): 408–412.

Ahemad, M., and M. S. Khan. 2012. Evaluation of plant growth promoting activities of rhizobacterium Pseudomonas putida under herbicide stress. *Ann. Microbiol.* 62(4):1531–1540.

Arunachalam, A. 2002. Dynamics of soil nutrients and microbial biomass during first year cropping in an 8-year jhum cycle. *Nutr. Cycl. Agroecosyst.* 64: 283–291.

Arunachalam, A., and H.N. Pandey. 2003. Ecosystem restoration of Jhum fallows in Northeast India: Microbial C and N along altitudinal and successional gradients. *Restor. Ecol.* 11(2): 168–173.

Ashraf, M., T. Mahmood, F. Azam, and R.M. Qureshi. 2004. Comparative effects of applying leguminous and non-leguminous green manures and inorganic N on biomass yield and nitrogen uptake in flooded rice (*Oryza sativa* L.). *Biol. Fertil. Soils* 40(2): 147–152.

Atkinson, D. 1990. Influence of root system morphology and development on the need for fertilizer and efficiency of use. In *Crops as Enhancers of Nutrient Use*, Eds. V.C Baligar, and R.R. Duncan, pp. 411–451. Academic Press, San Diego, CA.

Avasthe, R.K. 2002. Annual Report. ICAR Research Complex for NEH Region, Umiam, Meghalaya.

Azeez, J.O., and W. Van Averbeke. 2010. Nitrogen mineralization potential of three animal manures applied on a sandy clay loam soil. *Bioresour. Technol.* 101(14): 5645–5651.

Babu, S., R. Singh, R.K. Avasthe, G.S. Yadav, T.K. Chettri, and C.D. Phempunadi. 2015. Effect of organic nitrogen sources on growth, yield, quality, water productivity and economics of rice (*Oryza sativa* L.) under different planting methods in mid hills of Sikkim Himalayas. *Res. on Crops* 16(3): 389–400.

Babu, S., R. Singh, R.K. Avasthe, G.S. Yadav, T.K. Chettri, and D.J., Rajkhowa. 2016. Productivity, profitability and energetics of buckwheat (*Fagopyrum* sp.) cultivars as influenced by varying levels of vermicompost in acidic soils of Sikkim Himalayas, India. *Indian J. Agric. Sci.* 86(7): 844–848.

Baishya, L.K., M.A. Ansari, D. Sarkar, B.C. Deka, and N. Prakash. 2014. Differential responses in production efficiency of rice (*Oryza sativa* L.), energy budgeting and soil improvement to organic fertilization. *Eco. Env. Cons.* 20(4): 365–372.

Baishya, L.K., and D. Sarkar. 2015. Land–soil resources of north eastern region of India: Constraints and management options. In *Integrated Soil and Water Resource Management for Livelihood and Environmental Security,* Eds. D.J. Rajkhowa, A. Das, S.V. Ngachan, A.K. Sikka and M. Lyngdoh, pp. 20–33. ICAR Research Complex for NEH Region, Umiam, Meghalaya, India.

Baligar, V.C., N.K. Fageria, and Z.L. He. 2001. Nutrient use efficiency in plants. *Commun. Soil Sci. Plant. Anal.* 32:921–950.

Bargali, K. and S.S. Bargali, 2009. *Acacia nilotica*: A multipurpose leguminous plant. *Nat. Sci.* 7(4): 11–19.

Bargali, K., N.R. Maurya, and S.S. Bargali. 2015. Effect of a nitrogen-fixing Actinorhizal shrub on herbaceous vegetation in a mixed conifer forest of Central Himalaya. *Curr. World Environ.*10(3): 957–966.

Batabyal, K., B. Mandal, D. Sarkar, S. Murmu, A. Tamang, I. Das, G.C. Hazra, and P.S. Chattopadhyay. 2016. Comprehensive assessment of nutrient management technologies for cauliflower production under subtropical conditions. *Europ. J. Agron.* 79: 1–13.

Behera, S.N., and M. Sharma. 2010. Investigating the potential role of ammonia in ion chemistry of fine particulate matter formation for an urban environment. *Sci.Total Environ.* 408(17): 3569–3575.

Behera, U.K., A.R. Sharma, and H.N. Pandey. 2007. Sustaining productivity of wheat–soybean cropping system through integrated nutrient management practices on the Vertisols of central India. *Plant Soil* 297(1–2): 185–199.

Bhadauria, T., and P.S. Ramakrishnan. 1996. Role of earthworms in nitrogen cycling during the cropping phase of shifting agriculture (Jhum) in north-east India. *Biol. Fertil. Soils* 22 (4): 350–354.

Bharali, B., B. Haloi, J. Chutia, and P.C. Dey. 2012. Atmospheric deposition of Nitrogen in North-East India and productivity of winter cereals. 5th International Conference on Water, Climate and Environment, Web BALWOIS, Ohrid, the Republic of Macedonia.

Bhatt, D.K., V.K. Joshi, and R.K. Arora. 1999. Conodont biostratigraphy of the Lower Triassic in Spiti Himalaya, India. *J. Geol. Soc. India* 54(2): 153–167.

Bhatta, K.P., and O.R. Vetaas. 2016. Does tree canopy closure moderate the effect of climate warming on plant species composition of temperate Himalayan oak forest? *J. Veg. Sci.* 27(5): 948–957.

Bhattacharyya, P.N., and D.K. Jha. 2012. Plant growth-promoting rhizobacteria (PGPR): Emergence in agriculture. *World J. Microbiol. Biotechnol.* 28(4):1327–1350.

Bhuyan, S.I., O.P. Tripathi, and M.L. Khan. 2014. Soil N-mineralization dynamics under four major land use patterns in Eastern Himalaya. *J. Trop. Agric.* 52(2): 162–168.

Bujarbaruah, K.M. 2004. February. Organic farming: Opportunities and challenges in North Eastern region of India. In *Souvenir, International Conference on Organic Food*, February 14–17, 2004, ICAR Research Complex for NEH Region, Umiam, Meghalaya, India, pp 7–13.

Chase P., and O.P. Singh. 2014. Soil nutrients and fertility in three traditional land use systems of Khonoma, Nagaland, India. *Resour. Environ.* 4(4): 181–189.

Chen, J.H. 2006. The combined use of chemical and organic fertilizers and/or biofertilizer for crop growth and soil fertility. In *International Workshop on Sustained Management of the Soil-Rhizosphere System for Efficient Crop Production and Fertilizer Use*. October 16–20 (Vol. 16, p. 20). Land Development Department, Bangkok, Thailand.

Choudhary, V.K., and B. Choudhury. 2016. A staggered maize–legume intercrop arrangement influences yield, weed smothering and nutrient balance in the eastern Himalayan region of India. *Exp. Agric.* pp. 1–20. DOI:10.1017/S0014479716000144.

Choudhury, B.U., K.P. Mohapatra, A. Das, P.T. Das, L. Nongkhlaw, R.A. Fiyaz, S.V. Ngachan et al. 2013. Spatial variability in distribution of organic carbon stocks in the soils of North East India. *Curr. Sci.* 104: 604–614.

Collins, R., and A. Jenkins. 1996. The impact of agricultural land use on stream chemistry in the middle hills of the Himalayas, Nepal. *J. Hydrol.* 185(1): 71–86.

Cui, Z., X. Chen, and F. Zhang. 2010. Current nitrogen management status and measures to improve the intensive wheat–maize system in China. *Ambio* 39(5–6): 376–384.

Das, A., P.K. Ghosh, B.U. Choudhury, D.P. Patel, G.C. Munda, S.V. Ngachan, and P. Chowdhury. 2009a. Climate change in North East India: Recent facts and events—Worry for agricultural management. In: *ISPRS Archives XXXVIII-8/W3 Workshop Proceedings: Impact of Climate Change on Agriculture*, Eds., S. Panigrahy, S.S. Ray, J. J. Parihar, pp. 32–37.

Das, A., U.K. Hazarika, G.C. Munda, D.P. Patel, R. Kumar, J.M.S. Tomar, A. Chandra et al. 2006. Non-conventional source of organic nutrient supply in crop production. National seminar on standards and technologies of non-conventional source of organic inputs. April 8–9, 2006, PDCSR, Meerut, Modipuram, UP, India.

Das, A., Munda, G.C., Patel, D.P., Ghosh, P.K., Ngachan S.V. and Baiswar, Pankaj. 2009b. Productivity, nutrient uptake and post-harvest soil fertility in lowland rice as influenced by compost made from locally available plant biomass. *Archives of Agronomy and Soil Science* doi: 10.1080/03650340903207907.

Das, A., D.P. Patel, M. Kumar, G.I. Ramkrushna, A. Mukherjee, J. Layek, S.V. Ngachan, and J. Buragohain. 2016b. Impact of seven years of organic farming on soil and produce quality and crop yields in eastern Himalayas, India. *Agric. Ecosyst. Environ.* 236: 142–153.

Das, A., D.P. Patel, G.C. Munda, U.K. Hazarika, and J. Bordoloi. 2008. Nutrient recycling potential in rice—Vegetable cropping sequences under in-*situ* residue management at mid-altitude subtropical Meghalaya. *Nutr. Cycl. Agroecosys.* 82 (3): 251–258.

Das, A., D.P. Patel, G.C. Munda, G.I. Ramkrushna, M. Kumar, and S.V. Ngachan. 2014. Improving productivity, water and energy use efficiency in lowland rice (*Oryza sativa*) through appropriate establishment methods and nutrient management practices in the mid-altitude of north-east India. *Exp. Agric.* 50 (3): 353–375.

Das, A., G.I. Ramkrushna, B.U. Choudhury, G.C. Munda, D.P. Patel, S.V. Ngachan, P.K. Ghosh et al. 2012. Natural resource conservation through indigenous farming systems: Wisdom alive in North East India. *Indian J. Tradit. Know.* 11(3): 505–13.

Das, A., G.I. Ramkrushna, B. Makdoh, D. Sarkar, J. Layek, S. Mandal, and L. Rattan. 2016a. Managing soils of the lower himalayas. *Encyclopedia of Soil Science*, 3rd ed. DOI: 10.1081/E-ESS3-120053284.

Das, A., G.I. Ramkrushna, G.S. Yadav, J. Layek, C. Debnath, B.U. Choudhury, K.P. Mohaptara et al. 2015. Capturing traditional practices of rice based farming systems and identifying interventions for resource conservation and food security in Tripura, India. *Appl. Ecol. Environ. Sci.* 3(4): 100–107.

Das, A., J.M.S. Tomar, T. Ramesh, G.C. Munda, P.K. Ghosh, and D.P. Patel. 2010. Productivity and economics of lowland rice as influenced by N-fixing tree leaves under mid-altitude subtropical Meghalaya. *Nutr. Cycl. Agroecosys.* 87: 9–19.

Datt, C., Datta, M., and Singh, N.P. 2008. Assessment of fodder quality of leaves of multipurpose trees in subtropical humid climate of India. *J. For. Res.* 19(3): 209–214.

Deka, S.C., and R.C. Borah. 1993. *ORYZA-An International Journal on Rice.* 30: 219–225.

Delgado, J.A., and A.R. Mosier. 1996. Mitigation alternatives to decrease nitrous oxides emissions and urea-nitrogen loss and their effect on methane flux. *J. Environ. Qual.* 25: 1105–1111.

Dobermann, A., C. Witt, D. Dawe, S. Abdulrachman, H.C. Gines, R. Nagarajan, S. Satawathananont et al. 2002. Site-specific nutrient management for intensive rice cropping systems in Asia. *Field Crops Res.* 74(1): 37–66.

Erisman, J.W., M.A.J. Sutton, K.Z. Galloway, and W. Winiwarter. 2008. How a century of ammonia synthesis changed the world. *Nat. Geosci.* 1(10): 636–639.

Finzi, A.C., D.J. Moore, E.H. DeLucia, J. Lichter, K.S. Hofmockel, R.B. Jackson, H.S. Kim et al. 2006. Progressive nitrogen limitation of ecosystem processes under elevated CO2 in a warm-temperate forest. *Ecology* 87(1): 15–25.

Gangopadhayay, S.K., P.K. Das, S.S., Mukhopdhayay, and S.K. Bansal. 1990. Altitudinal pattern of soil characteristics under forest vegetation in Eastern Himalayan Region, *J. Indian Soc. Soil Sci.* 38: 93–99.

Ghosh, P.K., A. Das, E. Kharkarang, A.K. Tripathi, G.C. Munda and S.V. Ngachan. 2010. Conservation agriculture towards achieving food security in North East India. *Curr. Sci.* 99(7): 915–921.

Ghosh, P.K., R. Saha, J.J. Gupta, T. Ramesh, A. Das, T.D. Lama, G.C. Munda et al. 2009. Long-term effect of pastures on soil quality in acid soil of North-East India. *Aust. J. Soil. Res.* 47: 1–8.

Graham, P.H., and C.P. Vance. 2000. Nitrogen fixation in perspective: An overview of research and extension needs. *Field Crops Res.* 65: 93–106.

Grogan, P., F. Lalnunmawia, and S.K. Tripathi. 2012. Shifting cultivation in steeply sloped regions: A review of management options and research priorities for Mizoram state, Northeast India. *Agrofor. Syst.* 84(2): 163–177.

Gruhn, P., F. Goletti, and M. Yudelman. 2000. *Integrated Nutrient Management, Soil Fertility, and Sustainable Agriculture: Current Issues and Future Challenges.* Intl Food Policy Res Inst.

Gupta, R.D., K.R. Bherdwaj, B.C. Morwan, and B.R. Tripathi. 1986. Occurrence of phosphate dissolving bacteria in soils of North-West Himalayas under varying biosequence and climosequence. *J. Indian Soc. Soil Sci.* 34: 498–504.

Haileslassie, A., J. Priess, E. Veldkamp, D. Teketay, and J.P. Lesschen. 2005. Assessment of soil nutrient depletion and its spatial variability on smallholders' mixed farming systems in Ethiopia using partial versus full nutrient balances. *Agric. Ecosyst. Environ.* 108(1): 1–16.

Hazarika, U.K., G.C. Munda, K.M. Bujarbaruah, A. Das, D.P. Patel, K. Prasad, R. Kumar et al. 2006. Nutrient Management in Organic Farming. Technical Bulletin No. 30, ICAR RC for NEH Region, Umiam, Meghalaya.

Hazarika, U.K., N.P. Singh, D.C. Saxena, A.S. Panwar and G.C. Munda. 2003. Prospect of using Azolla as green manure in mid hills of Meghalaya. In *Biotechnology in Sustainable Organic Farming*, Eds. A.K. Yadav, S.R. Chaudhuri, and N.C. Talukdar, pp. 290–294. RBDC, Imphal, Manipur, India.

Hong, N., P.C. Scharf, J.G. Davis, N.R. Kitchen, and K.A. Sudduth. 2007. Economically optimal nitrogen rate reduces soil residual nitrate. *J. Environmental Qual.* 36(2): 354–362.

Javorekova, S., J. Makova, J. Medo, S. Kovacsova, I. Charousova, and J. Horak. 2015. Effect of bio-fertilizers application on microbial diversity and physiological profiling of microorganisms in arable soil. *Eurasian J. Soil Sci.* 4(1): 54.

Jodha, N.S., and S. Shrestha. 1994. Towards sustainable and more productive mountain farming. In *Proceedings of the International Symposium on Mountain Environment and Development: Constraints and Opportunities.* ICIMOD, Kathmandu.

Ju, X.T., C.L. Kou, P. Christie, Z.X. Dou, and F.S. Zhang. 2007. Changes in the soil environment from excessive application of fertilizers and manures to two contrasting intensive cropping systems on the North China Plain. *Environ. Pollut.* 145(2): 497–506.

Kennedy, I.R., A.T.M.A. Choudhury, and M.L. Kecskes. 2004. Non-symbiotic bacterial diazotrophs in crop-farming systems: Can their potential for plant growth promotion be better exploited? *Soil Biol. Biochem.* 36(8): 1229–1244.

Kumar, K., and K.M. Goh. 2002. Management practices of antecedent leguminous and non-leguminous crop residues in relation to winter wheat yields, nitrogen uptake, soil nitrogen mineralization and simple nitrogen balance. *Eur. J. Agron.* 16(4): 295–308.

Kumar, K., and K.V.P. Rao. 1992. N and P requirement of upland rice in Manipur. *ORYZA—An International Journal on Rice.* 29: 306.

Kumar, M. 2011. Northeast India: Soil and water management imperatives for food security in a changing climate. *Curr. Sci.* 101(9): 1119.

Kumar, M., M.H. Khan, P. Singh, S.V. Ngachan, D.J. Rajkhowa, A. Kumar, and M.H. Devi. 2012. Variable lime requirement based on differences in organic matter content of iso-acidic soils. *Indian J. Hill Farming* 25(1):26–30.

Lal, R. 1994. Tillage effects on soil degradation, soil resilience, soil quality, and sustainability. *Soil Till. Res.* 27: 1–8.

Laxminarayana, K., B.P. Bhatt, and T. Rai. 2006b. Soil fertility build up through hedgerow intercropping in integrated farming system: A case study. In *Agroforestry in North East India: Opportunities and Challenges*, Eds. B.P. Bhatt, and K.M. Bujarbaruah, pp. 479–490. ICAR Research Complex for NEH Region, Umiam, Meghalaya, India.

Laxminarayana K., M. Datta, Y.P. Sharma, U.C. Sharma, and B. Gopichand. 2006a. Soil microbes and bio-fertilizers. In *Soils and Their Management in North East India*, Eds. U.C. Sharma, M. Datta, and J. S. Samra. ICAR Research Complex for NEH Region, Umiam, Meghalaya, India, p. 502.

Layek, J., B.G. Shivakumar, D.S. Rana, S. Munda, K. Lakshman, A. Das, and G.I. Ramkrushna. 2014a. Soybean–cereal intercropping systems as influenced by nitrogen nutrition. *Agron. J.* 106 (7): 1–14.

Layek, J., B.G. Shivakumar, D.S. Rana, S. Munda, K. Lakshman, A.S. Panwar, A. Das, and G.I. Ramkrushna 2014b. Performance of soybean (*Glycine max*) intercropped with different cereals under varying levels of nitrogen. *Indian J. Agric. Sci.* 85 (12): 1571.

Lemaire, G., E. van Oosterom, J. Sheehy, M.H. Jeuffroy, A. Massignam, and L. Rossato. 2007. Is crop N demand more closely related to dry matter accumulation or leaf area expansion during vegetative growth? *Field Crops Res.* 100(1): 91–106.

Lenka, N.K., P.R. Choudhury., S. Sudhishri., A. Dass, and US Patnaik. 2012a. Soil aggregation, carbon build up and root zone soil moisture in degraded sloping lands under selected agroforestry based rehabilitation systems in eastern India. *Agric. Ecosyst Environ.* 150: 54–62.

Lenka, N.K., A. Dass, S. Sudhishri, and U.S. Patnaik. 2012b. Soil carbon sequestration and erosion control potential of hedgerows and grass filter strips in sloping agricultural lands of eastern India. *Agric. Ecosyst. Environ.* 158:31–40.

Mahanty, T., S. Bhattacharjee, M. Goswami, P. Bhattacharyya, B. Das, A. Ghosh, and P. Tribedi. 2016. Biofertilizers: A potential approach for sustainable agriculture development. *Environ. Sci. Pollut. Res.* 26:1–21. doi 10.1007/s11356-016-8104-0.

Maikhuri, R.K., D. Dangwal, V.S. Negi, and Rawat, L.S. 2016. Evaluation of symbiotic nitrogen fixing ability of legume crops in Central Himalaya, India. *Rhizosphere* 1: 26–28.

Majumder, M., A.K. Shukla and A. Arunachalam. 2010. Nutrient release and fungal succession during decomposition of crop residues in a shifting cultivation system. *Comm. Soil Sci. Plant Anal.* 41 (4): 497–515.

Malusa, E. and N. Vassilev. 2014. A contribution to set a legal framework for biofertilisers. *Appl. Microbiol. Biotechnol.* 98(15): 6599–6607.

Mandal, D., K.S. Dadhwal, O.P.S. Khola, and B.L. Dhyani. 2006. Adjusted T values for conservation planning in Northwest Himalayas of India. *J. Soil. Water. Conserv.* 61(6): 391–397.

Mandal, D., and V.N. Sharda. 2013. Appraisal of soil erosion risk in the Eastern Himalayan region of India for soil conservation planning. *Land Degrad. Dev.* 24(5): 430–437.

Manjunath, B.L., and V.S. Korikanthimath. 2009. Sustainable rice production through farming systems approach. *J. Sustain. Agric.* 33(3): 272–284.

Maskey, S.L., S. Bhattarai, M.B. Peoples, and D.F. Herridge. 2001. On-farm measurements of nitrogen fixation by winter and summer legumes in the Hill and Terai regions of Nepal. *Field Crops Res.* 70(3): 209–221.

Mazid, M., T.A. Khan, and F. Mohammad. 2011. Cytokinins, a classical multifaceted hormone in plant system. *J. Stress Physiol. Biochem.* 7(4): 347–368.

Mishra, B.K., and P.S. Ramakrishnan. 1983. Slash and burn agriculture at higher elevations in north-eastern India. II. Soil fertility changes. *Agric., Ecosyst. Environ.* 9 (1): 83–96.

Mishra, B.K., and P.S. Ramakrishnan. 1984. Nitrogen budget under rotational bush fallow agriculture (jhum) at higher elevations of Meghalaya in north-eastern India. *Plant Soil* 81(1):37–46.

Mishra, B.N., R. Prasad, B. Gangaiah, and B.G. Shivakumar. 2006. Organic manures for increased productivity and sustained supply of micronutrients Zn and Cu in a rice-wheat cropping system. *J. Sustain. Agric.* 28(1): 55–66.

Morrison, J.E., T.J. Gerik, F.W. Chichester, J.R. Martin, and J.M. Chandler. 1990. A no-tillage farming system for clay soils. *J. Produc. Agric.* 3(2): 219–227.

Mosier, A.R. 2002. Environmental challenges associated with needed increases in global nitrogen fixation. *Nutr. Cycl. Agroecosys.* 63:101–116.

Munda, G.C., D.P. Patel, A. Das, R. Kumar, and A. Chandra. 2006. Production potential of rice (*Oryza sativa*) under in situ fertility management as influenced by variety and weeding. *J. Eco-friendly Agric.* 1(1): 12–15.

NBSS. 1997. Soils of Manipur for optimizing land use. National Bureau of Soil Survey, Publ. 56b, Nagpur, India.

NEC (North Eastern Council). 2015. Basic statistics of North Eastern Region 2015, Govt. of India. http://
necouncil.gov.in/writereaddata/mainlinkFile/BasicStatistic2015.pdf.

Nitant, H.C., L.S. Bhushan, and I. Pal. 1992. Status of soil properties in Yamuna Ravines under forest cover.
*J. Trop. For.* 8(1): 31–37.

Palni, L.M.S. and R.S. Rawal. 2010. Conservation of Himalayan bioresources: An ecological, economical and
evolutionary perspective. In *Nature at Work: Ongoing Saga of Evolution.* Springer, India. pp. 369–402.

Pandey, A., A. Mathur, S.K. Mishra, and B.C. Mal. 2009. Soil erosion modeling of a Himalayan watershed
using RS and GIS. *Environ. Earth Sci.,* 59(2): 399–410.

Pandey, K.K. 1992. Sustainability of the environmental resource base and development priorities of a mountain community. ICI MOD Occasional Paper No. 19. Kathmandu.

Partap, T. 1999. Sustainable land management in marginal mountain areas of the Himalayan region. *Mt. Res.
Dev.* 19(3): 251–260.

Patel, A. 2013. Harnessing agricultural potential in North Eastern Region of India. http://indiamicrofinance
.com/agricultural-in-north-east-india.html (accessed on  December 4, 2016).

Patel, C.S. 1985. Fertilizer management in rice. In Annual Report ICAR Research Complex for NEH Region,
Umiam, Meghalaya, 1985–86, pp. 22–25.

Patel, C.S., and J. Singh 1990. Efficacy of form, timing and application method of urea and urea-based fertilizers in wetland rice (*Oryza sativa*) on Histosols and Haplaquents soils under mid-altitude of north-
eastern hills region. *Indian J. Agric. Sci.* 60(3): 169–173.

Patel, D.P., A. Das, M. Kumar, G.C. Munda, S.V. Ngachan, G.I. Ramkrushna, J. Layek et al. 2015. Continuous
application of organic amendments enhance soil health, produce quality and system productivity of
vegetable based cropping systems at subtropical eastern Himalayas. *Exp. Agric.* 51(1): 85–106.

Pilbeam, C.J., B.P. Tripathi, and G.P. Acharya. 1998. Estimation of nitrogen fixation by blackgram using 15N
techniques. *LARC Working Paper (Nepal).*

Prasad, R., and J.F. Power. 1995. Nitrification inhibitors for agriculture, health, and the environment. *Adv.
Agron.* 54: 233–281.

Prasad, R., G. Singh, Y.S. Shivay, and H. Pathak. 2016. Need for determining ecofriendly optimum fertilizer
nitrogen level for better environment and for alleviating hunger and malnutrition. *Indian J. Agron.* 61:
1–8.

Prasad, R.N., Patiram, R.C. Barroah, and M. Ram. 1981. Soil Fertility management in North Eastern Hills
Region, Research Bulletin No. 9, ICAR Research Complex for NEH Region.

Rai, S.C., and E. Sharma. 1998. Comparative assessment of runoff characteristics under different land use
patterns within a Himalayan watershed. *Hydrol. Process.* 12(13): 2235–2248.

Raines, C.A. and J.C. Lloyd. 2007. C3 carbon reduction cycle. *Encycl. Life Sci.* doi: 10.1002/9780470015902
.a0001314.pub2.

Rajurkar, M.P., U.C. Kothyari, and U.C. Chaube. 2004. Modeling of the daily rainfall-runoff relationship with
artificial neural network. *J. Hydrol.* 285(1): 96–113.

Ram, M., and B.P. Singh. 1993. Soil fertility management in farming systems. In Lectures notes, off campus
training on farming system, Aizawl, India.

Ramakrishnan, P.S. 1994. The jhum agroecosystem in North-Eastern India: A case study of the biological
management of soils in a shifting agricultural system. In *The Biological Management of Tropical Soil
Fertility,* Eds. P.L. Woomer, and M.J. Swift, pp. 189–207. Wiley, Chichester, UK.

Ramesh, T., K.M. Manjaiah, J.M.S. Tomar, and S.V. Ngachan. 2013. Effect of multipurpose tree species on
soil fertility and $CO_2$ efflux under hilly ecosystems of Northeast India. *Agroforest. Syst.* 87: 1377–1388.

Rao, D.L.N., D. Balachandar, and D. Thakuria. 2015. Soil biotechnology and sustainable agricultural intensification. *Indian J. Fert.* 11(10): 87–105.

Ravindranath, N.H., S. Rao, N. Sharma, M. Nair, R. Gopalakrishnan, A.S. Rao, S. Malaviya et al. 2011.
Climate change vulnerability profiles for North East, India. *Curr. Sci.* 101(3): 1–11.

Saha, S., A.K. Pandey, K.A. Gopinath, R. Bhattacharya, S. Kundu, and H.S. Gupta. 2007. Nutritional quality
of organic rice grown on organic composts. *Agron. Sustain. Dev.* 27: 223–229.

Sanwal, S.K., K. Laxminarayana, R.K. Yadav, N. Rai, D.S. Yadav, and M. Bhuyan. 2007. Effect of organic
manures on soil fertility, growth, physiology, yield and quality of turmeric. *Indian J. Hort.,* 64(4):
444–449.

Sarkar, D., C.B. Meitei, L.K. Baishya, A. Das, S. Ghosh, K.L. Chongloi, and D. Rajkhowa. 2015. Potential of
fallow chronosequence in shifting cultivation to conserve soil organic carbon in northeast India. *Catena*
135: 321–327.

Sarkar, D., C.B. Meitei, L.K. Baishya, and P. Dey. 2016. Fertilizer prescription for desired target yield of direct seeded rainfed upland rice in north eastern hill region. *ORYZA-An International Journal on Rice*. 53(3): 288–293.

Sarmah, A.C., and P.K. Bordoloi. 1994. Decomposition of organic matter in soils in relation to mineralization of carbon and nutrient availability. *J. Indian Soc. Soil Sci.* 42(2): 199–203.

Sen, T.K., G.S. Chamuah, A.K. Maji, and J. Sehgal. 1996. Soils of Manipur for optimising land use. NBSS publ 56b (Soils of India series), National Bureau of Soil Survey and Land Use Planning, Nagpur, India, 52 p + 1 sheet map.

Sharma, H.R. 1996. Mountain agricultural development processes and sustainability micro-level evidence from Himachal Pradesh. Discussion Paper, Series No. MFS 96/2. ICIMOD, Kathmandu.

Sharma, J.P., and U.C. Sharma. 1991. Effect of nitrogen and phosphorus on the yield and severity of turcicum blight in maize Nagaland. *Indian Phytopathol.* 44: 383–385.

Sharma, M.P., S.V. Bali and D.K. Gupta. 2001. Soil fertility and productivity of rice (*Oryza sativa*)-wheat (*Triticum aestivum*) cropping system in an Inceptisol as influenced by integrated nutrient management. *Ind. J. Ag. Sci.* 71(2): 82–86.

Sharma, P.D., T.C. Baruah, A.K. Maji, and Patiram. 2006. Management of acid soils of NEH region. Natural Resource Management Division (ICAR), Krishi Anusandhan Bhawan-II, Pusa Campus, New Delhi. Technical Bulletin, p. 14.

Sharma, U.C. 1999. Loss of nitrogen through leaching and runoff from two potato land-use systems on different soils. In *Impact of Land Use Change on Nutrient Loads from Diffuse Sources,* Ed. L. Heathwaise, pp. 27–32. IAHS Publ. 257. IAHS Press, Wallingford, UK.

Sharma, U.C., M. Datta, and J.S. Samra. 2006. Soils and their management in north east India. ICAR research Complex for NEH region, Umiam, Meghalaya, India, pp. 122–170.

Sharma, U.C., and K.A. Singh. 1994. Agriculture with alder—A potential indigenous farming system of Nagaland. In *Proceedings of the National Seminar on Developments in Soil Science*, pp. 570–571. ICAR, New Delhi, India.

Shrestha, S. 1992. Mountain agriculture: Indicators of unsustainability and options for Reversai MFS Discussion Paper No. 32, ICIMOD, Kathmandu.

Singh, A., and S. Sharma. 2002. Composting of a crop residue through treatment with microorganisms and subsequent vermicomposting. *Bioresour. Technol.* 85(2): 107–111.

Singh, A., and M.D. Singh. 1980. Effect of various stages of shifting cultivation on soil erosion from steep hill slopes. *Indian For.* 108(2): 116–121.

Singh A., and M.D. Singh. 1981. Effect of various stages of shifting cultivation on soil erosion from steep hill slopes. *Indian For.* 106(2): 115–121.

Singh, A.K. 2015. Advances in Indian coldwater fisheries and aquaculture. *J. Fisheries Sciences.com.* 9(3): 48.

Singh, A.K., A. Arunachalam, S.V. Ngachan, K.P. Mohapatra, and J.C. Dagar. 2014. From shifting cultivation to integrating farming: Experience of agroforestry development in the Northeastern Himalayan Region. In *Agroforestry Systems in India: Livelihood Security & Ecosystem Services.* Springer, India, pp. 57–86.

Singh, B. 2014. Nitrogen management issues and strategies. *Indian J. Fert.* 10: 30–38.

Singh, B.P., M. Das, B.S. Dwibedi, M. Ram, and R.N. Prasad. 1993. *Annals Agric. Res.* 14(1): 100–103.

Singh, N.P., O.P. Singh, and N.S. Jamir. 1996. In *Sustainable Agriculture Development Strategy for North Eastern Hill Region of India*, Eds. S.P. Shukla and N. Sharma. Mittal, New Delhi, India.

Singh, O.P., and B. Datta. 1987. Phosphorus status of some hill soils of Mizoram in relation to pedegenic properties. *J. Indian Soc. Soil Sci.* 34: 600–705.

Singh, O.P., and B. Datta. 1988. Organic carbon and nitrogen status of some soils of Mizoram occurring at different altitudes. *J. Indian Soc. Soil Sci.* 36: 414–420.

Singh, T.C., and E.J. Singh. 2014. Effect of traditional fire on n-mineralization in the oak forest stand of Manipur, north east India. *Int. J. Sci. Res. Pub.* 4 (11):1–11.

Smith, B.E., R.L. Richards, and W.E. Newton (Eds.). 2013. Catalysts for nitrogen fixation: Nitrogenases, relevant chemical models and commercial processes. *Springer Science & Business Media* 1:340.

Tawnenga, U.S., and R.S. Tripathi. 1996. Evaluating second year cropping on jhum fallows in Mizoram, north-eastern India—phytomass dynamics and primary productivity. *J. Biosci.* 21 (4): 563–575.

Tomar, A., N. Kumar, R.P. Pareek, and A.K. Chaube. 2001. Synergism among VA mycorrhiza, phosphate solubilizing bacteria and Rhizobium for symbiosis with black gram (*Vigna mungo* L.) under field conditions. *Pedosphere* 11(4): 327–332.

Tomar, J.M.S., A. Annaduari, A. Das, and R. Singh. 2007. Agroforestry a viable component of organic farming. In *Advances in Organic Farming Technology in India*, Eds. G.C. Munda, P.K. Ghosh, A. Das, S.V. Ngachan, and K.M. Bujarbaruah. ICAR Research Complex for NEH Region, Umiam, Meghalaya, India.

Tripathi, A.K., A. Das, D.P. Patel, U.K. Hazarika, and G.C. Munda. 2007. Organic rice production—package of practices for North East India. Technical bulletin no. 2. KVK, Ri-Bhoi, ICAR Research Complex for NEH Region, Umiam, Meghalaya, India.

Vance, C.P. 2001. Symbiotic nitrogen fixation and phosphorus acquisition. Plant nutrition in a world of declining renewable resources. *Plant Physiol.* 127(2): 390–397.

Venkatesh, M.S., K. Kumar, M. Raychaudhuri, M. Datta, and U.C. Sharma. 2006. Soil fertility and crop responses to nutrients. In *Soils and Their Management in North East India,* Eds. U.C. Sharma, M. Datta, and J.S. Samra. ICAR Research Complex for NEH Region, Umiam, Meghalaya, India, pp. 502.

Vitousek, P.M., R. Naylor, T. Crews, M.B. David, L.E. Drinkwater, E. Holland, P.J. Johnes et al. 2009. Nutrient imbalances in agricultural development. *Science* 324(5934): 1519–1520.

Walter, C., P. Merot, B. Layer, and G. Dutin. 2003. The effect of hedgerows on soil organic carbon storage in hillslopes. *Soil Use Manage.* 19(3): 201–207.

Williams, A., P.M. Ewing, N.R. Jordan, A.S. Davis, A.S. Grandy, R.G. Smith, D.A. Kane et al. 2016. Enhanced control of soil nitrogen cycling through soil functional zone management. *Crops and Soils* 49(6): 42–45.

Xu, Q.X., T.W. Wang, C.F. Cai, Z.X. Li, and Z.H. Shi. 2012. Effects of soil conservation on soil properties of citrus orchards in the Three-Gorges Area, China. *Land Degrad. Dev.* 23(1): 34–42.

Yadav, G.S., M. Datta, S. Babu, C. Debnath, and P.K. Sarkar. 2013. Growth and productivity of lowland rice (*Oryza sativa*) as influenced by substitution of nitrogen fertilizer by organic sources. *Indian J. Agric. Sci.* 83(10): 1038–1042.

Yadav, S.K., and A.K. Srivastava. 2015. A review on agronomical aspects of potato production in northeastern region of India. *Int. J. Appl. Pure Sci. Agric.* 1(6): 26–34.

Yawalkar, K.S., J.P. Agrawal, and S. Bokade. 2002. *Manures and Fertilizers*, 9th ed. Agri-Horticulture Publishing House, Nagpur, p. 29.

Zhang, X., E.A. Davidson, D.L. Mauzerall, T.D. Searchinger, P. Dumas, and Y. Shen. 2015. Managing nitrogen for sustainable development. *Nature* 528(7580): 51–59.

# 11 Merits and Limitations of Enhanced Efficiency Fertilizers

*Matthew D. Ruark, Rogério P. Soratto, and Carl J. Rosen*

## CONTENTS

## 11.1 INTRODUCTION

The value of nitrogen (N) fertilizer to agricultural production is clear. The environmental effects of N fertilizer use are also clear: high levels in groundwater, export to Gulf of Mexico, and contribution to $N_2O$ emissions. The N use efficiency (NUE) of applied fertilizer is generally improving, but there is still plenty of room for improvement (Ladha et al. 2016). Controlling the availability of applied N to crops during periods of peak uptake is the key to improving NUE. However, achievement of maximum efficiency can be laborious as it requires multiple passes across a field, or expensive specific equipment (e.g., irrigation systems or high clearance applicators) to apply N to a growing crop. One way to offset these issues is to use fertilizer technologies, generally referred to as enhanced efficiency fertilizers (EEFs), to meter out N or inhibit part of the N cycle. These EEFs are those fertilizer products that can promote improvements in the agronomic effectiveness, minimize the potential of N loss to the environment from N fertilizer, or generate some economic benefit, as compared to conventional fertilizer, which in this context would be unprotected urea, ammonium sulfate, ammonium nitrate, anhydrous ammonia, or urea-ammonium nitrate (UAN) solution (Chien et al. 2009; Timilsena et al. 2015).

When discussing EEFs, often the first step required is to provide standardized definitions to provide clarity. The most cited reference for these definitions is the Association of American Plant Food Control Officials (AAPFCO) Official Publication 57 (AAPFCO 2014). One of the purposes of the AAPFCO is to provide uniform definitions of products and provide specific definitions for slow- or controlled-release fertilizers, stabilized N fertilizers, nitrification inhibitors, and urease inhibitors. Here, EEF was adopted as the term to describe these types of fertilizers and use the AAPFCO definitions, along with the further explanations of Trenkel (2010). The goal of any of the EEFs is to improve NUE through either greater uptake of the fertilizer per unit applied or by maintaining or increasing yields with less fertilizer applied.

The objective of this chapter is to summarize the EEF fertilizer types and the state of knowledge on each product's ability to improve yields or NUE and reduce N losses. To date, there are many review papers that explain the nature and process of these N fertilizers; the works of Shaviv (2001) and Trenkel (2010) provide thorough descriptions of these fertilizers and the mechanisms of the N release and are heavily referenced in this chapter. There have also been several meta-analyses conducted to determine the effect of EEFs on yield and N losses (Table 11.1). This chapter provides a review of the reviews as well as recent papers that build upon our understanding of the benefits of these products, specifically those published after Trenkel (2010) (Tables 11.2 and 11.3). This review

## TABLE 11.1
### Recent Publications Using Meta-Analysis to Determine Crop and Environmental Effects of Enhanced Efficiency Fertilizers (EEFs)

| Reference | Geographic Scope | Crops or Land Use Type | EEFs | Response Variables |
|---|---|---|---|---|
| Wolt (2004) | USA | Grain, forage | Nitrapyrin | Yield, soil N retention, N leaching, greenhouse gas emissions |
| Akiyama et al. (2010) | World | Upland field, grassland, paddy field | Urease inhibitors, nitrification inhibitors, PCF[a] | $N_2O$ and NO emissions |
| Linquist et al. (2013) | World | Rice | Urease inhibitors, nitrification inhibitors, slow- and controlled-release | Yield, N uptake |
| Abalos et al. (2014) | World | Grain, forage, vegetable, cotton | DMPP,[b] DCD,[c] NBPT,[d] DCD+NBPT | Yield, nitrogen use efficiency |
| Burzaco et al. (2014) | North America | Corn | Nitrapyrin | Yield, N uptake, nitrogen use efficiency |
| Decock (2014) | North America | Corn | DCD+NBPT, PCU[e] | $N_2O$ emissions |
| Hu et al. (2014) | Germany | Grain, potato, silage corn | Nitrification inhibitors | Yield, crude protein (wheat), starch (potato) |
| Huang et al. (2016) | China | Annual crops | Urease inhibitors, nitrification inhibitors, slow- and controlled-release | Yield, $NH_3$ emissions |
| Gilsanz et al. (2016) | World | Cropland, grassland, upland, paddy | DMPP, DCD | $N_2O$ emissions |
| Thapa et al. (2016) | World | Corn, rice, wheat | Urease inhibitors, nitrification inhibitors, PCU | Yield, $N_2O$ emissions |
| Yang et al. (2016) | World | Grain, forage, vegetable and industrial crops | DMPP, DCD | Yield, N uptake, soil inorganic N, N leaching, $NH_3$ and greenhouse gas emissions |
| Silva et al. (2016) | World | Grain, forage, sugarcane, cotton | NBPT | Yield, $NH_3$ emissions |

*Note:* The EEFs are listed by category if using many different products in the analysis.
[a] PCF: polymer-coated fertilizers (as used by Akiyama et al., 2010)
[b] DMPP: 3,4-dimethyldimepyrazole phosphate
[c] DCD: dicyandiamide
[d] NBPT: N-(n-butyl) thiophosphoric triamide
[e] PCU: polymer-coated urea

TABLE 11.2
Summary of Yield and Nitrogen Use Efficiency (NUE) Effects of Polymer, Resin, or Sulfur-Coated Urea from Peer-Reviewed Publications Published Post-Trenkel (2010)

| Crop | Coated Urea | | Conventional Fertilizer | | N Rate kg ha⁻¹ | Relative Yield Difference % | Relative NUE Difference % | Reference |
|---|---|---|---|---|---|---|---|---|
| | Source | Application Timing | Source | Application Timing | | | | |
| Barley | PCU[a] | At planting | Urea | At planting | 10–150 | 0.1 | – | Blackshaw et al. (2011) |
| Bermudagrass | PCU | Topdress | Urea | Topdress | 336 | –7.4 | – | Connell et al. (2011) |
| Corn | PCU | Post-emergence | Urea | Post-emergence | 202–246 | –3.1 | – | Halvorson et al. (2010) |
| | PCU | Post-emergence | Urea | Post-emergence | 202–246 | –3.9 | – | |
| | PCU | Post-emergence | Urea | Post-emergence | 202 | 1.5 | – | Halvorson et al. (2011) |
| | PCU | Pre-planting | Urea | Split-applied | 180 | –4.0 | – | Rubin et al. (2016) |
| | PCU | Pre-planting | Urea | Split-applied | 225 | –1.6 | – | |
| | PCU | Pre-planting | Urea | Split-applied | 180 | –10.0 | – | Maharjan et al. (2014) |
| | PCU | Pre-planting | Urea | Pre-planting | 0–336 | 7.3 | 13.6 | Halvorson and Bartolo (2014) |
| | PCU | At planting | Urea | At planting | 225 | 7.5 | 63.7 | Li et al. (2015) |
| | PCU | Topdress | Urea | Topdress | 225 | 0.3 | – | |
| | PCU | After planting | UAN[b] | After planting | 168 | 59.0 | 153.0 | Maharjan et al. (2016a) |
| | PCU | After planting | UAN | After planting | 280, 199 | 53.4 | 219.2 | |
| | PCU | Pre-planting | Urea | Pre-planting | 84 | 2.6 | – | Nelson et al. (2014) |
| | PCU | Pre-planting | Urea | Pre-planting | 168 | 7.3 | 29.2 | |
| | PCU | After planting | UAN | After planting | 56–224 | 5.9 | 83.4 | Shapiro et al. (2016) |
| | PCU | Pre-planting (broadcast) | Urea | Pre-planting (broadcast) | 168 | –2.4 | – | Kaur et al. (2017) |
| | PCU+PSCU[c]+urea | At planting | Urea | Split-applied | 150 | 6.6 | 39.1 | Zheng et al. (2016) |
| | PCU+PSCU+urea | At planting | Urea | Split-applied | 225 | 15.4 | 19.5 | |
| Corn (silage) | PCU | Pre-planting | Urea | Pre-planting | 202 | –6.0 | – | Nash et al. (2015a) |
| Cotton | PCU | After planting | Urea | After planting | 101 | –0.7 | – | Watts et al. (2014) |

(Continued)

**TABLE 11.2 (CONTINUED)**

**Summary of Yield and Nitrogen Use Efficiency (NUE) Effects of Polymer, Resin, or Sulfur-Coated Urea from Peer-Reviewed Publications Published Post-Trenkel (2010)**

| | Coated Urea | | Conventional Fertilizer | | | | | |
|---|---|---|---|---|---|---|---|---|
| Crop | Source | Application Timing | Source | Application Timing | N Rate kg ha$^{-1}$ | Relative Yield Difference % | Relative NUE Difference % | Reference |
| Rice | PCU | Pre-planting | Urea | Split-applied | 255 | −1.4 | – | Wang et al. (2016) |
| | TRCU[d] | At planting | Urea | Split-applied | 240 | −4.8 | – | Ji et al. (2013) |
| | TRCU | At planting | Urea | Split-applied | 240 | 4.6 | – | Wang et al. (2015) |
| Potato | PCU | Pre-planting | Urea+AN[e] | Split-applied | 225 | −1.0 | – | Hyatt et al. (2010) |
| | TRCU | Pre-planting | Urea+AN | Split-applied | 225 | −0.8 | – | |
| | PCU | At planting | DAP[e]+AN | Split-applied | 193 | 4.3 | – | Zebarth et al. (2012) |
| | PCU | At planting | DAP+AN | At planting | 193 | −0.3 | – | |
| | PCU | At emergence | AS[f]+AN | Split-applied | 280 | 3.1 | – | Bero et al. (2014) |
| | PCU | At planting | Urea | At planting | 150 | 18.2 | 147.7 | Gao et al. (2015) |
| | PSCU | At planting | Urea | At planting | 150 | 11.3 | 95.7 | |
| | PCU | At planting | AN | Split-applied | 60 | 8.2 | 17.5 | Cambouris et al. (2016) |
| | PCU | At planting | AN | Split-applied | 120 | −1.4 | – | |
| | PCU | At planting | AN | Split-applied | 200 | 1.1 | – | |
| | PCU | At planting | AN | Split-applied | 280 | −0.3 | – | |
| Rice | PCU | Pre-planting | Urea | Split-applied | 255 | −1.4 | – | Wang et al. (2016) |
| | TRCU | At planting | Urea | Split-applied | 240 | −4.8 | – | Ji et al. (2013) |
| | TRCU | At planting | Urea | Split-applied | 240 | 4.6 | – | Wang et al. (2015) |

*(Continued)*

**TABLE 11.2 (CONTINUED)**

**Summary of Yield and Nitrogen Use Efficiency (NUE) Effects of Polymer, Resin, or Sulfur-Coated Urea from Peer-Reviewed Publications Published Post-Trenkel (2010)**

| Crop | Coated Urea | | Conventional Fertilizer | | N Rate kg ha$^{-1}$ | Relative Yield Difference % | Relative NUE Difference % | Reference |
|---|---|---|---|---|---|---|---|---|
| | Source | Application Timing | Source | Application Timing | | | | |
| Wheat | PCU | At planting | Urea | At planting | 0–210 | 2.2 | – | McKenzie et al. (2010) |
| | PCU | Spring broadcast | Urea | Spring broadcast | 0–210 | –5.5 | – | |
| | PCU | Pre-planting | Urea | Split-applied | 225 | –1.0 | – | Wang et al. (2016) |
| | PCU + PSCU + urea | At planting | Urea | Split-applied | 150 | 10.6 | 59.8 | Zheng et al. (2016) |
| | PCU + PSCU + urea | At planting | Urea | Split-applied | 225 | 9.1 | 36.6 | |
| | TRCU | At planting | Urea | Split-applied | 225 | 10.9 | 25.4 | Yang et al. (2011) |

*Note:* Only papers where yields were directly reported were included. Relative yield and relative NUE differences were calculated as the gain or decrease relative to conventional fertilizer. Calculations were also based on the average across all site years. The NUE was calculated as the yield difference between fertilized and unfertilized treatments divided by N applied. Positive numbers for relative yield and relative NUE difference indicate a positive effect of the coated urea fertilizer.

a  PCU: polymer-coated urea
b  UAN: urea-ammonium-nitrate
c  PSCU: polymer sulfur-coated urea
d  TRCU: thermoplastic resin-coated urea
e  AN: ammonium nitrate
f  DAP: diammonium phosphate
g  AS: ammonium sulfate

**TABLE 11.3**

**Summary of Yield and Nitrogen Use Efficiency (NUE) Effects of Urease and Nitrification Inhibitors from Peer-Reviewed Publications Published Post-Trenkel (2010)**

| Inhibitor | Application Timing | Conventional Fertilizer | Application Timing | Crop | N rate kg ha⁻¹ | Relative Yield Difference % | Relative NUE Difference % | Reference |
|---|---|---|---|---|---|---|---|---|
| NBPT[a] | Spring broadcast | Urea | Spring broadcast | Wheat | 0–210 | 0 | – | McKenzie et al. (2010) |
| | Topdress | Urea | Topdress | Bermudagrass | 336 | 11.0 | 21.7 | Connell et al. (2011) |
| | Topdress | UAN[b] | Topdress | Bermudagrass | 336 | −2.3 | – | |
| Nitrapyrin | Pre-emergence | UAN | Pre-emergence | Corn | 90 | 6.5 | – | Burzaco et al. (2014) |
| | Pre-emergence | UAN | Pre-emergence | Corn | 180 | 0.8 | – | |
| | Post-emergence | UAN | Post-emergence | Corn | 90 | 4.0 | – | |
| | Post-emergence | UAN | Post-emergence | Corn | 180 | 0.8 | – | |
| | Pre-planting | Urea | Pre-planting | Corn | 168 | −1.7 | – | Kaur et al. (2017) |
| DCD[c] | Split-applied | Urea | Split-applied | Corn (pot study) | 100 | −4.8[d] | – | Mahmood et al. (2011) |
| | Post-emergence | Urea | Post-emergence | Corn (pot study) | 100 | 2.9[d] | – | |
| | Split-applied | Urea | Split-applied | Wheat (pot study) | 100 | −5.3 | – | |
| | Post-emergence | Urea | Post-emergence | Wheat (pot study) | 100 | 5.2 | – | |
| | Pre-planting | Urea | Pre-planting | Cotton (pot study) | 155 | −8.5[e] | – | |
| | Pre-flood | Urea | Pre-flood | Flood-rice | 84 | 14.6 | 54.4 | Fitts et al. (2014) |
| | Pre-flood | Urea | Pre-flood | Flood-rice | 168 | 16.6 | 36.4 | |
| DMPP[f] | Split-applied | AS[g] | Split-applied | Wheat | 180 | 2.7 | | Huérfano et al. (2016) |
| | Tillering | AS | Tillering | Wheat | 180 | −10.5 | | |
| DMPSA[h] | Split-applied | AS | Split-applied | Wheat | 180 | 1.6 | | |
| | Tillering | AS | Tillering | Wheat | 180 | −10.8 | | |

(Continued)

**TABLE 11.3 (CONTINUED)**

**Summary of Yield and Nitrogen Use Efficiency (NUE) Effects of Urease and Nitrification Inhibitors from Peer-Reviewed Publications Published Post-Trenkel (2010)**

| Inhibitor | Application Timing | Conventional Fertilizer | Application Timing | Crop | N rate kg ha$^{-1}$ | Relative Yield Difference % | Relative NUE Difference % | Reference |
|---|---|---|---|---|---|---|---|---|
| DCD+NBPT | Topdress | Urea | Topdress | Bermudagrass | 336 | 6.5 | 12.7 | Connell et al. (2011) |
| | Topdress | UAN | Topdress | Bermudagrass | 336 | -0.2 | – | |
| | Post-emergence | Urea | Post-emergence | Corn | 202–246 | -6.6 | – | Halvorson et al. (2010) |
| | Post-emergence | UAN | Post-emergence | Corn | 202–246 | 6.1 | – | |
| | Post-emergence | Urea | Post-emergence | Corn | 202 | 4.8 | – | Halvorson et al. (2011) |
| | Post-emergence | UAN | Post-emergence | Corn | 202 | -3.6 | – | |
| | Pre-planting | Urea | Pre-planting | Corn | 0–336 | 0.8 | – | Halvorson and Bartolo (2014) |
| | Pre-planting | Urea | Split-applied | Corn | 180 | -10.9 | – | Maharjan et al. (2014) |
| | Pre-planting | Urea | Split-applied | Corn | 180 | -4.8 | – | Rubin et al. (2016) |
| | After planting | Urea | After planting | Cotton | 101 | -2.1 | – | Watts et al. (2014) |

*Note:* Only papers where yields were directly reported were included. Relative yield and relative NUE differences were calculated as the gain or decrease relative to conventional fertilizer. Calculations were also based on the average across site years. The NUE was calculated as the yield difference between fertilized and unfertilized treatments divided by N applied. Positive numbers for relative yield and relative NUE difference indicate a positive effect of the inhibitor.

a  NBPT: N-(n-butyl) thiophosphorictriamide
b  UAN: urea-ammonium-nitrate
c  DCD: dicyandiamide
d  Shoot dry matter yield
e  Seed + lint yield
f  DMPP: 3,4-dimethylpyrazol phosphate
g  AS: ammonium sulfate
h  DMPSA: 3,4-dimethylpyrazole succinic

focuses on manufactured, N-based EEF products, rather than naturally occurring sources of N (such as manure or plant material), which could be considered a slow-release product based on decomposition and mineralization rates. This review also focuses on EEF use in field-based agricultural systems, rather than on greenhouse crops or turfgrass systems, where these products are more widely used. As often stated in nearly all studies on EEF, there is an added cost to these products and they need to be evaluated for consistency of benefit before adoption in lower-margin agricultural ventures, such as grain production (as compared to vegetable, fruit, or ornamental production). This review does not attempt to list all types of commercial products; the most recent and complete reference for products can be found in Trenkel (2010), who summarizes all available products at the time of that publication, but additional products are continuously emerging.

## 11.2    SLOW- AND CONTROLLED-RELEASE FERTILIZER

Generally, N fertilizers that are slow- or controlled-release fertilizers are those that are coated or rely on N compounds more complex then urea [$(NH_2)_2CO$]. Interestingly, the AAFPCO uses both slow and controlled as interchangeable terms, and include coated urea products and products that rely on urea-organic compounds. This has led to some confusion. Trenkel (2010) explains the differences as this: controlled-release fertilizers have known release rates (i.e., the process of release is controlled by the mechanism), while slow-release fertilizers also have slower release than unprotected commercial fertilizer, but the mechanism of release is either too complex to be predicted or relies on a mechanism that cannot be predicted. Under this definition, encapsulated products [e.g., polymer-coated urea (PCU)] are controlled-release and urea-formaldehyde (UF) products are slow-release. Ultimately, slow- and controlled-release fertilizers attempt to provide the same function: the prevention of the full amount of applied N existing in the soil solution where losses can occur. This not only provides environmental benefits, but if placed near the seed can reduce stress and toxicity to the plant (Trenkel 2010).

### 11.2.1    Polymer-Coated Urea

Polymer-coated urea [sometimes referred to as polyolefin coated urea, organic polymers, or thermoplastic polymer-coated fertilizers (Shaviv 2001)] is the most commonly available controlled-release fertilizer. Polymer-coated urea is a fertilizer product in which each urea prill is individually coated with a polymer (or plastic) coating. When the urea dissolves inside the coating it slowly diffuses through the plastic membrane into the soil over time. Shaviv (2001) provides a thorough description of the conceptual model of N release from PCU and provides the history of the research to develop release prediction models. Shaviv (2001) also describes N release from PCU as a three-phase process: lag phase, constant release phase, and release decay phase. During the lag phase, water is absorbed inside the coating through the pores of the polymer. Little, if any, N is released into the soil during this phase. During the constant release phase, N release from PCU is controlled by the rate the dissolved urea diffuses through the polymer coating. Nitrogen release increases as temperature increases as the coating slightly expands making the pores in the plastic larger. During the release decay phase, the release rate of N through the polymer coating slows after the urea is completely dissolved within the polymer. This is also referred to as the tailing effect (Shaviv 2001). The release pattern has been described as "generally sigmoidal" (Shaviv et al. 2003), but will vary depending on granule size and polymer thickness and properties. The kinetics of N release are discussed in detail by Dave et al. (1999) and Shaviv (2001), with more recent work suggesting that both temperature and moisture should be considered to accurately predict the N release rate of PCU (Fujinuma et al. 2009). Kenawy (1998), Dave et al. (1999), Azeem et al. (2014), Timilsena et al. (2015), and Naz and Sulaiman (2016) provide thorough reviews of the coating products (organic polymers and mineral-based inorganic coating) as well as the coating process. Chien et al. (2009) indicates the future of PCU products will rely on the tailoring of release relative to farmer need. Since these products are considered controlled-release, they should be somewhat predictable in the release of N and as such, many studies have been published that describe models that predict the release. The product specific N release pattern has been highlighted by Shaviv (2001) as a key component of future work to increase adoption of these N fertilizers. Trenkel (2010) also notes the lack of prediction of N release as a limitation of these products. This is especially important as it will need to be fit into future modeling-based approaches for N management (Yuan et al. 2017).

The majority of the PCU use occurs for non-agricultural crops, with the remainder used for high value agricultural crops (e.g., tree crops, vegetables, and fruits) (Trenkel 2010). Because of cost, PCU remains only a small portion of the N fertilizer used in grain production. Each product's polymer is company specific and proprietary, different PCU products will have different release rates of N based on the molecular characteristics and thickness of the polymer, and products are often marketed for specific crops. The thickness of the coating reduces the N concentration of the fertilizer product. For example, urea is 46% N by weight while PCUs often range from 42–44% N. Polymer-coated urea is more expensive than urea fertilizer although the difference in price will vary year to year; a survey of fertilizer retailers in Wisconsin indicates it can be as much as 30–60% more per metric ton (Ruark, personal communication). Thus, the advantage of using PCU over urea may not be economical every year. The benefits of PCU are only realized when there are environmental conditions causing substantial N loss. Deciding to use PCU means that the famer is accepting an additional cost every year to mitigate a risk that may not occur in every year.

The advantage of PCU over unprotected urea is that it prevents large amounts of ammonium $\left(NH_4^+\right)$ or nitrate $\left(NO_3^-\right)$ from existing in the soil early in the growing season, reducing the likelihood that it could be leached, denitrified, or volatilized, which, depending on the rate of N applied, can lead to yield increases. Trenkel (2010) suggests a 20–30% reduction in fertilizer rate is achievable with PCU and still maintain yields. However, there is not enough wide evidence to support this practice. But, in general, PCU has been shown to have a positive effect on a variety of agronomic and environmental factors across the globe (Chien et al. 2009). In a review by Wakimoto (2004), a single application of N fertilizer containing 30–70% PCU achieved the same yields as split-applications of other fertilizers in transplanted rice (*Oryza sativa* L.) systems. Although in direct seeded rice systems, 85–100% of the single application is needed for PCU to have an advantage over split-application. Wang et al. (2015) showed that PCU increased N recovery, and in some cases yield, in rice in southeast China. There is some suggestion that progress in technology can help; Lyu et al. (2015) showed that an experimental coated urea delayed release compared to a commercial PCU product and led to yield increase of rice in pot studies. While PCU appears promising in rice production, it is interesting to note that in the most comprehensive analysis to date, Linquist et al. (2013) did not obtain enough individual studies on PCU to conduct meta-analysis to quantify its individual effect on rice yield.

Use of PCU will have the clearest benefit on sandy soils where $NO_3^-$ leaching can be problematic. In studies on sandy soil (often irrigated), PCU is typically tested against split applications of conventional N (Table 11.2). Interestingly, research has shown that the effect may be crop specific. For example, potato (*Solanum tuberosum* L.) has shown more consistent benefits on irrigated sands compared to corn (*Zea mays* L.). Zvomuya et al. (2003) showed the clearest effect of the benefit of PCU, with polyolefin-coated urea having greater potato yields compared to split applications of conventional and a 34–49% reduction in $NO_3^-$ leaching. Wilson et al. (2009) compared various timings and rates of PCU against conventional fertilizer and showed single application of PCU resulted in similar yields to those under split application (up to five splits). Bero et al. (2014) found similar results in Wisconsin. A simple economic analysis showed that net returns were similar as the PCU had higher cost, but saved in energy costs associated with additional applications. Wilson et al. (2010) also showed that $NO_3^-$ leaching was lower with PCU compared to split applications of conventional N. Cambouris et al. (2016) showed that a single application of PCU was equivalent to the split application in Quebec Province, Canada, but did demonstrate greater plant N uptake in a year with a lot of leaching; Shoji et al. (2001) also showed that PCU produced equivalent potato yields compared to split applications. Venterea et al. (2011) showed that $NO_3^-$ leaching was less with PCU in potato, but only in 1 of 3 years, and only with a thicker-coated urea product (42% N vs. 44% N), although PCU only served to maintain yields in this trial (Hyatt et al. 2010). The controlled-release fertilizer may also prevent the need for late season, rescue application of N, although this has not been directly studied. In addition, it is not clear if use of PCU will alter interpretations of plant tissue diagnostic tests in potato (e.g., petiole $NO_3^-$) as it will change the timing of N availability in the soil. In contrast to the potato research, use of PCU for corn production on irrigated, sandy soils has not shown consistent benefits. In research conducted in Minnesota at the same location as Wilson et al. (2009), PCU applied preplant resulted in lower corn yields compared to split application of conventional N and PCU did not lead to a reduction in $NO_3^-$ leaching losses (Maharjan et al. 2014). Rubin et al. (2016) also reported a slight corn yield decrease (4%) on an irrigated, sandy soil comparing preplant PCU with split-applied urea, although Struffert et al. (2016) showed similar $NO_3^-$ leaching losses between treatments on similar soils. However, research in Nebraska on an irrigated, loamy sand has shown PCU to consistently increase yield compared to UAN when applied after planting (Maharjan et al., 2016a). In addition, when preplant applied PCU is compared to preplant applied urea on irrigated, sandy soil, the PCU has been shown to be beneficial (Maharjan et al. 2016b). Thus, PCU can be beneficial when limited to one, early season application, and one-time application may offset the fuel costs of additional application. Further research is needed to optimize PCU with corn production on irrigated, sandy soil to consistently maintain similar yields with a split application approach.

Polymer-coated urea can also be beneficial on wet or poorly drained soils where denitrification losses (conversion of $NO_3^-$ to nitrous oxide [$N_2O$] or $N_2$ gas) can be substantial (e.g., Halvorson et al. 2011; Noellsch et al. 2009). The PCU can also reduce N loss through ammonia ($NH_3$) volatilization compared to non-coated urea when surface applied, as demonstrated by Connell et al. (2011) on bermudagrass [*Cynodon dactylon* (L.) Pers]. Noellsch et al. (2009) showed that PCU outyielded urea in low-lying areas prone to waterlogging (although PCU was relatively equivalent to anhydrous ammonia), but did not observe the same effect at summit or sideslope landscape positions. However, on claypan soils in Missouri near the same location, PCU was not shown to be beneficial over urea on poorly drained across irrigated, drained, and waterlogged treatments (Nelson et al. 2009; Nash et al. 2015a; Kaur et al. 2017). Conversely, Nelson et al. (2014) reported that under no-till, PCU resulted in greater corn yields when compared to urea on these same soils. Interestingly, there does not appear to be much research on benefits of PCU on tile drained soils. Nash et al. (2015c) found no significant corn yield benefit of PCU with free or managed tile drainage on claypan soils, and no difference in $NO_3^-$ export was determined (Nash et al. 2015b).

It is likely that the benefit of PCU will be variable outside of flooded, waterlogged, and sandy soils (Table 11.2). For example, Cahill et al. (2010) found no benefit of PCU over UAN for corn grain, but corn stover yields were greater in 3 of 6 years. Shapiro et al. (2015) found the PCU had greater corn yields when compared to UAN when broadcast applied in 2 of 3 site years in Nebraska. In Iowa, Hatfield and Parkin (2014) reported both negative, neutral, and positive benefits of PCU to corn yield across 3 site-years compared to many other N sources and timings. Li et al. (2015) found PCU to improve corn yields on a clay loam in China as did Halvorson and Barolo (2014) in Colorado, although previous work by Halvorson et al. (2010, 2011) did not find PCU to lead to greater yields in irrigated corn. Benefits to small grains have also been shown to be variable. McKenzie et al. (2010) found no positive effect of PCU to yield on winter wheat (*Triticum aestivum* L.), while Nyborg et al. (1999), Blackshaw et al. (2011), Yang et al. (2011), and Nash et al. (2012) showed benefits to winter wheat, canola (*Brassica napus* L.), or barley (*Hordeum vulgare* L.) in some years. Cahill et al. (2014) showed PCU resulted in lower wheat yields in North Carolina compared to UAN. Research in Minnesota showed a negative yield effect on wheat with PCU compared to urea, but greater protein concentrations (Farmaha and Sims 2013). No advantage of PCU (on non-sandy soils) has been reported on cotton (*Gossypium hirsutum* L.) in Alabama (Watts et al. 2014), on broccoli (*Brassica oleracea* L. var. *italica* Plenck) in the Salinas Valley of California (Hartz and Smith 2009), nor on potato in Canada (Zebarth et al. 2012). In contrast, Gao et al. (2015) found that PCU improved potato yield on a silt loam indicating the effect may also be potato specific, especially when compared to urea also applied at the same time (at planting). These variable effects demonstrate clearly that on non-sandy, non-flooded soils, there can be benefits, limitations, or no effect of the controlled-release aspect of the fertilizer depending on if weather conditions were such to drive significant enough N loss and method of application. Much more work is needed to explain and understand all the interaction effects with application method to identify scenarios where negative effects will occur. Interestingly, mixing of PCU (rather than complete use) has not been well studied. Mixtures of PCU applied at one timing outyielded split applications of conventional N in wheat and corn in the North China Plain (Zheng et al. 2016). In addition, what most of these studies are lacking are comparisons of PCU with conventional fertilizer across enough N rates to identify optimum N rates based on plateau curves using nonlinear regression.

Generally, controlled-release fertilizer reduces gaseous losses of N, but the effect has been shown to be affected by texture and tillage. Akiyama et al. (2010) conducting meta-analysis across all crops and global locations determined that PCU reduced $N_2O$ and NO emissions compared to conventional fertilizers by 35 and 46%, but the reduction was much greater on wetland soils compared to upland soils. A meta-analysis by Decock (2014) showed that PCU did not significantly reduce $N_2O$ emissions in the Midwestern United States. These results were confirmed by a meta-analysis by Abalos et al. (2014), who showed that PCU did not reduce $N_2O$ emissions (nor affect corn yield) across the United States and Canada. In contrast, taking a global approach, Thapa et al. (2016) conducted a meta-analysis across corn, rice, and wheat studies and showed that PCU reduced $N_2O$ emissions by 19% compared to conventional fertilizer. They also showed no significant reduction in yield (across all crops and soil conditions). However, the reduction in $N_2O$ from PCU was greater under coarse textured soil compared to medium and fine textured soil, and in tilled fields compared to no-till. In contrast, in irrigated, dryland conditions of Colorado, PCU was shown to significantly reduce $N_2O$ emissions, but only under reduced and no-till management (Halvorson et al. 2014). Thermoplastic resin-coated urea was shown to reduce $N_2O$ emissions by 13% (although not statistically significant) compared to urea in rice production in South China (Ji et al. 2013), although Wang et al. (2016) showed significant reductions in $NH_3$ emissions with the same product in eastern China. In North Dakota, PCU reduced both $N_2O$ and $NH_3$ emissions, but only on a sandy loam soil, as opposed to a silty clay soil (Awale and Chatterjee 2017). Parkin and Hatfield (2014) indicate that the benefit of PCU in reducing $N_2O$ emissions in rainfed corn systems is limited due to the $N_2O$ being driven by timing of rainfall events.

Damage to PCU's polymer coating can result in faster N release rates and reduce its effectiveness as a controlled-release fertilizer. Once the polymer is cracked, water can easily enter and the release of N will be similar to uncoated urea. Damage can occur because of handling practices, blending or mixing the PCU with other fertilizers, or the method of application. The most severe damage has been seen during handling of PCU, especially when transferring in equipment with abrasive surfaces (Beres et al. 2012). Transporting with a belt conveyer instead of a steel auger is suggested to reduce damage (Beres et al. 2012). Having the PCU be the last component mixed if blended can reduce the time in contact with other fertilizers and reduce damage as well. Changing the way the fertilizer is applied can also reduce damage. For example, use of airboom spreaders to apply the PCU tends to increase polymer damage. However, as long as the PCU is not damaged prior to loading into the airboom truck, the amount of damage with the airboom spreader is usually less than 15% of the prills (Rosen unpublished). Spinner spreaders or drop spreaders will result in less damage than airboom spreaders. Polymer-coated urea will work the best (and most predictively) when not mixed with other fertilizers and with avoidance of unnecessary handling prior to application.

Damage to the polymer coating can change the release pattern of N during the growing season. Research conducted at the Sand Plain Research Farm in Becker, MN showed that damaged PCU (damaged via applicator) released 60% of its N after 8 days, while undamaged PCU only released 12% (Bierman et al. 2015). Since there is extra cost of the fertilizer N with the polymer coating, it is important that PCU is not damaged during application as well as transport. Since damage to the prill is not visible to the naked eye, a simple test can be conducted to test the damage to the PCU. The test, referred to as the 24-hour water test, is where PCU is weighed before and after submersion in water for 24 hours to calculate the amount of N released. This is a simple test and can be used to test the quality of the PCU during different stages of transport and application. Knowing if PCU is damaged and may release over a shorter period of time is important when interpreting mid- and late-season plant N levels especially if tissue results are low. Carson and Ozores-Hampton (2012) provide a review of all laboratory, greenhouse, and field-based methods for PCU assessments; Shaviv (2001) and Trenkel (2010) also provide details on evaluation methods. Overall, more research and development are needed to develop coating products that are resistant to damage, but still provide release in a predictable, timely manner (Shaviv 2001). Additionally, there is some concern about the coatings not decomposing in soil after use (Trenkel 2010).

### 11.2.2 SULFUR-COATED UREA

Before the invention of PCU, sulfur-coated urea (SCU) was a popular slow/controlled release fertilizer. Trenkel (2010) puts SCU products into the slow-release category because the cracking of the coating immediately ends the controlled-release nature of the fertilizer. Shaviv (2001), Chien et al. (2009), and Trenkel (2010) provide a history and thorough description of SCU and suggest positive yield results have occurred across the globe on a variety of crops. However, there are some important limitations to SCU, specifically the lack of predictable release. As summarized by Trenkel (2010), one-third of the urea may become plant available upon contact with soil water (referred to as the burst effect) and one-third of the urea-N may not release during the time period of N uptake by the plant (referred to as the lock-off effect). Liegel and Walsh (1976) showed that on potato, SCU was beneficial in years with high precipitation, but resulted in lower yields in below average rainfall years. In concurrent trials, SCU was not beneficial in corn, even in high rainfall years (Liegel and Walsh 1976). Polymer-sulfur-coated urea (PSCU), which has polymer coating over sulfur coating, can prolong release beyond that of PCU alone. The PSCU may be more beneficial than urea, as shown by Gao et al. (2015), where it increased potato yields, but not necessarily more beneficial than PCU. Although, examples of contrasting effects have been found (Fitts et al. 2014; Sanderson and Fillmore 2012).

### 11.2.3 OTHER SLOW-RELEASE PRODUCTS

Trenkel (2010) describes slow-release N fertilizers as those that are "low solubility compounds with a complex/high molecular weight" and result in slow microbial degradation to achieve the slow-release effect. Common products are UF and isobutlyedene-diruea (IBDU). Description of these types of products can be found in Shaviv (2001), Trenkel (2010), and Timilsena et al. (2015). Chien et al. (2009) consider the UF products as the most important in this category and indicate that the release of N from these products is complex, but also acknowledge that there are clear benefits to reducing $NH_3$ volatilization. In general, studies are lacking on both UF and IBDU products. Cahill et al. (2007) showed that a UF polymer provided a neutral to negative effect on wheat yields in North Carolina compared to split applications of UAN. In slight contrast, UF was shown to produce slightly (but not statistically) greater rice grain and straw yields compared to urea (Lyu et al. 2015). Sanderson and Fillmore (2012) showed no benefit of either UF or IBDU on carrot (*Daucus carota* L.) yield on Prince Edward Island, Canada. However, IBDU performed nearly as well as PCU and above urea in flooded rice in Spain (Carreres et al. 2003) and there is evidence that IBDU has slow-release properties that can reduce N leaching on sandy soil (Wang and Alva 1996). In general, these products are not widely used in large-scale agriculture, which is a function of the increased cost, lack of published field research trials, and lack of clarity on the predictability of N release.

## 11.3  NITROGEN STABILIZERS

Stabilized fertilizers are those that incorporate substances, in small quantities, which stabilize the N in urea and ammonium-based fertilizer by inhibiting their transformations by chemical or microbial action. Stabilization is one of the most efficient and cost effective methods of enhancing the efficiency of fertilizers (Timilsena et al. 2015). Nitrogen stabilizers is the general term used to describe products that include nitrification and urease inhibitors, which respectively inhibit the biological oxidation of ammoniacal-N to nitrate-N and the hydrolytic action on urea by the enzyme urease (Trenkel 2010; Guelfi 2017).

As opposed to controlled-release fertilizers, which are beneficial on a soil and crop basis based on release patterns, nitrification stabilizers stop part of the N cycle from occurring and prevent N loss only for a short period of time following timing of application. Evaluation of N stabilizers is concerned with the ability of the product to inhibit bacterial or enzymatic transformation of N. Products reviewed here are those that have been shown to stabilize N (through inhibition). There is an abundance of review papers on N stabilizers including Subbarao et al. (2006), Chien et al. (2009), Trenkel (2010), and Di and Cameron (2016). There are other key review papers that cover modes of action (McCarthy 1999), behavior in soils (Wolt 2000), and the relationship between inhibitors and $N_2O$ and methane emissions (Sahrawat 2004; Ruser and Shutlz 2015). There are products that are promoted as inhibitors without scientific validation. For example, maleic-itaconic acid copolymer products do not appear to function as a urease inhibitor as reviewed by Chien et al. (2014). These products will not be discussed.

### 11.3.1  NITRIFICATION INHIBITORS

Nitrification is a term used to define the biological oxidation of $NH_4^+$ to nitrite $\left(NO_2^-\right)$ and then to $NO_3^-$. It is carried out mainly and respectively by two groups of chemolithotropic bacteria, *Nitrosomonas* sp. and *Nitrobacter* sp., which are ubiquitous in soil (Subbarao et al. 2006; Di and Cameron 2016). Recently, the same ammonia monooxygenase gene, as found in *Nitrosomonas* spp, was discovered in the ammonia-oxidizing archaea population. The presence of archaea has been reported in most soils, especially acid soils, but their contribution to nitrification is not well known (Subbarao et al. 2012). With moderate temperature and soil water content, nitrification occurs in most soils within a few days or weeks after application. The rapid conversion of $NH_4^+$ to $NO_3^-$ in the soil limits the effectiveness of much of the applied N fertilizer, because $NH_4^+$, but not $NO_3^-$, is retained by negatively charged soil particles, thus $NO_3^-$ is much more mobile and prone to leaching than $NH_4^+$. Under saturated conditions, $NO_3^-$ is converted via denitrification to N oxides ($NO_x$), $N_2O$ [a powerful greenhouse gas (GHG)], and $N_2$. Thus, application of nitrification inhibitors will be the most beneficial when N is applied early in the growing season to prevent losses when there is little crop N uptake. However, the maintenance of N in the $NH_4^+$ form by the addition of nitrification inhibitors can cause higher N losses by $NH_3$ volatilization than that of conventional urea (Zaman et al. 2008; Trenkel 2010; Soares et al. 2012; Frame 2017). This is even more concerning when urea application with nitrification inhibitors is carried out under conditions favorable for the occurrence of volatilization of $NH_3$, like topdressed applications of urea especially where reduced or no-till is practiced (Trenkel 2010; Guelfi 2017). So, most nitrification inhibitor coated fertilizer needs to be incorporated into soil at least 5 to 10 cm during or immediately after application mechanically or through rainfall or irrigation (Subbarao et al. 2006; Trenkel 2010).

Nitrification inhibitors are compounds that are added to N fertilizers during the production process or added or mixed immediately prior to N fertilizer application. There are many compounds with potential to be used as a nitrification inhibitor. These include nitrapyrin [2-chloro-6-(trichloromethyl-pyridine)], DCD (dicyandiamide), DMPP (3,4-dimethylpyrazole phosphate), ATC (4-amino-1,2,4-6-triazole-HCl), CI-1580 (2,4-diamino-6-trichloro-methyltriazine), TU (thiourea), MT (1-mercapto-1,2,4-triazole), AM (2-amino-4-chloro-6-methyl-pyramidine),

ASU (1-amide-2-thiourea), ATS (ammonium-thiosulphate), HPLC (1H-1,2,4-triazole), Terrazole (5-ethylene oxide-3-trichloro-methlyl,2,4-thiodiazole), 3-MP (3-methylpyrazole), CMP (1-carbamoyle-3-methyl-pyrazole), and Neem (oil or cake from seeds of *Azadirachta indica* A. Juss) (Subbarao et al. 2006; Trenkel 2010; Thapa et al. 2016). Sometimes mixtures of more than one of the aforementioned products may be used (Trenkel 2010; Hu et al. 2014). According to Di and Cameron (2016), nitrification inhibitors are one of the few environmental technologies that provide an economic benefit by increasing NUE and crop yield to offset the cost of implementing the technology. However, only a very few of these nitrification inhibitors have been adopted extensively in the agriculture (Subbarao et al. 2006; Trenkel 2010). Nitrapyrin, DCD, and DMPP are the main compounds now extensively used (Chien et al. 2009) (Table 11.3).

Nitrapyrin is widely used, especially in North America (Subbarao et al. 2006), and the chemical nature is thoroughly described in Wolt (2000). Nitrapyrin kills the bacteria *Nitrosomonas*, leading to inhibition of nitrification until the *Nitrosomonas* repopulates or the N is transported to other areas of the soil (Trenkel 2010); nitrapyrin does not kill or inhibit *Nitrobacter*. Nitrapyrin typically degrades within 30 days or less if applied during the crop growing season (Subbarao et al. 2006; Trenkel 2010; Burzaco et al. 2014), and many field studies show inhibition up to 35 days after application, although environmental conditions dictate how effective the product will be (Wolt 2000). Regular fertilizer rates can be reduced when applying nitrapyrin in this situation, although use of nitrapyrin without reducing N rates may also lead to greater yields (Wolt 2004). The effectiveness of nitrapyrin is longer with cooler soils, and thus can be useful with fall or winter N fertilizer applications (Subbarao et al. 2006; Trenkel 2010).

Nitrapyrin can be added to any ammonium-based fertilizer such as ammonium sulfate, ammonium nitrate, urea, UAN solutions, anhydrous ammonia, and liquid animal manures. However, the incorporation of nitrapyrin into conventional N fertilizers, such as impregnating into urea during development, is difficult due to nitapyrin's high vapor pressure, but decreasing the vapor pressure will reduce its nitrification inhibiting properties (Trenkel 2010). Frye et al. (1981) found that nitrapyrin was effective at inhibiting nitrification when directly sprayed onto urea or ammonium nitrate granules. Nitrapyrin is also available in an encapsulated formulation, which prevents nitrapyrin losses by evaporation when surface applied, allowing for more time to occur between application and incorporation without risk of loss (Trenkel 2010). A review by Subbarao et al. (2006) indicates that nitrapyrin is stable between pH values of 2.7 to 11.9; the authors also suggest that nitrapyrin may lose effectiveness when pH is above 6.5 (due to hydrolysis) or in soils with high organic matter (due to sorption). However, there does not appear to be a clear relationship between nitrapyrin inhibition effectiveness and soil surface pH reported in the literature. It should be noted that in the United States, nitrapyrin application is only allowed for corn, sorghum [*Sorghum bicolor* (L.) Moench], and wheat.

DCD is very soluble in water and non-volatile (unlike nitrapyrin), and while it inhibits *Nitrosomonas*, unlike nitrapyrin, it does not outright kill the bacteria, but instead suppresses their biological activity (Subbarao et al. 2006; Trenkel 2010; Di and Cameron 2016). DCD does not have an inhibitory effect on other soil microbial communities, such as methanotrophs, and does not adversely affect other important enzymatic activities after multi-year or long-term use (Subbarao et al. 2006; Di and Cameron 2016). Depending on the amount of mineral N applied and the temperature, water content, soil organic matter (SOM), and pH of the soil, the inhibitory effect of DCD on nitrification lasts for several weeks (4 to 10) (Subbarao et al. 2006; Trenkel 2010). Because DCD is non-volatile, it is relatively more suitable for tropical climates than products like nitrapyrin (Subbarao et al. 2006).

DCD is suitable for use as coatings on solid N fertilizers such as urea and ammonium sulfate or for incorporation with solid, liquid, or suspension N fertilizers such as UAN or liquid manure (Subbarao et al. 2006; Trenkel 2010). A major advantage of DCD over other potential nitrification inhibitors is its low vapor pressure and high water solubility (Di and Cameron 2016). These properties make DCD an ideal nitrification inhibitor for application to soils in a liquid or slurry form (Di and Cameron 2002, 2005). For incorporation of DCD into anhydrous ammonia, special high-pressure equipment is necessary (Trenkel 2010). One of the major limitations of DCD is that it is quite mobile in soil and readily leaches away from the fertilizer and potentially contaminating aquifers (Subbarao et al. 2006). For example, DCD has been shown to increase $NO_3^-$ leaching losses in a pot study on alkaline calcareous soil (Mahmood et al. 2011).

DMPP is a relatively new nitrification inhibitor, produced as a powder and is soluble in water (Trenkel 2010; Di and Cameron 2016). DMPP will generally inhibit nitrification for 4–10 weeks after application, but the inhibition time is temperature dependent (Zerulla et al. 2001) and it can be used on solid or liquid fertilizers, or in a slurry (Di and Cameron 2016). Like other nitrification inhibitors such as nitrapyrin and DCD, DMPP is persistent and effective in inhibiting nitrification at 5°C, but at 20°C the inhibitory effect from DMPP lasts only for 40 days (Zerulla et al. 2001). However, according to Trenkel (2010), satisfactory efficiency of DMPP under high temperature (25°C) was also found in an experiment under laboratory conditions. It is a more potent nitrification inhibitor than DCD and thus an application rate of 0.5–1.5 kg ha$^{-1}$ is sufficient to provide effective inhibition (Zerulla et al. 2001). DMPP also has lower mobility than DCD and nitrapyrin, and has almost the same mobility of $NH_4^+$ (and sorbs readily to clay particles) and thus will generally not leach away from the N fertilizer (Subbarao et al. 2006; Trenkel 2010). However, application with animal urine has been shown to increase the leaching of DMPP on organic soils (Marsden et al. 2016). An even newer nitrification inhibitor recently tested is 3,4-dimethylpyrazole succinic acid (DMPSA), which is similar to DMPP except for the succinic group (Huérfano et al. 2016; Pacholski et al. 2016). The succinic acid portion needs to be degraded to release the reactive compound, which may result in longer availability of the inhibitory compound in the soil. Further testing of this inhibitor under a range of conditions is needed to determine its efficacy.

Neem cake or neem oil, extracted from seed from a neem tree, is used to coat the urea prill. The neem product functions as a nitrification inhibitor (Vyas et al. 1991; Kashiri and Kumar 2017), and functions under both aerobic and anaerobic conditions. A meta-analysis by Linquist et al. (2013) showed that neem led to greater rice yields compared to unprotected N (22 observations and 9 studies). It appears that most of neem research exists in non-peer reviewed publications. For example, Neem Foundation (www.neemfoundation.org) hosts the World Neem Conference and publishes a book that includes research reports for each conference.

Nitrification inhibitors have been the subject of many meta-analysis papers to determine their overall value to production and the environment (Table 11.1). A meta-analysis from Yang et al. (2016) clearly shows that DCD and DMPP were effective at increasing soil NH$_4$-N content in soil compared to unprotected N fertilizer. Wolt (2004) showed that nitrapyrin increased soil N retention as well, by 28% compared to unprotected N fertilizer. In addition, Wolt (2004) found that nitrapyrin decreased GHG emissions by 51% relative to N fertilization without nitrapyrin. Akiyama et al. (2010) showed nitrapyrin, DCD, and DMPP were all effective at reducing $N_2O$ emissions in upland and grassland soils and both DCD and neem have been shown to increase yield (Linquist et al. 2013) and reduce $N_2O$ emissions (Akiyama et al. 2010) in rice production systems. Thapa et al. (2016) also found that nitrapyrin, DCD, DMPP, and neem reduced $N_2O$ emissions in grain production systems, but found that only DCD and DMPP significantly increased grain yields. In furthering the work by Akiyama et al. (2010), Gilsanz et al. (2016) showed that both DCD and DMPP were still effective in reducing $N_2O$ emissions, with reductions of 42.3 and 40.2%, respectively. Soares et al. (2015) found that the addition of DCD and DMPP to urea similarly reduced $N_2O$ emissions by more than 90%, which did not differ from those plots without N. DCD has also been shown to be effective in both reducing $NO_3^-$ leaching and $N_2O$ emissions in grazed pastures (Luo et al. 2010; Di and Cameron 2016). Although in contrast, DCD containing products were not effective in reducing seasonal $N_2O$ emissions in Iowa (Parkin and Hatfield 2014).

The reduction of $N_2O$ emissions by DMPP is attributed to reduced nitrification rates and a reduced $NO_3^-$ pool available for denitrification (Hatch et al. 2005; Ruser and Schulz 2015). The reduction of $NO_3^-$ concentrations by DMPP also causes a shift in the $N_2$:$N_2O$ ratio toward $N_2$ (Hatch et al. 2005). Wu et al. (2017) observed that the application of DMPP significantly reduced both $N_2O$ and NO emissions at three soil moisture conditions (50, 65, and 80% soil water-filled pore space) and increased $N_2$ emissions at 80% soil water-filled pore space. However, Friedl et al. (2017) found that DMPP reduced these $N_2$ losses by more than 70% in pasture field when used with urea, but had no effect on $N_2O$ emissions. According to Lam et al. (2017), nitrification inhibitors decrease both direct emissions of $N_2O$, and indirect emissions related to the decrease in $NO_3^-$ leaching, but can lead to an increase in indirect $N_2O$ emissions associated with increased $NH_3$ emissions and subsequent deposition. Furthermore, Weiske et al. (2001) reported that DMPP stimulated methane ($CH_4$) oxidation in soil. In contrast, Hatch et al. (2005) and Maris et al. (2015) reported that $CH_4$ emissions were larger when DMPP was applied leading to contradictory results regarding the effect of DMPP on $CH_4$ emissions.

### 11.3.2 Urease Inhibitors

Urea is a ubiquitous, organic N compound that exists in animal urine and soils (Fisher et al. 2017). Urea-N based products are the most commonly used conventional N fertilizer in world agriculture, especially because urea has a high N content (46% N), low cost, high water solubility, and is easy to handle and apply (Chien et al. 2009; Trenkel 2010). Urea is rapidly converted by the enzyme urease in the presence of water, creating ammonium bicarbonate, which then reacts with hydrogen ions to form $NH_4^+$. Ammoniacal-N in soil exists as an equilibrium between $NH_4^+$ and $NH_3$ depending on the soil pH. If this reaction takes place at the soil surface, significant losses of $NH_3$ due to volatilization will occur. Soils with low cation exchange capacity (CEC) and low SOM have a high potential for $NH_3$ volatilization losses. Ammonia losses are normally higher under conditions of high wind, moist soil surface, high temperature, low plant canopy, and the presence of plant residue (Chien et al. 2009). According to Ferm (1998), worldwide $NH_3$ losses range from 10 to 19% (average 14%) of the applied N fertilizer. Cantarella et al. (2008) found that $NH_3$ volatilization losses from the urea varied from 1% (rainy days after fertilization) to 25% of the applied N. However, in some cases, as observed by Lara Cabezas et al. (1997) who applied urea over humid crop residues in a no-till system without significant rainfall in the days following fertilizer application, N losses by $NH_3$ volatilization can reach 78% of the applied N. Thus, the agronomic disadvantage of urea is related to high potential N losses by $NH_3$ volatilization when urea is applied over the soil surface or crop residues, especially in neutral, alkaline, and flooded soils; when rainfall does not occur soon after application; and during the early stage of plant growth (Sangoi et al. 2003; Chien et al. 2009; Upadhyay 2012).

Urease exists in soil as an extracellular enzyme produced by bacteria, fungi, archaea, and plants and is sorbed to and stabilized by clay particles (Krajewska 2009; Fisher et al. 2017). The activity of urease in the soil is a function of temperature, humidity, CEC, soil pH, and if crop residues are present on the soil surface (Frankenberger Jr. and Tabatabai 1982; Watson et al. 2008; Frame 2017; Silva et al. 2017). The use of urease inhibitors to reduce $NH_3$ volatilization from urea hydrolysis has been considered a cost-effective strategy to increase NUE of urea-based N products (Chien et al. 2009). According to Timilsena et al. (2015), the intent of using urease inhibitors is to reduce the conversion of urea to $NH_4$-compounds by reducing soil urease activity until adequate water is available through rain or irrigation. Many chemical compounds have been tested for urease inhibition, but few meet the criteria listed by Trenkel (2010) as being nontoxic, effective at low concentrations, degradable in soil, and inexpensive. Thus, only three widely studied urease inhibitors are N-(n-butyl) thiophosphoric triamide (NBPT), hydroquinone (HQ), and phenylphosphorodiamidate (PPD or PPDA) (Chien et al. 2009; Trenkel 2010).

The consensus is that NBPT is the most effective urease inhibitor (Trenkel 2010; Chien et al. 2009; Soares et al. 2015; Silva et al. 2017). NBPT has solubility and diffusivity properties similar to those of urea, which is important as movement of both compounds will be similar in soil (Chien et al. 2009). Inhibition occurs due to the formation of chemical bonds between the urease inhibitor

(O and $NH_2$) atoms and the active site of the enzyme (Guelfi 2017). According to Manunza et al. (1999), NBPT coordinates both nickel atoms of the urease active site and binds the oxygen atom of the urea-derived carbamate. NBPT is a non-toxic compound, its use has not been shown to have environmental or personal safety risks, and it readily degrades into its constituent elements (Trenkel 2010). The liquid formulation of NBPT can be injected into molten urea before granulation, applied to the surface granules or prills in batch- or continuous-processes, or added to a UAN solution (Trenkel 2010). Watson et al. (2008) indicates that incorporating NBPT in the urea melt produces a more homogeneous product with superior stability than coating the urea granule.

Nearly all studies have shown that NBPT is efficient in reducing N losses by $NH_3$ volatilization. Soares et al. (2012) showed that use of NBPT both delayed peak $NH_3$ volatilization losses and reduced loss by 54–78% compared to untreated urea. Evaluating several studies published between 1990 and 2014, Silva et al. (2017) found that on average the use of NBPT reduced $NH_3$ volatilization by 52% when compared to untreated urea and delayed peak $NH_3$ loss by 3.5 days. The authors also note that the range in $NH_3$ loss reduction with NBPT was wide, ranging from 0 to 94%. In a meta-analysis focused only on Chinese studies, urease inhibitors were shown to reduce $NH_3$ emissions by over 70% on average (Huang et al. 2016). Silva et al. (2017) also determined that NBPT-treated urea reduced $NH_3$ volatilization loss across all soil pH classes, soil texture classes, SOM contents, N rates, and NBPT concentrations. Previous research has shown the importance of soil property interactions. For example, soil with low organic matter content and high pH has a greater response to NBPT application rate compared to other soils (Watson et al. 1994).

Silva et al. (2017) reported an average crop yield increase of 5.3% for NBPT-treated urea compared with urea, but yield increases were only found on fine textured soils. Other studies have shown similar yield increases with use of NBPT: >10% increase (Abalos et al. 2014), ~5% increase (Huang et al. 2016); ~1% increase in grain crops (Thapa et al. 2016), and >5% increase in rice (Linquist et al. 2013). Abalos et al. (2014) also observed that pH did not affect NBPT yield increases. Some studies have reported phytotoxic effects in plants treated with urea plus NBPT with the development of leaf scorches and necrotic leaf margins, which could be associated with an excessive uptake of N, but these effects are short-lived (Trenkel 2010; Artola et al. 2011; Cruchaga et al. 2011). Zanin et al. (2015) found that the presence of NBPT around roots inhibited urea uptake and assimilation. It is unclear if use of NBPT in agricultural settings, where it is typically applied to the soil surface would result in similar issues, as opposed to the hydroponically grown plants in the Zanin et al. (2015) study.

Urease and nitrification inhibitors can also be used together with N applications; the most common use is DCD with NBPT, but the effects have been variable (Table 11.3). Maharjan et al. (2016b) reported that under irrigated conditions on a sandy soil, a preplant application of DCD with NBPT (labeled as IU in Figure 11.1) improved corn yield compared with a preplant application of uncoated urea. However, yields with the stabilizers were lower than yields when urea was split-applied, but was similar to yield with preplant applied PCU (Figure 11.1). Of interest, in that same study, yields were not affected by N source or timing under water-stressed conditions, presumably due to less nitrate leaching when irrigation was limited (Figure 11.1). Jantalia et al. (2012) found that nitrification inhibitors could improve the beneficial effect of NBPT in reducing the loss of $NH_3$ due to volatilization when applied together. Clay et al. (1990), Lee et al. (1999), and Dillon et al. (2012) found that DCD did not affect $NH_3$ volatilization when compared to urea plus NBPT. However, Soares et al. (2012) found that the addition of DCD to urea with NBPT greatly increased $NH_3$ volatilization. The authors suggest that the DCD created greater $NH_4^+$ concentrations and soil pH at the soil surface for a longer period of time compared to urea without DCD, causing the NBPT to be less effective. Interestingly, the dual use still resulted in less $NH_3$ emissions when compared to DCD or urea alone. Zaman et al. (2008) determined that NBPT alone was more effective in reducing $NH_3$ emissions from urine than the addition of NBPT plus DCD. Frame (2017) found the application of NBPT together a nitrification inhibitor, DCD or nitrapyrin, continued to minimize the loss of $NH_3$ from urea, but N losses were still above levels observed when NBPT was applied alone; similar

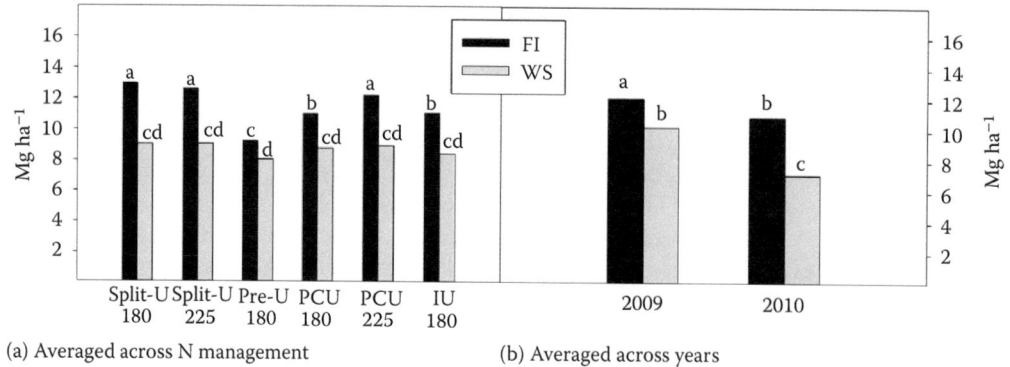

(a) Averaged across N management                    (b) Averaged across years

**FIGURE 11.1**  Mean corn grain yield for fully irrigated (FI) and water-stressed (WS) treatments in 2009 and 2010 under different N management systems, which included split-applied urea (Split-U), pre-plant applied urea (Pre-U), polymer-coated urea (PCU), and urea with inhibitors (IU, which included both the urease inhibitor NBPT and the nitrification inhibitor DCD) each at rates of 180 kg N ha$^{-1}$. Split-U and PCU were also applied at 225 kg N ha$^{-1}$. Results by N treatment are averaged across years (a) and results by year are averaged across N treatments (b). Means with the same letter are not significantly different ($P < 0.05$). (Originally published in Maharjan, B.C.J. Rosen, J.A. Lamb, and R.T. Venterea. 2016. Crop response to nitrogen management under fully-irrigated vs. water-stressed conditions. *Agron. J.* 108:2089–2098 and republished with permission from *Agronomy Journal.*)

results were found by Gioacchini et al. (2002) and Nastri et al. (2000). Gioacchini et al. (2002) hypothesized that the DCD primes soil organic mineralization through the maintenance of greater $NH_4^+$ concentration in soil solution, thus further promoting more $NH_4^+$ at the soil surface.

The dual use of DCD and NBPT may also have positive effects in reducing $N_2O$ emissions. Decock (2014) identified, using meta-analysis, that dual use of NBPT and DCD was the only management practice that consistently reduced $N_2O$ emissions but also acknowledges the limited number of studies used to draw this conclusion. Thapa et al. (2016) found that dual use of NBPT and DCD led to only slight yield gains in grain systems, but significantly reduced $N_2O$ emissions. Furthermore, dual use was particularly beneficial in reducing $N_2O$ in coarse textured soils, alkaline soils, and irrigated cropping systems. Linquist et al. (2013) determined, through meta-analysis, that dual use of urease and nitrification inhibitors significantly increase corn yields. In contrast, Parkin and Hatfield (2014) did not find a reduction in seasonal $N_2O$ emissions with dual use of NBPT and DCD compared to urea or UAN.

Other urease inhibitors exist, but are not as widely used as NBPT. Hydroquinone has been shown to be a urease inhibitor (Bremner and Douglas 1971). However, HQ has some disadvantages: apparent toxicity, mutagenic and carcinogenic action, negative effect on germination, and is photosensitive, which must be considered when urea is treated with HQ (Trenkel 2010). Some studies also have demonstrated that other inhibitors such as PPD and NBPT are more effective than HQ in retarding the hydrolysis of urea in the soil (Bremner 1995; Chien et al. 2009). PPD was considered a potent urease inhibitor that received extensive investigation. However, PPD showed inconsistent results in field tests, having high efficiency in some cases, but not in others (Watson 2000). In addition, PPD decomposes rapidly in the soil (Bremner 1995).

According to Modolo et al. (2015), many organic compounds, especially plant derivatives, have potential to be used as urease inhibitors. However, the identification and isolation of the major constituents of promising plant extracts, evaluation of the mechanism of action of the pure natural compounds, and production of the promising compounds in large scale when the availability is limited in nature are some challenges. Some inorganic elements like boron (B), copper (Cu), and zinc (Zn) are able to inhibit urease activity when added to urea at rates compatible with nutrient recommendations (Chien et al. 2009; Upadhyay 2012; Guelfi 2017). In a laboratory experiment,

Junejo et al. (2011) found that the addition of 5% Cu or 5% Cu plus 5% Zn in urea significantly reduced $NH_3$ loss compared to untreated urea. Guelfi (2017) reported that urea should be treated with boric acid and copper sulfate in the proportion of 1.5–2.4% and 0.6–1.5%, respectively. The principle of reducing urease activity is explained by the effect of uncompetitive and competitive inhibition of B and Cu, respectively. The inhibition promoted by $Cu^{2+}$ can be explained by the reaction of this ion with a sulfhydryl urease group, blocking the active site of the enzyme. Cancellier et al. (2016) found that urea with 0.15% Cu (copper sulfate) and 0.4% B (boric acid) decreased $NH_3$ volatilization compared to untreated urea.

## 11.4 SUMMARY AND CONCLUSION

There is a vast array of EEFs and they each can lead to an improvement in NUE if used under specific environmental conditions. In general, PCU will have the best return on investment on sandy or waterlogged soils, especially if fertilizer can only be applied early in the growing season. Use of PCU on sandy soils can also reduce the need for split application. The N stabilizer products reviewed here, specifically nitrapyrin, DCD, DMPP, and NBPT, are all well-studied products with clear inhibitory effects. The use of NBPT should be considered for all surface applicants of urea. Nitrification inhibitors use ensures against N loss in years when weather conditions promote faster nitrification and leaching. Unfortunately, there is no guarantee that there will be a return on investment related to the additional cost of any EEF product. The additional cost of the product compared to conventional fertilizers and lack of predictable benefit has limited the widespread adoption of EEFs products. This occurs despite clear benefits being reported by multiple meta-analysis studies, which show the EEFs can lead to yield increases, reduce $NO_3^-$ leaching, reduce gaseous losses of N, and reduce the contribution of agricultural systems to global GHG emissions. These meta-analyses help to point us to the broad conclusion that these products are beneficial, but it is likely that more regional research is needed to help farmers and land-owners to decide if they should invest in these products. The types of studies that are missing from the literature are those that compare EEFs to conventional fertilizer across several N application rates. These studies would allow the use of regression analysis to identify differences in agronomically optimum N rates, and thus identify if N fertilizer rates could be lowered with the use of EEF which would help offset the additional costs associated with these products.

## REFERENCES

AAPFCO. 2014. Official publication No. 67. Association of American Plant Food Control Officials (AAPFCO), Inc. West Lafayette, Indiana.

Abalos, D., S. Jeffery, A. Sanz-Cobena, G. Guardia, and A. Vallejo. 2014. Meta-analysis of the effect of urease and nitrification inhibitors on crop productivity and nitrogen use efficiency. *Agric. Ecosyst. Environ.* 189:136–144.

Akiyama H., X. Yan, and K. Yagi. 2010. Evaluation of effectiveness of enhanced-efficiency fertilizers as mitigation options for $N_2O$ and NO emissions from agricultural soils: Meta-analysis. *Global Change Biol.* 16:1837–1846.

Artola E., S. Cruchaga, I. Ariz, J.F. Moran, M. Garnica, F. Houdusse, J.M.G. Mina, I.I.B. Lasa, and P.M. Aparicio-Tejo. 2011. Effect of N-(n-butyl) thiophosphoric triamide on urea metabolism and the assimilation of ammonium by *Triticum aestivum* L. *Plant Growth Regul.* 63:73–79.

Awale, R. and A. Chatterjee. 2017. Enhanced efficiency nitrogen products influence ammonia volatilization and nitrous oxide emission from two contrasting soils. *Agron. J.* 109:47–57.

Azeem, B., K. KuSharri, Z.B. Man, A. Basit, and T.H. Thanh. 2014. Review on materials and methods to produce controlled release coated urea fertilizer. *J. Controlled Release* 181:11–21.

Beres, B.L., R.H. McKenzie, R.E. Dowbenko, C.V. Badea, and D.M. Spaner. 2012. Does handling physically alter the coating integrity of ESN urea fertilizer? *Agron. J.* 104:1149–1159.

Bero, N.J., M.D. Ruark, and B. Lowery. 2014. Controlled-release fertilizer effect on potato and groundwater nitrogen in sandy soil. *Agron. J.* 106:359–368.

Bierman, P.M., J.E. Crants, and C.J. Rosen. 2015. Evaluation of a quick test to assess polymer-coated urea prill damage. *Agron. J.* 107:2381–2390.

Blackshaw, R.E., X. Hao, K.N. Harker, J.T. O'Donovan, E.N. Johnson, and C.L. Vera. 2011. Barley productivity response to polymer-coated urea in a no-till production system. *Agron J.* 103:1100–1105.

Bremner, J.M. 1995. Recent research on problems in the use of urea as a nitrogen fertilizer. *Fertil. Res.* 42:312–329.

Bremner, J.M., and L.A. Douglas. 1971. Inhibition of urease activity in soils. *Soil Biol. Biochem.* 3:297–307.

Burzaco, J.P., I.A. Ciampitti, and T.J. Vyn. 2014. Nitrapyrin impacts on maize yield and nitrogen use efficiency with spring-applied nitrogen: Field studies vs. meta-analysis comparison. *Agron. J.* 106:753–760.

Cahill, S., D. Osmond, C. Crozier, D. Israel, and R. Weisz. 2007. Winter wheat and maize response to urea ammonium nitrate and a new urea formaldehyde polymer fertilizer. *Agron. J.* 99:1645–1653.

Cahill, S., D. Osmond, R. Weisz, and R. Heiniger. 2010. Evaluation of alternative nitrogen fertilizers for corn and winter wheat production. *Agron. J.* 102:1226–1236.

Cambouris, A.N., M. St. Luce, B.J. Zebarth, N. Ziadi, C.A. Grant, and I. Perron. 2016. Potato response to nitrogen sources and rates on an irrigated sandy soil. *Agron. J.* 108:391–401.

Cancellier, E.L., D.R.G. Silva, V. Faquin, B.A. Gonçalves, L.L. Cancellier, and C.R. Spehar. 2016. Ammonia volatilization from enhanced-efficiency urea on no-till maize in Brazilian Cerrado with improved soil fertility. *Cienc. Agrotec.* 40:133–144.

Cantarella, H., P.C.O. Trivelin, T.L.M. Contin, F.L.F. Dias, R. Rossetto, R. Marcelino, R.B. Coimbra, and J.A. Quaggio. 2008. Ammonia volatilisation from urease inhibitor-treated urea applied to sugarcane trash blankets. *Sci. Agric.* 65:397–401.

Carreres, R., J. Sendra, R. Ballesteros, E.F. Valiente, A. Quesada, D. Carrasco, F. Leganes, and J.G. de la Cuadra. 2003. Assessment of slow release fertilizers and nitrification inhibitors in flooded rice. *Biol. Fertil. Soils* 39:80–87.

Carson, L.C. and M. Ozores-Hampton. 2012. Methods for determining nitrogen release from controlled-release fertilizers used for vegetable production. *HortTech.* 22:20–24.

Chien, S.H., D. Edmeades, R. McBride, and K.L. Sahrawat. 2014. Review of maleic-itaconic acid copolymer purported as urease inhibitor and phosphorus enhancer in soils. *Agron. J.* 106:423–430.

Chien, S.H., L.I. Prochnow, and H. Cantarella. 2009. Recent developments of fertilizer production and use to improve nutrient efficiency and minimize environmental impacts. *Adv. Agron.* 102:267–322.

Clay, D.E., G.L. Malzer, and J.L. Anderson. 1990. Ammonia volatilization from urea as influenced by soil temperature, soil water content, and nitrification and hydrolysis inhibitors. *Soil Sci. Soc. Am. J.* 54:263–266.

Connell, J.A., D.W. Hancock, R.G. Durham, M.L. Cabrera, and G.H. Harris. 2011. Comparison of enhanced efficiency nitrogen fertilizers for reducing ammonia loss and improving bermudagrass forage production. *Crop Sci.* 51:2237–2248.

Cruchaga S., E. Artola, B. Lasa, I. Ariz, I. Irigoyen, J.F. Moran, and P.M. Aparicio-Tejo. 2011. Short term physiological implications of NBPT application on the N metabolism of *Pisum sativum* and *Spinacea oleracea*. *J. Plant Physiol.* 168:329–336.

Dave, A.M., M.H. Mehta, T.M. Aminabhavi, A.R. Kulkarni, and K.S. Soppimath. 1999. A review on controlled release of nitrogen fertilizers through polymeric membrane devices. *Polymer-Plastics Technol. Engineer.* 38:675–711.

Decock, C. 2014. Mitigating nitrous oxide emissions from corn cropping systems in the Midwestern U.S.: potential and data gaps. *Environ. Sci. Technol.* 48:4247–4256.

Di, H.J., K.C. Cameron. 2002. The use of a nitrification inhibitor, dicyandiamide (DCD), to reduce nitrate leaching and nitrous oxide emissions in a simulated grazed and irrigated grassland. *Soil Use Manage.* 18:395–403.

Di, H.J., K.C. Cameron. 2005. Reducing environmental impacts of agriculture by using a fine particle suspension nitrification inhibitor to decrease nitrate leaching from grazed pastures. *Agric. Ecosyst. Environ.* 109:202–212.

Di, H.J. and K.C. Cameron. 2016. Inhibition of nitrification to mitigate nitrate leaching and nitrous oxide emissions in grazed grassland: A review. *J. Soils Sediments* 16:1401–1420.

Dillon, K. A., T. W. Walker, D. L. Harrell, L. J. Krutz, J. J. Varco, C. H. Koger, and M. S. Cox. 2012. Nitrogen sources and timing effects on nitrogen loss and uptake in delayed flood rice. *Agron. J.* 104:466–472.

Farmaha, B.S. and A.L. Sims. 2013. Yield and protein response of wheat cultivars to polymer-coated urea and urea. *Agron. J.* 105:229–236.

Ferm, M. 1998. Atmospheric ammonia and ammonium transport in Europe and critical loads: A review. *Nutr. Cycl. Agroecosyst.* 1:5–17.

Fisher, K.A., S.A. Yarwood, and B.R. James. 2017. Soil urease activity and bacterial urea gene copy numbers: Effect of pH. *Geoderma* 285:1–8.

Fitts, P.W., T.W. Walker, L.J. Krutz, B.R. Golden, J.J. Varco, J. Gore, J.L. Corbin, and N.A. Slaton. 2014. Nitrification and yield for delayed-flood rice as affected by a nitrification inhibitor and coated urea. *Agron. J.* 106:1541–1548.

Frame, W. 2017. Ammonia volatilization from urea treated with NBPT and two nitrification inhibitors. *Agron. J.* 109:378–387.

Frankenberger Jr., W.T. and M.A. Tabatabai. 1982. Amidase and urease activities in plants. *Plant Soil* 64:153–166.

Friedl, J., C. Scheer, D.W. Rowlings, M.T. Mumford, and P.R. Grace. 2017. The nitrification inhibitor DMPP (3,4-dimethylpyrazole phosphate) reduces $N_2$ emissions from intensively managed pastures in subtropical Australia. *Soil Biol. Biochem.* 108:55–64.

Fujinuma, R., N.J. Balster, and J.M. Norman. 2009. An improved model of nitrogen release for surface-applied controlled-release fertilizer. *Soil Sci. Soc. Am. J.* 73:2043–2050.

Frye, W.W., R.L. Blevins, L.W. Murdock, K.L. Wells, and J.H. Ellis. 1981. Effectiveness of nitrapyrin with surface-applied fertilizer nitrogen in no-tillage corn. *Agron. J.* 73:287–289.

Gao, X., C. Li, M. Zhang, R. Wang, and B. Chen. 2015. Controlled release urea improved the nitrogen use efficiency, yield, and quality of potato (*Solanum tuberosum* L.) on silt loamy soil. *Field Crops Res.* 181:60–68.

Gilsanz, C., D. Báez, T.H. Misselbrook, M.S. Dhanoa, and L.M. Cárdenas. 2016. Development of emission factor and efficiency of two nitrification inhibitors, DCD and DMPP. *Agric. Ecosyst. Environ.* 216:1–8.

Gioacchini, P., A.C. Marzadori, L.V. Antisari, and C. Gessa. 2002. Influence of urease and nitrification inhibitors on N losses from soils fertilized with urea. *Biol. Fertil. Soils* 36:129–135.

Guelfi, D. 2017. Fertilizantes nitrogenados estabilizados, de liberação lenta ou controlada. *Informações Agronômicas*, 157:1–14.

Halvorson, A.D. and M.E. Bartolo. 2014. Nitrogen source and rate effects on irrigated corn yields and nitrogen use efficiency. *Agron. J.* 106:681–693.

Halvorson, A.D., S.J. Del Grosso, and F. Alluvione. 2010. Nitrogen source effects on soil nitrous oxide emissions from irrigated no-till corn. *J. Environ. Qual.* 39:1554–1562.

Halvorson, A.D., S.J. Del Grosso, and C.P. Jantalia. 2011. Nitrogen source effects on soil nitrous oxide emissions from strip-till corn. *J. Environ. Qual.* 40:1775–1786.

Halvorson, A.D., C.S. Synder, A.D. Blaylock, and S.J. Del Grosso. 2014. Enhanced-efficiency nitrogen fertilizers: Potential role in nitrous oxide emission mitigation. *Agron. J.* 106:715–722.

Hartz, T.K. and R.F. Smith. 2009. Controlled-release fertilizer for vegetable production: The California experience. *HortTech.* 19:20–22.

Hatch, D., H. Trindade, L. Cardenas, J. Carneiro, J. Hawkins, D. Scholefield, and D. Chadwick. 2005. Laboratory study of the effects of two nitrification inhibitors on greenhouse gas emissions from a slurry-treated arable soil: Impact of diurnal temperature cycle. *Biol. Fertil. Soils* 41:225–232.

Hatfield, J.L. and T.B. Parkin. 2014. Enhanced efficiency fertilizers: Effect on agronomic performance of corn in Iowa. *Agron. J.* 106:771–780.

Hu, Y., M. Schraml, S. von Tucher, F. Li, and U. Schmidhalter. 2014. Influence of nitrification inhibitors on yields of arable crops: A meta-analysis of recent studies in Germany. *Int. J. Plant Prod.* 8: 33–50.

Huang, S., W. Lv, S. Bloszies, Q. Shi, X. Pan, and Y. Zeng. 2016. Effects of fertilizer management practices on yield-scaled ammonia emissions from croplands in China: A meta-analysis. *Field Crops Res.* 192:118–125.

Huérfano, X., T. Fuertes-Mendizábal, K. Fernández-Diaz, J.M. Estavillo, C. González-Murua, and S. Menéndez. 2016. The new nitrification inhibitor 3,4-dimethylpyrazole succinic (DMPSA) as an alternative to DMPP for reducing N2O emissions from wheat crops under humid Mediterranean conditions. *Eur. J. Agron.* 80:78–87.

Hyatt, C.R., R.T. Ventera, C.J. Rosen, M. McNearney, M.J. Wilson, and M.S. Dolan. 2010. Polymer-coated urea maintains potato yields and reduces nitrous oxide emissions in a Minnesota loamy sand. *Soil Sci. Soc. Am. J.* 74:419–428.

Jantalia, C.P., A.D. Halvorson, R.F. Follett, B.J.R. Alves, J.C. Polidoro, and S. Urquiaga. 2012. Nitrogen source effects on ammonia volatilization as measured with semi-static chambers. *Agron. J.* 104: 1595–1603.

Ji, Y., G. Liu, J. Ma, G. Zhang, H. Xu, and K. Yagi. 2013. Effect of controlled-release fertilizer on mitigation of $N_2O$ emission from paddy field in South China: A multi-year field observation. *Plant Soil* 371: 473–486.

Junejo, N., M.Y., Khanif, M.M., Hanfi, K.A. Dharejo, and Z.W.Y. Wan. 2011. Reduced loss of $NH_3$ by coating urea with biodegradable polymers, palm stearin and selected micronutrients. *African J. Biotech.* 10:10618–10625.

Kashiri, H.O. and D. Kumar. 2017. Coating of essential oils onto prilled urea retards its nitrification in soil. *Arch. Agron. Soil Sci.* 63:96–105.

Kaur, G., B.A. Zurweller, K.A. Nelson, P.P. Motavalli, and C.J. Dudenhoeffer. 2017. Soil waterlogging and nitrogen fertilizer management effects on corn and soybean yields. *Agron. J.* 109:97–106.

Kenway, E. 1998. Recent advances in controlled release of agrochemicals. *J. Macromol. Sci.* 38:365–390.

Krajewska, B. 2009. Ureases I. Functional, catalytic and kinetic properties: A review. *J. Mol. Catal. B Enzym.* 59:9–21.

Ladha, J.K., A. Tirol-Padre, C.K. Reddy, K.G. Cassman, S. Verman, D.S. Powlson, C. van Kessel et al. 2016. Global nitrogen budgets in cereals: A 50-year assessment for maize, rice, and wheat production systems. *Sci. Reports* 6:19355.

Lam, S.K., H. Suter, A.R. Mosier, and D. Chen. 2017. Using nitrification inhibitors to mitigate agricultural $N_2O$ emission: A double-edged sword? *Global. Change Biol.* 23:485–489.

Lara Cabezas, W.A.R., G.H. Korndörfer, and S.A. Motta. 1997. Volatilização de N-$NH_3$ na cultura de milho: II. avaliação de fontes sólidas e fluidas em sistema de plantio direto e convencional. *Rev. Bras. Cienc. Solo* 21:489–496.

Lee, J.H, H.J. Lee, and B.W. Lee. 1999. Effects of urease inhibitor, nitrification inhibitor, and slow-release fertilizer on nitrogen fertilizer loss in direct-seeding rice. *Korean J. Crop Sci.* 44:230–235.

Li, N., T. Ning, Z. Cui, S. Tian, Z. Li, and R. Lal. 2015. $N_2O$ emissions and yield in maize field fertilized with polymer-coated urea under subsoiling or rotary tillage. *Nutr. Cycl. Agroecosyst.* 102:397–410.

Liegel, E.A. and L.M. Walsh. 1976. Evaluation of sulphur coated urea (SCU) applied to irrigated potatoes and corn. *Agron. J.* 68:457–463.

Linquist, B.A., L. Liu, C. van Kessel, and K.J. van Groenigen. 2013. Enhanced efficiency nitrogen fertilizers for rice systems: Meta-analysis of yield and nitrogen uptake. *Field Crops Res.* 154:246–254.

Luo, J., C.A.M. de Klein, S.F. Ledgard, and S. Sagger. 2010. Management options to reduce nitrous oxide emissions from intensively grazed pastures: A review. *Agric. Ecosyst. Environ.* 136:282–291.

Lyu, X., Y. Yang, Y. Li, X. Fan, Y. Wan, Y. Geng, and M. Zhang. 2015. Polymer-coated tablet urea improved rice yield and nitrogen use efficiency. *Agron. J.* 107:1837–1844.

Maharjan, B., R.B. Ferguson, and G.P. Slater. 2016a. Polymer-coated urea improved corn response compared to urea-ammonium-nitrate when applied on a coarse-textured soil. *Agron. J.* 108:509–518.

Maharjan, B. C.J. Rosen, J.A. Lamb, and R.T. Venterea. 2016b. Crop response to nitrogen management under fully-irrigated vs. water-stressed conditions. *Agron. J.* 108:2089–2098.

Maharjan, B., R.T. Venterea, and C.J. Rosen. 2014. Fertilizer and irrigation management effects on nitrous oxide emissions and nitrate leaching. *Agron. J.* 106:703–714.

Mahmood, T., R. Ali, Z. Latif, and W. Ishaque. 2011. Dicyandiamide increases the fertilizer N loss from an alkaline calcareous soil treated with [15]N-labelled urea under warm climate and under different crops. *Biol. Fertil. Soils.* 47:619–631.

Manunza, B., S. Deiana, M. Pintore, and C. Gessa. 1999. The binding mechanism of urea, hydroxamic acid and N-(n-butyl)-phosphoric triamide to the urease active site. A comparative molecular dynamics study. *Soil Biol. Biochem.* 31:789–796.

Maris, S.C., M.R. Teira-Esmatges, A. Arbonés, and J. Rufat. 2015. Effect of irrigation, nitrogen application, and a nitrification inhibitor on nitrous oxide, carbon dioxide and methane emissions from an olive (*Olea europaea* L.) orchard. *Sci. Total Environ.* 15:966–978.

Marsden, K.A., A.J. Marín-Martínez, A. Vallejo, P.W. Hill, D.L. Jones, and D.R. Chadwick. 2016. The mobility of nitrification inhibitors under simulated ruminant urine deposition and rainfall: A comparison between DCD and DMPP. *Biol. Fertil. Soils* 52:491–503.

McCarthy, G.W. 1999. Modes of action of nitrification inhibitors. *Biol. Fertil. Soils* 29:1–9.

McKenzie, R.H., A.B. Middleton, P.G. Pfiffner, and E. Bremer. 2010. Evaluation of polymer-coated urea and urease inhibitor for winter wheat in southern Alberta. *Agron. J.* 102:1210–1216.

Modolo, L.V., A.X. Souza, L.P. Horta, D.P. Araújo, and A. Fátima. 2015. An overview on the potential of natural products as ureases inhibitors: A review. *J. Adv. Res.* 6:35–44.

Nash, P.R., K.A. Nelson, and P.P. Motavalli. 2015a. Corn response to drainage and fertilizer on a poorly drained, river bottom soil. *Agron. J.* 107:1801–1808.

Nash, P.K. Nelson, and P. Motavalli. 2015b. Reducing nitrogen loss with managed drainage and polymer-coated urea. *J. Environ. Qual.* 44:256–264.

Nash, P.R., K.A. Nelson, P.P. Motavalli, and S.H. Anderson. 2015c. Corn yield response to managed drainage and polymer-coated urea. *Agron J.* 107:435–441.

Nash, P.R., K.A. Nelson, P.P. Motavalli, and C.G. Meinhardt. 2012. Effects of polymer-coated urea application ratios and dates on wheat and subsequent double-crop soybean. *Agron. J.* 104:1074–1084.

Nastri, A., G. Toderi, E. Bernati, and G. Govi. 2000. Ammonia volatilization and yield response from urea applied to wheat with urease (NBPT) and nitrification (DCD) inhibitors. *Agrochimica* 44:231–239.

Naz, M.Y. and S.A. Sulaiman. 2016. Slow release coating remedy for nitrogen loss from conventional urea: A review. *J. Controlled Release* 225:109–120.

Nelson, K.A., P.P. Motavalli, and C.J. Dudenhoeffer. 2014. Cropping system affects polymer-coated urea release and corn yield response in claypan soils. *J. Agron. Crop Sci.* 200:54–65.

Nelson, K.A., S.M. Paniagua, and P.P. Motavalli. 2009. Effect of polymer coated urea, irrigation, and drainage on nitrogen utilization and yield of corn in a claypan soil. *Agron. J.* 101:681–687.

Noellsch, A.J., P.P. Motavalli, K.A. Nelson, and N.R. Kitchen. 2009. Corn response to conventional and slow-release nitrogen fertilizers across a claypan landscape. *Agron. J.* 101: 607–614.

Nyborg, M., S.S. Malhi, E.D. Solberg, and M.C. Zhang. 1999. Influence of polymer-coated urea on mineral nitrogen release, nitrification, and barley yield and nitrogen uptake. *Comm. Soil Sci. Plant Anal.* 30:1963–1974.

Pacholski, A., N. Berger, I. Bustamante, R. Ruser, G. Guardia, T. Mannheim. 2016. Effects of the novel nitrification inhibitor DMPSA on yield, mineral N dynamics and $N_2O$ emissions. Proceedings of the 2016 International Nitrogen Initiative Conference. Victoria, Australia, December 4–8. Available online: http://www.ini2016.com/pdf-papers/INI2016_Pacholski_Andreas.pdf (July 25, 2017).

Parkin, T.B. and J.L. Hatfield. 2014. Enhanced efficiency fertilizers: Effect on nitrous oxide emissions in *Iowa*. *Agron. J.* 106:694–702.

Rubin, J.C., A.M. Struffert, F.G. Fernandez, and J.A. Lamb. 2016. Maize yield and nitrogen use efficiency in upper Midwest irrigated sandy soils. *Agron. J.* 108:1681–1691.

Ruser, R., and R. Schulz. 2015. The effect of nitrification inhibitors on the nitrous oxide ($N_2O$) release from agricultural soils—A review. *Plant Nutr. Soil Sci.* 178:171–188.

Sahrawat, K.L. 2004. Nitrification inhibitors for controlling methane emissions from submerged rice soils. *Current Sci.* 87:1084–1087.

Sanderson, K.R. and S.A.E. Fillmore. 2012. Slow-release nitrogen fertilizer in carrot production on Prince Edward Island. *Can. J. Plant Sci.* 92:1223–1228.

Sangoi, L., P.R. Ernani, V.A. Lech, and C. Rampazzo. 2003. Volatilization of $N\text{-}NH_3$ influenced by urea application forms, residue management and soil type in lab conditions. (In Portuguese, with English abstract.) *Cienc. Rural* 33:687–692.

Shapiro, C., A. Attia, S. Ulloa, and M. Mainz. 2015. Use of five nitrogen source and placement systems for improved nitrogen management of irrigated corn. 80:1663–1674.

Shaviv, A. 2001. Advances in controlled-release fertilizers. *Adv. Agron.* 71:1–49.

Shaviv, A., S. Raban, and E. Zaidel. 2003. Modeling controlled nutrient release from polymer coated fertilizers: Diffusion release from single granules. *Environ. Sci. Technol.* 37:2251–2256.

Shoji, S., J. Delgado, A. Mosier, and Y. Miura. 2001. Use of controlled release fertilizers and nitrification inhibitors to increase nitrogen use efficiency and to conserve air and water quality. *Comm. Soil Sci. Plant Anal.* 32:1051–1070.

Silva, A.G.B., C.H. Sequeira, R.A. Sermarini, and R. Otto. 2017. Urease inhibitor NBPT on ammonia volatilization and crop productivity: A meta-analysis. *Agron. J.* 109:1–13.

Soares, J.R., H. Cantarella, and M.L.C Menegale. 2012. Ammonia volatilization losses from surface-applied urea with urease and nitrification inhibitors. *Soil Biol. Biochem.* 52:82–89.

Soares, J.R., H. Cantarella, V.P. Vargas, J.B. Carmo, A.A. Martins, R.M. Sousa, and C.A. Andrade. 2015. Enhanced-efficiency fertilizers in nitrous oxide emissions from urea applied to sugarcane. *J. Environ. Qual.* 44:423–430.

Struffert, A.M., J.C. Rubin, F.G. Fernandez, and J.A. Lamb. 2016. Nitrogen management for corn and groundwater quality in upper Midwest irrigated sands. *J. Environ. Qual.* 45:1557–1564.

Subbarao, G.V., K.L. Sahrawat, K. Nakahara, T. Ishikawa, M. Kishii, I. M. Rao, C.T. Hash et al. 2012. Biological nitrification inhibition—A novel strategy to regulate nitrification in agricultural systems. *Adv. Agron.* 114:249–302.

Subbarao, G.V., O. Ito, K.L. Sahrawat, W.L. Berry, K. Nakahara, T. Ishikawa, T. Watanabe et al. 2006. Scope and strategies for regulation of nitrification in agricultural systems—Challenges and opportunities. *Crit. Rev. Plant Sci.* 25:303–335.

Thapa, R., A. Chatterjee, R. Awale, D.A. McGranahan, and A. Daigh. 2016. Effect of enhanced efficiency fertilizers on nitrous oxide emissions and crop yields: A meta-analysis. *Soil Sci. Soc. Am. J.* 80:1121–1134.

Timilsena, Y.P., R. Adhikari, P. Casey, T. Muster, H. Gilla, and B. Adhikaria. 2015. Enhanced efficiency fertilisers: A review of formulation and nutrient release patterns. *J. Sci. Food. Agric.* 95:1131–114.

Trenkel, M.E. 2010. *Slow- and controlled-release and stabilized fertilizers: An option for enhancing nutrient use efficiency in agriculture.* 2nd ed. International Fertilizer Industry Association, Paris, France. 163 p. Available online: http://www.fertilizer.org/imis20/images/Library_Downloads/2010_Trenkel_slow%20release%20 book.pdf?WebsiteKey=411e9724-4bda-422f-abfc-8152ed74f306&=404%3bhttp%3a%2f%2fwww .fertilizer.org%3a80%2fen%2fimages%2fLibrary_Downloads%2f2010_Trenkel_slow+release+book .pdf (July 26, 2017).

Upadhyay, L.S.B. 2012. Urease inhibitors: A review. *Indian J. Biotech.* 11:381–388.

Vyas, B.N., N.B. Godrej, and K.B. Mistry. 1991. Development and evaluation of neem extract as a coating for urea fertilizer. *Fert. News* 36:19–25.

Venterea, R.T., C.R. Hyatt, and C.J. Rosen. 2011. Fertilizer management effects on nitrate leaching and indirect nitrous oxide emissions in irrigated potato production. *J. Environ. Qual.* 40:1103–1112.

Wakimoto, K. 2004. Utilization advantages of controlled release nitrogen fertilizer on paddy rice cultivation. *Japan Agricultural Res. Quart.* 38:15–20.

Wang, F.L. and A.K. Alva. 1996. Leaching of nitrogen from slow-release urea sources in sandy soils. *Soil Sci. Soc. Am. J.* 60:1454–1458.

Wang, H., A.M. Hegazy, X. Jiang, Z. Hu, J. Lu, J. Mu, X. Zhang, and X. Zhu. 2016. Suppression of ammonia volatilization from rice-wheat rotation fields amended with controlled-release urea and urea. *Agron. J.* 108:1214–1224.

Wang, S., Z. Zhao, G. Xing, Y. Yang, M. Zhang, and H. Chen. 2015. Improving grain yield and reducing N loss using polymer-coated urea in southeast China. *Agron. Sustain. Dev.* 35:1103–1115.

Watson, C.J. 2000. Urease activity and inhibition—principles and practice. The International Fertilizer Society Meeting, November 28, 2000. London, The International Fertilizer Society Proceedings, n. 454, 39 p.

Watson, C.J., H. Miller, P. Poland, D.J. Kilpatrick, M.D.B. Allen, M.K. Garrett, and C.B. Chistianson. 1994. Soil properties and the ability of the urease inhibitor N-(n-butyl) thiophosphoric triamide (nBTPT) to reduce ammonia volatilization from surface-applied urea. *Soil Biol. Biochem.* 26:1165–1171.

Watson, C.J., N.A. Akhonzada, J.T.G. Hamilton, and D.I. Mathews. 2008. Rate and mode of application of the urease inhibitor N-(n-butyl) thiophosphoric triamide on ammonia volatilization from surface-applied urea. *Soil Use Manage.* 24:246–253.

Watts, D.B., G.B. Runion, K.W.S. Nannenga, and H.A. Torbert. 2014. Enhanced-efficiency fertilizer effects on cotton yield and quality in the coastal plains. *Agron. J.* 106:745–752.

Weiske, A., G. Benckiser, T. Herbert, and J.C.G. Ottow. 2001. Influence of the nitrification inhibitor 3,4-dimethylpyrazole phosphate (DMPP) in comparison to dicyandiamide (DCD) on nitrous oxide emissions, carbon dioxide fluxes and methane oxidation during 3 years of repeated application in field experiments. *Biol. Fertil. Soils* 34:109–117.

Wilson, M.L., C.J. Rosen, and J.F. Moncrief. 2009. Potato response to a polymer-coated urea on an irrigated, coarse-textured soil. *Agron. J.* 101:897–905.

Wilson, M.L., C.J. Rosen, and J.F. Moncrief. 2010. Effects of polymer-coated urea on nitrate leaching and nitrogen uptake by potato. *J. Environ. Qual.* 39:492–499.

Wolt, J.D. 2000. Nitrapyrin behavior in soils and environmental considerations. *J. Environ. Qual.* 29:367–379.

Wolt, J.D. 2004. A meta-evaluation of nitrapyrin agronomic and environmental effectiveness with emphasis on corn production in the Midwestern USA. *Nutr. Cycl. Agroecosys.* 69: 23–41.

Wu, D., L.M. Cárdenas, S. Calvet, N. Brüggemann, N. Loick, S. Liu, R. Bol. 2017. The effect of nitrification inhibitor on $N_2O$, NO and $N_2$ emissions under different soil moisture levels in a permanent grassland soil. *Soil Biol. Biochem.* 113:153–160.

Yang, M., Y. Fang, D. Sun, and Y. Shia. 2016. Efficiency of two nitrification inhibitors (dicyandiamide and 3, 4-dimethypyrazole phosphate) on soil nitrogen transformations and plant productivity: A meta-analysis. *Sci. Rep.* 6: 22075.

Yang, Y.C., M. Zhang, L. Zheng, D.D. Cheng, M. Liu, and Y.Q. Geng. 2011. Controlled release urea improved nitrogen use efficiency, yield, and quality of wheat. *Agron. J.* 103:479–485.

Yuan, M.W., M.D. Ruark, and W.L. Bland. 2017. Adaption of the AmaizeN model for nitrogen management in sweet corn (Zea mays L.). *Field Crops Res.* 209:27–38.

Zaman, M., M. Nguyen, J. Blennerhassett, and B. Quin. 2008. Reducing $NH_3$, $N_2O$ and $NO_3$–N losses from a pasture soil with urease or nitrification inhibitors and elemental S-amended nitrogenous fertilizers. *Biol. Fertil. Soils* 44:693–705.

Zanin, L., N. Tomasi, A. Zamboni, Z. Varanini, and R. Pinton. 2015. The urease inhibitor NBPT negatively affects DUR3-mediated uptake and assimilation of urea in maize roots. *Front. Plant Sci.* 6:1007.

Zebarth, B.J., E. Snowdon, D.L. Burton, C. Goyer, and R. Dowbenko. 2012. Controlled release fertilizer product effects on potato crop response and nitrous oxide emissions under rain-fed production on a medium-textured soil. *Can. J. Soil Sci.* 92:759–769.

Zerulla, W., T. Barth, J. Dressel, K. Erhardt, K.H. von Locquenghien, G. Pasda, M. Radle, and H. Wissemeier. 2001. 3,4-Dimethylpyrazole phosphate (DMPP)—A new nitrification inhibitor for agriculture and horticulture. *Biol. Fertil. Soils* 34:79–84.

Zheng, W., M. Zhang, Z. Liu, H. Zhou, H. Lu, W. Zhang, Y. Yang, C. Li, and B. Chen. 2016. Combining controlled-release urea and normal urea to improve the nitrogen use efficiency and yield under wheat-maize double cropping system. *Field Crops Res.* 197:52–62.

Zvomuya, F., C.J. Rosen, M.P. Russelle, and S.C. Gupta. 2003. Nitrate leaching and nitrogen recovery following application of polyolefin-coated urea to potato. *J. Environ. Qual.* 32:480–489.

# 12 Economic and Policy Implications of Nitrogen Management

*Otto C. Doering III, Benjamin M. Gramig, and Dawoon Jeong*

## CONTENTS

## 12.1 INTRODUCTION

Nitrogen (N) has played an increasingly important role in agricultural productivity and is essential to support the food needs of the world's population. Guano, the excrement of seabirds or cave dwelling bats, was a major initial source of N for nineteenth-century agriculture. Von Humboldt was the first European to investigate guano at the beginning of the nineteenth century and Peruvian guano soon became a major export to Europe and the United States. After 1870, Peruvian guano was increasingly replaced by sodium nitrate ($NaNO_3$) from the Atacama Desert in Peru or from Chile. Utilization was stimulated by Von Liebig's popularization of the Law of the Minimum. The $NaNO_3$ trade grew until replaced in the early twentieth century through Haber and Bosch's discovery for producing ammonia from the air. While agriculture may have used animal manures, Guano, or $NaNO_3$ as major sources of N for crop production in the past, the vast amounts of anthropogenic N used in crop production today are made possible by the Haber–Bosch process.

One critical factor about N is its ability to move over time in different forms and to different places. This cycling of N has been called the "nitrogen cascade" (Galloway et al. 2003). N applied

315

to the land for crop growth may be transformed and end up in aquatic ecosystems. It can move from the land or water to atmospheric systems where it may result in ozone depletion, transform into a potent greenhouse gas (GHG), or be redeposited on land or water. This makes the management of N difficult. Once N is in the ecosystem, the major challenge lies in how to manage it to reap the most benefit from its essential contributions to plant nutrition or control it to reduce the environmental and other damage it can cause when released as part of the N cycle in an uncontrolled form.

When gaining economic benefit or reducing economic cost from N, metrics become important in defining the problem and in identifying goals for policy (Birch et al. 2011). In agricultural production, the N inputs and production benefits of N are often considered in physical rather than in dollar value terms. Thus, the metrics for improving the primary goal of water quality in the Chesapeake Bay in the United States has been in terms of units like tons of N removed or prevented from getting into the Bay. Because agriculture is a large contributor in terms of physical units introduced into the Bay's ecosystem, its use of N is thus being constrained, and intensive management is being prescribed. If there were a different metric, that is, the economic value of the damage actually caused by N in the Chesapeake Bay area, then the problem is defined differently and the solution would be altered. The largest economic damage from N in the Chesapeake Bay area is from health impacts related to atmospheric N. That new economic metric changes the strategy to one of reducing atmospheric N rather than the flows of N from the terrestrial sources into the aquatic environment. In addition, it is initially less expensive to control the N from power plants and transportation equipment causing health problems due to air quality. Institutions and individuals do what they measure. So, we need to understand the impact of what is measured and its consequences, whether increasing our agricultural N utilization efficiency is our goal or whether increasing economic returns to N is our goal. Both are important, and both may be necessary. The argument here is that society may have to have goals for both and be able to understand and explain the trade-offs between them.

The focus of this chapter is on economic implications and policy with respect to N in agriculture. Decision making about N, given the "nitrogen cascade" and other factors, is a wicked problem that requires a different approach from normal science. Both deficits and excesses of N are of concern. There is a review of the economics of N use and such things as the crop to N price ratio. There are important instances of N deficiency where N use efficiency (NUE) and economic efficiency both become important goals and also have critical trade-offs. Important issues are involved when there are N deficiencies and many alternative policy instruments are possible for dealing with these deficits. The same is true where there is excess N. Some of these policy instruments are described and assessed. Finally, an example is provided of farm level management approaches that can improve economic returns and N use efficiency in the context of trade-offs with farmer's risk from weather.

## 12.2 DEFICIT OR EXCESS, BOTH WICKED PROBLEMS

### 12.2.1 WICKED CHARACTERISTICS

When people deal with a problem, they usually approach it in terms of the normal "tame" problem. They arrive at a clear definition of the problem, which often appears obvious. This straightforward definition of the problem then points them toward a solution or some set of clear alternative solutions. As they proceed along the solution path, the outcome is usually something like true, false, solved, or not solved. While they are working on the problem, it does not change over time and it can be conceptualized through the linear progression of the scientific method. Many of the basic scientific problems people address are tame problems and they can be structured into distinct steps to be taken that lead to a definitive solution. This is not true of the "wicked" problem. At the outset, there may not be agreement about what the problem is with a wicked problem. In addition, attempts to create a solution to the problem may change the nature of the problem as people work toward a solution. There is no definitive outcome with a wicked problem. Often, the best that can be done is to assess whether things are "better" or "worse" off than they were before or decide that the attempted solution is "good enough." While that is being done, the problem has likely changed over time.

The role of stakeholders for a wicked problem is different as well. For a tame problem, the cause of the problem is determined primarily by experts—often using scientific data. For a wicked problem, there are likely to be many different stakeholders who have different ideas about what the real problem is and what its causes might be. The nature of these problems is also different. In a tame problem, scientifically based protocols guide the choice of solutions. Tame problems have low uncertainty about system components and outcomes. There are also shared values as to the desirability of the various outcomes. The acceptable solution to a wicked problem may depend on judgments, values, and perceptions of multiple stakeholders. There is usually a high degree of uncertainty about system components and outcomes. It is also not likely that there are shared values with respect to societal goals. The important factor is that values and perceptions become important in all aspects of dealing with the problem. Therefore, experts are not allowed to dictate the problem definition, solution process, or outcome. These are negotiated, not determined analytically by experts. For the wicked problem, there is no stopping rule. Stakeholders, political forces, and resource availability become the determining factors as to the extent of the problem-solving process. Wicked problems' higher degree of uncertainty relates to the value conflicts and trade-offs that must be negotiated. There is uncertainty about causes and effects as well as system components and outcomes (Batie 2008). Wicked problems involve systems that are complex and interconnected. Sending a man to the moon is a tame problem. Such things as national healthcare, protecting the environment, and solving the negative impacts of N deficits or excesses represent truly wicked problems.

### 12.2.2  How a Nitrogen Deficit or Excess Can Be Wicked

When thinking of N as a nutrient in the context of agriculture, it is considered as a relationship between the nutrient and its role in plant growth. Yield trials in controlled situations on field stations provide data on basic relationships between yield changes from different amounts of N under given conditions. Additional considerations include such things as NUE and how different application methods and fertilizer types might affect yield. These kinds of questions are often relatively "tame." However, these tame questions become part of larger wicked problems when dealing with broader concerns about the impacts of N deficits or excesses on food production, sustainability, and agriculture's impact on the environment. Now policies must be considered that are designed to correct nutrient imbalances driven by broader sets of goals. It is here that one must consider multiple trade-offs and stakeholders with different definitions of the problem and objectives. China presents an excellent example of the wicked problem policy dilemma of excess N in agricultural production and Africa gives us the counter case of N deficits.

## 12.3  ECONOMICS OF NITROGEN USE

### 12.3.1  The Nitrogen Production Function

The economic importance of N to crop yields is most easily illustrated using a graph of crop yield ($y$) as a function of N application ($x$), holding all other inputs and conditions constant to focus on yield response to N. Figure 12.1 depicts yield on the vertical axis and N on the horizontal axis. As one moves from the origin to the right, increasing the quantity of N added to the crop as fertilizer, yield increases rapidly even at low levels of N. These yield increases are increasing at an increasing rate over some range of fertilization levels before the marginal gain in crop productivity slows, eventually leveling off (and potentially decreasing) at very high levels of N. This general yield-N relationship is common among many agricultural crops (Grassini et al. 2013). It simultaneously illustrates how important N is to: (1) improving crop yields—indeed, closing persistent yield gaps (Mueller et al. 2012) between more and less developed regions of the world—and (2) how consequential the gains from increasing NUE

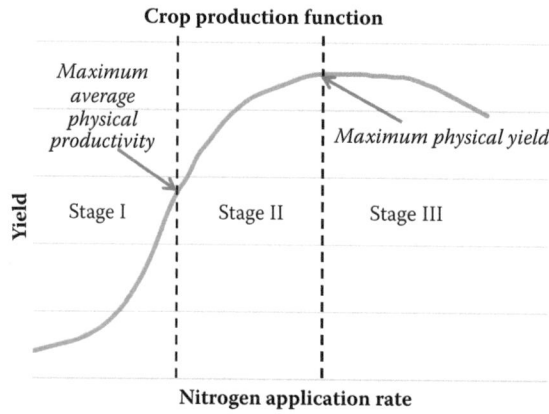

**FIGURE 12.1**   Crop yield response to nitrogen.

through management can be whenever and wherever N is a limiting input to yield. On land with very low soil organic matter (SOM) and inadequate N, yield is limited by insufficient amounts of N; crops demand more N than is available; and are unable to achieve their yield potential.

Economic analysis of a crop production function traditionally separates a yield response curve like that depicted in Figure 12.1 into three distinct stages. The rightmost section of the curve where the plot becomes flat over a range and may decline is referred to as Stage III of the production function, and begins at the level where one additional unit of N results in no increase in yield. Economically speaking, there is no marginal return from fertilizing at levels beyond where the curve becomes flat at the boundary between stages II and III; additional expenditures on fertilizer do not result in any increase in yield. Stage II encompasses the region from the maximum attainable physical yield identified in Figure 12.1, moving to the left along the curve down to the level of N where the average physical productivity of N is maximized. This happens at the level of N that corresponds to the maximum average physical productivity identified in Figure 12.1, beyond which the average change in yield from an increase in N begins to decline. Stage II is the production stage that is normally the focus of economically optimal management when input use is not constrained by, say, the inability to buy more fertilizer. Stage I is the region of Figure 12.1 from the origin to the beginning of Stage II, where the average physical productivity of increasing levels of N is *always* increasing.

Figure 12.1 does not include input and output prices that ultimately determine profitability. The economically optimal N rate (EONR) is the amount of N applied per unit of land area (kg N/ha) such that an incremental increase in the N rate costs the same amount as the increase in income from the resulting increase in yield. In other words, the economically optimal fertilization rate is the one where the net economic return from an incremental increase in N applied is zero. The EONR changes depending on the ratio of N input price to crop output price (holding all other input prices constant): a higher [lower] price ratio implies a lower [higher] EONR. N can have extremely high value in use if crop prices are high and N costs are low in relative terms.

N is often described as a "risky" input from an economic standpoint. It tends to be a high cost input to crop production. However, despite the risk of N loss to the environment before crops take up the N, the economic gain from increased yield that can be realized can be significant. Thus, if a farm is already operating in Stage II, the economic gains from increasing N inputs may be small but worth it relative to the risk of losing that N to the environment. A second important risk consideration in the economic management of N fertilizer relates to the timing of fertilizer application. Agronomic recommendations generally prescribe application of N as close as possible in time to the stages of crop development when the largest amount of the nutrient is demanded and taken up by a developing crop. Typically, this means applying the majority of N after planting by side dressing the soil adjacent to the crop already growing in the field. If soils become too saturated to allow fieldwork to be performed during the limited amount of time before the crop becomes too tall to side-dress fertilizer from the ground (as opposed to aerially, as from a plane), then the crop will suffer from potentially severe N deficiency and associated economic losses relative to applying all N before planting (Gramig et al. 2017). Both types of risk are driven by weather. Risk of N loss is greatest when warm, wet conditions occur *and* N is available as nitrate in the soil following pre-plant N application; and excessively wet conditions that are persistent post-planting will constrain days suitable for field working days to side-dress any required N not applied before planting.

### 12.3.2 Optimizing Nitrogen Use

The economic problem of optimizing the amount of N applied (subject to budget constraints) and when to apply it is fundamentally about NUE. This concept is rooted in plant biology and agronomic processes dependent on management, weather, and soil characteristics. Situations where there is either deficit *or* excess N can be addressed through management that improves NUE. In practice, N's high value in use and low cost relative to yield gain profit effectively creates incentives to over-apply N relative to the EONR as a form of insurance against loss of N before a crop can utilize it—to try to make sure there is enough N available when the plant needs it to avoid yield loss. This is contrary to NUE and is relevant mainly in the context of the excess N problem.

The wicked problem posed by *excess N* pertains especially to farms and regions of the world operating in Stage III of the crop production function where N is being applied at levels that are neither privately nor publicly economically efficient. Privately, farms are wasting money on N (or perhaps governments are wasting money on fertilizer subsidies) that will never contribute to yield increases that translate into increased revenue. Publicly, crops cannot take up excess N and there is a larger quantity of N available to be lost to the environment from the soil through leaching, surface runoff, and denitrification in gaseous forms. This negative public externality can manifest itself in water quality problems (e.g., eutrophication) or may contribute to health concerns or climate change.

The wicked problem of *deficit N* primarily pertains to farms operating in Stage I of a crop's production function where the gains in total crop productivity are increasing at the highest possible rates in the amount of N, but farmers are financially or otherwise constrained and cannot afford N to fertilize their crops. Producers who are financially constrained and cannot afford the amount of fertilizer required to achieve the economically optimal yield level that maximizes profit are behaving in an economically rational way by operating in Stage I of the production function. The global research community has called for sustainable or ecological intensification (e.g., Cassman 1999) of agricultural management and food production, in food insecure regions of the world in particular, to close gaps between current yields and yield potential by increasing crop productivity, part of which is reducing the N deficit (Mueller et al. 2012).

In summary, the core of the economic management problem lies in improving NUE (in the context of the EONR). For situations and areas of the world where there are insufficient private resources to buy enough N fertilizer to optimize production, it is even more important to utilize limited amounts of N in the most efficient way possible. This may require education or agricultural extension to help farmers utilize N in ways that minimize loss to the environment, potentially by adjusting timing, placement, and/or form of N fertilizer. Even where public subsidies can provide fertilizer to low-income farmers, farmers need training to take full advantage of this opportunity to improve yields that provide food security and income to households. In situations where N is being over-applied far beyond the level where the economic returns to additional N cover its cost, the same education or training could benefit society where subsidized fertilizer use is resulting in an excess N problem. In circumstances of excess N where there are no budget constraints and farms are operating in Stage II of their production function, alternative policy or market-based incentives are necessary to induce reduced or redistributed (over space and/or time) N use that reduces loss to the environment and the associated environmental and economic damages.

## 12.4  NUE AND ECONOMIC EFFICIENCY

### 12.4.1  The Case of Nitrogen Excess: China

In the 1950s, crop consumption in China outweighed its production level causing continual famine. Faced with a growing population, China was in need of supplying enough food to meet growing demand and deal with food shortages. Over the next five decades, China put significant effort into agriculture by implementing food-increasing policies, in particular N fertilizer subsidies and crop price increases, aimed at increasing food production and boosting rural agricultural economies. The Chinese government subsidized fertilizer industries in various ways that include manufacturing, transport, storage, and distribution. Through these subsidy programs, the fertilizer industry in China received a total subsidy of $0.95 billion in 2003, which later increased to $18.76 billion in 2010 (Li et al. 2014). In addition, the government supported corn (*Zea mays*) price in China was roughly double the United States corn price in early 2016 (USDA 2016). The low-priced input and sustained high output price both effectively resulted in increasing N fertilizer use and crop yields in China from an artificially high EONR. These policies also drove the nationwide trend of fertilizer overuse that farmers now consider a normal farming practice to ensure higher yield.

With N being applied at much more than its optimal level (for either NUE or EONR), production is taking place in stage III of the production function illustrated in Figure 12.1. Table 12.1 reports N overuse rates in various regions in China from the peer-reviewed literature. This over-application of N does not increase yield because Chinese producers are already operating in Stage III. At this point, N over-application in China is only contributing to massive N losses to the environment without yield increases or improved economic returns. Chinese NUE is much lower than that of other countries and the total amount of surplus N in China is the highest in the world (Table 12.2). Recently, the Chinese government has recognized this issue and has responded through agricultural policy changes such as lowering fertilizer subsidies and crop price supports.

### TABLE 12.1

**Optimal Nitrogen Rate and Farmer Nitrogen Rate (kg of N/ha) Used in China**

| Locations and Years | Crop | Optimal N Rate | Farmer N Rate Used | Reference |
|---|---|---|---|---|
| Nationwide (2005–2010) | Maize | 174 | 220 | Wu et al. (2014) |
| Near Beijing (1999–2003) | Summer maize | 191.5 | 300 | Zhao et al. (2006) |
| North China Plain (2003–2005) | Summer maize | 158 | 263 | Cui et al. (2008) |

### TABLE 12.2

**N Budget and NUE in Crop Production by Region in 2010**

| | Harvest N (Tg N yr$^{-1}$) | Input N (Tg N yr$^{-1}$) | NUE | Surplus N (Tg N yr$^{-1}$) |
|---|---|---|---|---|
| China | 13 | 51 | 0.25 | 38 |
| U.S. and Canada | 14 | 21 | 0.68 | 7 |
| Europe | 7 | 14 | 0.52 | 7 |
| Sub-Saharan Africa | 4 | 5 | 0.72 | 2 |
| Other OECD countries | 1 | 2 | 0.52 | 1 |

*Source:* Zhang, X., E.A. Davidson, D.L. Mauzerall, T.D. Searchinger, P. Dumas, and Y. Shen. 2015. Managing nitrogen for sustainable development. *Nature* 528: 51–59.

## 12.4.2   THE CASE OF NITROGEN DEFICIT: AFRICA

The population of the Sub-Saharan Africa (SSA) region is growing annually by 19 million people, which simultaneously necessitates growing food production at the same rate. To produce enough crops to feed 19 million additional people with the minimum diet of 2500 calories per day recommended by FAO, an additional 90,569 tons of N fertilizer is needed (Daniel et al. 2009). Despite this, N fertilizer consumption in SSA, one of the most essential nutrients for crop growth, has actually decreased over the last decade. Because of this very low N application rate, SSA's NUE is the highest in the world, exceeding 70% when the global average is around 40% (Table 12.2). For SSA countries in the Stage I production level as in Figure 12.1, more N input would dramatically increase yield. However, increasing N use is a highly complex issue since it is multi-constrained by such things as transportation, logistics, high cost of fertilizer relative to crop prices, and often risk avoidance behavior by farmers.

## 12.5   IMPLICATIONS AND POLICY INSTRUMENTS FOR NITROGEN DEFICITS

### 12.5.1   SUBSIDIES AND PRICING

For developing countries, one of the biggest concerns is food shortages and this problem heavily stems from the N deficits. As food production declined relative to population, many countries in SSA pursued large-scale "universal subsidy programs" from the 1960s through the 1980s. Those programs were characterized by government-controlled input and output marketing systems. In many cases, these programs succeeded in raising fertilizer use and increasing crop yield. However, these subsidy programs could not be sustained over a longer period because the budgetary costs for the governments were too high and they crowded out other investments such as roads and education (Demeke et al. 2012). In addition, the subsidy program itself was not working effectively facing two major challenges. The first challenge was that the fertilizer prices were high since many countries in SSA are landlocked and infrastructure is poor. Fertilizers must be shipped from distant overseas manufacturing centers causing additional costs of US$50 up to US$100 per metric ton (Mg). Additional in-country transportation costs resulted from the poor port and road infrastructure that could double the farm-level fertilizer price compared to the import price (Gregory and Bumb 2006; Morris et al. 2007). The second challenge was that the benefits of a fertilizer subsidy did not successfully reach smallholder farmers. The monetary transfer through subsidies programs were often used to tactically target political goals rather than maximize economic or social welfare (Banful 2010). After decades of inefficient subsidy programs, many countries in SSA phased out subsidies policy in the 1980s through 1990s due to the high fiscal burden on the government. However, unlike the China case, the private fertilizer firms could not successfully compete with the absence of government subsidies when still having the same fundamental problems such as poor transportation and insufficient market development. In recent years, large-scale subsidy programs have re-emerged renamed "smart subsidy programs" supposedly avoiding the same mistakes as in the past but these subsidy programs still may not have solved the structural problems (Baltzer and Hansen 2011). The smart subsidy programs can perform better with well-targeted beneficiaries and clear objectives if there is less political interference. Integrated approaches are needed in SSA.

## 12.6   IMPLICATIONS AND POLICY INSTRUMENTS FOR EXCESS NITROGEN

The developed world suffers the problem of excess N. The primary concerns for excess reactive N (Nr) are environmental and health on the one hand and the decline in NUE on the other. Both entail economic costs of production and downstream impacts that have been discussed. Excess Nr causes a spectrum of different impacts that can include decreased visibility, biodiversity loss, forest decline, crop yield loss, acidification of surface waters, hypoxia of coastal waters, harmful algal blooms, and human mortality (U.S. EPA 2011). For each of these impacts there are costs of mitigation or control and costs of damages that are inflicted that together make a part of the economic case for taking action or not.

These issues are referred to as "externalities." The party that causes the problem is not affected by the consequences of its own action. This might be a manufacturing firm that releases toxic materials into a river or toxic gas into the air, it might be a farmer whose excess N enters drain tiles and contributes to excess nutrients in a stream or lake. The party that causes the problem does not compensate those affected by the damage nor suffer the damage, so there is no incentive for it to control its activities. The free market does not solve this problem, so government may be asked to intervene. Control strategies for Nr might involve improved practices and conservation, product substitution, transformation of Nr, source limitation, removal, or improved use or reuse efficiency. Effective management of Nr likely requires combinations of these different approaches (U.S. EPA 2011). The policy actions that might be taken tend to involve regulations, voluntary approaches, or market-based approaches. Each of these has advantages and disadvantages. Each has characteristics that may make them effective or ineffective in particular situations to control the entry of Nr into the environment or mitigate its damage once it is there. Each of these is described below along with the characteristics and implications of each that makes them less than complete solutions to solve clearly wicked problems.

### 12.6.1  REGULATIONS

Most regulations are a form of command and control—mandating something or requiring its control. A regulatory body sets rules for the discharge of a pollutant over time and space. In the United States, this approach is primarily based on the authority of the Clean Water Act and the Clean Air Acts passed in the 1970s. The U.S. EPA was created during the Nixon Administration to enforce these acts with the goal for water that it be swimmable, fishable, and drinkable. A key distinction in the Clean Water Act is whether the source is point, emanating from a particular place, or non-point, being widely dispersed. Primarily point sources are targeted under the Clean Water and Clean Air Acts. The bulk of agriculture's contribution to aquatic Nr is non-point, so agriculture has not felt the regulatory action that has been felt by manufacturing plants or municipal wastewater treatment plants that release excess Nr or other pollutants. The Clean Water Act and the Clean Air Act have had substantial success in cleaning up point sources of pollution. For non-point sources, the major regulatory lever in the United States has been the Total Maximum Daily Load (TMDL) process. If a water body is deemed impaired, ultimately state and federal governments can move toward the source of that waterbody (upstream in a river) and require reduced TMDLs from non-point as well as point sources. There are a limited number of water bodies where this has occurred, affecting agriculture and other sources. This is similar to what has occurred in the Chesapeake Bay. Both point and non-point control efforts originally were based largely on applying pollution limiting technology. This would be best management practices (BMPs) for agriculture or best available technology in other cases for point sources. These practices or technology standards are not the same as a performance standard where requirements are met only when the required result is achieved.

Conservation compliance is a good example of a regulatory approach applied to non-point sources of pollution from agriculture, though it is somewhat indirect in its operation. As it stands now, conservation compliance only covers highly erodible land, wetlands, and grasslands. Under the farm bill, farmers who do not apply the proscribed conservation measures on such lands will lose federal benefits from farm-bill commodity programs, crop insurance, and other programs to benefit farmers. The farmer is free not to apply the proscribed conservation measures, but the consequence is not being able to take advantage of the benefits of the federal programs. It is estimated that a little over 100 million acres (40.5 million hectares) of U.S. cropland is highly erodible. This represents somewhat less than a third of U.S. cropland (Claassen 2012). While conservation compliance does not have the direct power to force the control of run-off from highly erodible land, swamp busting, or sod-busting, most farmers try to comply with BMPs—given the value of farm program benefits which can be very high when commodity prices are low (Doering and Smith 2012). The focus of U.S. control of excess non-point nutrients is not through direct regulatory processes.

Europe has taken an approach that is driven by government involvement. One reason for this is the much greater population density in Europe with a smaller land base to act as environmental sink for waste or excess Nr. Excess Nr is perceived as more of a problem than it is in the United States. From a European standpoint, there are four principal paths through with societies organize and govern themselves that are somewhat different from what has been described for the United States. These are culture (human values and traditions), market power (the free market), public policy (regulation and pressure by government), and civic society pressure (public opinion, NGOs, and lobby groups). In Europe, there is more utilization of policy instruments to control excess Nr, which includes regulatory instruments, economic instruments, and communications instruments (Oenema 2011). The umbrella for action on N is the European Union's Nitrates Directive of 1991. Under this, member states are to take specific actions to reduce excess Nr, and they have done so over time with those states having severe problems, like the Netherlands, taking the lead. One characteristic of the European approach is extensive monitoring across the European Union (EU) both to identify areas of concern and identify whether progress is being made. The individual states have drawn up action programs to reduce excess nitrates ($NO_3$) that include things like a limit of N from manure per unit of land. Under Europe's common agricultural policy, cross compliance is the main vehicle for enforcing N rules or best practices. Under this, a farmer would give up European Union agricultural and rural development payments if the N reduction obligations were not met. Compliance requires that the farmer maintain land in good agricultural and environmental condition—a basic standards requirement that does not exist in the United States. There are established codes of good agricultural BMPs that are to be adopted voluntarily and actions required of farmers in $NO_3$ vulnerable zones on a compulsory basis. Where there are nutrient vulnerable zones, farmers may find it difficult or impossible to operate as they did before and have to change crops, farming systems, or cease farming. While reductions in excess Nr have been slow to achieve, they have occurred. In the 27 EU states, fertilizer nutrient consumption has declined by almost 30% since the late 1980s. Between the periods 2004–2007 and 2008–2011, $NO_3$ concentrations in surface water remained stable or fell at 80% of monitored sites (European Commission 2013).

### 12.6.2  VOLUNTARY APPROACHES

Voluntary programs for farmers in the United States to tackle conservation and nutrient problems began in the administration of Franklin Roosevelt. This was partially because the Supreme Court declared key requirements in the Agricultural Adjustment Act of 1933 unconstitutional and this influenced the course that government programs for farmers would be voluntary. The Soil Conservation and Domestic Allotment Act of 1936 became the basic instrument for providing incentives to farmers to adopt voluntary conservation practices. It set the pattern under which efforts on the ground are undertaken today to encourage practices that reduce negative and promote positive environmental externalities. The early programs were aimed at reducing erosion and taking fragile farmland out of production by putting it into conserving uses. The programs were also a primary means of getting cash into rural areas during the depression (Benedict 1953). During the 1930s, the bulk of the money distributed to farmers was in the form of conservation payments. Farmers would terrace erodible land, plant windbreaks, replace grass on eroded land, take fragile land out of production and would be paid to do this by the federal government. Today the bulk of the federal agricultural expenditure is in commodity or price support programs. Nutrient management is becoming more prominent in the USDA's incentive driven voluntary conservation programs. An example of this would be the initiative in the Mississippi River Basin where the Natural Resource Conservation Service (NRCS) has concentrated programs that address nutrients. The approach to reduce nutrient externalities in the United States has been to encourage voluntary activities through education, persuasion, and incentives. The long tradition of payments for voluntary conservation programs replaced any development of basic environmental standards that all farmers would be expected to meet in farming—which is similar to the European approach. Today the expectation is that the payments offered by the government should at least cover the cost of the conservation or nutrient reduction effort and possibly give an incentive beyond that. The question is how can adequate participation and the environmental results be achieved cost-effectively? There is concern that voluntary approaches by themselves may not be capable of achieving meaningful new improvements as well as maintaining the efforts over time. Critics cite such things as the loss of stream buffers and declines in other conservation efforts despite continuing large federal payments (Rundquist and Cox 2016). Only 35% of U.S. cropland is estimated to be managed with BMPs for reducing Nr pollution. Extending the adoption of such BMPs would require a substantial increase in the federal budget for increased incentives for nutrient management (Ribaudo et al. 2011). Another observation is whether something like performance-based standards might be an improvement over the current BMPs based metric for U.S. voluntary programs. Performance based standards are more cost-effective than practice based standards (Antle et al. 2003). They also give a strong incentive for the development and use of the most effective and least expensive practices to meet the performance metric. In addition, a threat of regulatory action on the horizon can also induce increased voluntary cooperation. This has occurred where a threat of a TMDL has given parties the incentive to reduce pollution levels on their own.

### 12.6.3 Market-Based Incentives

There are many potential market-based incentives. However, in many cases the effective operation of the incentive depends on some initial government action toward the negative externality trying to be controlled. If the negative externality were naturally amenable to being controlled by market forces, this would occur. However, the problem arises when dealing with externalities that are not automatically controlled by the market. If market forces are going to be brought to bear, there must be a government action that stimulates that—like a cap on a pollution level. The kinds of market-based instruments available include water quality trading, auction-based contracting, individual transferable quotas, risk indemnification for specified behavior, and conservation easements (U.S. EPA 2011). There is a lot of activity around water quality trading and this is described below as an example of a market-based approach. This is similar in principle to the emissions trading under the Clean Air Act. To start the process there must be a pollution cap established. Then those who are polluting are issued permits to pollute at a level where the totality of permits issued brings about a desired reduction in the overall pollution. The ability of those discharging pollutants to trade their permits increases the cost effectiveness of the expenditure needed to reduce pollution to the new cap level. Envision that there are a number of firms that have been issued permits, and all of them are going to have to reduce their pollution. Given a number of firms, their individual costs for controlling the pollution will not be the same. Some firms will be able to do it at lower cost. A market price can develop for the pollution permits that is somewhere between the costs for firms who have high costs of abatement and the costs for firms with low costs of abatement. This allows those firms with high costs to buy permits from those firms with low costs of abatement. The high-cost firms save the control expenditure (and may continue to pollute at the same level) and the low-cost firms reduce their pollution below their required level at low cost and make a profit doing so. Remember, there was a cap set to reduce the overall amount of pollution, so the overall pollution reduction still takes place. It is just that the cost of control is reduced because those firms with a cost advantage in reducing pollution do a higher proportion of the control.

For agriculture, this market approach has taken the form of nutrient trading that might exist between agricultural producers and entities like sewage treatment plants. The increasingly stringent discharge standards for sewage treatment plants over the years has resulted in increasingly stringent and expensive controls on their part to meet water quality standards. The lower cost treatments have already been in place and more and more expensive efforts are required, as standards have been tightened. Within the treatment plant's watershed, there may be a number of farmers who have done little in the way of BMPs to control excess N (or phosphorus, P). In this case, the farmers could adopt BMPs that would reduce the flows of the critical nutrient in the watershed at relatively low cost. The treatment plant could pay the farmers to do so through tradable permits or some other device that would save the plant additional expenditure on control. Such trading is a way to encourage non-point pollution control through a market mechanism. One drawback is that trading is limited by lack of information. Measurement is critical to the process. Sulfur dioxide ($SO_2$) air quality trading was facilitated by the fact that the emitters, like power plants, were all point source polluters and there were accurate measurements and measurement systems in place for the pollutants as they were discharged. In the case of farmers and sewage treatments plants, the plant is a point source where measurement data are reliable and available. For the farmer who applies a BMP, it is still not certain exactly what amount of nutrient reduction will take place. The trading is then based on an assumption that the expected reduction occurs—but we know this is not always the case (Stephenson and Shabman 2011). Another drawback with nutrient trading between point and non-point entities is that they may not be sufficiently co-located geographically. They would likely have to be in the same or connected watersheds. Only a portion of the non-point agricultural community is located in watersheds where there are industrial scale point source nutrient polluters with whom to trade. So, trading is not likely to be physically able to encourage the majority of non-point sources of excess N to reduce their flows through trading with a nearby point source.

The issue of measurability is important not just for market-based instruments. This limitation applies equally to regulation or to voluntary approaches where the achievement of goals on the one hand and linking those achievements to specific actions on the other hand becomes critical to success. Individuals and institutions do what they measure. In addition, incentives, positive or negative, play a big role in regulation, voluntary action, or market instruments. The U.S. conservation programs are based on incentives. Governments have the ability to tax negative externalities and incentivize positive efforts to control them. Positive or negative incentives can be used in concert with regulations and in essence allow a combination of carrot and stick. It is within this broader policy realm that agricultural producers have to make decisions about production and the extent to which they can or cannot afford to control excess Nr. The kinds of trade-offs farmers face in decisions to improve NUE and reduce excess Nr can be illustrated with a farm decision model.

## 12.7  FARM LEVEL DECISIONS

### 12.7.1  Risk Factors and Trade-Offs

Underneath the broader umbrella of policies and prices, the decisions that are made at the farm level by farmers ultimately determine NUE, the situation with respect to excess Nr, and farm incomes. Farmers may be risk-averse toward making changes in cropping decisions or adopting new agricultural practices, or might have very conservative attitudes toward technology or lower or higher levels of concern for the natural environment (Bowman and Zilberman. 2013). With annual profits dependent on the crop yield in any one year, farmers tend to change farming practices only when the extent of risk is acceptable relative to potential reward. Thus, new agricultural practices have to be explained in the perspective of economic attractiveness and risk management.

The practice of "splitting" N fertilizer can be a way for improving NUE, increasing profit, and reducing excess Nr. In some instances, especially with fall or previous season fertilization, farmers may tend to apply excessive N considering the possible N loss via leaching, runoff, and denitrification between when N fertilizer is applied and when the plants need it the most. To reduce unnecessary N loss, splitting the application of N with one application before or around the time of planting and a second application after planting can be undertaken. Through this practice, farmers can reduce total N use resulting in economic savings and potentially reduce N loss to the environment. Such a split N application can result in reduced negative externalities to society. However, there are risks involved with a split N application strategy as well. One is that an incomplete second N application due to the weather factors that determine fieldwork days may result in lower yields. The trade-off here is between weather variability and the capacity the farmer has at hand to make the split N application. This capacity involves labor availability and the capacity of the machinery to apply N. On one extreme, a farmer's N application capacity could be limited enough so that perfect weather would be required for the entire ideal application period to get the job done. On the opposite extreme, the farmer could have such excess labor and machinery capacity that only a few good days would suffice to complete the post-planting N application in a timely manner. The latter case would involve not only excess labor, but also a large capital investment in machinery that was generally underutilized.

### 12.7.2 Example of the Split N Calculator

In the illustration of the trade-off described above, the economic costs and benefits of splitting N application are estimated and take account of weather risk with an online decision support tool. Corn Split NDST (available at: splitn.agclimate4u.org) is the decision tool utilized. With this tool, farmers can estimate net benefits (or loss) for split N application for a given number of acres based on historical days suitable for fieldwork (DSFW) for major corn areas in the United States. The suitable fieldwork days are driven by variable weather conditions. When fields are too wet, N application is restricted. The trade-off results provided are the result of scenario analysis on a hypothetical farm in Tippecanoe County, Indiana. The farm chosen for the case study has 608 ha of corn and a machinery capacity of 6.5 ha per hour. Thus, 94 hours would be needed to cover all 608 ha. The labor availability is up to 15 hours per day, 6 days a week. With planting May 15, one would like to get the N applied between the V4 and V8 stages of corn growth—a period that might average about 11 days between June 13 and 23. Over the 35 years of weather experience, the average hours suitable for fieldwork are only 92 hours. This is slightly less than the required 94 hours. (The application constraint could be relaxed by increasing machinery capacity or labor availability.) Fertilizer application involved an initial basal dose of 22.7 kg/ha and then enough split application of N to meet the yield goal of 10.6 Mg/ha in Tippecanoe County. The yield penalty for incomplete side dress application is 2.8 Mg/ha. Choosing the best time physiologically for the corn plant to side dress provides a yield benefit from side dressing of 0.31 Mg/ha. This is even though 16.8 kg/ha less fertilizer is used in the side dress than would be required in total pre-plant application. In terms of economic factors, the side dress operation costs $37 per ha, the nitrogen price is $0.88 per kg, and the corn price is $148 per Mg.

Starting with the best case, there are 18 of the 35 years with excess or sufficient good field working days so that the entire crop can be side dressed. The fertilizer savings plus yield enhancement minus side dressing costs yields the farmer an increase of $20,000 in profits. In 21 out of the 35 years the farmer gets at least 597 ha harvested and gains at least $15,000. The breakeven point is getting 567 ha side dressed, and this occurs in at least 23 of the 35 years. Thus, two-thirds of the time the farmer will break even or be ahead economically by side dressing in addition to likely improving NUE and reducing excess Nr. In the worst of the 35 years, only 69.6 ha could be side dressed, and the loss amounted to $253,000 (Gramig et al. 2017). The worst-case risk could be reduced by increasing labor availability, machinery capacity, or through such things as improved drainage or cover crops that improve soils' ability to dry out after rainfall. Getting improved N use practices on the ground will have to involve understanding the farmer's risk parameters and finding ways to lesson risk or change perception that is often based on the concern with the catastrophic risk.

## 12.8  CONCLUDING THOUGHTS: ECONOMIC AND POLICY IMPLICATIONS

Achieving technical efficiency in agricultural N use (improving the NUE) while achieving economic optimality (EONR) is a truly wicked problem even without adding the constraint of trying to minimize excess Nr—all in the difficult biophysical context of the "nitrogen cascade." The seeming simplicity and achievability of the EONR is confounded by many factors including such things as prices that may be distorted on the economic side and less than ideal timing and weather from the farm example on the production side. There is something of a tradeoff function between EONR and NUE where results for one or the other may be less than desired and cases where apparent achievement of either or both still does not meet broader societal goals. Many Chinese farmers likely met their own EONR under low fertilizer and high crop prices, but were tempted to exceed N applications into Stage III because the fertilizer/crop price ratio was so much in favor of using more fertilizer. This has led to large-scale Nr pollution. In SSA, farmers may assess their risk and fail to reach the EONR that confronts them because of poor logistics or weather/timing difficulties. Meeting the economic optimum and having an extremely high NUE still may not gain farmers an adequate livelihood and their country adequate food supplies for many reasons outside the N use decision. The challenge from this is finding better means of utilizing the powerful concepts of EONR and NUE to better address problems of overuse and underuse of N, recognizing the kinds of risk factors farmers face—weather being one illustrated by the split nitrogen tool.

From a policy standpoint in the United States, the concern is excess Nr even without the high degree of overuse we see in China. The same is true in Europe, and governments are relying primarily on regulation as the policy tool to reduce levels of excess Nr. When thinking in economic terms, it is important to recognize that excess Nr is an externality in that the producer of the excess Nr does not suffer the "downstream" damages caused by the excess Nr. Two options to deal with this are to build the cost of the external damages into the production costs of the producer of excess Nr so there will be a market incentive for the producer to do something about it, or to regulate and simply prohibit the producer from releasing excess Nr. The market does not take care of this problem automatically. For point source problems, like $SO_2$ from power plants, the regulatory route has been taken and greatly reduced pollution levels from these sources. Many economists would favor intervention to improve efficiency or some form of regulation under these circumstances. The societal choice for excess non-point Nr in the United States has been to use subsidies and voluntary participation with few exceptions. There are concerns that this approach is not adequate (Rundquist and Cox 2016). One conclusion might be that if U.S. farmers are not too far off their EONR then, despite the economics, fertilizer applications need to be scaled back some to improve the NUE which can also be enhanced through better placement and timing of application. This also brings the discussion back into the realm of the "nitrogen cascade" and begs the question whether it is best to reduce the amount of N added to the environment or try to interdict and stabilize it later in the cascade through such things as managed drainage or field bioreactors. A broad mix of technical and practical approaches is often recommended for reducing Nr. The policy approaches must also be broadly based. If the limit of what is possible to achieve through practice and behavioral changes from voluntary action has been reached, are there actions that should be taken on the regulatory front and through the establishment of market mechanisms? For the latter, it is important to remember that the development of a market mechanism like nutrient trading depends upon a regulatory cap that creates the market, that is, a mixed approach.

Nitrogen use and control remains one of the grand challenges for the world—both ensuring its beneficial use and preventing damage from its excess in the environment. No single branch of science has the key to this challenge. Economics provides some guidance for appropriate use, but this must be balanced with understanding of what goes on biophysically in the "nitrogen cascade" as well as understanding the economic position, resources, knowledge base, and risk trade-offs of those farmers who use N for agriculture on the ground.

## REFERENCES

Antle, J., S. Capalbo, S. Mooney, E. Elliott, and K. Paustian. 2003. Spatial heterogeneity, contract design, and the efficiency of carbon sequestration policies for agriculture. *Journal of Environmental Economics and Management* 46(4): 231–250.

Baltzer, K., and H. Hansen. 2011. *Evaluation study: Agricultural input subsidies in Sub-Saharan Africa.* Danida Evaluation Study. Copenhagen: International Development Cooperation (DANIDA), Ministry of Foreign Affairs of Denmark.

Banful, A.B. 2010. *Old problems in the new solutions? Politically motivated allocation of program benefits and the "new" fertilizer subsidies.* IFPRI Discussion Paper No. 01002. Washington, DC: International Food Policy Research Institute.

Batie, S.S. 2008. Wicked problems and applied economics. *American Journal of Agricultural Economics* 90(5): 1176–1191.

Benedict, M.R. 1953. *Farm policies of the United States, 1750–1950.* New York: The Twentieth Century Fund.

Birch, M.B.L., B.M. Gramig, W.R. Moomaw, O.C. Doering III, and C.J. Reeling. 2011. Why metrics matter: Evaluating policy choices for reactive nitrogen in the Chesapeake Bay Watershed. *Environmental Science and Technology* (45)(1): 168–174.

Bowman, M.S., and D. Zilberman. 2013. Economic factors affecting diversified farming systems. *Ecology and Society* 18(1): 33.

Cassman, K.G. 1999. Ecological intensification of cereal production systems: Yield potential, soil quality, and precision agriculture. *Proceedings of the National Academy of Sciences* 96: 5952–5959.

Claassen, R. 2012. *The future of environmental compliance incentives in U.S. agriculture: The role of commodity, conservation and crop insurance programs.* EIB-94. Washington, DC: Economic Research Service, U.S. Department of Agriculture.

Cui, Z., X. Chen, and Y. Miao et al. 2008. On-farm evaluation of the improved soil N min–based nitrogen management for summer maize in North China Plain. *Agronomy Journal* 100(3): 517–525.

Daniel E.E., S.L. Abreu, A. West, D.R. Caasi, and T.O. Conley et al. 2009. Cereal nitrogen use efficiency in Sub Saharan Africa, *Journal of Plant Nutrition,* 32: 2107–2122.

Demeke, M., D. Dawe, J. Tefft, T. Ferede, and W. Bell. 2012. *Stabilizing price incentives for staple grain producers in the context of broader agricultural policies—Debates and country experiences.* ESA Working paper No. 12-05. Rome: Agricultural Development Economics Division, Food and Agriculture Organization of the United Nations.

Doering, O.C., and K. Smith. 2012. *Examining the relationship of conservation compliance and farm program incentives.* Washington, DC: The Council on Food, Agricultural and Resource Economics, CFARE.

European Commission. 2013. *Report from the commission to the council and the European parliament on the implementation of Council Directive 91/676/EEC concerning the protection of waters against pollution caused by nitrates from agricultural sources based on member state reports for the period 2008–2011,* Luxenberg: The European Commission.

Galloway, J.N., J.D. Aber, and J.W. Erisman et al. 2003. The nitrogen cascade. *Bioscience* 53: 341–356.

Gramig, B.M., R. Massey, and S.D. Yun. 2017. Nitrogen application decision-making under climate risk in the U.S. Corn Belt. *Climate Risk Management* 15: 82–89. http://dx.doi.org/10.1016/j.crm.2016.09.001.

Grassini, P., K.M. Eskridge, and K.G. Cassman. 2013. Distinguishing between yield advances and yield plateaus in historical crop production trends. *Nature Communications* 4: 2918. doi:10.1038/ncomms3918.

Gregory, D.I., and B.L. Bumb. 2006. *Factors affecting supply of fertilizer in Sub-Saharan Africa.* Agriculture and Rural Development Discussion Paper 24. Washington, DC: The World Bank.

Li, Y., W. Zhang, and L. Ma et al. 2014. An analysis of china's fertilizer policies: Impacts on the industry, food security, and the environment. *Journal of Environmental Quality* 42: 972–981.

Morris, M., V.A. Kelly, R.J. Kopicki, and D. Byerlee. 2007. *Fertilizer use in African agriculture lessons learned and good practice guidelines.* Directions in Development. Washington, DC: Agriculture and Rural Development, the World Bank.

Mueller, N.D., J.S. Gerber, M. Johnston, D.K. Ray, N. Ramankutty, and J.A. Foley. 2012. Closing yield gaps through nutrient and water management. *Nature* 490: 254–257.

Oenema, O. 2011. Nitrogen in current European politics. In *The European nitrogen assessment,* Eds. M.A. Sutton et al., pp. 62–87. Cambridge: Cambridge University Press.

Ribaudo, M., J. Delgado, L. Hansen, M. Livingston, R. Mosheim, and J. Williamson. 2011. *Nitrogen in agricultural systems: Implications for conservation policy.* Economic Research Report No. 127. Washington, DC: U.S. Department of Agriculture.

Rundquist, O., and C. Cox. 2016. *Fooling ourselves: Voluntary programs fail to clean up dirty water.* Washington, DC: Environmental Working Group.

Stephenson, K., and L. Shabman. 2011. Rhetoric and reality of water quality trading and the potential for market-like reform. *Journal of the American Water Resources Association* 47(1): 15–28.

U.S. Department of Agriculture. 2016. *China—Peoples Republic of grain and feed update huge stocks challenge grain policy.* Global Agricultural Information Network GAIN Report: 16002. Washington, DC: USDA Foreign Agricultural Service.

U.S. Environmental Protection Agency. 2011. *Reactive nitrogen in the United States: An analysis of inputs, flows, consequences and management actions.* Washington, DC: EPA Science Advisory Board.

Wu, L., L.X. Chen, Z. Cui, W. Zhang, and F. Zhang. 2014. Establishing a regional nitrogen management approach to mitigate greenhouse gas emission intensity from intensive smallholder maize production, *PloS ONE* 9(5), e98481. http://dx.doi.org/10.1371/journal.pone.0098481.

Zhao, R.F., X.P. Chen, F.S. Zhang, H. Zhang, J. Schroder, and V. Römheld. 2006. Fertilization and nitrogen balance in a wheat–maize rotation system in North China. *Agronomy Journal* 98(4): 938–945.

Zhang, X., E.A. Davidson, D.L. Mauzerall, T.D. Searchinger, P. Dumas, and Y. Shen. 2015. Managing nitrogen for sustainable development. *Nature* 528: 51–59.

# 13 Formulations of Slow Release Fertilizers for Enhancing N Use Efficiency

*Amit Roy*

## CONTENTS

## 13.1 INTRODUCTION

While the global population is projected to grow by about 30% by 2050 (UN 2017), food demand is predicted to increase by nearly 70% as emerging regions urbanize, develop economically, and consume a richer diet. However, the absolute increase in food required by 2050 will be as large as the increase since the Green Revolution was launched in the 1960s—as available arable land and water become scarcer. Inorganic fertilizers have been essential in increasing the food production in the past and will be so in the future but within the context of environmental and social sustainability. The negative environmental impact of fertilizers, particularly nitrogen (N), has been under scrutiny but has accelerated in recent years due to health issues related to excess nitrates $\left(NO_3^-\right)$ in water bodies. Hence, there is greater emphasis in improving fertilizer nutrient use efficiency (NUE), reducing greenhouse gas (GHG) emission from agriculture, improving water use efficiency (WUE), and maintaining a healthy soil for improved productivity.

Of the 17 nutrients essential for plant growth, N, phosphorus (P), and potassium (K) are the most likely to be deficient and needed in the largest quantities. All crops require N, P, and K. The only exceptions are legume crops that have symbiotic bacterial colonies associated with roots and thus have the ability to meet plant N needs through biological N fixation (BNF). However, legumes do require large amounts of P and K. Of the primary nutrients, N and P fertilizers are subjected to more chemical- and energy-intensive production processes. These nutrients are available in many formulations and forms (solid blends and liquids) and often "carry" much smaller (often minute) amounts of secondary and micronutrients that are essential for the maximum economic yields and improved nutritional value. While large amounts of NPK application have increased production, it also has raised issues related to the uptake efficiencies by crops. Of the three primary nutrients (N, P, K), N is the most problematic. Urea is the most common N source used for crop production and its uptake by plants varies but rarely exceeds 50% under a very well-managed system. N nutrient use efficiency (N-NUE) rarely exceeds 50%, but in developing regions, where most fertilizer use occurs, farmers (especially smallholder farmers) struggle to achieve a N-NUE of 20–30%. The unused N is lost to the environment as GHG, contributing to climate change; as leachate in water, causing eutrophication; or it accumulates in the soils. Further, the lower efficiency is an economic loss for the farmers and waste of energy.

Improving N-NUE has been the focus of researchers and several products have been commercialized but most are used for high-value crops, lawn, golf courses, and ornamental plants. Most of these products are not economically viable for field crops. With increasing demand for producing food for ever-growing population and reducing environmental footprint of fertilizer use, there is a renewed drive to develop urea-based efficient fertilizers for field crops. These products, referred to as "specialty fertilizers," promise to be transformational akin to "leaded to unleaded gasoline."

This chapter reviews the future food demand and corresponding fertilizer needs, and articulates the rationale for developing more efficient fertilizers that are economically viable, particularly for developing countries. It also reviews the current development trajectories for "specialty" products by the researchers, entrepreneurs, and the private sector.

## 13.2  FUTURE POPULATION AND FOOD DEMAND

Since the 1960s, population has more than doubled and is estimated to increase up to 9.8 billion by 2050 (Figure 13.1). While the population increases in developed regions are expected to remain virtually unchanged at 1.3 billion, developing regions will see increases from 5.7 billion (2010) to 8.3 billion by 2050. In addition, the developing regions, driven by massive urbanization, will see a shift of the urban:rural ratio from 0.8 to 2.0 (Figure 13.2) (UN 2017).

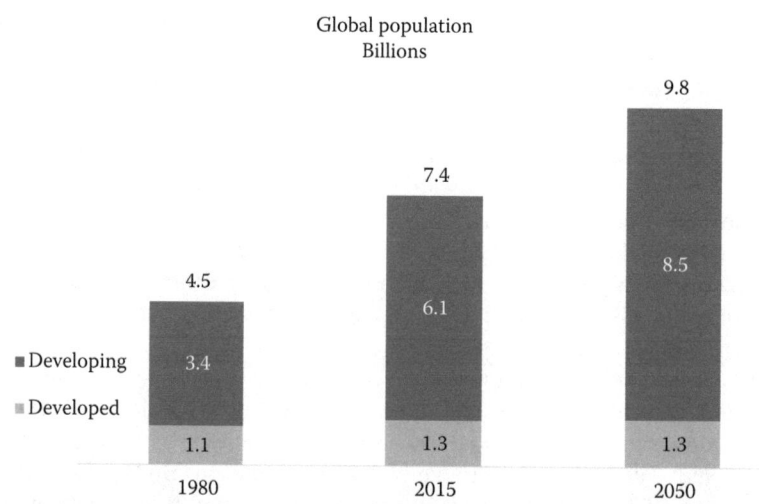

**FIGURE 13.1**  Population growth in developed and developing countries. (Adapted from Roy, A. H. (2015). Global fertilizer industry: transitioning from volume to value—The 29th Francis New Memorial Lecture. *Proceedings International Fertiliser Society* 769.) The population stat is from UN. (2017). United Nations Department of Economic and Social Affairs, Population Division. *World Population Prospects: The 2017 Revision, Key Findings and Advance Tables.*

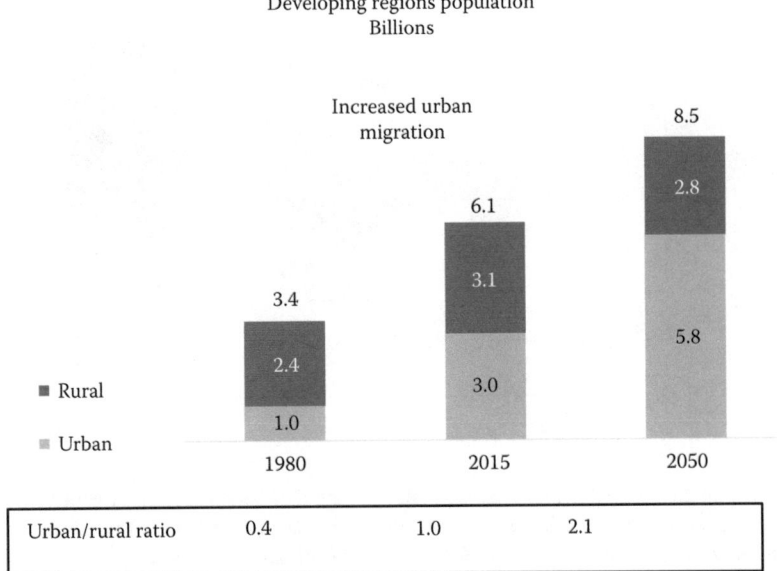

**FIGURE 13.2** Developing regions population growth segregated by rural and urban increases. (Adapted from Roy, A.H. (2015). Global fertilizer industry: transitioning from volume to value—The 29th Francis New Memorial Lecture. *Proceedings International Fertiliser Society* 769.)

As with population growth, most (87%) of the increased food demand will occur in developing regions, with the developed economies experiencing only a slight increase (6%). Globally, yields will need to increase at a yearly compound rate of about 1.4%, but in developing regions, the increase is estimated to be 1.8%. Further, a significant infrastructure expansion will be required to transport up to 2.5 times more food from the rural areas of production to the urban markets (Figures 13.3 and 13.4). However, developing regions are dominated by farmers whose land holdings generally range from 0.5 to 2.0 hectares. To achieve required yield increases, these smallholder farmers would need suitable seeds, the right type of fertilizers, and best management practices (BMPs) supported by government policies to overcome economic and infrastructure challenges to access inputs and participate more fully in post-harvest markets.

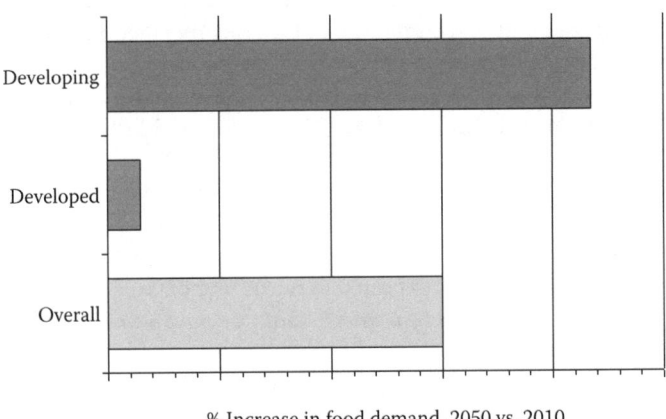

% Increase in food demand, 2050 vs. 2010

**FIGURE 13.3** Overall estimated global food demand increase segregated by developed and developing regions. (Adapted from Roy, A. H. (2015). Global fertilizer industry: transitioning from volume to value—The 29th Francis New Memorial Lecture. *Proceedings International Fertiliser Society* 769.)

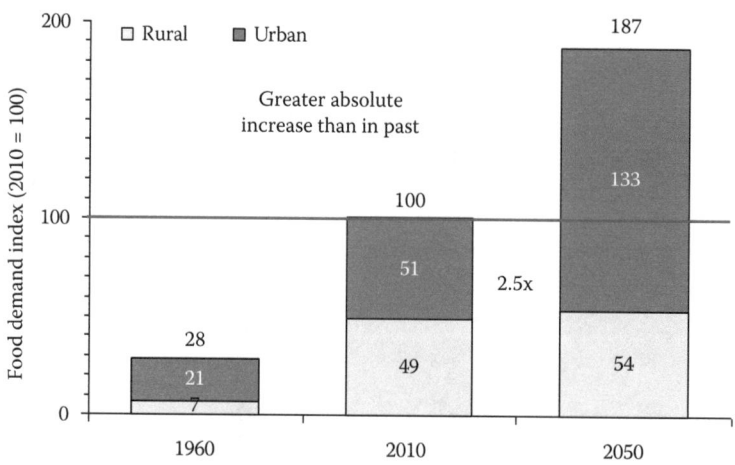

**FIGURE 13.4** Estimated food demand increase in developing regions segregated by rural and urban demand. (Adapted from Roy, A. H. (2015). Global fertilizer industry: transitioning from volume to value—The 29th Francis New Memorial Lecture. *Proceedings International Fertiliser Society* 769.)

## 13.3   CONVENTIONAL FERTILIZERS AND THEIR SHORTCOMINGS

Conventional fertilizers, mostly containing N, P, and K, have helped to more than double world cereal production since 1975. These fertilizers, however, are largely unchanged from the formulations and forms manufactured in the 1950s, 1960s, and 1970s. Their characteristics, in the absence of BMPs, result in significant loss of nutrients to the ecosystem. These losses produce negative economic and environmental impacts. Even under the most efficient agro-production systems that are highly knowledge intensive, the apparent NUE of the macro-elements (N, P, K) range from only 20% to over 80% (Baligar et al. 2001). Among N fertilizers, urea is the most common, which is also the most troublesome. Under a very well-managed system, the nitrogen NUE (N-NUE) from urea rarely exceeds 50%, but in developing regions, where most of the urea fertilizer use occurs, farmers (especially smallholder farmers) struggle to achieve an N-NUE of 20–30% (Bindraban et al. 2014). The unused N is lost to the environment as a GHG, contributing to climate change; as leachate in water, causing eutrophication; or it accumulates in the soils. Further, the lower NUE is economic loss for the farmers and waste of energy.

The economic impact of low N-NUE is significant. For example, assuming an annual urea consumption of 130 million tons (Mt) and N-NUE of 50%, up to $25 billion of the total price paid by farmers is lost annually. In addition, considering the associated environmental and energy costs, the losses could be as high as $15 billion per year if GHG taxes and water remediation costs were assessed (VFRC 2012).

Any recovery of these losses would significantly benefit farmers in the developing regions, where most of the increase in food demand is expected to occur. For example, a 50% improvement in the average N-NUE (25–30%) in developing regions would capture some $10 billion per year of the current loss and improve yields as well.

From 2009 to 2014, the global urea capacity increased by 33 Mt at a total investment cost of U.S. $40 billion (Prud'homme 2015). If urea-N NUE had increased by 50% yearly, the need for investment in new capacity would have been reduced significantly.

Long-term studies in India have shown that while crop production increased with increasing fertilizer usage (specifically N but also P and K), the productivity per unit of N fertilizer applied has fallen (Figure 13.5).

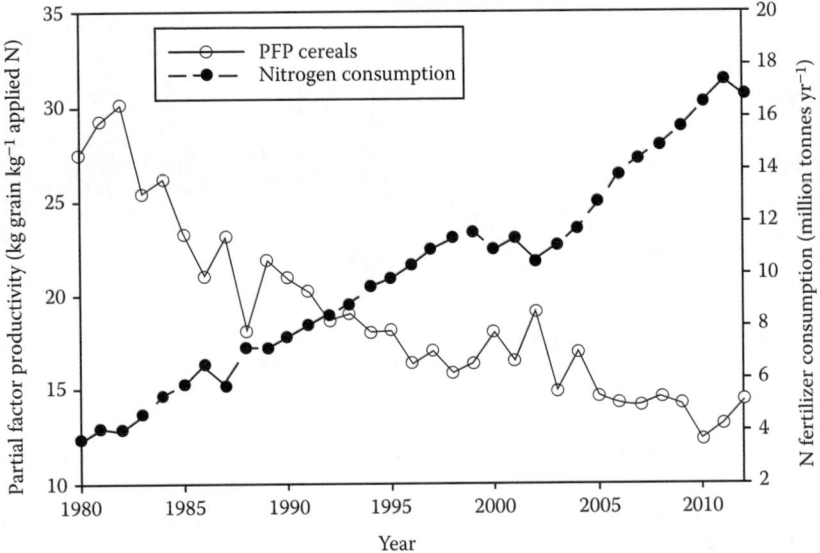

**FIGURE 13.5** Partial factor productivity of nitrogen (N) for cereal production and fertilizer N consumption in India. (From FAOSTAT. (2015). Food and Agriculture Organization of the United Nations, statistics website: http://faostat3.fao.org/home/E.)

Against these environmental and economic factors, and the need to produce food for ever-increasing population, the farmers use more fertilizers with its associated environmental factors or use an alternate path that can produce more from same or less amount of fertilizer. So, the slow, controlled-release and stabilized fertilizers once targeted for horticulture, golf courses, and lawns, are now transitioning to commodity markets from niche markets.

## 13.4   SLOW- AND CONTROLLED-RELEASE AND STABILIZED FERTILIZERS

The term "slow-controlled-release and stabilized fertilizers," as used in the following discussion, refers to fertilizers that release, either by design or naturally, their nutrient content over an extended period of time to match nutrient requirement of the crop (Figure 13.6). Because of economics and environmental considerations, the slow, controlled-release and stabilized N fertilizers are much more important than phosphate (or potash), particularly under certain soil and climatic conditions. In most cases unutilized phosphate and potash remain available for subsequent crops. In contrast, there is seldom much carryover of fertilizer N from one crop to the next.

Potential advantages of these fertilizers are increased efficiency of uptake by plants; minimization of losses by leaching, fixation, or decomposition; a reduction in the release of nutrients and by-products to air and water; and avoidance of burning of vegetation or damage to seedlings. Despite this impressive list of potential advantages, the slow- and controlled-release fertilizers have mainly been applied to high-value crops and/or ornamental plants and golf courses. However, there has been progressive increase in use of these fertilizers in Japan, the United States, Western Europe, and China. Trenkel (2010) provides more detailed discussions about slow- and controlled-release and stabilized fertilizers.

The Association of American Plant Food Control Officials (AAPFCO) defines:

1. *Slow- and controlled-release fertilizers* (SCRF) are those containing a plant nutrient in a form which delays its availability for plant uptake and use after application, or which extends its availability to the plant significantly longer than ammonium nitrate or urea,

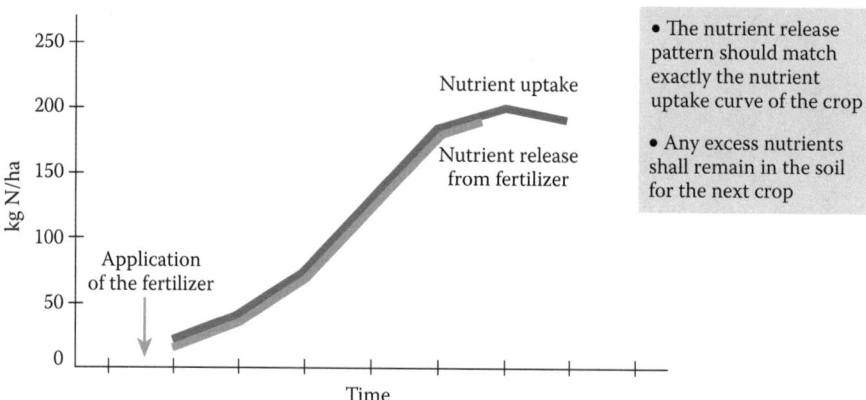

**FIGURE 13.6**  The ideal slow and controlled fertilizer matches the crop's nutrient requirements. (Adapted from Lammel, J. (2005). Cost of different options available to farmers: Current situation and prospects. IFA International Workshop on Enhanced Efficiency Fertilizers, Frankfurt. International Fertilizer Industry Association, Paris, France.)

ammonium phosphate or potassium chloride. Such delay of initial availability or extended time of continued availability may be achieved by a variety of mechanisms. These include controlled water solubility of the material by semi-permeable coatings, occlusion, protein materials, or other chemical forms by slow hydrolysis of water-soluble low molecular weight compounds.

2. *Stabilized fertilizers* are those containing a substance (nitrogen stabilizer) which delays the nitrification of ammonia or ammonification of urea. These substances are grouped under urease inhibitors and nitrification inhibitors.

### 13.4.1  SLOW- AND CONTROLLED-RELEASE FERTILIZERS

#### 13.4.1.1  Slow-Release Nitrogen Fertilizers (SRNFs)

Slow-release nitrogen fertilizers release their nutrient at a slower rate than urea and ammonium nitrate, but the rate, pattern, and duration of release are controlled by soil and climatic conditions and cannot be predicted with any accuracy. Examples of SRFs are organic-N low-solubility compounds such as urea formaldehyde (UF) and isobutylidene-diurea (IBDU).

#### 13.4.1.2  Controlled-Release Fertilizers (CRFs)

Controlled-release fertilizers are those whose rate, release pattern, and duration of release are predictable and controllable through a physical barrier applied onto tablets or granules. These coatings can be hydrophobic polymers or as matrices in which the soluble active material restricts the dissolution of the fertilizer. The coated fertilizers can be further divided into fertilizers with organic polymer-coatings—that are either thermoplastic or resins—and fertilizers coated with inorganic materials such as sulfur (S) or mineral-based coatings (Figure 13.7). Additional information regarding various coating materials is discussed by Shaviv (2005).

The sulfur-coated urea (SCU), one of the early CRFs, was developed by the Tennessee valley Authority (TVA) in the early 1960s and involved spraying molten sulfur (S) onto urea granules in a rotating drum. The applied sulfur oxidized over time and was an important secondary nutrient, but the coating had many imperfections that hindered prediction of precise release rate and additional paraffin coating was applied to cover the imperfections. The relatively high price of SCU compared with conventional fertilizers has made it uneconomical for use in crop production but has seen acceptance in non-agricultural markets such as golf courses and high-value specialty crops.

FIGURE 13.7    (See color insert.) Polymer coated and uncoated commercial urea.

The SCU provided the clue for the industry to develop improved coatings using cross-linked thermosetting and thermoplastic polymers resulting in a polymer coated commercial product known as Osmocote. Subsequently, the recent technology enhancements have resulted in new slow-release products that are cost-effective for corn (*Zea mays*), wheat (*Triticum aestivum*), potatoes (*Solanum tuberosum*), rice (*Oryza staiva*), and others. For example, Agrium Inc. manufactures a product sold mainly in North America under the trade name ESN®, which is a polymer-coated urea N product whose release mechanism is temperature and moisture controlled (Figure 13.8) (Blaylock 2010). Typically, the ESN shows a yield increase of about 10% compared with conventional fertilizers (Figure 13.9). Although not currently practical for use in developing economies, the concept of polymer coating as used for ESN has spurred other companies to research alternative chemicals to produce an affordable slow-release product for use in developing economies.

## 13.4.2  Stabilized Fertilizers

### 13.4.2.1  Urease Inhibitors

These compounds prevent or suppress the transformation of amide-N in urea to ammonium hydroxide and ammonium through the hydrolytic action of the enzyme urease. By slowing down the rate at which urea is hydrolyzed in the soil, volatilization losses of ammonia to the air (as well as further leaching losses of nitrate) are either reduced or avoided. Thus, the efficiency of urea and of N fertilizers containing urea (e.g., urea ammonium nitrate solution) is increased and any adverse environmental impact from their use is decreased. There has been considerable research to develop

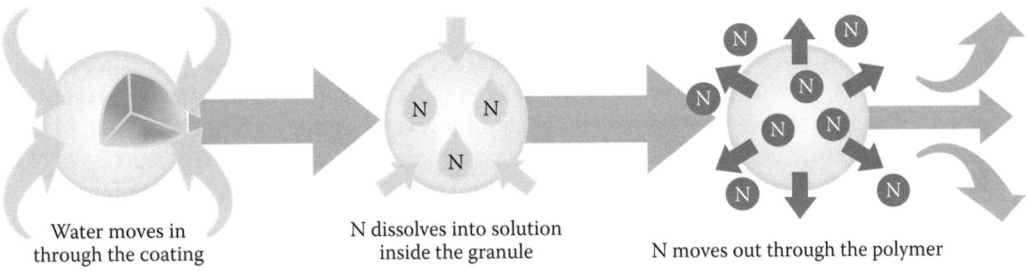

Water moves in through the coating          N dissolves into solution inside the granule          N moves out through the polymer

FIGURE 13.8   Release of N from polymer coated urea. (Adapted from Blaylock, A. (2010). Enhanced Efficiency Fertilizers. Colorado State University Soil Fertility Lecture.)

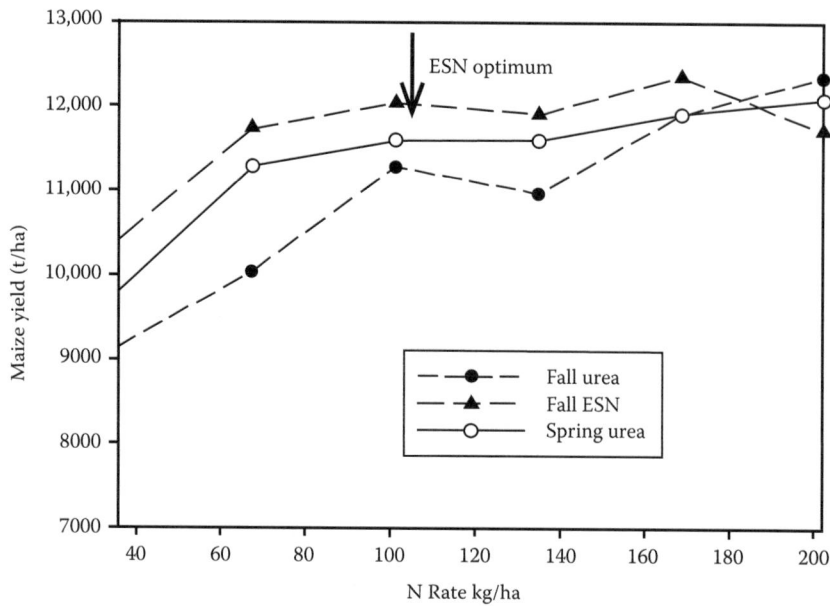

**FIGURE 13.9**  Maize response to autumn and spring urea versus ESN (polymer coated urea). (From Iowa State University, Kanawha 2003.)

an effective urease inhibitor. Among many, N-(n-butyl) thiophosphoric triamide (NBPT) is the only urease inhibitor, marketed under the trade name Agrotain®, that has gained practical and commercial importance in agriculture.

### 13.4.2.2  Nitrification Inhibitors

These compounds delay the oxidation of the ammonium ion (NH4+) by the *Nitrosomonas* bacteria in the soil. These bacteria transform ammonium ions into nitrite (NO2$^-$), which is further transformed into NO3$^-$ by *Nitrobacter* and *Nitrosolobus* bacteria. Thus, the nitrification inhibitors control the loss of NO3$^-$ by leaching or the production of nitrous oxide (N$_2$O) by denitrification thus increasing N-NUE. Nitrification inhibitors include dicyandiamide (DCD), 3,4-dimethylpyrazole phosphate (DMPP), and 2-chloro-6-(trichloromethyl)pyridine (nitrapyrin), which are commercially produced and used in certain markets, mainly in developed economies.

The neem (*Azadipachta indica*) tree is indigenous to many tropical and semi-arid countries. Meliacins, a component of the neem oil, has the characteristic of retarding the nitrification process (Mangat and Narang 2004). Neem cake has been used to coat urea to improve N-NUE. The efficacy of neem-coated urea (NCU) has been confirmed through several years of field trials in India. Compared with commercial urea, NCU results in 2–10% higher rice yields. In some cases, the rice yields with 80% NCU are comparable to that by application of 100% commercial urea. In the initial two to three years, commercial production of NCU in India was quite limited since the companies were not allowed to recover costs of neem coating. However, because of farmer demand, in 2008 the Government of India (GOI) allowed companies to convert up to 20% of their capacity to NCU and sell it at a 5% higher price than uncoated urea. On January 7, 2015, in a major shift in policy, the GOI directed domestic urea manufacturers to convert 100% of their installed capacity to produce NCU. In general, the farmers apply the same amount of NCU as uncoated urea but realize 5–10% yield increases and subsequently profits.

In flooded or waterlogged soils, as for example, in a continuously flooded rice paddy, anaerobic (reducing) conditions exist in the soil below the surface, and hence nitrification of ammoniacal-N

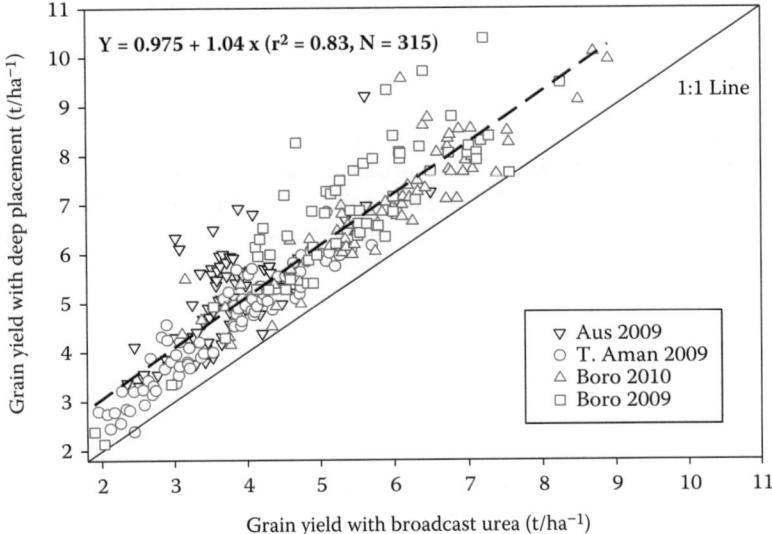

**FIGURE 13.10** Comparison of grain yield with urea briquette deep placement and broadcast split urea application, demonstration plots in Bangladesh.

does not occur in this layer. Therefore, when ammonia, ammonium salts, or urea (which hydrolyzes in the soil to ammonia and carbon dioxide) is placed in the reducing zone of the soil, it remains in the ammonium form and is more or less immobilized by sorption on the clay. Some of the advantages of controlled release may be attained by placement of ammoniacal-N in the reducing zone of the rice paddy soil. However, this placement is difficult since applicators quickly become clogged with mud and wet fertilizer. To facilitate subsurface placement, large granules or briquettes of urea ranging in weight from 1 to 2.7 g are "deep" placed 9–10 cm below the soil surface. Experiments in Bangladesh involving deep placement of urea briquettes (guti) (referred to as urea deep placement) have shown substantially increased efficiency of nitrogen utilization by rice as compared with the usual broadcast placement (Figure 13.10). On average, rice yield increased by 15% while applying 35% less urea. This technology is being introduced and scaled up in several African countries.

## 13.5 FUTURE OF SLOW-CONTROLLED-RELEASE AND STABILIZED FERTILIZERS

From a consumption of about 350,000 tons in the early 1980s mainly in the United States, West Europe, and Japan, the current market size for slow and controlled release N fertilizers globally is about 4.3 Mt which is projected to reach 7.74 Mt by 2024/2025. During this period, China will account for 2.6 Mt of the increase bringing its projected total demand to 5.4 Mt or about 70% of the total slow- and controlled-release market (Pursell 2017). This growth in China is attributed to the government policy encouraging production and use of more nutrient efficient fertilizers with lower environmental footprint. This policy was further strengthened in 2015 (Reuters 2015) when the government announced that it will target zero fertilizer growth by 2020.

The market size for stabilized fertilizers containing either nitrification or urease inhibitors was about 8 million tons in 2015/2016 and is projected to reach about 18.5 million tons in 2025 with North America accounting for about 8.5 million tons and China about 6.5 million tons.

The demand for slow-controlled release and stabilized fertilizers is determined by several factors including incremental price over conventional urea, increased production from its use, and prevailing market price of the crop. The price of conventional urea is 1.5–2 times cheaper so there is a tradeoff for both consumers and producers. For consumers, the tradeoff is paying more for efficient fertilizer and thus buying less of conventional fertilizer. For producers, the tradeoff is between receiving higher price and selling less volume.

The cost of producing CRFs mostly depends on capacity and cost of the plant, cost and thickness of coating materials, release rate, and so on. The present thrust of research into CRFs is to test alternate coating materials that are cheaper than and as effective as the current materials and reduce the coating thickness without compromising release rate and improve technology of coating. Also, there is on-going research to add micronutrients during coating of granules to not only control the release rate but also overcome soil micronutrient deficiencies thereby increasing yield and nutrient content of crops.

## 13.6  WAY FORWARD

The need for sustainably producing more nutritious food for an expanding population, particularly in developing economies, necessitates the increased use of fertilizers with higher N-NUE. The current suit of controlled-release nitrogen fertilizers, because of their costs and other factors, is mainly used for high value crops except for some products, for example, ESN, which are mainly used in the United States. But the recent policy in China that will target zero fertilizer growth by 2020 (Reuters 2015) to reduce wasted resources and air and water pollution and the Indian government's support of neem-coated urea adds to the urgency for fertilizer with higher N-NUE for cereal crops in developing economies.

These new CRFs are based primarily on existing knowledge of systems and chemistry. Hence, in developing the next generation of fertilizers with higher N-NUE, it is important to utilize knowledge of plant physiology, including mineral nutrient uptake mechanisms, their translocation and metabolism in identifying the composition, amount, and timing of nutrient release to align with plant needs. Other avenues, particularly advances in nanotechnology, could be leveraged to produce or deliver fertilizers in nanoparticle form. Furthermore, ecological processes, including interactions and symbioses with microorganisms, could be exploited for efficient N delivery and uptake. Finally, it is important to look beyond agriculture-related sectors to others, such as the pharmaceutical sector for clues to develop new products. Drugs that deliver medication directly to intended cells or regions might aid in determining how to develop fertilizers that only release nutrients when plants need them.

Such a transformative approach must be a joint public-private sector endeavor. The public sector develops the fundamental knowledge forming the basis for the private sector to develop new products to optimize resource use, reduce the negative impact on the environment, improve productivity, and increase farmer profits. Revolutionizing global food production with this new era of volume to value will only prevail when everyone is fully engaged and supported.

## REFERENCES

Baligar, V.C., Fageria, N.K. and He, Z.L. (2001). Nutrient use efficiency in plants. *Commun. Soil Sci. Plant Anal.* **32**, 921–950.

Bindraban, P.S., Dimkpa, C.O., Nagarajan, L., Roy, A.H. and Rabbinge, R. (2014). Towards fertilisers for improved uptake by plants. *Proceedings International Fertiliser Society* **750**.

Blaylock, A. (2010). Enhanced Efficiency Fertilizers. Colorado State University Soil Fertility Lecture.

FAOSTAT. (2015). Food and Agriculture Organization of the United Nations, statistics website: http://faostat3.fao.org/home/E.

Lammel, J. (2005). Cost of different options available to farmers: Current situation and prospects. IFA International Workshop on Enhanced Efficiency Fertilizers, Frankfurt. International Fertilizer Industry Association, Paris, France.

Mangat, G.S. and Narang, J.K. (2004). Agronomical trial for efficacy of NFL neem-coated urea. *Fertiliser Marketing News* **35**(11), 1–3, 5, 7.

Prud'homme, M. (2015). International Fertilizer Association, Paris, France.

Pursell, T. (2017). Pursell Agri-Tech, Birmingham, AL.

Reuters. (2015). China targets zero growth in chemical fertilizer use in 2020. http://www.producer.com/daily/china-targets-zero-growth-in-chemical-fertilizer-use-in-2020/.

Roy, A.H. (2015). Global fertilizer industry: Transitioning from volume to value—The 29th Francis New Memorial Lecture. *Proceedings International Fertiliser Society* **769**.

Shaviv, A. (2005). Controlled release fertilizers. IFA International Workshop on Enhanced Efficiency Fertilizers, Frankfurt. International Fertilizer Association, Paris, France.

Trenkel, M.E. (2010). Slow- and controlled-release and stabilized fertilizers: An option for enhancing nutrient use efficiency in agriculture. International Fertilizer Association, Paris, France.

UN. (2017). United Nations Department of Economic and Social Affairs, Population Division. *World Population Prospects: The 2017 Revision, Key Findings and Advance Tables.*

VFRC. (2012). Virtual Fertilizer Research Center, part of International Fertilizer Development Center. *A Blueprint for Global Food Security.*

# 14  Managing the Soil Nitrogen Cycle in Agroecosystems

*Richard A. Ferrieri, Eliot Herman,*
*Benjamin Babst, and Michael J. Schueller*

## CONTENTS

## 14.1  INTRODUCTION

### 14.1.1  Agriculture Is Challenged by a Growing Global Population

Agriculture in the twenty-first century faces many formidable challenges with the growing and economically empowered global population placing increasing demands on the planet's natural resources and energy supplies. The increased demand for food and fiber to feed and clothe this growing population is already taxing existing agricultural practices. Projections indicate that feeding a world population of 9.77 billion people by 2050 (UN DESA Report 2017) will require further increases in food production by more than 60% (Baldwin et al. 2013). Production in developing countries will need to almost double. Furthermore, as we attempt to increase reliance on a bioenergy-based market, agriculture will need to increase outputs of biomass and chemical feedstocks to keep future energy and industrial resources in balance. These challenges to agriculture are compounded by the fact that business and governmental responses to climate change are placing additional pressure on the agricultural industry to make these rapid gains in crop improvement and production without increasing agricultural inputs.

Since the Green Revolution, breeding programs have delivered a successful record of accomplishments by producing new, environmentally sustainable cultivars across many cropping systems that can deliver high yields in the face of environmental stressors. Progress in crop tree breeding has also made progress, although at a much slower pace due to the long generation time of trees (Neale 2007; Holliday et al. 2017). The limiting factors to trait selection in crop breeding have been the speed and cost at which phenotypes and genotypes can be measured accurately and precisely. The recent infusion of new technologies such as unmanned aerial drones for improving high-throughput phenotyping in the field as well as automation and improvements in data analytics has been a major driver for reducing the turnaround time for breeding programs, and will contribute to future gains. Even so, there are many examples where plateaus in genetic gain are now being realized despite these improvements in the trait selection process. This is not to imply that crop breeding is exhausted, and progress in breeding will in fact be aided by both high-throughput phenotyping technologies and genomics/systems biology approaches. There are thousands of plant genes whose roles in metabolism and/or regulation have yet to be discovered. However, to connect the dots in the scientific roadmap between plant genotypes and plant phenotypes, the path to new gene discovery requires a more in-depth analysis of the plant's transcriptomics, proteomics, metabolomics, and ionomics. In the meantime, as a compensatory measure to optimize crop yields in the present day, farmers around the world are deploying intensive high-density cultivation, often requiring excessive amounts of fertilizers to sustain yields. However, there are environmental consequences to this practice.

### 14.1.2 Consequences of Too Much Fertilizer

Nitrogen (N) management is the most critical component for most agronomic systems. Nitrogen fertilizers common to crop production usually contain N in one or more of the following forms including nitrate, ammonium, or urea. Agriculture reflects the single largest geo-engineering initiative that humans have initiated on this planet, largely through the introduction of unprecedented amounts of reactive nitrogen into ecosystems through crop fertilization. Worldwide, chemical fertilizer consumption has increased fourfold during the last 50 years (FAO 2011); at present, 70% of all global N input to the biosphere is from human activities. With the continued growth in the global population it is expected that chemical fertilizer consumption will increase another fourfold by 2050 (Schlesinger 2009).

Once administered, nitrates in fertilizer dissolve readily in water and, therefore, can move about in the soil with the movement of soil water. Through leaching and run-off, rainfall can wash nitrates into local water systems and onward to coastal waterways, significantly impacting both local and distant aquatic ecosystems. While water-soluble, ammonium attaches readily to clay and organic matter particles in soil and, thus, it is less likely to leach away from the site of application. However, much ammonium can be lost through volatilization as ammonia gas. Furthermore, during the growing season, soil microorganisms convert ammonium to nitrate, which is the main form taken up by plants (Haynes and Goh 1978; Salsac et al. 1987; Schlesinger 2009; Boudsocq et al. 2012). Nitrogen from urea becomes available to plants, when it is converted to ammonium by soil bacteria. However, urea, like nitrates, readily dissolves in water and thus can be lost through the leaching process causing similar effects to the surrounding ecosystem.

Agricultural N fertilization can have numerous negative planetary effects. For example, $N_2O$ and NO emissions from microbial denitrification increases greenhouse gas (GHG) levels in the atmosphere and accounts for 70% and 30% of atmospheric emissions, respectively (Bremner and Blackmer 1978; Smith et al. 1997, 2007; Hofstra and Bouwman 2005). Last year, approximately 19 M tons of $N_2O$ was released into the atmosphere from fertilizer sources. This gas has a global warming potential that is equivalent to 265–298 times that of $CO_2$ for a 100-year timescale. $N_2O$ emitted today remains in the atmosphere for more than 100 years, on average (IPCC 2012). Furthermore, the planet's protective ozone layer can be damaged by NOs that reach the stratosphere (Crutzen and Ehhalt 1977). Additionally, nitrite in drinking water can cause human health problems and has been linked to methemoglobin anemia. Finally, eutrophication, or the surface run-off of nitrate, can influence ecosystem biodiversity. There have been numerous documented cases of algal blooms forming in freshwater estuaries and in coastal waters due to nitrate run-off (Landsberg 2002; Gilbert et al. 2005; Heisler et al. 2008). As these algae die, bacteria feed on their decaying biomass, consuming dissolved oxygen during respiration, creating severe hypoxic conditions, and in turn, killing off major fisheries resulting in economic loss. The ecological damage alone is considerable and far-reaching on the landscape. In and of its own, global nitrate run off can add up to $81B in losses annually just for the cost of manufacturing these fertilizer resources.

### 14.1.3  ECONOMICS AND ENVIRONMENTAL CONSEQUENCES OF FERTILIZER PRODUCTION

Currently, industrial synthesis of ammonia via the Haber–Bosch process is the starting point to all N fertilizer production worldwide. This process uses iron oxide as a catalyst coupled with other oxide promoters to react $N_2$ and $H_2$ in the gas phase. Overall, the net reaction is thermodynamically exothermic:

$$N_{2(g)} + 3H_{2(g)} \leftrightarrows 2NH_{3(g)} \quad \Delta H^\circ = -92 \text{ kJ}$$

However, the reaction kinetics are sluggish owing to high activation energy barriers for breaking chemical bonds, hence the process requires high temperature between 400 and 600°C, and elevated pressures between 200 and 400 atmospheres to drive it. Both inputs are costly commodities. Additionally, the process requires $H_2$ which derives from steam reforming of methane (Schrock 2006). The total methane steam reforming process, including the gas-shift reduction of CO, releases $CO_2$ into the atmosphere. Even so, the Haber–Bosch process is relatively efficient, typically operating at 8–15% conversion rates (Modak 2002; Schlogl 2004) producing 1800 tons of ammonia gas per day globally, with roughly one-third going to ammonia fertilizers at a net energy cost of 28 GJ/ton $NH_3$ produced (Patyk 1996; Kongshaug 1998; Davis and Haglund 1999). With these facts, the question to ask is how renewable and non-polluting is this process? Consuming roughly 2–3% of the global energy budget, reliance solely on this process may not be sustainable in the long-term. In addition, the production of nitric acid from ammonia, the feedstock used for synthesis of complex N-fertilizers, results in additional $N_2O$ emissions to the atmosphere and energy costs (IPCC 2012).

### 14.1.4  NITROGEN CYCLE IN INTENSIFIED AGRICULTURAL SYSTEMS

Biological N fixation (BNF), ammonification, nitrification, and denitrification are all important components of the soil N cycle which regulate the status of N in terrestrial ecosystems (Figure 14.1). The soil microbiome can be biologically diverse. A tablespoon of soil can contain 50 billion microorganisms with a biodiversity numbering in the thousands.

The diazotrophic microorganisms have evolved with the capacity to produce reduced forms of N as ammonium utilizing biocatalytic processes. These include: (1) nitrate/nitrite reductase enzymes operating on nitrate ($NO_3^-$) and nitrite ($NO_2^-$) substrates, respectively; and (2) nitrogenase enzyme operating on $N_2$ substrate from the atmosphere. *In vivo*, these enzymes depend upon several biochemical processes for recycling redox cofactors, generating substrates, and cycling electron carriers (Figure 14.2).

The biological oxidation of ammonium ($NH_4^+$) to $NO_2^-$ and $NO_3^-$ evolved roughly 2.5 billion years ago (Berner 2006). Two groups of chemo-lithotrophic microorganisms (*Nitrosomonas* spp. and *Nitrobacter* spp.) are ubiquitous in the soil and actively participate in ammonium oxidation under aerobic conditions (Leninger et al. 2006; Taylor et al. 2010). Additionally, other soil bacterial spp. including *Nitrosocytus* and *Nitrosospira*, as well as some heterotrophic fungi including *Aspergillus flavis* can be important in the nitrification process of certain ecosystems (Sommer et al. 1976).

In intensified agricultural ecosystems, rapid and unregulated nitrification can often result in inefficient N use by crops and increased nitrate leaching to the surrounding environment. Higher plants possess the ability to assimilate and use either $NH_4^+$ or $NO_3^-$ as their N source. In fact, assimilation of $NH_4^+$ is energetically more efficient for the plant than that of $NO_3^-$ where 20 moles of ATP are consumed by the plant per mole of $NO_3^-$ assimilated, whereas only 5 moles of ATP are consumed per mole of $NH_4^+$ assimilated (Salsac et al. 1987). Moreover, assimilation of $NO_3^-$, but not of $NH_4^+$, results in the direct emission of $N_2O$ from crop canopies (Smart and Bloom 2001). Unfortunately, the state of the soil N cycle is such that 95% of N uptake in non-leguminous cropping systems is through $NO_3^-$ assimilation (Haynes and Goh 1978; Salsac et al. 1987; Boudsocq et al. 2012).

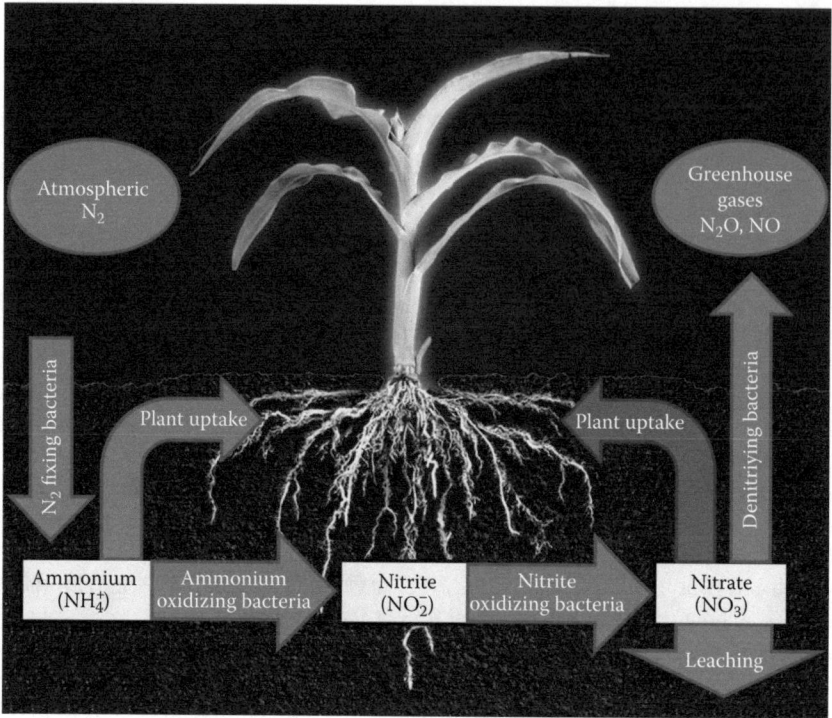

**FIGURE 14.1** The soil nitrogen cycle reflects the actions of microorganisms involved in biological nitrogen fixation, ammonification, nitrification, and denitrification.

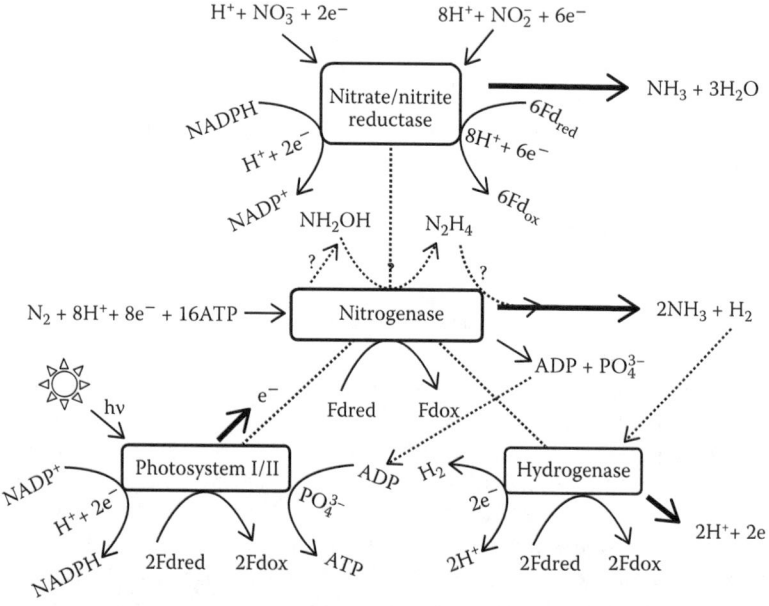

**FIGURE 14.2** Illustration of the interrelated reaction centers in the cyanobacterium *Anabaena variablis*. Nitrogenase is one target enzyme for its ability to generate $NH_3$ from $N_2$ in this proposed project. Note that peripheral reaction centers have many shared mediators, co-factors, reactants, and products in the integrated system. Solid black lines reflect known pathways; gray lines reflect possible pathways that have not yet been verified.

Consequently, mitigation strategies that focus on managing the soil N cycle can have far reaching effects on the future bioeconomy.

## 14.2 STRATEGIES FOR ATTAINING A GLOBAL NITROGEN-NEUTRAL FOOTPRINT

### 14.2.1 BIOLOGICAL NITROGEN FIXATION

Rather than relying purely on applications of N fertilizer, alternative N sources are needed to help develop more sustainable farming systems. Nodule-forming legumes have the potential to fulfill this requirement due to the ability of their nodules to fix N biologically from the atmosphere, benefiting not only the legumes themselves but also intercropped plants or subsequent crops. The ability to fix atmospheric N comes from the intimate symbiotic relationship between legumes and rhizobia, bacteria in soil, in which the legume supplies the products of photosynthesis, energy, and carbon (C) resources, to rhizobia, and rhizobia provide the legume with N, mainly in the form of ammonium (Howard and Rees 1996). The symbiosis initiates from the infection of legume roots by rhizobium, forming root nodules where N fixation occurs (Figure 14.3). There are many instances in agriculture where such symbiotic relationships have been incorporated into farming of legumes with 40–90% total N accumulated through BNF (Figure 14.3) (Rumjanek et al. 2005; Mendonca et al. 2017).

There are also a few examples of nodule formation in non-leguminous plants (for a review, see Santi et al. 2013). Actinorhizal plants tend to be woody shrubs and trees. They represent a diverse group of about 220 species belonging to eight plant families distributed in the three orders, Fagales (*Betulaceae, Casuarinaceae and Myricaceae*), Rosales (*Rosaceae, Eleagnaceae* and *Rhamnaceae*), and Cucurbitales (*Datiscaceae* and *Coriariaceae*) (Wall 2000; Pawlowski 2009; Franche and Bogusz 2011). Actinorhizal plants can develop an endosymbiosis with the N-fixing soil actinomycete *Frankia*, and in this establishment, they readily form root nodules much like legumes in which *Frankia* provides fixed N to the host plant in exchange for reduced carbon.

Staple grains account for roughly 43% of caloric intake by the global human population. This statistic is higher for certain developing countries. While staple grains derived from grass cropping

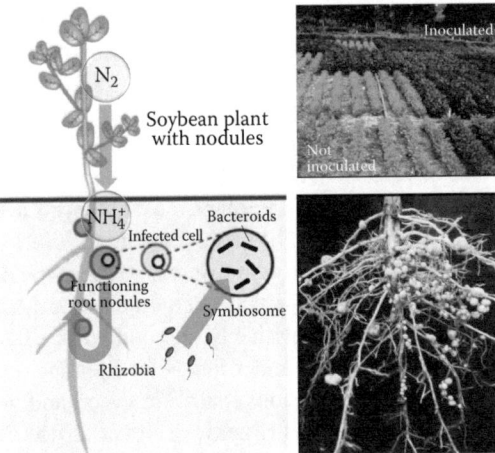

**FIGURE 14.3** Symbiosis between rhozbia and legume root cells results in the formation of root nodules, or tiny root appendages that contain bacteroids that fix $N_2$ from the atmosphere. After years of research the means to establish this symbiosis is well understood and has been cultivated to a commercialized practice with many leguminous cropping systems like soybean (*Glycine max*) where high crop yields can be obtained with minimal nitrogen fertilization.

systems are not known for their ability to form conspicuous symbioses with rhizobia like legumes, there is a large body of literature that attests to the effects of certain plant growth promoting bacteria (PGPB) in grasses (see *Plant and Soil*, Special Focus Issue, 2012; Döbereiner 1968, 1970, 1973, 1976, 1979; Hurek et al. 1994a; Yanni et al. 1997; James et al. 1998), including significant increases in crop yield (Charyulu et al. 1985; Okon et al. 1994; Dobbelaere et al. 2001; Pedraza et al. 2009; Farrar et al. 2014). PGPB have been documented to colonize many grass species relevant to both bioenergy and agriculture (Pacovsky 1990; Eckert et al. 2001; Bashan et al. 2004). In some cases, it has been suggested that BNF may contribute to this plant growth enhancement in grass systems (Carroll and Sommerville 2009; Tjepkema and Burris 1976; Christian et al. 2008). However, in most cases, the underlying mechanisms of plant growth promotion are unknown and have been attributed to a variety of sources, including antagonism toward phytopathogens (Raaijmakers et al. 2009), induction of plant resistance (Verhagen et al. 2004), or phytostimulation through the transfer of phytohormones to their host (Richardson et al. 2009).

Several $N_2$-fixing PGPB have been identified as endophytes of grass species, including *Azoarcus* spp. in Kallar grass (*Leptochloa fusca* L. Kunth) and rice (*Oryza sativa*) (Reinhold et al. 1986; Hurek et al. 1994b), *Herbaspirillum seropedicae* in sugarcane (*Saccharum officinarum*) (James et al. 1998) and sorghum (*Sorghum bicolor*) (James et al. 1997), and *Gluconacetobacter diazotrophicus* in sugarcane (James et al. 1994). Others have been identified as epiphytes, including *Azospirillum brasilense* and *A. lipoferum*, which have been commercialized as crop inoculants for maize and wheat (Okon et al. 1994; Dobbelaere et al. 2001; Hungria et al. 2010) and are gaining increasing acceptance in agriculture as PGPB inoculants (e.g., http://votivo.us/). Unlike rhizobia that form an intracellular symbiosis with their legume hosts (Figure 14.3), PGPB do not induce the formation of observable plant structures (nodules). They are also usually not major components of the soil microflora (James et al. 1998; Reinhold-Hurek and Hurek 1998). These $N_2$-fixing bacteria infect at the emergence of lateral roots and in the zone of elongation and differentiation above the root tip (Pankievicz et al. 2015). After infection, the bacteria colonize the outer root cell layers and the root cortex, but can also access the vascular tissue. These associations are strictly defined by the lack of any evidence of intracellular infection (James et al. 1998; Reinhold-Hurek and Hurek 1998). Even so, very high numbers of PGPB in roots have been reported (i.e., $\leq 10^8$/gram root dry weight) and with no observable disease symptoms (Reinhold et al. 1986; Barraquio et al. 1997).

Early work in green foxtail grass (*Setaria viridis*) established it as a model C4 grass for exploring BNF by *H. seropedicae* and *A. brasilense* bacteria (Pankievicz et al. 2015). For the first time, direct evidence for provision of N to grass by BNF in these PGPB was provided using the radiotracer $^{13}N_2$. These studies firmly established $N_2$ as a nitrogen source for plants inoculated with these bacteria, including observation of $^{13}N$ transport to the shoot, and incorporation into $^{13}N$-ribulose-1,5-bisphosphate carboxylase oxygenase (RUBISCO), a protein unique to plants which is involved in leaf photosynthesis and the capture of $CO_2$. Perhaps most notably, the optimal plant and bacterial genotypes showed evidence that sufficient BNF was occurring to fulfill 100% of the N needs of the plant (Figure 14.4). Specifically, the A10.1 growth responsive genotype of *S. viridis* inoculated with a hyper $N_2$-fixing strain of *A. brasilense* (HM053) fixed on average 12,230.7 ± 5922.3 ppt $N_2$ on a dry root mass basis which equates to a cumulative fixation rate of 2 ± 0.6 µmol $N_2$ d$^{-1}$. This level of BNF was sufficient to allow plants to thrive under low N fertilization conditions. Even so, this was the only genotype out of 32 *S. viridis* accessions that were tested and verified to exhibit a positive growth response under N-limiting conditions (Pankievicz et al. 2015). More research is needed to understand why the A10.1 genotype was so responsive.

Unlike agricultural fields, forests generally tend to accumulate N over time, given a long-enough rotation between harvesting. This is partly because harvest of food crops removes nutrients from the site, but it is also partly due to the nurturing environment that trees create for N-fixing microbes in forest soils. The above studies suggest that N-fixing microbes could also potentially be co-cultivated with agricultural crops to supplement or replace manufactured N inputs.

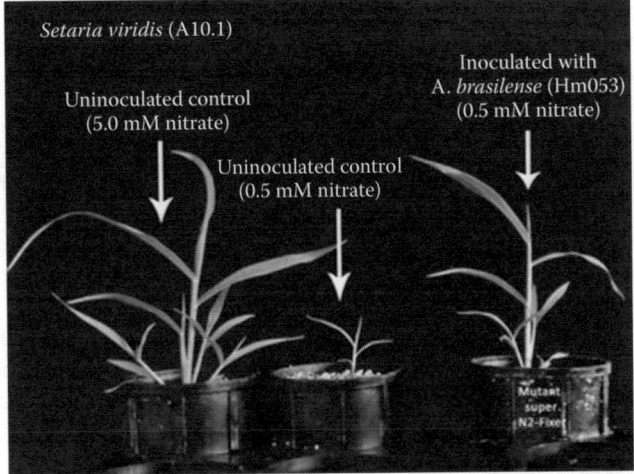

**FIGURE 14.4**  Work using *Setaria viridis* as a model C4 grass has shown that inoculation of the A10.1 genotype with a hyper N$_2$-fixing mutant strain of *Azospirillum brasilense* (HM053), a root epiphyte, results in sufficient biological nitrogen fixation and transference of nitrogen to the host to enable healthy plant growth under limiting soil nitrate levels.

### 14.2.2  DECOUPLING OF PLANT CARBON ASSIMILATION FROM NITROGEN DEMANDS

In a biogeographic and ecological context, plants are highly competitive when N availability is limiting. Plants employ various mechanisms in competition to acquire N such as enhanced root growth to enlarge the surface area for uptake, sequestration of N in constituents of the plant, adaptations to minimize N needs and maximize N use efficiency (NUE), and, in legumes and a few other plant types, to reduce N with the aid of symbiotic bacteria.

Typically, the uptake and metabolism of N in higher plants is kept in homeostasis with the plant's C input or photosynthetic capacity (Cooke et al. 2003; Nunes-Nesi et al. 2010; Schmollinger et al. 2014). This behavior was demonstrated in a series of radiotracer experiments conducted by the authors where a young birch sapling was tested for radioactive $^{13}$N-nitrate acquisition using positron emission tomography (Figure 14.5), an imaging technique used to quantify plant nutrient uptake and allocation (Karve

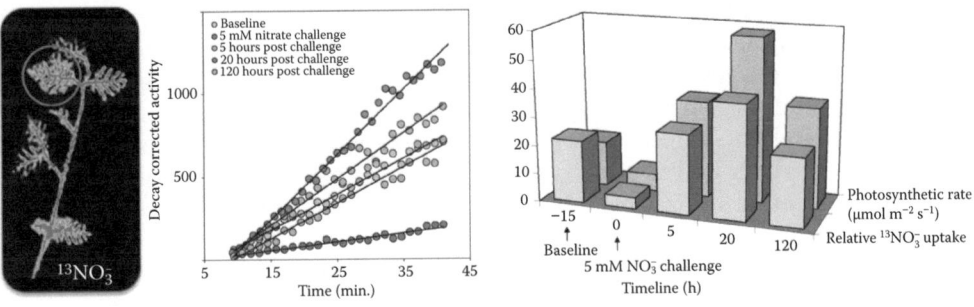

**FIGURE 14.5**  **(See color insert.)** Radiotracer imaging using $^{13}$N-nitrate coupled with positron emission tomography shows changes in nitrate acquisition of a young poplar sapling before and after a nitrate challenge. The kinetic uptake curves reflect the amount of $^{13}$N accumulation in the red circled mature leaf. The short half-life of $^{13}$N ($t_{1/2}$ 9.97 min) also affords opportunity to reimage the sapling over time after repeated administration of new doses of tracer to the roots. At administration, nitrate acquisition is initially shut down, but quickly increases in the subsequent hours. By 120 hours, nitrate acquisition is seen to return to baseline response. An infrared gas exchange cell placed on the same mature leaf maps changes in leaf photosynthetic capacity ($\mu$mol mm$^{-1}$ s$^{-1}$). Carbon input to the sapling, reflected by changes in leaf photosynthetic capacity, is seen to correlate with changes in nitrate acquisition.

et al. 2015; Qu et al. 2016). In these studies, nitrate acquisition was altered due to a nitrate addition treatment. Nitrate acquisition initially decreased after treatment, and then increased for several hours before returning to baseline kinetics at 120 hours. Leaf photosynthesis, measured using infrared gas exchange showed changing trends in C input that mapped to the plants nitrate acquisition patterns indicating that plant C and N acquisition are directly coupled. Similarly, root development is responsive to the levels of C input through photosynthesis (Brun et al. 2010). What if the two could be decoupled?

The plant $P_{II}$ protein is involved in perception of C-N balance, through ATP, 2-oxoglutarate, and glutamine binding (Leigh et al. 2007; Uhrig et al. 2009; Osanai et al. 2007; Fokina et al. 2010; Forchhammer et al. 2008; Huergo et al. 2013; Zeth et al. 2014). *Arabidopsis* is the only plant where $P_{II}$ has been studied in detail (Ferrario-Mery et al. 2005; Ferrario-Mery et al. 2006; Mizuno et al. 2007; Feria Bourrellier et al. 2009). Although metabolism was altered in *Arabidopsis* mutants with knocked out $P_{II}$ protein, the mutants had no obvious morphological or developmental phenotype (Ferrario-Mery et al. 2005). Grown in hydroponics with low N availability, $P_{II}$ mutants accumulated slightly more carbohydrates and slightly fewer amino acids than wild type plants, but key N assimilation activities (nitrate reductase and glutamine synthetase) were not altered (Ferrario-Mery et al. 2005). Although these effects on C-N balance were slight in *Arabidopsis* $P_{II}$ mutants, plants of the *Brassicaceae* family are missing the C-terminal Gln binding loop (Chellamutha et al. 2014), so it is possible the $P_{II}$ pathway might be a good target for altering or decoupling carbon-nitrogen balance in non-*Brassicaceae* species.

Additionally, lessons learned from observing natural evolution can provide us with guidance on how to decouple plant C-N demands in agroecosystems. With competition for N being high in natural ecosystems, there is selection pressure for plants to be more broadly tolerant of limiting N, and for plants to tolerate shifts in composition to a higher C:N ratio than might occur under conditions of N surplus. In some biogeographic regions, the soil and environmental conditions can limit microbial N fixation, and these environments are relatively N-poor. Alpine regions are an example of such poor soil environments and in these environments plant diversity and density is much reduced to plants that can adapt to lower N and likely other limiting nutrient conditions (Shaheen et al. 2011; Körner et al. 1986; Haselwandter et al. 1983). It is of selective advantage for plants to be able to thrive with a higher C-N composition in low N-environments. For example, many flowering plants of the *Umbelliferae* family that are commonly found in alpine pastures tend to have low asparagine content (McKee 2013). We took this as a clue to an engineering approach to create N efficient crops.

As a primary vehicle of N flux, the concentration of free asparagine in plant cells is a determinant of N status that comprises one of the primary regulators of plant growth and development and of N metabolism (Joy 1988; Fischer et al. 1998; Lam et al. 1996; Lea et al. 2007; Lima and Sodek 2003). Low carbohydrate content stimulates asparagine synthesis that is the inverse of the other amino acids. Asparagine has a feedback inhibition of nitrate reduction when systemic sugar content is depleted (Lamm et al. 1996). The role of asparagine in regulating metabolism suggests that plants can sense its content. Therefore, we hypothesized that reducing asparagine levels would result in rebalancing the systemic C flux in plants to favor increased carbohydrate and its polymers paralleling the observations on N-limited plants in the high alpine pastures. By reducing asparagine content and the resulting cascade of metabolic and transcriptomic changes would alter the plant in ways favoring a high C fixation and high biomass.

Asparagine content can be manipulated by either inhibiting its biosynthesis, or by reducing cellular free asparagine through increased *in situ* hydrolysis. Hydrolysis of free asparagine does not impair any major metabolic pathway, and the Asparaginase enzyme (ASNase) does not completely eliminate free asparagine. Rather, the content of free asparagine is a sum of complex factors, including competition for asparagine substrate and regulation of compensatory synthesis. As shown in the results described herein, enhanced hydrolysis of asparagine by overexpressing ASNase significantly reduces the concentration of free asparagine, resulting in useful plant phenotypes, as well as increased C storage with less demand on soil N.

Transgenic methods were used to transfer a nucleic acid sequence encoding the Asnase from potato and soybean (*Glycine max.*) as donor plants for generating ASNase overexpressing lines

(35S:PotASNase, lines 1–4; 35S:SoyASNase, lines 2–38) in *Nicotiana tobaccum* (Herman 2014). In addition to growth phenotypic and metabolic comparisons with wild-type plants, C and N demands were measured in a subset of transgenic plants using radioactive tracers ($^{11}CO_2$ and $^{13}NO_3^-$). As expected, free steady state asparagine levels were significantly reduced in 35S:SoyASNase (2–38) overexpressing plants to about 15% of control non-transgenic plant levels. The transgenic tobacco plants exhibited dwarfism as a growth phenotype with shorter stems and fewer leaves and a short-ened life cycle (Figure 14.6). However, leaf biomass was significantly increased in transgenic plants although there was a lower leaf count. This observation was attributed to increased specific leaf mass. No differences were observed in root growth between wild-type and the 35S:SoyASNase transgenic line. Altogether, the overall biomass of the ASNase over-expressing plants was compa-rable to the wild type controls or slightly higher (not statistically significant), but with an altered morphology and the production of biomass over a truncated time frame.

Physiological measurements of leaf function using $^{11}CO_2$ revealed that there were no differences in C fixation or allocation between wild type and 35S:SoyASNase transgenic plants (Figure 14.7).

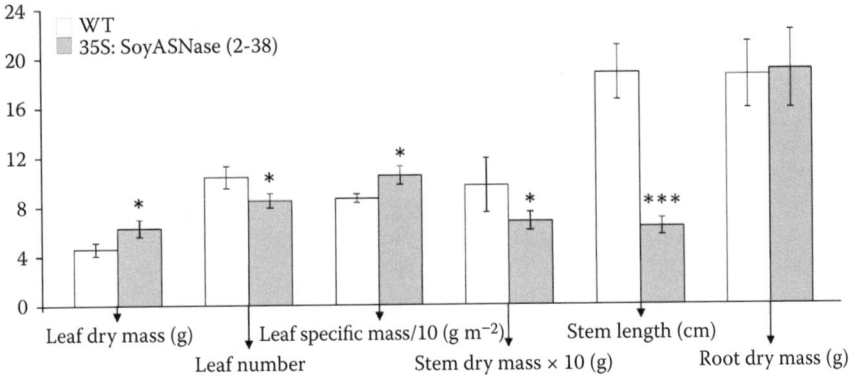

**FIGURE 14.6**  Leaf, stem, and root growth traits were compared between WT and 35S:SoyASNase (2–38) transgenic tobacco plants.

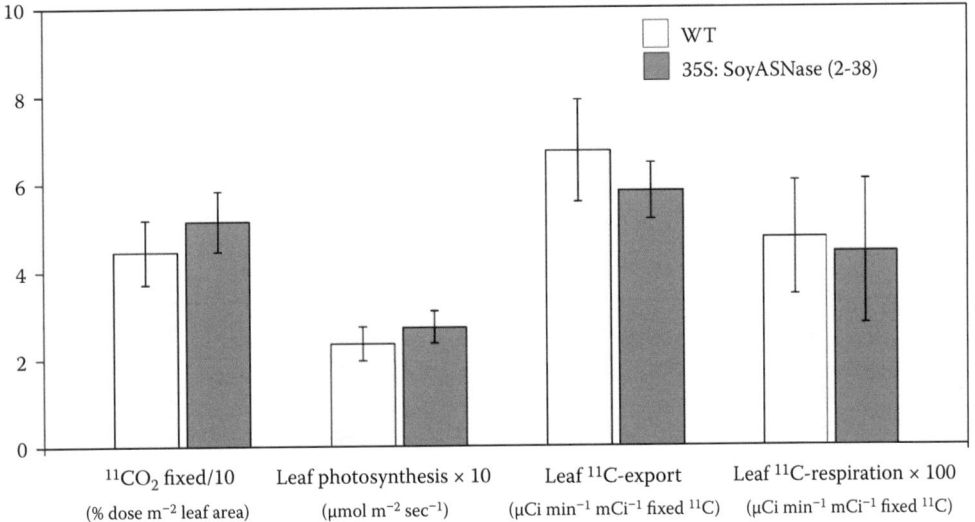

**FIGURE 14.7**  Leaf physiological functions, including $CO_2$ fixation, photosynthetic capacity, leaf export of photoassimilates, and leaf respiration were assessed using $^{11}CO_2$ tracer administered to WT and 35S:SoyASNase (2–38) transgenic tobacco plants.

However, we found that there was a reprogramming of leaf C metabolism into storage substrates such as starch (Figure 14.8) and cell-wall cellulose (Figure 14.9). Finally, leaf total protein content was significantly reduced in both 35S:PotASNase and 35S:PotASNase transgenic lines relative to wild-type by about 33% (Figure 14.10). This major alteration in leaf metabolism was reflected by a lower nitrogen demand in 35S:PotASNase, as seen by $^{13}NO_3^-$ uptake studies (Figure 14.11). Taken together, the 35S:PotASNase transgenic plants exhibited a higher C/N value of 27, as determined by elemental C/N analysis, than the wild type which had a C/N value of 12, demonstrating that enhanced C content can be engineered in plants with an overall lower N demand by decoupling C and N metabolism. For applications in which high C/N is acceptable, plants with decoupled C and

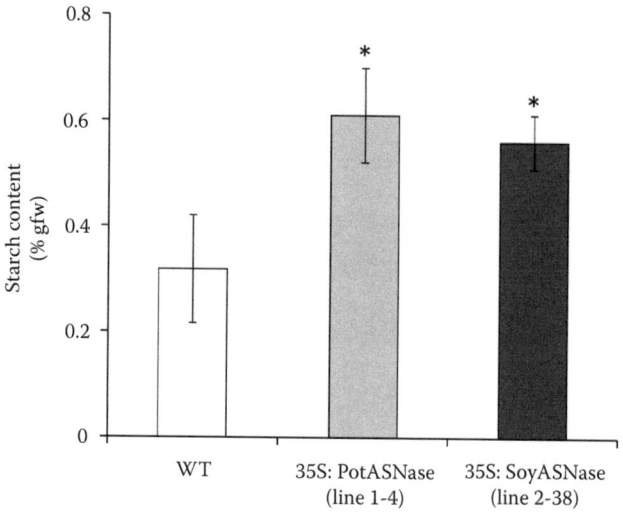

**FIGURE 14.8**   Leaf starch content was compared between WT and S35:PotASNase (1–4) and S35:SoyASNase (2–38) transgenic tobacco plants.

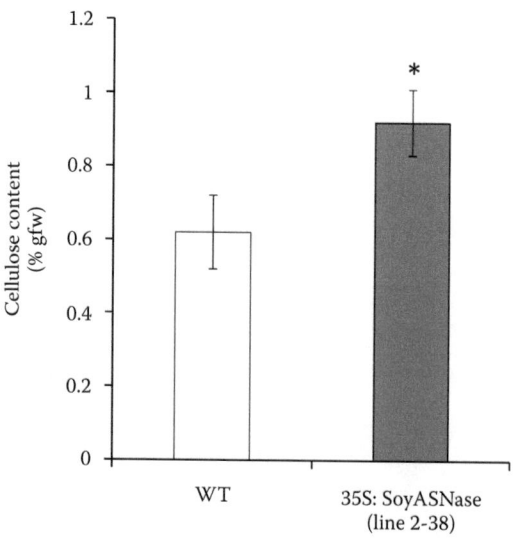

**FIGURE 14.9**   Leaf cellulose content was compared between WT and S35:SoyASNase (2–38) transgenic tobacco plants.

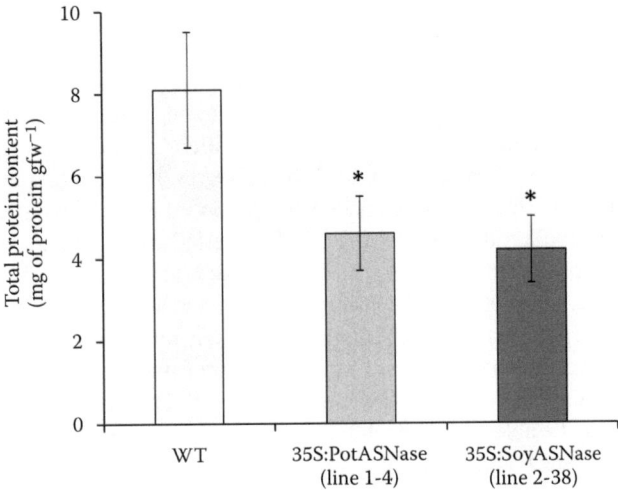

**FIGURE 14.10** Leaf total protein content was compared between WT and S35:PotASNase (1–4) and S35:SoyASNase (2–38) transgenic tobacco plants.

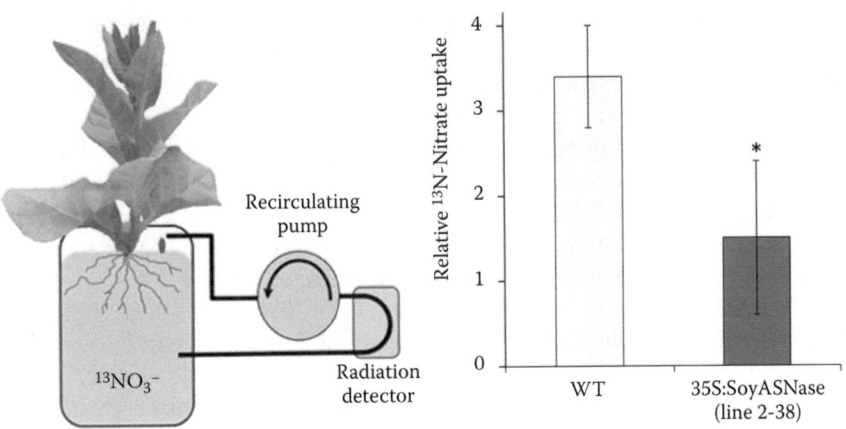

**FIGURE 14.11** Relative uptake of nitrate was compared between WT and S35:SoyASNase (2–38) transgenic tobacco plants using hydroponically grown plants and $^{13}NO_3^-$ tracer.

N metabolism may require less extrinsic N input, decreasing the economic and environmental costs of fertilizer. Future work is needed to see if the observed short stem stature can be rescued, while retaining the desirable high biomass, low nitrogen demand. For crops that are produced for biomass or other C-based feedstocks, reducing input costs could provide a new strategy to engineer crops for reduced life-cycle N footprint, enhancing overall sustainability.

### 14.2.3 Striving Toward Low Nitrifying Agroecosystems

Rapid microbial conversion of $NH_4^+$ to $NO_3^-$ in the soil sets the stage for inefficient use of natural soil N resources, applied N fertilizers, as well as organically bound forms of N in the soil (Raun and Johnson 1999). All are subject to oxidative conversion in the nitrification processes involving microorganisms which, as already discussed, can have far-reaching effects on the environment and human health, as well as result in huge economic losses. It is clear that N availability and crop utilization of this resource are key agronomic factors to crop performance and crop yield in any agroecosystem.

So, what can be done to deliberately reshape the balance of power in the soil microbiome to suppress soil nitrification? Use of synthetic antioxidant chemical treatments that inhibit the bacterial oxidation of $NH_4^+$ in principle should improve N availability and N utilization in agroecosystems (Hendrickson et al. 1978; Bremner et al. 1981; Rodgers 1986; Dinnes et al. 2002; Liao et al. 2004). In fact, several chemical inhibitors have been proposed for use in agroecosystems including nitrapyrin, DCD (dicyandiamide), and DMPP (3, 4-dimethyl pyrazole phosphate), which have all reached the agronomic evaluation stage (Guthrie and Bomke 1980; Weiske et al. 2001; Zerulla et al. 2001; Subbarao et al. 2006a, 2012). Nitrapyrin has been adopted for certain niche production systems such as winter wheat in North America. However, nitrification inhibitors are not widely used in production agriculture due to their limited biological stability in the soil, and the high costs to apply these chemicals to the soil (Sahrawat and Keeney 1985; Subbarao et al. 2012).

However, plants may be able to do the job for us. Many higher plants have evolved with the capacity to produce and excrete specialized chemicals from their roots that can suppress nitrification in the soil. For example, tropical pasture grasses that are adapted to low-N-environments, in particular *Brachiaria* spp., have extremely high biological nitrification inhibition (BNI) activity in root systems (Subbarao et al. 2007a). In contrast, *Panicum* spp. adapted to high N-environments have relatively weak BNI-activity in their root systems (Subbarao et al. 2007a, b). Among field crops, sorghum spp. adapted to low N input environments appear to have stronger BNI-capacity than crops adapted to high N input environments such as wheat (*Triticum aestivum*) and maize (*Zea mays*) (Subbarao et al. 2007a). In various sorghum accessions, a diverse group of lipobenzoquinones that possess BNI-activity have been identified in root exudates. These lipobenzoquinones are continuously excreted from the root hairs where they are biosynthesized. In particular, root hairs of *Sorghum* spp. secrete large amounts (roughly 20 µg of exudate $mg^{-1}$ fresh weight) of an oily exudate (Hess et al. 1992; Nimbal et al. 1996; Czarnota et al. 2003) that is comprised of between 80 and 95% sorgoleone, 2-hydroxy-5-methoxy-3-[(Z,Z)-8′,11′,14′-pentadecatriene]-p-benzoquinone; 4-methoxy-3,6-dioxo-2-[(8Z,11Z)-pentadeca-8,11,14-trien-1-yl]cyclohexa-1,4-dien-1-olate, a lipid benzoquinone, and several structural congeners which differ in their substitution pattern on the ring and in the length and level of unsaturation in the tail (Netzly et al. 1988; Fate and Lynn 1996; Kagan et al. 2003).

Nitrification inhibition by plants is likely an adaptive mechanism to retain and use N efficiently in N-limited natural systems (Subbarao et al. 2006; Subbarao et al. 2007a; Lata et al. 2004). However, the natural ability of plants to suppress nitrification is not widely recognized or utilized in agricultural practices even though it has been shown to strongly influence the microbiome (Subbarao et al. 2006). For example, brachialactone secreted by *B. humidicola* roots inhibited growth of nitrifying bacteria in the soil microbiome, while having no adverse effects on other plant growth promoting microorganisms such as *Azospirillum lipoferum*, *R. leguminosarum*, and *Azotobacter chroococcum* (Gopalakrishnan et al. 2007).

Advances in genome editing capabilities hold promise for engineering desirable biochemical traits into cropping systems that could improve their capabilities for biological nitrification inhibition. This approach would be far superior to chemical treatments since there would be a continual supply of chemical inhibitors to the soil microbiome at no cost to the farmer.

## 14.3   CONCLUDING REMARKS

Never in the history of this planet has there been such a major upset and perturbation to the planetary nitrogen balance. Annual fertilizer input ($N + P_2O_5 + K_2O$) into agroecosystems has been increasing at an alarming rate since 2008 (Galloway et al. 2008) and has reached close to 197M metric tons per year (FAO, 2015), a level 1.5 times greater than Earth's nitrogen-fixing capacity. These actions have nearly doubled the reactive nitrogen capacity of the planet in just over 50 years. While societal efforts abound to try to attain a net neutral global carbon footprint, it seems apparent that an equal effort must be applied to achieve a neutral global nitrogen footprint. We have discussed many of the environmental, health, and economic concerns with present agricultural practices, and have proposed several strategies that could mitigate this global problem including the following:

1. Biological nitrogen fixation in non-leguminous cropping systems,
2. Use of genetic engineering approaches to decoupling plant C/N demands creating high carbonaceous feedstocks, and
3. Implementation of management practices that can yield low nitrifying agroecosystems.

Likely, no single strategy will solve this global problem. As a case in point, genetic engineering of high carbonaceous cropping systems by metabolically decoupling plant carbon-nitrogen demands will likely not be useful for food resources, as the nutritional value of crop products will be low. But for diverse industrial crops produced for wood, fiber, oil, and carbohydrate polymers, this strategy could have great benefits to humankind for sustainable future fuel and fiber resources, as well as for future chemical feedstocks. Finally, public opinion on the general use of GMOs in agroecosystems must evolve with broader acceptance of these practices if we are to be successful at implementing some of the mitigation strategies described in this chapter. The strategy to create low nitrifying agroecosystems would benefit most from use of GMOs in farming practices, but until then, intercropping practices using plants that have evolved with natural capabilities to reshape the balance of power of the soil microbiome should be considered.

## ACKNOWLEDGMENTS

This chapter has been authored under contracts DE-SC0012704 in support of the Radiochemistry Program with the U.S. Department of Energy (DOE), which supported RAF, BAB, and MJS for the experiments conducted at Brookhaven National Laboratory; and in part from a USDA National Institute of Food and Agriculture award # 2017-67013-26216 (to RAF and MJS), a AR/MO NSF EPSCoR Track 2 award #s IIA-1430427 and IIA-1430428 through a Plant Imaging Consortium (PIC) seed grant (to BAB); and from the USDA National Institute of Food and Agriculture, McIntire Stennis project 1009319 (to BAB).

## REFERENCES

Baldwin I.T.C., Benning C.A., Burke A., Caicedo A., Carpita N., Dilworth M., Horsch R. et al. 2013. *Unleashing a decade of innovation in plant science—A vision for 2015–2025*. American Society of Plant Biology, Howard Hughes Medical Institute, National Science Foundation, U.S. Department of Agriculture, U.S. Department of Energy.

Barraquio, W.L., Revilla L., and Ladha, J.K. 1997. Isolation of endophytic diazotrophic bacteria from wetland rice. *Plant and Soil* **194**: 15–24.

Bashan, Y., Holguin, G., and de-Bashan L.E. 2004. *Azospirillum*-plant relationships: Physiological, molecular, agricultural, and environmental advances (1997–2003). *Canadian Journal of Microbiology* **50**: 521–577.

Berner, R.A. 2006. GEOCARBSULF: A combined model for phanerozoic atmospheric $O_2$ and $CO_2$. *Geochimica et Cosmochimica Acta* **70**: 5653–5664.

Boudsocq, S., Niboyet, A., Lata, J.C., Raynaud, X., Loeuille, N., Mathieu, J., Blouin, M. et al. 2012. Plant preference for ammonium versus nitrate: A neglected determinant of ecosystem functioning? *American Naturalist* **180**: 60–69.

Bremner, J.M. and Blackmer, A.M. 1978. Nitrous oxide: Emission from soils during nitrification and denitrification of fertilizer nitrogen. *Science* **199**: 295–296.

Bremner, J.M., Breitenbeck, G.A., and Blackmer, A.M. 1981. Effect of nitrapyrin on emission of nitrous oxide from soils fertilized with anhydrous ammonia. *Geophysical Research Letters* **8**: 353–356.

Brun, F., Richard-Molard, C., Pagès, L., Chelle, M., and Ney, B. 2010. To what extent may changes in the root system architecture of *Arabidopsis thaliana* grown under contrasted homogenous nitrogen regimes be explained by changes in carbon supply? A modelling approach. *J Exp Bot* **61**: 2157–2169.

Carroll, A. and Somerville, C. 2009. Cellulosic biofuels. *Annual Review of Plant Biology* **60**: 165–182.

Charyulu, P.B.B.N., Foucassie, F., Barbouche, A.K., RondroHarisoa, L., Omar, A.M.N., Marie, R., and Balandreau J. 1985. Field inoculation of rice using *in vitro* selected bacterial and plant genotypes, in *Azospirillum III: Genetics, Physiology, Ecology,* Klingmuller W. (Ed.), Springer-Verlag, Heidelberg, Germany, pp. 163–179.

Chellamuthu, V.R., Ermilova, E., Lapina, T., Lüddecke, J., Minaeva, E., Herrmann, C., Hartmann, M.D., and Forchhammer, K. 2014. A widespread glutamine-sensing mechanism in the plant kingdom. *Cell* **159**: 1188–1199.

Christian, D.G., Riche, A.B., and Yates, N.E. 2008. Growth, yield and mineral content of *Miscanthus* × *giganteus* grown as a biofuel for 14 successive harvests. *Industrial Crops and Products* **28**: 320–327.

Cooke, J.E.K., Brown, K.A., Wu, R., and Davis, J.M. 2003. Gene expression associated with N-induced shifts in resource allocation in poplar. *Plant Cell and Environment* **26**: 757–770.

Crutzen, P.J. and Ehhalt, D.H. 1977. Effects of nitrogen fertilizers and combustion on the stratospheric ozone layer. *Ambio* **6**: 112–116.

Czarnota, M.A., Rimando, A.M., and Weston, L.A. 2003. Evaluation of seven sorghum (Sorghum sp.) accessions. *J Chem Ecol* **29**: 2073–2083.

Davis, J. and Haglund, C. 1999. Life cycle inventory (LCI) of fertilizer production. Fertilizer products used in Sweden and Western Europe. SIK-Report No. 654. Master's Thesis, Chalmers University of Technology, Sweden.

Dinnes, D.L., Karlen, D.L., and Jaynes, D.B. 2002. Nitrogen management strategies to reduce nitrate leaching in tile drained Mid-Western soils. *Agronomy Journal* **94**: 153–171.

Dobbelaere, S., Croonenborghs, A., Thys, A., Ptacek, D., Vanderleyden, J., Dutto, P., Labandera-Gonzalez, C. et al. 2001. Responses of agronomically important crops to inoculation with *Azospirillum. Functional Plant Biology* **28**: 871–879.

Döbereiner, J. 1968. Non-symbiotic nitrogen fixation in tropical soils. *Pesq Agropec Bras* **3**: 1–6.

Döbereiner, J. 1970. Further research on *Azotobacter paspali* and its variety specific occurrence in the rhizosphere of *Paspalum notatum* Flügge. *Zentralb Bakteriol Parasint Infektion Hyg* **124**: 224–230.

Döbereiner, J. 1973. Fixação de nitrogênio atmosférico na rizosfera de gramíneas tropicais. In: Conferência Internaciona; Sobre Os Impactos Globais Da Microbiologia Aplicada, 4, São Paulo, SP, Brasil, pp. 461–481.

Döbereiner, J. 1976. Fixação de nitrogênio atmosférico em gramíneas tropicais. In: Congresso Brasileiro De Ciência Do Solo, 15, Campinas, SP, Brasil, pp. 593–602.

Döbereiner, J. 1977. Biological nitrogen fixation in tropical grasses—Possibilities for partial replacement of mineral N fertilizers. AMBIO; *J Human Environ Res Manag* **6**: 174–177.

Eckert, B., Weber, O.B., Kirchhof, G., Halbritter, A., Stoffels, M., and Hartmann, A. 2001 *Azospirillum doebereinerae* sp. nov. a nitrogen-fixing bacterium associated with the $C_4$-grass *Miscanthus. International Journal of Systematic & Evolutionary Microbiology* **51**: 17–26.

FAO. 2011. FAOSTAT Database—Agriculture production. Rome: Food and Agriculture Organization of the United Nations.

FAO. 2015. World fertilizer trends and outlook to 2018. Rome: Food and Agriculture Organization of the United Nations.

Farrar, K., Bryant, D., and Cope-Selby, N. 2014. Understanding and engineering beneficial plant–microbe interactions: Plant growth promotion in energy crops. *Plant Biotechnology Journal* **12**: 1193–1206.

Fate, G.D. and Lynn, D.G. 1996. Xenognosin methylation is critical in defining the chemical potential gradient that regulates the spatial distribution in Striga pathogenesis. *Journal of the American Chemical Society* **118**: 11369–11376.

Feria Bourrellier, A.B., Ferrario-Méry, S., Vidal, J., and Hodges, M. 2009. Metabolite regulation of the interaction between Arabidopsis thaliana PII and N-acetyl-l-glutamate kinase. *Biochemical and Biophysical Research Communications* **387**: 700–704.

Ferrario-Méry, S., Besin, E., Pichon, O., Meyer, C., and Hodges, M. 2006. The regulatory PII protein controls arginine biosynthesis in Arabidopsis. *FEBS Letters* **580**: 2015–2020.

Ferrario-Méry, S., Bouvet, M., Leleu, O., Savino, G., Hodges, M., and Meyer, C. 2005. Physiological characterization of Arabidopsis mutants affected in the expression of the putative regulatory protein PII. *Planta* **223**: 28–37.

Fischer, W.N., André, B., Rentsch, D., Krolkiewicz, S., Tegeder, M., Breitkreuz, K., and Frommer, W.B. 1998. Amino acid transport in plants. *Trends Plant Sci.* **3**: 188–195.

Fokina, O., Chellamuthu, V.-R., Forchhammer, K., and Kornelius, Z. 2010. Mechanism of 2-oxoglutarate signaling by the *Synechococcus elongatus* PII signal transduction protein. *Proc Nat Acad Sci* (USA) **107**: 19760–19765.

Forchhammer, K. 2008. PII signal transducers: Novel functional and structural insights. *Trends in Microbiol* **16**: 65–72.

Franche, C. and Bogusz, D. 2011. Signalling and communication in actinorhizal symbiosis. In: Perotto, S. and Baluska, F., Eds. *Signalling and communication in plant symbiosis*. Berlin: Springer, pp. 73–92.

Galloway, J.N., Townsend, A.R., Erisman, J.W. et al. 2008. Transformation of the nitrogen cycle: Recent trends, questions and potential solutions. *Science* **320**: 889–892.

Glibert, P.M., Anderson, D.M., Gentien, P., Graneli, E., and Sellner, K.G. 2005. The global complex phenomena of harmful algal blooms. *Oceanography* **18**: 136–147.

Gopalakrishnan, S., Watanabe, T., Pearse, S.J., Ito, O., Hossain, Z.A.K.M., and Subbarao, G.V. 2009. Biological nitrification inhibition by *Brachiaria humidicola* roots varies with soil type and inhibits nitrifying bacteria, but not other major soil microorganisms. *Soil Sci. Plant Nutr.* **55**: 725–733.

Guthrie, T.T. and Bomke, A.A. 1980. Nitrification inhibition by N-Serve and ATC in soils with varying texture. *Soil Science Society of America Journal* **44**: 314–320.

Haselwandter, K., Hofmann, A., Holzmann, H.-P., and Read, D.J. 1983. Availability of nitrogen and phosphorous in the nival zone of the Alps. *Oecologia* (Berlin) **57**: 266–296.

Haynes, R.J. and Goh, K.M. 1978. Ammonium and nitrate nutrition of plants. *Biological Reviews* **53**: 465–510.

Heisler, J., Glibert, P., Burkholder, J., Anderson, D., Cochlan, W., Dennison, W., Gobler, C. et al. 2008. Eutrophication and harmful algal blooms: A scientific consensus. *Harmful Algae* **8**: 3–13.

Hendrickson, L.L., Keeney, D.R., Walsh, L.M., and Liegel, E.A. 1978. Evaluation of nitrapyrin as a means of improving N use efficiency in irrigated sands. *Agronomy Journal* **70**: 699–708.

Herman, E.M. 2014. Methods and materials for producing enhanced sugar, starch, oil and cellulose output traits in crop plants. U.S. Patent, No. 20170037419 A1.

Hess, D., Ejeta, G., and Butler, L.G. 1992. Selecting sorghum genotypes expressing a quantitative biosynthetic trait that confers resistance to Striga. *Phytochemistry* **31**: 493–497.

Hofstra, N. and Bouwman, A.F. 2005. Denitrification in agricultural soils: Summarizing published data and estimating global annual rates. *Nutrient Cycling Agroecosystems* **72**: 267–278.

Holliday, J.A., Aitken, S.N., Cooke, J.E.K., Fady, B., González-Martínez, S.C., Heuertz, M., Jaramillo-Correa, J.P. et al. 2017. Advances in ecological genomics in forest trees and applications to genetic resources conservation and breeding. *Molecular Ecology* **26**: 706–717.

Howard, J.B. and Rees, D.C. 1996. Structural basis of biological nitrogen fixation. *Chem. Rev.* **96**: 2965–2982.

Huergo, L.F., Chandra, G., and Merrick, M. 2013. PII signal transduction proteins: Nitrogen regulation and beyond. *FEMS Microbiology Reviews* **37**: 251–283.

Hungria, M., Campo, R., de Souza, E.M. et al. 2010. Inoculation with selected strains of *Azospirillum brasilense* and *A. lipoferum* improves yields of maize and wheat in Brazil. *Plant and Soil* **331**: 413–425.

Hurek, T., Reinhold-Hurek, B., Van Montagu, M., and Kellenberger, E. 1994a. Root colonization and systemic spreading of *Azoarcus* sp. strain BH72 in grasses. *J Bacteriol* **176**: 1913–1919.

Hurek, T., Reinhold-Hurek, B., Turner, G.L., and Bergersen, F.J. 1994b. Augmented rates of respiration and efficient nitrogen fixation at nanomolar concentrations of dissolved $O_2$ in hyperinduced *Azoarcus* sp. strain BH72. *J Bacteriol* **176**: 4726.

IPCC (Intergovernmental Panel on Climate Change). 2012. Fifth assessment report (AR5). XXXV/Doc. 7. Geneva, Switzerland.

James, E.K. and Olivares, F.L. 1998. Infection and colonization of sugar cane and other Graminaceous plants by endophytic diazotrophs. *Critical Reviews in Plant Sciences* **17**: 77–119.

James, E.K., Olivares, F.L., Baldani, J.I., and Döbereiner J. 1997. *Herbaspirillum*, an endophytic diazotroph colonizing vascular tissue in leaves of *Sorghum bicolor* L. Moench. *J Exp Bot* **48**: 785–797.

James, E.K., Reis, V.M., Olivares, F.L., Baldandi, J.I., and Döbereiner, J. 1994. Infection of sugar cane by the nitrogen-fixing bacterium *Acetobacter diazotrophicus*. *J Exp Bot* **45**: 757–766.

Joy, K.W. 1988. Ammonia, glutamine and asparagine: A carbon–nitrogen interface. *Can. J. Bot.* **66**: 2103–2109.

Kagan, I.A., Rimando, A.M., and Dayan, F.E. 2003. Chromatographic separation and in vitro activity of sorgoleone congeners from the roots of Sorghum bicolor. *J Agricultural and Food Chemistry* **51**: 7589–7595.

Karve, A., Alexoff, D., Kim, D., Schueller, M.J., Ferrieri, R.A., and Babst, B. 2015. *In vivo* quantitative imaging of photoassimilate transport dynamics and allocation in large plants using a commercial positron emission tomography (PET) scanner. *BMC Plant Biology* **15**: 273–284.

Kongshaug, G. 1998. Energy consumption and greenhouse gas emissions in fertilizer production. IFA Technical Conference, Marrakech, Morocco, September 28 to October 1, 1998.

Körner, Ch., Bannister, P., and Mark, A.F. 1986. Altitude variation in stomatal conductance, nitrogen content and leaf anatomy in different plant life forms in New Zealand. *Oecologia* (Berlin) **69**: 577–588.

Lam, H.-M., Coschigano, K.T., Oliveira, I.C., Melo-Oliveira, R., and Coruzzi, G. 1996. The molecular-genetics of nitrogen assimilation into amino acids in higher plants. *Annu Rev Plant Physiol Plant Mol Biol* **47**: 569–593.

Landsberg, J.H. 2002. The effects of harmful algal blooms on aquatic organisms. Reviews in *Fisheries Sci* **10**: 113–390.

Lata, J.C., Degrange, V., Raynaud, X., Maron, P.-A., Lensi, R., and Abbadie, L. 2014. Grass populations control nitrification in savanna soils. *Funct Ecol* **18**: 605–611.

Lea, P.J., Sodek, L., Parry, M.A.J. et al. 2007. Asparagine in plants. *Ann Appl Biol* **150**: 1–26.

Leigh, J.A. and Dodsworth, J.A. 2007. Nitrogen regulation in bacteria and archaea. *Ann Rev Microbiol*, **61**: 349–377.

Leininger, S., Urich, T., Schloter, M. et al. 2006. Archaea predominate among ammonia-oxidizing prokaryotes in soils. *Nature* **442**: 806–809.

Liao, M., Fillery, I.R.P., and Palta, J.A. 2004. Early vigorous growth is a major factor influencing nitrogen uptake in wheat. *Functional Plant Biology* **31**: 121–129.

Lima, J.D. and Sodek, L. 2003. N-stress alters aspartate and asparagine levels of xylem sap in soybean. *Plant Science*, **165**: 649–656.

Mckee, H.S. 2013. Nitrogen metabolism of seedlings, in *Der Stickstoffumsatz/Nitrogen Metabolism*, Springer-Verlag, pp. 478–489.

Mendonça, E.S., Lima, P.C., Guimarães, G.P., Moura, W.M., and Andrade, F.V. 2017 Biological nitrogen fixation by legumes and N uptake by coffee plants. *Rev Bras Cienc Solo.* 41: e0160178.

Mizuno, Y., Moorhead, G.B.G., and Ng, K.K.-S. 2007. Structural basis for the regulation of N-acetylglutamate kinase by PII in *Arabidopsis thaliana*. *J Biological Chemistry* **282**: 35733–35740.

Modak, J.M. 2002. Haber process for ammonia synthesis. *Resonance* **7**: 69–77.

Neale, D.B. 2007. Genomics to tree breeding and forest health. *Current Opinion in Genetics & Development* **17**: 539–544.

Netzly, D.H., Riopel, J.L., Ejeta, G., and Butler, L.G. 1988. Germination stimulants of witchweed (*Striga asiatica*) from hydrophobic root exudates of sorghum (Sorghum bicolor). *Weed Sci* **36**: 411–446.

Nimbal, C.I., Yerkes, C.N., Weston, L.A., and Weller, S.C. 1996. Herbicidal activity and site of action of the natural product sorgoleone. *Pesticide Biochemistry and Physiology* **54**: 73–83.

Nunes-Nesi, A., Fernie, A.R., and Stitt, M. 2010. Metabolic and signaling aspects underpinning the regulation of plant carbon nitrogen interactions. *Molecular Plant* **3**: 973–996.

Okon, Y. and Labandera-Gonzalez, C.A. 1994. Agronomic applications of *Azospirillum*: An evaluation of 20 years worldwide field inoculation. *Soil Biology and Biochemistry* **26**: 1591–1601.

Osanai, T. and Tanaka, K. 2007. Keeping in touch with PII: PII-interacting proteins in unicellular cyanobacteria. *Plant and Cell Physiol* **48**: 908–914.

Pacovsky, R. 1990. Development and growth effects in the *Sorghum–Azospirillum* association. *Journal of Applied Bacteriology* **68**: 555–563.

Pankievicz, V.C.S., Amaral, F.P., Santos, K.F.D.N., Agtuca, B., Xu, Y., Schueller, M.J., Arisi, A.C.M. et al. 2015. Robust biological nitrogen fixation in a $C_4$ model grass, *Setaria viridis*. *Plant Journal* **81**: 907–919.

Patyk, A. 1996. Balance of energy consumption and emissions of fertilizer production and supply. Reprints from the International Conference of Life Cycle Assessment in Agriculture, Food and Non-Food Agro-Industry and Forestry: Achievements and Prospects. Brussels, Belgium, April 4–5, 1996, pp. 73–87.

Pawlowski, K. 2009. Induction of actinorhizal nodules by Frankia. In Pawlowski, K. (Ed.), *Prokaryotic symbionts in plants*. Berlin: Springer, pp. 127–154.

Pedraza, R.O., Bellone, C.H., Carrizo de Bellone, S., Boa Sorte, P.M.F., and Teixeira, K.R.D.S. 2009. *Azospirillum* inoculation and nitrogen fertilization effect on grain yield and on the diversity of endophytic bacteria in the phyllosphere of rice rainfed crop. *European Journal of Soil Biology* **45**: 36–43.

Qu, W., Robert, C.A.M., Erb, M., Hibbard, B.E., Paven, M., Gleede, T., Riehl, B. et al. 2016. Dynamic precision phenotyping reveals mechanisms of crop tolerance to root herbivory. *Plant Physiology* **172**: 776–788.

Raaijmakers, J., Paulitz, T., Steinberg, C., Alabouvette, C., and Moënne-Loccoz, Y. 2009. The rhizosphere: A playground and battlefield for soilborne pathogens and beneficial microorganisms. *Plant and Soil* **321**: 341–361.

Raun, W.R. and Johnson, G.V. 1999. Improving nitrogen use efficiency for cereal production. *Agronomy Journal* **91**: 357–363.

Reinhold, B., Hurek, T., Niemann, E.-G., and Fendrik, I. 1986. Close association of *Azospirillum* and diazotrophic rods with different root zones of Kallar grass. *Applied and Environmental Microbiology* **52**: 520–526.

Reinhold-Hurek, B. and Hurek, T. 1998. Life in grasses: Diazotrophic endophytes. *Trends in Microbiology* **6**: 139–144.

Richardson, A., Barea, J.-M., McNeill, A., and Prigent-Combaret, C. 2009. Acquisition of phosphorus and nitrogen in the rhizosphere and plant growth promotion by microorganisms. *Plant and Soil* **321**: 305–339.

Rodgers, G.A. 1986. Nitrification inhibitors in agriculture. *Journal of Environmental Science and Health* **A21**: 701–722.

Rumjanek, N.G., Martins, L.M.V., Xavier, G.R., and Neves, M.C.P. 2005. Fixação biológica de nitrogênio. In: Freire Filho, F.R., Lima, J.A.A., and Ribeiro, V.Q., Eds. Feijão-caupi: Avanços tecnológicos. Brasília, DF: Embrapa/Informação Tecnológica; pp. 281–335.

Sahrawat, K.L. and Keeney, D.R. 1985. Perspectives for research on development of nitrification inhibitors. *Communications in Soil Science and Plant Analysis* **16**: 517–524.

Salsac, L., Chaillou, S., Morot-Gaudry, J., and Lesaint, C. 1987. Nitrate and ammonium nutrition in plants. *Plant Physiology and Biochemistry* **25**: 805–812.

Santi, C., Boguscz, D., and Franche, C. 2013. Biological nitrogen fixation in non-legumes. *Annals Bot* **111**: 743–767.

Schlesinger, W.H. 2009. On the fate of anthropogenic nitrogen. *Proc Nat Acad Sci, USA* **106**: 203–208.

Schlogl, R. 2004. Catalytic synthesis of ammonia—A never-ending story. *Agnew Chem Int Ed* **42**: 2004–2008.

Schmollinger, S., Mühlhaus, T., Boyle, N.R., Blaby, I.K., Casero, D., Mettler, T., Moseley, J.L. et al. 2014. Nitrogen-sparing mechanisms in *Chlamydomonas* affect the transcriptome, the proteome, and photo-synthetic metabolism. *Plant Cell* **26**: 1410–1435.

Schrock, R.R. 2006. Reduction of nitrogen. *Proc Natl Acad Sci* (USA) **103**: 17087–17089.

Shaheen, H., Khan, S.M., Harper, D.M., Ullah, Z., and Allem, R. 2011. Species diversity, community structure, and distribution patterns in Western Himalayan alpine pastures of Kashmir, Pakistan. *Mountain Research and Development* **31**: 153–159.

Smart, D.R. and Bloom, A.J. 2001. Wheat leaves emit nitrous oxide during nitrate assimilation. *Proc Natl Acad Sci* (USA) **98**: 7875–7878.

Smith, K.A., McTaggart, I.P., and Tsuruta, H. 1997. Emissions of $N_2O$ and NO associated with nitrogen fertilization in intensive agriculture, and the potential for mitigation. *Soil Use Management* **13**: 296–304.

Smith, P., Martino, D., Cai, Z., Gwary, D., Janzen, H., Kumar, P., McCarl, B. et al. 2007. Agriculture. In: Metz, B., Davidson, O.R., Bosch, P.R., Dave, R., and Meyer, LA. (Eds.), *Climate change 2007: mitigation*. Contribution of working group III to the fourth assessment report of the Intergovernmental Panel on Climate Change. Cambridge: Cambridge University Press, pp. 497–540.

Sommer, K., Mertz, M., and Rossig, K. 1976. Stickstoff zu weizen mit ammoniumnitrificiden, mitteilung 2: Proteingehalte, proteinfraktionen und backfahigkeit. *Landwirtschaftliche Forschung* **29**: 161–169.

*Special Issue*: The role of biological nitrogen fixation by non-legumes in the sustainable production of food and biofuels. *Plant and Soil*, 2012. **356**: issues 1–2: 1–417.

Subbarao, G.V., Ishikawa, O.I., Nakahara, K. et al. 2006b. A bioluminescence assay to detect nitrification inhibitors released from plant roots: A case study with *Brachiaria humidicola*. *Plant Soil* **288**: 101–112.

Subbarao G.V., Ito, O., Sahrawat, K.L., Berry, W.L., Nakahara, K., Ishikawa, T., Watanabe, T. et al. 2006a. Scope and strategies for regulation of nitrification in agricultural systems—Challenges and opportunities. *Critical Reviews in Plant Sciences* **25**: 303–335.

Subbarao, G.V., Rondon, M., Ito, O., Ishikawa, T., Rao, I.M., Nakahara, K., Lascano, C., and Berry, W.L. 2007a. Biological nitrification inhibition (BNI)—Is it a widespread phenomenon? *Plant Soil* **294**: 5–18.

Subbarao, G.V., Sahrawat, K.L., Nakahara K., Ishikawa, T., Kishii, M., and Rao, I.M. 2012. Biological nitrification inhibition—A novel strategy to regulate nitrification in agricultural systems. *Advances in Agronomy* **114**: 249–302.

Subbarao, G.V., Wang, H.Y., Ito, O., and Berry, W.L. 2007b. $NH_4^+$ triggers the synthesis and release of biological nitrification inhibition compounds in *Brachiara humidicola* roots. *Plant Soil* **290**: 245–257.

Taylor, A.E., Zeglin, L.H., Dooley, S., Myrold, D.D., and Bottomley, P.J. 2010. Evidence for different contributions of archaea and bacteria to the ammonia-oxidizing potential of diverse Oregon soils. *Applied Environmental Microbiology* **76**: 7691–7698.

Tjepkema, J.D. and Burris, R.H. 1976. Nitrogenase activity associated with some Wisconsin prairie grasses. *Plant and Soil* **45**: 81–94.

United Nations. Department of Economic and Social Affairs 2017 Report, "World Population Prospects: The 2015 Revision."

Uhrig, R.G., Ng, K.K.S., and Moorhead, G.B.G. 2009. PII in higher plants: A modern role for an ancient protein. *Trends in Plant Science* **14**: 505–511.

Verhagen, B.W., Glazebrook, J., Zhu, T., Chang, H.S., van Loon, L.C., and Pieterse, C.M. 2004. The transcriptome of rhizobacteria-induced systemic resistance in *Arabidopsis*. *Molecular Plant-Microbe Interactions* **17**: 895–908.

Wall, L.G. 2000. The actinorhizal symbiosis. *J Plant Growth Regulation* **19**: 167–182.

Weiske, A., Benckiser, G., and Ottow, J.C.G. 2001. Effect of new nitrification inhibitor DMPP in comparison to DCD on nitrous oxide ($N_2O$) emissions and methane ($CH_4$) oxidation during 3 years of repeated applications in field experiments. *Nutrient Cycling in Agroecosystems* **60**: 57–64.

Yanni, Y.G., Rizk, R.Y., Corich, V., Squartini, A., Ninke, K., Philip-Hollingsworth, S., Orgambide, G. et al. 1997. Natural endophytic association between *Rhizobium leguminosarum* bv. trifolii and rice roots and assessment of its potential to promote rice growth. *Plant and Soil*, **194**: 99–114.

Zerulla, W., Barth, T., Dressel, J., Erhardt, K., von Locquenghien, K.H., Pasda, G., Rädle, M., and Wissemeier, A. 2001. 3, 4-Dimethylpyrazole phosphate (DMPP)—A new nitrification inhibitor for agriculture and horticulture. *Biology and Fertility of Soils* **34**: 79–84.

Zeth, K., Fokina, O., and Forchhammer, K. 2014. Structural basis and target-specific modulation of ADP sensing by the Synechococcus elongatus PII signaling protein. *Journal of Biological Chemistry* **289**: 8960–8972.

# 15  Nitrogen: Managing the Necessary Evil

*Rattan Lal*

## CONTENTS

## 15.1  INTRODUCTION

The global nitrogen (N) cycle, similar to and strongly coupled with that of the carbon (C) cycle, is a critical component of the Earth's biogeochemistry. Anthropogenic alterations of the global N cycle have strong environmental impacts with regard to eutrophication and acidification of water along with drastic changes in atmospheric concentration of $N_2O$, a highly potent greenhouse gas (GHG) with the relative global warming potential of 310. The atmospheric concentration of $N_2O$ has increased by 121% compared with the preindustrial level of 270 ppb to 328 ppb in 2015, and is increasing at the average annual rate of 1.0 ppb with the mean annual absolute increase during the last 10 years (2005–2014) of 0.89 ppb/yr (WMO 2016). In addition to the adverse impacts on the environment (water and air quality) and on both the terrestrial and aquatic biodiversity, there are also strong impacts on human health and wellbeing.

There has been unprecedented increase in the global food production, both in absolute and per capita basis, since the 1960s. As an essential plant nutrient, input of reactive N ($N_r$) from organic and/or inorganic sources is critical to obtaining higher grain yield of major cereals including corn (*Zea mays*), rice (*Oryza sativa*), and wheat (*Triticum aestivum*). The achieved increase in food production would not have been possible without the use of N fertilizer. The global N fertilizer input increased from <10 Tg (Tg = teragram = $10^{12}$g = million metric ton) in 1950 to 123 Tg in 2012, and is projected to be 135 Tg in 2020 and 236 Tg in 2050. Global livestock productions, and the manure generated, which are a bigger secondary source of N and P than the chemical fertilizers, are also increasing (Smith 1991; Van Horn et al. 1996; Wilkinson et al. 1997; Bouwman et al. 2013a,b).

Heavy demands of growing and increasingly affluent human population (e.g., food, energy, water, raw materials) have strongly impacted the environment through changes in biogeochemical and biogeophysical cycling especially those related to water, C, N, and P (Lal 2010a). Addition of $N_r$ into the ecosystems to increase food production has polluted water, increased radiative forcing of the atmosphere and exacerbated global warming, and affected species composition and the overall biodiversity. Increasing input of $N_r$ is a necessary evil: it is needed for achieving food production of the growing world population but has severe adverse environmental consequences. The world population (billion) was 2.54 in 1950, 7.55 in 2017, and is projected to be 8.55 in 2030, 9.77 in 2050, and 11.18 in 2100 (U.N. 2017). Thus, the necessary evil must be managed judiciously to advance food production while minimizing its adverse environmental impacts on soil, air, and water resources.

The impact of anthropogenic activities on the global N cycle (GNC) has been discussed in several chapters of this volume. It is argued that the Green Revolution of the 1960s has stalled in South Asia and elsewhere, air and water resources are severely polluted, and the world has researched "the end of plenty" (Bourne Jr., 2015). The critical issue is the challenge of reconciling the need of achieving high agronomic production with that of restoring the environment and minimizing additional risks of environmental pollution. Judicious management of $N_r$ is the focus and central theme of this concluding chapter. Therefore, a critical and an objective appraisal is needed toward judicious and sustainable management of the $N_r$.

## 15.2   THE GLOBAL NITROGEN CYCLE

Anthropogenic activities have drastically altered the GNC by adding $N_r$ into the atmosphere. During the preindustrial level, the fixation of total $N_r$ from the atmosphere was 203 Tg/yr and is comprised

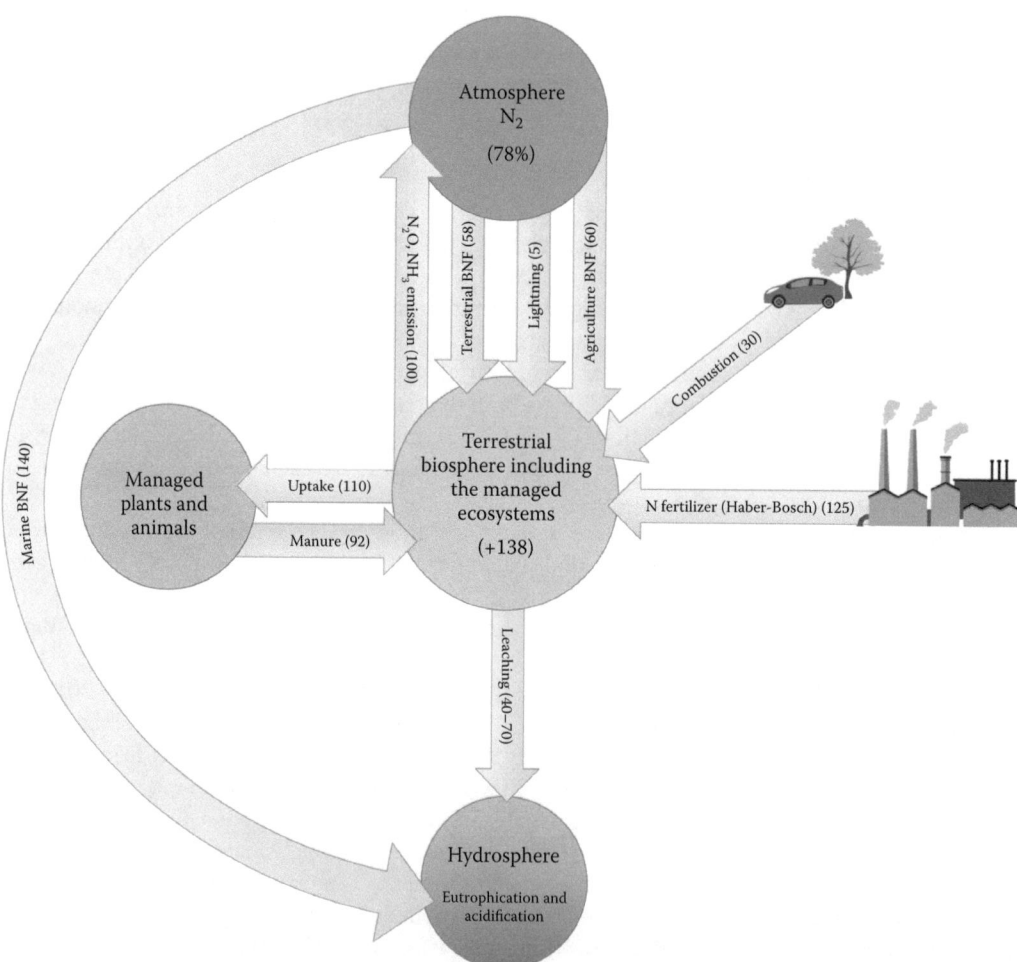

**FIGURE 15.1**   The contemporary global nitrogen cycle (GNC). Fluxes are shown as arrows and the value within the arrow is in Tg N/yr. The data are compiled from FAO (2016), Bouwman et al. (2016), and Fowler et al. (2017). In comparison with preindustrial fixation of 203 Tg N/yr, the contemporary fixation is 418 Tg N/yr of which 215 Tg (51.4%) is of the anthropogenic origin.

of: terrestrial BNF (58 Tg N/yr), marine BNF (140 Tg N/yr), and lightning fixation of 5 Tg N/yr (Fowler et al. 2017). However, the data on the critical components of the current GNC outlined in Figure 15.1 show drastic changes since the preindustrial era. The global N fixation of 203 Tg in the preindustrial era has increased to 418 Tg at present. It includes 140 Tg N/yr of BNF at present in marine ecosystems (Canfield et al. 2010), 5 Tg by lightning (Levy et al. 1996; Tie et al. 2002), and 58 Tg (range of 40–100 Tg) in terrestrial ecosystems (Vitousek et al. 2013; Fowler et al. 2017). BNF by agricultural crops and grazed savannas is estimated at 50–70 Tg N/yr with an average of 60 Tg N/yr (Herridge et al. 2008). Of the terrestrial BNF, phytoautrophic communities including cryptogamic covers (e.g., cyanobacteria, algae, fungi, lichens, and bryophytes) can fix both C and N (Freiberg 1999; Büdel 2002; Belnap and Lange 2003). Elbert et al. (2012) estimated the global annual fixation of 3.9 Pg C/yr and 49 Tg N/yr by cryptogamic covers.

The $N_r$ produced by anthropogenic processes (215 Tg N/yr, or 51.4% of the total $N_r$) is a cause of environmental pollution including eutrophication and acidification of water and increase in atmospheric concentration of $N_2O$. Some argue that "fertilizers are good for the father and bad for the son," because it may also destroy SOC stock, undermine soil health (Philpott 2010; Khan et al. 2007), and jeopardize the health of rivers (Ranade 2010) and aquatic ecosystems (algal bloom, hypoxia, etc.). Thus, there is a need for the controlled evolution of the biosphere (Yablokov and Levechenko 2015). Others argue that there is no evidence that fertilizer N causes loss of SOC stock (Grove et al. 2009). Debate aside, a judicious management of reactive $N_r$ is essential to reducing the environmental impact while improving and sustaining productivity of the agroecosystems.

## 15.3 GLOBAL AND REGIONAL FERTILIZER CONSUMPTION

Global fertilizer nutrient consumption ($N + P_2O + K_2O$) increased from 162.4 Tg in 2008 to 194.8 Tg in 2016 (FAO 2012) with an average rate of growth of 2.5%/yr. However, there are large regional differences in the fertilizer consumption (Table 15.1). In 2014, the rate (kg/ha of arable land) was 138 for the world, compared with 565 for China, 218 for Germany, 175 for Brazil, 165 for India, 158 for South Asia (SA), and only 16 for Sub-Saharan Africa (SSA). In the United States, total use of

**TABLE 15.1**
**Total Fertilizer Consumption in Some Countries**

| | kg/ha of Arable Land | | | | | | | |
|---|---|---|---|---|---|---|---|---|
| Year | China | Germany | Brazil | India | South Asia | SSA | Nigeria | World |
| 2002 | 337 | 220 | 121 | 100 | 100 | 12 | 5 | 105 |
| 2003 | 394 | 220 | 152 | 105 | 102 | 11 | 6 | 110 |
| 2004 | 414 | 215 | 151 | 115 | 112 | 13 | 5 | 111 |
| 2005 | 423 | 209 | 120 | 128 | 124 | 11 | 7 | 114 |
| 2006 | 452 | 194 | 124 | 136 | 131 | 13 | 10 | 118 |
| 2007 | 469 | 222 | 165 | 143 | 135 | 13 | 4 | 124 |
| 2008 | 483 | 160 | 142 | 153 | 144 | 12 | 6 | 120 |
| 2009 | 499 | 181 | 118 | 167 | 158 | 12 | 5 | 120 |
| 2010 | 515 | 212 | 156 | 179 | 166 | 15 | 12 | 129 |
| 2011 | 533 | 191 | 183 | 181 | 168 | 14 | 7 | 134 |
| 2012 | 549 | 199 | 182 | 165 | 155 | 16 | 11 | 134 |
| 2013 | 557 | 203 | 176 | 158 | 152 | 17 | 16 | 135 |
| 2014 | 565 | 218 | 175 | 165 | 158 | 16 | 11 | 138 |

*Source:* World Bank. 2016. Fertilizer Consumption, Washington, DC. (http://date.worldbank.org/indicated/AG.CON .FERT.ZS.)

fertilizer nutrients increased from 5.1 Tg/yr in 1960 to 20.3 Tg/yr in 2011. Of this, a fraction of N fertilizer ranged from 48.3% in 1960 to 57.4% in 2011 (Table 15.2). Rates of fertilizer consumption remain less in all regions of SSA (Table 15.3), and average crop yields (especially of cereals) in SSA are also among the lowest in the world. There is also a strong contrast in total fertilizer use in South Asia versus SSA. Differences in total fertilizer consumption in SA and SSA were approximately threefold in 1961 compared with 11-fold in 2014 (Table 15.4). There is a strong relationship between total agricultural production and fertilizer use in both regions. Similar to total fertilizer use, there are also differences in regional N fertilizer use in SA and SSA, which differs by a factor of 12 over the period of 2002–2018 (Table 15.5), and these differences are also reflected in yield per unit area and total production.

**TABLE 15.2**
**Use of Fertilizer Nutrients in the United States**

| Year | Nutrients (Tg/yr) | | | | |
|------|------|------|------|-------|----------------|
|      | N    | P    | K    | Total | N (% of total) |
| 1960 | 2.48 | 1.03 | 1.62 | 5.13  | 48.3 |
| 1965 | 4.20 | 1.41 | 2.13 | 7.74  | 54.3 |
| 1970 | 6.77 | 3.00 | 3.04 | 12.81 | 52.8 |
| 1975 | 7.80 | 3.46 | 3.35 | 14.61 | 53.4 |
| 1980 | 10.35 | 4.59 | 4.70 | 19.64 | 52.7 |
| 1985 | 10.42 | 4.63 | 4.18 | 19.23 | 54.2 |
| 1990 | 10.05 | 4.46 | 3.92 | 18.43 | 54.5 |
| 1995 | 10.62 | 4.72 | 3.86 | 19.20 | 55.3 |
| 2000 | 11.19 | 4.97 | 3.74 | 19.90 | 56.2 |
| 2005 | 11.19 | 4.97 | 3.89 | 20.05 | 55.8 |
| 2010 | 11.14 | 4.95 | 3.36 | 19.45 | 57.3 |
| 2011 | 11.65 | 5.17 | 3.46 | 20.28 | 57.4 |

*Source:* Recalculated from (Recalculated from World Bank. 2016. Fertilizer Consumption, Washington, DC. (http://date.worldbank.org/indicated/AG.CON.FERT.ZS.)

**TABLE 15.3**
**Average Global Fertilizer Nitrogen Use on Cropland**

| Region | N Use (kg/ha) | | | |
|--------|------|------|------|------|
|        | 2002 | 2005 | 2007 | 2010 |
| World | 57.0 | 58.9 | 65.8 | 68.7 |
| Middle Africa | 1.0 | 1.1 | 1.2 | 0.8 |
| Northern Africa | 33.9 | 42.5 | 34.1 | 36.4 |
| Southern Africa | 30.7 | 23.0 | 30.5 | 27.4 |
| Western Africa | 2.5 | 3.9 | 2.6 | 2.9 |
| Southern Asia | 65.2 | 76.6 | 83.8 | 92.2 |

*Source:* FAO. 2016. World Fertilizer Trends and Outlook to 2019. Food and Agriculture Organization of the United Nations (FAO), Rome, Italy.

**TABLE 15.4**
**Fertilizer Consumption in SSA and SA**

| Year | Fertilizer Use ($10^6$ Mg) | |
| | South Asia | Sub-Saharan Africa |
| --- | --- | --- |
| 1961 | 0.49 | 0.16 |
| 1965 | 1.02 | 0.26 |
| 1970 | 2.83 | 0.44 |
| 1975 | 4.44 | 0.71 |
| 1980 | 7.37 | 0.96 |
| 1985 | 11.22 | 1.06 |
| 1990 | 15.21 | 1.25 |
| 1995 | 18.07 | 1.07 |
| 2002 | 21.03 | 1.38 |
| 2005 | 27.53 | 1.80 |
| 2010 | 35.08 | 2.68 |
| 2011 | 35.15 | 2.66 |
| 2012 | 32.33 | 3.05 |
| 2013 | 31.75 | 3.25 |
| 2014 | 33.07 | 3.02 |

*Source:* IFDC 2003, FAOSTAT. 2016. Faostat database collections. Data retrieved October 12, 2016. Food and Agriculture Organization of the United Nations, Rome. (http://faostat3.Fao.Org/home/e.)

**TABLE 15.5**
**Use of Nitrogen Fertilizer (Tg) in Sub-Saharan Africa and South Asia**

| Year | South Asia | Sub-Saharan Africa |
| --- | --- | --- |
| 2002 | 15.00 | 1.11 |
| 2003 | 15.55 | 1.03 |
| 2004 | 16.48 | 1.23 |
| 2005 | 17.69 | 1.12 |
| 2006 | 19.17 | 1.26 |
| 2007 | 19.34 | 1.18 |
| 2008 | 20.38 | 1.25 |
| 2009 | 21.17 | 1.23 |
| 2010 | 21.50 | 1.56 |
| 2011 | 22.39 | 1.51 |
| 2012 | 21.57 | 1.72 |
| 2013 | 21.66 | 1.97 |
| 2014 | 21.98 | 1.87 |
| 2015 | 23.11 | 1.80 |
| 2016 | 23.50 | 1.88 |
| 2017 | 23.88 | 1.97 |
| 2018 | 24.22 | 2.07 |

*Source:* FAOSTAT 2016.

**TABLE 15.6**

**Global N Budget for Principal Cereals**

| Cereal | Tg N, 1961–2010 | | |
| | Inputs | Output | Change in Soil N |
| --- | --- | --- | --- |
| Maize | 956.2 | 988.2 | −32.0 |
| Rice | 1018.5 | 992.8 | +25.7 |
| Wheat | 1147.3 | 1209.1 | −61.8 |
| Total | 3122.0 | 3190.1 | −68.1 |

*Source:* Adapted from Ladha, J.K., A. Tirol-Padre, C.K. Reddy, K.G. Cassman, S. Verma, D.S. Powlson, C. van Kessel et al. 2016. Global nitrogen budgets in cereals: a 50-year assessment for maize, rice, and wheat production systems. *Scientific Reports* 6:19355.

## 15.4 MANAGING NITROGEN IN AGROECOSYSTEMS

Global N fertilizer use, despite large regional differences, has increased steadily (Heffer and Prud'homme 2016). There are large temporal and spatial variations in global use of N and P (Bouwman et al. 2017). There are also regional variations in environmental consequences especially with regard to emission of $N_2O$ and eutrophication of surface and groundwater. Whereas the N use efficiency (NUE) has gradually increased, especially in China and India, more needs to be done to limit its leakage into the environment. Change in the ratio of N:P input into soils of agroecosystems raises important questions: (1) do the regional and national differences in fertilizer consumption reflect differences in stages of development in agroecosystems, (2) are changes in productivity related to changes in nutrient budget, (3) do differences in inputs of N and P reflect differences in environmental quality, and (4) can the environmental impacts indicate the need for change in policy to minimize the environmental risks while enhancing production?

Ladha et al. (2016) computed the global N budgets in cereals over a 50-year period from 1961 to 2010 (Table 15.6). All three cereals harvested a total of 1551 Tg N over the 50-year period. Of this, 48% was supplied by fertilizer N and 4% through soil depletion. Uptake of 48% of crop N is equivalent to 29, 38, and 25 kg/ha.yr for maize, rice, and wheat, respectively. Non-symbiotic fixation of N contributed 13, 22, and 13 kgN/ha.yr for maize, rice, and wheat, respectively (Ladha et al. 2016).

## 15.5 MANAGING N USE

An efficient management of N fertilizer is critical to sustaining productivity and improving environment quality. Indiscriminate use of N causes contamination of groundwater by leaching and eutrophication of surface waters by transport of nitrates in runoff. Further, excess N in agroecosytems is released into the atmosphere through volatilization as ammonia ($NH_3$) and nitrous oxide ($N_2O$) and the latter occurs through denitrification and nitrification processes. In addition to acidification of soil, input of N can also acidify water.

Crop yield response to N follows a sigmoid curve. Therefore, understanding the response function under soil/crop specific situations is important to optimize productivity and minimize losses into the environment. In this context, reducing input of N per ha of cropland in China but increasing that in Sub-Saharan Africa would increase global production and minimize the environmental pollution. Mueller et al. (2014) reported that the current levels of cereal production in the world can be achieved even with 50% less N application. The goal is to manage N in such a way as to increase its use efficiency from about 30% at present to 75% and more under optimal conditions. The hotspots of fertilizer use shifted from the United States to Western Europe in the 1960s and to China in the early 2000s (Lu and Tian 2017). Rather than emphasizing the need to increase the fertilizer input per unit of agricultural land, the goal is to enhance the nutrient use efficiency (NUE) in the high input regions (e.g., China, India) while reducing losses into the water and the atmosphere. The strategy is to prudently achieve balance between crop demand and N supply without excess or deficit (Cassman et al. 2002). Further, it is important to discard the anthropocentric philosophy and reduce the global environmental crisis of the biosphere (Yablokov and Levecheko 2015). An excessive and indiscriminate use of N and other agrochemicals is an example of the anthropocentricism, which cannot and must not continue.

## 15.6  THE COUPLED CYCLING OF NITROGEN AND CARBON

There is a strong $H_2O$-C-N-P nexus, and the biogeochemical and biogeophysical cycles are strongly interconnected. Systems of sustainable management of soil, water, and crops of these elements must be identified and adopted (Lal 2010a). Thus, the eco-efficiency can be greatly enhanced by a judicious management of C inputs (Lal 2010b). Use efficiency of N and other inputs depend on concentration of soil organic carbon (SOC) in soils of temperate climate in the root zone, which should be above the threshold level of 2% (Loveland and Webb 2003). Restoring and sustaining SOC also improves soil aggregation and tilth, increases plant available water holding capacity, decreases risks of run off and erosion, and enhances soil resilience against the extreme climatic events (e.g., heat wave, pedologic and agronomic drought). Through high surface area and increase in charge density (cation and anion exchange capacity), an optimal level of SOC concentration would reduce leaching losses of $NO_3$ and emission of $N_2O$ into the atmosphere through nitrification and denitrification. A high soil biodiversity (of macro, meso, and microfauna) can create disease-suppressive soils, thereby reducing the dependence on pesticides for the control of pests and pathogens. Because of the strong nexus, a wholistic approach is needed to sustain management of Nr along with managing the stock and concentration of SOC, P, S, Ca, and Mg. The wholistic approach would involve the use of conservation agriculture (CA), mulch farming, cover cropping, BNF, mycorrhizal inoculation, complex rotations, agroforestry, and integration of crops with tree plantations and livestock achieving an optimal rather than maximum crop yield ensuring a stable yield rather than high yield during the bad seasons would be the best strategy. The wholistic approach is also based on the concept of sustainable intensification—producing more from less use of N, irrigation water energy, and other inputs (Figure 15.2).

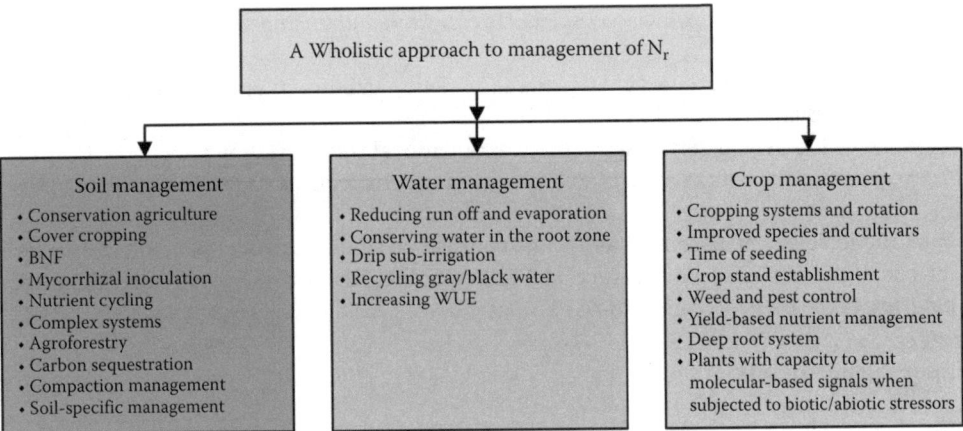

**FIGURE 15.2**  A Wholisitic approach to managing the reactive N (Nr) by decreasing losses and increasing the coefficiency. (BNF = biological N fixation, WUE = water use efficiency, $N_r$ = reactive N.)

## 15.7   CONCLUSION

Conversion of the inert nitrogen ($N_2$) in the atmosphere into reactive N ($N_r$) as $NH_3$ was the most important scientific discovery that revolutionized agriculture and increased grain production from the same land area. The so-called "Green Revolution" of the 1960s saved hundreds of millions from starvation because of the adoption of yield-responsive (dwarf) varieties of wheat and rice (with a high harvest index) grown with high rates of $N_r$ (chemical fertilizers) under irrigated conditions. However, the indiscriminate use of $N_r$ (along with other agrochemicals comprising of herbicides and pesticides) have adversely affected the quality of soil, water, and air. Soils are eroded and depleted of SOC stocks and are prone to salinization and other degradation processes (i.e., compaction, crusting), surface and ground waters are contaminated (eutrophication, non-point source pollution, leaching of $NO_3$ and pesticides), and air is polluted with ever-increasing concentration of greenhouse gases not only from the fossil fuel combustion but also from the soil (especially $N_2O$, NO, and $CH_4$). Yet, rather than a problem, a judiciously managed soil can be a sink of the greenhouse gases ($CO_2$, $CH_4$, $N_2O$). Enhancing SOC concentration to above the critical level (2% in the root zone) can adapt to and mitigate climate change while also improving the quality of water and air.

Agriculture has a bright future, and the twenty-first century is an exciting era until the population peaks at 11.2 billion by 2100 and begins to decrease and eventually stabilize at a manageable number (6, 5, 4 billion or less). Hopefully, the lower stable population will be reached because the people want it rather than Mother Nature imposes it. Therefore, soil scientists and those engaged in sustainable management of natural resources especially that of the GNC and the $N_r$ must actively undertake the following:

1. Engage with the public: in discussion of popular topics (e.g., food security, climate change, water quality, biofuel, biodiversity).
2. Speak with journalists and media: regarding implications of their research.
3. Meet with members of the public: to resolve inconsistencies, discuss issues, and address public concerns.
4. Communicate with policymakers: to translate science into action.
5. Discuss with educators: to change curricula and include soil science in K2–12.
6. Discuss with other disciplines: engineers, climatologist, hydrologists, plant scientists, and social scientists issues of mutual interest.
7. Create environment for soil science to contribute to the solutions of societal problems.
8. Help soil scientists to become better public communicators.
9. Identify fora: through which soil scientists can work with members of the public on identifying solutions.
10. Build a societal culture that champions soil science and thrives on the outcome of a strong pedology-public relationship.
11. Work with policymakers to seek approval of national soil protection policy.
12. Work with educators to include soil science in the curricula at all levels.

With the judicious management and with stabilization of the human population at a lower level by the end of the twenty-second century (at 4 billion or less), the necessary evil of $N_r$ (the genie) can be put back in the bottle. It can be done.

## REFERENCES

Belnap, J. and Lange, O.L. 2003. in *Biological Soil Crusts: Structure, Function, And Management* (Eds. Belnap, J. and Lange, O.L.), pp. 471–479 (Springer).
Bourne, Jr. J.K. 2015. *The End of Plenty: The Race to Feed the Crowded World*. W.W. Norton & Co., New York.

Bouwman, A., A. Beusen, J. Griffioen, J. Van Groenigen, M. Hefting, O. Oenema, P. Van Puijenbroek et al. 2013a. Global trends and uncertainties in terrestrial denitrification and $N_2O$ emissions. *Philosophical Transactions of the Royal Society B-Biological Sciences* 368(1621).

Bouwman, L., K. Goldewijk, K. Van Der Hoek, A. Beusen, D. Van Vuuren, J. Willems, M. Rufino and E. Stehfest. 2013b. Exploring global changes in nitrogen and phosphorus cycles in agriculture induced by livestock production over the 1900–2050 period. *Proceedings of the National Academy of Sciences of the United States of America* 110(52): 20882–20887.

Bouwman, A., A. Beusen, L. Lassaletta, D. Van Apeldoorn, H. Van Grinsven, J. Zhang, and M. Van Ittersum. 2017. Lessons from temporal and spatial patterns in global use of n and p fertilizer on cropland. *Scientific Reports* 7.

Büdel, B. 2002. In *Progress in Botany* (Eds., Lüttge, U. E., Beyschlag, W., Büdel, B., and Francis, D.,) pp. 386–404 (Springer).

Canfield, D.E., Glazer, A.N., and Falkowski, P.G. 2010 The evolution and future of Earth's nitrogen cycle. *Science* 330, 192–196.

Cassman, K.G., A.R. Dobermann, and D.T. Walters. 2002. Agroecosystems, nitrogen-use efficiency, and nitrogen management. *Ambio* 31:132–140.

Elbert, W., B. Webber, S. Burrows, J. Steinkamp, B. Budel, M.O. Andreae, and U. Poschl. 2012. Contributions of cryptogamic covers to the global cycles of carbon and nitrogen. *Nat. Geosci.* 5: 459–462.

FAOSTAT. 2016. Faostat database collections. Data retrieved October 12, 2016. Food and Agriculture Organization of the United Nations, Rome. (http://faostat3.Fao.Org/home/e).

FAO. 2016. World Fertilizer Trends and Outlook to 2019. Food and Agriculture Organization of the United Nations (FAO), Rome, Italy.

FAO. 2012. Current World Fertilizer Trends and Outlook to 2016. FAO, Rome, Italy.

Fowler, D., M. Coyle, U. Skiba, M.A. Sutton, J.N. Cape, S. Reis et al. 2013. The global nitrogen cycle in the 21st century. *Philip Traur. R. Soc (B)* 369: (http://dx.doi.org/10.1098/rstb.2013.016u).

Freiberg, E. 1999. Influence of microclimate on the occurrence of cyanobacteria in the phyllosphere in a pre-montane rain forest of Costa Rica. *Plant Biol.* 1: 244–252.

Grove, J.H., E.M. Pena-Yewtukhiw, M. Diaz-Zorita, and R.L. Blevins. 2009. Does fertilizer nitrogen burn up soil organic matter? *Better Crops* 93(4): 6–8.

Heffer, P., and M. Prud'homme. 2016. Global nitrogen fertilizer demand and supply: Trend, current level and outlook. Proceedings of the 2016 International Nitrogen Initiative Conference. December 4–8, 2016, Melbourne, Australia.

Herridge, D.F., Peoples, M.B., and Boddey, R.M. 2008 Global inputs of biological nitrogen fixation in agricultural systems. *Plant Soil* 311: 1–18.

IFDC. 2004. Global and regional data on fertilizer production and consumption; 1961/62-2002/03, Muscle Shoals, AL, 73 pp.

Khan, S., R. Mulvaney, T. Ellsworth, and C. Boast. 2007. The myth of nitrogen fertilization for soil carbon sequestration. *Journal of Environmental Quality* 36(6): 1821–1832.

Ladha, J.K., A. Tirol-Padre, C.K. Reddy, K.G. Cassman, S. Verma, D.S. Powlson, C. van Kessel et al. 2016. Global nitrogen budgets in cereals: A 50-year assessment for maize, rice, and wheat production systems. *Scientific Reports* 6:19355.

Lal, R. 2010a. Managing soils and ecosystems for mitigating anthropogenic carbon emissions and advancing global food security. *BioScience.* 60(9): 708–721.

Lal, R. 2010b. Enhancing eco-efficiency in agroecosystems through soil C sequestration. *Crop Sci.* 50.

Levy, H., Moxim, W.J., and Kasibhatla, P.S. 1996 A global three-dimensional time-dependent lightning source of tropospheric NOx. *J. Geophys. Res. Atmos.* 101: 22911–22922.

Loveland, P. and J. Webb. 2003. Is there a critical level of organic matter in the agricultural soils of temperate regions: A review. *Soil & Tillage Research* 70(1): 1–18.

Lu, C., and H. Tian. 2017. Global nitrogen and phosphorus fertilizer use for agriculture production in the past half century: Shifted hot spots and nutrient imbalance. *Earth Systems Science Data* 9:181–192.

Mueller, N.D., P.C. West, J.S. Gerber, G.K. MacDonald, S. Polasky, and J.A. Foley. 2014. A tradeoff frontier for global nitrogen use and cereal production. *Environ. Res. Lett.* 9:1–8.

Philpott, T. 2010. New research: Synthetic nitrogen destroys soil carbon, undermines soil health. *Grist Magazine*, February 24, 2010.

Ranade, V. 2010. Human activities and health of rivers: Case study of a river basin in the peninsular India. Proceedings of the 4th International Yellow River Forum, 213–217. S120–S131.

Smith, L.W. 1991. Production systems. In *Handbook of Animal Science*, Putnam P.A. (Ed.), Academic, San Diego, pp. 280–291.

Tie, X.X., Zhang, R.Y., Brasseur, G., and Lei, W.F. 2002 Global NOx production by lightning. *J. Atmos. Chem.* 43, 61–74.

Van Horn, H.H., G.L. Newton, and W.E. Kunkle. 1996. Ruminant nutrition from an environmental, perspective: Factors affecting whole-farm nutrient balance. *J. Anim. Sci.* 74: 3082–3102.

Vitousek, P.M., Menge, D.N.L., Reed, S.C., and Cleveland, C.C. 2013. Biological nitrogen fixation: Rates, patterns, and ecological controls in terrestrial ecosystems. *Phil. Trans. R. Soc. B* 368, 20130119. doi:10.1098/rstb.2013.0119.

Yablokov, A. and Levchenko, V. 2015. The decision exists: Transition to controlled evolution of the biosphere. *Philosophy & Cosmology* 14: 92–118.

WMO. 2016. Greenhouse Gas Bulletin: The State of Greenhouse Gases in the Atmosphere Based on Global Observations Through 2015. World Meteorological Organization, Geneva, Switzerland.

Wilkinson, J. M. 2011. Redefining efficiency of feed use by livestock. *Animal* 5: 1014–1022.

U.N. 2017. World Population Prospects: Key Findings of Advance Tables, 2017 Revision. Economic and Social Affairs. United Nations, New York.

World Bank. 2016. Fertilizer Consumption, Washington, DC (http://date.worldbank.org/indicated/AG.CON .FERT.ZS).

USDA. 2016. Fertilizer Use and Markets. Washington, DC (www.ens.uda,gov/topics/farm-practices-management /chemical-inputs/fetilizer-use-markets).

# Index